Axoplasmic Transport
and Its Relation to Other
Nerve Functions

AXOPLASMIC TRANSPORT AND ITS RELATION TO OTHER NERVE FUNCTIONS

Sidney Ochs

Department of Physiology
Indiana University School of Medicine
Indianapolis, Indiana

1807 1982

A Wiley-Interscience Publication

John Wiley & Sons

New York/Chichester/Brisbane/Toronto/Singapore

Library of Congress Cataloging in Publication Data:

Ochs, Sidney.
  Axoplasmic transport and its relation to
other nerve functions.

  "A Wiley-Interscience publication."
  Bibliography: p.
  Includes index.
  1. Axonal transport.   I. Title.
QP363.027   599.01′88   82-1978
ISBN 0-471-65255-5   AACR2

Printed in the United States of America

10 9 8 7 6 5 4 3 2 1

Preface

Nearly 25 years ago when I and my colleagues first began our work on axoplasmic transport, the question at issue then was whether the phenomena did in fact occur, whether a regular rate of movement of materials in nerve fibers could be convincingly established and related to other neural functions. It is now generally recognized that proteins and other materials are synthesized in the nerve cell bodies and are continuously being transported in their nerve fibers and dendrites to maintain structure and function; resting and action potentials, neurotransmission at the nerve terminals of motor fibers, and transduction of physical and chemical stimuli into nerve impulses at the sensory nerve terminals. Additionally, trophic materials are transported and released, which explains the influence of nerve over contiguous neurons, muscles, and other cells.

The recognition that axoplasmic transport (also termed axonal flow, axonal transport, neuroplasmic transport, or simply transport) has such a fundamental role in the function of the nervous system, is evidenced by the large and ever-increasing number of published papers, symposia, and reviews devoted to the subject. However, no text or monograph has as yet appeared which gives an integrated and systematic account of axoplasmic transport as a whole. When this book was first begun, it appeared that a small volume, one essentially outlining the field, would serve that purpose. However, the number of new advances made, even in the last 6 years, has necessitated a larger work. Some of those advances were basic and allowed a clearer and more unified presentation to be made. Other advances made have thrown more light on closely related fields which depend on transport for their full understanding—such topics as nerve degeneration and regeneration, neurotrophic control, and neurotropism. These are dealt with in this book.

The very large literature which has grown up based on the use of transport as a method to trace various nerve tracts in the CNS has only been briefly noted insofar as it has become a special branch of neuroanatomy in its own right, chiefly in studies made in the central nervous system. However, some basic principles revealed by those studies made in the CNS have been briefly delineated in the book and transport in one part of the CNS, the hypothalamo-neurohypophysial system, has been described in detail. In recent years, it has become clear that transport in this system shares the same properties as those found in the peripheral nervous system.

A sketch of the historical development of the concept of transport from its ancient shadowy past to our modern understanding is outlined in the Introduction. Additionally, attention was given to historical matters at various points throughout the book because an understanding of how the subject developed in the last several decades is necessary for a clear view of its present state. A number of earlier findings based on the limited techniques then available have remained in the literature to clash with results more recently attained, giving rise to obscurity and conflicts of interpretation. A critique of those findings and the various models advanced to account for transport is undertaken in this book as a service to the uninitiated. Such an analysis of current theories has been carried out in the spirit that the positive and negative aspects of each model will lead to a deeper appreciation of the phenomena by the reader. The aim of a generally accepted model is to account for all the known phenomena and to meet the criterion of non-falsification. This applies as well to the transport filament model, which is considered to be the one best able to explain the known phenomena at the present time. However, it, as any other model, must be subjected to further tests. The principal aim of this book is to provide a thorough understanding of the properties of transport, one which will serve as a foundation for further advances in the field.

The obligation to make this book as clear and trustworthy a guide as possible is a responsibility deeply felt. Where differences in interpretation exist, the various alternative positions have been set out as fairly as possible to allow the reader to make the appropriate judgements for himself. For additional details of methodology, results, or interpretations, an extensive bibliography has been given. Even at this relatively early stage in the development of this field, the literature has grown beyond the point where a complete bibliography can be given in a book of this size. However, a substantial selection of the primary literature is given, and with the general reviews cited the reader should be able to trace most, if not all, the available literature on transport.

In the course of organizing the various themes presented in this book, I have had the opportunity to gain much from the responses to lectures in neurophysiology and to seminars given on transport at various institutions in this country and abroad, and at symposia. I am especially appreciative of my appointment as Visiting Professor in the Centro de Investigación y de Estudios Avanzados in Mexico City during the fall of 1976, where the general outline of this text was used in my lectures.

Above all, I would especially like to take this opportunity to thank my colleagues and students, who by their collaboration with me on studies of axoplasmic transport over the years have contributed so greatly to the experimental basis on which this book is based. We are indebted to the National Institute of Health, the National Science Foundation, and the Muscular Dystrophy Association for their support which made our work possible. The librarians of the School of Medicine Library have been unfailingly helpful and the assistance of the Illustration and Photographic services at the medical school most appreciated. My colleagues, Drs. Nira Ben-Jonathan, Luis Gaziri, Koert Gerzon, Bernardino Ghetti, Richard Haak, Zafar Iqbal, Ralph Jersild, Jans Muller, Richard Peterson, and Robert Worth were kind enough to read gallies of several of the chapters and suggest corrections. Tim Breen and Diana Jackson also examined several of the chapters in galley form for errors as did William Rochlin and Lance McKitrick at an earlier stage.

Not least, my thanks are given to those who over the years have so patiently and conscientiously assisted in the typing of this text: Linda Cheatham, Erika Ramsey, Jennie Parker, Sherry Cotton, Susan Ochs, Karen Viewegh, Debbie Walters, Linda Gruver, and Kandy Golladay, who also patiently checked the accuracy of the references.

SIDNEY OCHS

Indianapolis, Indiana
April 1982

Acknowledgments

The author is grateful to the following publishers for their permission to use illustrations which originally appeared in their publications in this book:

Academic Press, New York; The American Association for the Advancement of Science, Washington, D.C.; The American Physiological Society, Bethesda; The Biochemical Society, London; Brain Research Institute, Los Angeles; The British Council, London; Chapman and Hall, Hampshire; Churchill Livingstone, Edinburgh; Cold Spring Harbor Laboratory, Cold Spring Harbor; Consejo Superior de Investigaciones Cientificas, Madrid; Elsevier/ North Holland Biomedical Press, Amsterdam; John Wiley and Sons, New York; Karolinska Institutet, Stockholm; Lippincott/Harper and Row, Philadelphia; Longman Group, LTD, Edinburgh; MIT Press, Cambridge, Massachusetts; Macmillan Journals, New York; New York Academy of Sciences, New York; Oxford University Press, New York; Pergamon Press, Elmsford; Pergamon Press, Oxford; Physiological Laboratory, Cambridge; Raven Press, New York; Rockefeller University Press, New York; The Royal Society, London; S. Karger AG, Basel; Sigma Xi, The Scientific Research Society, New Haven; Springer-Verlag, New York; The Williams and Wilkins Co., Baltimore; The Wistar Institute, Philadelphia.

S.O.

Contents

Contents

12. Agents and Pathological States Affecting Transport   278

13. Nerve Terminal Processes in Peripheral and Hypothalamo-Neurohypophysial Systems   312

1

A Brief Historical Introduction

A. DEVELOPMENT OF THE CONCEPT OF MATERIAL TRANSPORT IN NERVE

1. The Older Conception of Nerve Function

While evidence definitively establishing the movement of materials in nerve fibers has only been gained within the last several decades, the idea of some kind of an influence flowing in nerve is very old. It can be traced back to the animistic belief of primitive man in a spirit (pneuma) drawn into the body in the act of respiration. This concept of an animal spirit which serves to animate the body, together with a more or less explicit relation to mental attributes, was incorporated in ancient Greek physiological explanations from the 6th century B.C. onward. The meaning of the term "spirit" has undergone a series of changes in ancient times from Plato to Aristotle and their contemporaries (Peters, 1967), but in the main it includes motion and sensation, and to varying degrees, intellectual processes. Historical changes in the concept have occurred which are too complex for discussion here. In any case, a clear identification of animal spirit with the brain and nerves was made by the Alexandrian Erasistratus (c. 331 B.C.), who appears from fragments of his work to have been a great pioneer investigator of the nervous system (Wilson, 1959). It was Galen (130–200) who, through his clinical and experimental investigations, transformed ancient thought and produced a physiological system which survived for over 1400 years (Siegel, 1968; Temkin, 1973). In brief, in his schematization "vital spirit" produced by the action of respiration on the blood in the heart is carried by the arteries to the body to vivify it. The arterial blood is also carried by the carotids to the base of the brain where the rete mirabilis (and choroid plexus) abstracts the finer animal spirits from the arterial blood and passes it to the brain. Various psychic functions were subsequently related to the movement of animal spirits in the ventricles (Clarke and Dewhurst, 1972). The nerves of the body take origin from the brain and under higher (voluntary) control, animal spirits move from the brain into the nerves whence they are distributed through the

1

body to control the muscles. Both sensations and motor control were considered to be subserved by the animal spirits. The optic nerves were explicitly analogized to hollow tubes and implicitly all the nerves as well (Siegel, 1973). The concept of hollow nerves and the movement in them of animal spirits had a long-enduring effect on neurological thought through the succeeding centuries (Clarke, 1968), with Galen remaining the guiding authority for physiology and medicine until Harvey's discovery (1628) of the circulation of the blood (Rothschuh, 1973; Harris, 1973). While the concept of a blood circulation eventually replaced Galenic physiology, the idea of animal spirits flowing in hollow nerves continued to exert an influence for a considerable time afterwards, with traces found even in the 19th century.

Descartes (1664) radically altered the relation of animal spirits to the body. The animal spirits were conceived of as the "liveliest and subtlest" of particles which, when separated from the blood in the vessels at the base, enter the ventricles of the brain and serve not only "to nourish and sustain its substance, but also and principally to produce there a certain very subtle wind, or rather a very lively and pure flame" (Hall, 1972). In the ventricles of the brain the flow of animal spirits was pictured as being under the control of the pineal gland, which Descartes viewed as a readily moveable structure in which the soul is located. Thus, higher functions were clearly separated from the more mechanical operations of the nervous system. Animal spirits were projected into the ventricles by sensory nerves and reflected by the moveable pineal gland onto the inner walls of the ventricle where the animal spirits can "enter the pores in its substance and from these conduits proceed to the nerves." By determining which pore was entered, the pineal gland controlled the pattern of animal spirits flowing down the hollow motor nerves to the individual muscles thus causing the various parts of the body to move in an appropriate fashion.

In Descarte's schematization, sensory information is utilized by the pineal to direct animal spirits to the appropriate nerve tubes in the ventricles. When the animal spirits flow from the nerve down into the muscles it causes them to swell, bringing their ends together and thus giving rise to muscle contraction. This concept of an inflation of muscle by animal spirits received a decisive setback by plethysmographic studies of muscle contraction (Swammerdam, 1752). In his work carried out in 1663, but published only later, Swammerdam found that when the nerve of an isolated frog muscle preparation was stimulated, there was no increase in muscle volume during the resulting contraction. At about this same time, human arm plethysmographic studies were carried out by Steno (1664) and Glisson (1677), with a similar lack of increased muscle volume seen on contraction (Bastholm, 1950; Keele, 1974). Thus, it became generally apparent that some other process than an inflation causes muscle contraction.

In the 17th and 18th centuries various theories were advanced to account for nerve and muscle action, most of them invoking special properties for the animal spirits with the nerves for the most part still considered to be hollow tubes (Bastholm, 1950). The nerve influence was believed to be a thin watery-like fluid (Boerhaave, 1743) or a thicker fluid, something like egg albumin (Malpighi, 1667). Croone (1664) considered that the substance which moved in nerve into the muscle brings about an explosive-like reaction with a resulting contraction. Willis (1664) conceived two spirits to be present, a vital or "flamy" spirit, "inkindled" in the blood, and an animal spirit produced in the brain related to mind. Specifically, the animal spirits were elaborated from the blood in the cerebral cortex. As Willis put it, in the grey of the cortex the "subtil liquor" received from the blood is acted on "like a ferment" by a "volatile salt" to exalt it into a very "pure spirit." The medullary substance (the white matter) with its hollow nerve fibers serves to dispense the animal spirits to the various parts of the brain and thence to the spinal cord and nerves of the body. The ventricles of the brain, so far from elaborating or processing animal spirits as conceived of from the time of antiquity and by Descartes, were considered by Willis to simply contain the watery humors excreted from the brain, the "effete" matter discharged from the solid parts of the brain.

The idea of a movement of a nerve fluid in hollow nerve fibers was further supported by the microscopic studies of Van Leeuwenhoek (1717). He placed freshly cut cross-sections of optic nerve under his simple microscopes and found the fibers to be circular. Within a minute or so, a pearly fluid was seen to have exuded from the center of each of the circular canals to form a little mound which passed off as a vapor, the fibers subsequently collapsing to form flattened bands. Van Leeuwenhoek concluded from those observations that normally the nerve fluid within the fibers distends their walls causing them to assume a cylindrical form. On cutting the nerves, the nerve fluid escapes, the pressure on the walls decreases, and the fibers collapse.

In other more "physical" conceptions of nerve action, the fibers were viewed as solid "capillary" structures, transmitting a vibratory influence something like light; in Newton's (1730) view, a vibration of the "ether." A mechanical effect of sound vibrations was proposed by Baglivi (1723). He considered that music could, by "dissolving the venomous *coagulum* of spirits and humours," allow the spirits to recover their motion and "enter with greater facility and agility into the little tubes of the nerves and fibres." Baglivi thus explains the therapeutic value of music in the curious disease caused by the tarantula spider (cf. Sigerist, 1948).

When the phenomenon of electricity came under scientific study in the late 18th century, Galvani (1791) conceived of it as a fluid of a "lively" and peculiar kind which he assimilated to a nerve spirit or fluid. Galvani wrote that, "the electric fluid is produced by the activity of the cerebrum, that it is extracted in all probability from the blood and that it enters the nerves and circulates within them in the event that they are hollow and empty, or, as seems more likely, they are carriers for a very fine lymph or other similarly subtle fluid which is secreted from the cortical substance of the brain, as many believe. If this be the case, perhaps at last the nature of animal spirits, which has been hidden and vainly sought after for so long will be brought to light with clarity.''

2. Early Microscopy and Nerve Form

Following van Leenwenhoek, others carried out microscopic studies to determine whether spirits or some material agency could flow in the hollow nerve fibers. However, progress was hampered by the errors in visualization brought about by chromatic and other optical distortions inherent in the lenses then available. In spite of these difficulties, Fontana (1778) was able to picture the individual nerve fibers. After the introduction of the achromatic compound lens in the early part of the 19th century, the clearer view then made possible brought about rapid advances in the study of nerve. Fixation and staining methods were not yet developed, and to better visualize the individual nerve fibers, Ehrenberg (1838) used a compressorium. This instrument consisted of a hinged glass cover which was screwed down over nerves placed on the stage glass plate in order to spread out and separate the individual fibers. The pressure caused the fibers to assume an abnormal beaded shape, which resulted in their being confused with cell bodies. Purkinje and Valentin identified the true form of the cell body, showing it to contain a nucleus and nucleolus (Kruta, 1971a,b). However, they unfortunately also concluded that the cell bodies and nerve fibers constituted two separate and distinct neural entities (Van der Loos, 1967; Ochs, 1975b). The connection between the nerve cell body and the fibers in the dorsal root ganglion is difficult to visualize by light microscopy and we can understand how this led to a conception of cells and fibers as two ''elements,'' closely opposed but not integrally connected.

The cell body was believed by Valentin and Purkinje to be the ''active'' element which in some fashion causes the nerve fluid contained in the nerve fibers, the ''passive'' element, to circulate. The fibers thus constituted continuous loops with their central ends in the ganglia or in the CNS apposed to the cell bodies and peripherally terminating as closed loops in sensory endings. This resembles to some extent an older concept of a circulation of nerve fluid in fibers proposed by Henrici in 1646, who apparently was influenced in his theory by Harvey's demonstration of the circulation of blood (Clarke, 1978).

Remak (1838) was the first to advance the idea that the nerve fibers and cell bodies were connected and in fact that they constituted a single neural entity (Ochs, 1975b, 1979). Remak wrote that, ''the organic fibers originate from the substance of the nucleated globules itself. In spite of the fact that this observation is very difficult and requires great dexterity in preparation as well as in observation, it is so well founded that it already would not be possible to doubt [it]'' (Kisch, 1954). This ''early neuron doctrine'' of Remak's was, however, not generally accepted until the late 19th and early 20th century when it was strongly supported by the work of His (1889) and Forel

(1887). The concept was formally christened the "neuron doctrine" by Waldeyer (1891) in his review of the then extensive evidence which had accumulated in its favor (cf. Barker, 1899). With the studies of Ramon y Cajal (1894, 1909, 1911, 1954) and the convincing evidence obtained in tissue culture studies by Harrison (1910), the neuron doctrine was conclusively established.

3. Cell Body-Fiber Dependence

In spite of the uncertain status of the form and nature of neural elements and the lack of acceptance of the neuron doctrine at the time, Waller (1852a) presciently inferred from his experiments on nerve fiber degeneration that the viability of the nerve fiber depends on its connection with the cell body. He considered that some "trophic influence" passes from the cell body into the fiber to maintain its viability. This Waller inferred from a classic study where he transected the roots and nerves of the 2nd cervical ganglia of kittens with the resulting various patterns of fiber degeneration shown in Fig. 1.1.

Those portions of the fibers of the nerves or roots connected to the dorsal root ganglion retain their normal form, while the part of the fiber separated from the ganglion undergoes degeneration. From his experimental observations, Waller (1852b) concluded that, "the ganglion corpuscles (cell bodies) present in the dorsal root ganglion exert a trophic influence necessary to maintain the form and function of the fibers ascending in the dorsal root fibers as well as on the sensory fibers descending in the peripheral nerve." And further, "as long as the influence of the ganglion over the nerve fiber

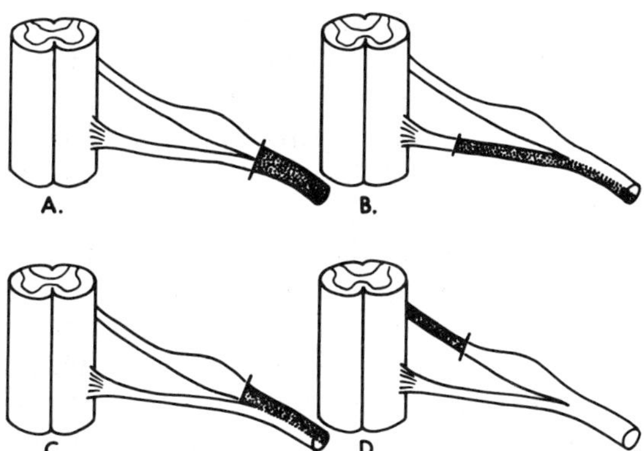

Fig. 1.1 Patterns of fiber degeneration. Sections made: (A) distal to the ganglion, (B) ventral root, (C) sensory nerve, and (D) dorsal root of the 2nd cervical segment of kittens at the levels indicated by the transverse lines. Fiber degeneration is shown by the shading of those portions of the nerve fibers separated from their cell bodies. From Waller (1861) and A.D. Waller (1891).

occurs, this equilibrium [forces of renewal as opposed to those of degeneration] is maintained, but as soon as the connection of the ganglion corpuscle with the nerve fiber is destroyed, its peripheral [severed] end …is subjected to forces of destruction [degeneration]." As an analogy, Waller (1861) presented the view that, "a ganglion therefore was to the fibers connected to it what a river was to the rivulet that trickled from it, a source of nutritive energy." He also inferred that there were cell bodies in the spinal cord which have the same trophic influence on the motor fibers, this at a time when the connection of motoneurons to ventral root fibers had not yet in fact been established.

4.   Fibrillary Elements and Their Role in the Fiber

In conflict with the possibility of a movement of fluid in hollow nerve fibers was the pioneer observation made by Remak (1838) that fibrillary material exists in the "primitive nerve fiber," the "nerve band," as he called it. This was verified by Purkinje, who renamed it the "nerve axis" or axon, the term by which it is referred to today. Schultze (1871), using more advanced staining methods, clearly established the presence of fibrillary material in nerve fibers and considered that its function is to conduct the nerve impulse. By then the electrical nature of the nerve impulse had been established.

When the axon membrane was recognized as the seat of the nerve action potential (Bernstein, 1912), it was clear that some other role had to be assigned to the neurofibrils. Parker (1929) proposed that the fibrillary elements are the means by which the cell body exerts a metabolic control over its fiber. Considering the T-shaped neurons of the dorsal root ganglion (Ranvier, 1875), Parker wrote, "In the ordinary sensory neurons, nerve impulses originate at the peripheral end, make their way centrally over the neurite, and without entering the body of the cell, pass on to discharge at the central end of the neuron. The metabolic influences on the other hand originate in the region of the nucleus of the cell body, pass down its neck to the tract of nervous transmission where they separate into two streams, one flowing peripherally over the neurite and the other centrally over the central nerve fiber process. And, thus, the course of the neurofibrils does not follow that of the nerve impulses but does duplicate exactly that of the metabolic influences. I conclude therefore that the neurofibrillar system in the neuron is concerned specifically with the distribution of the metabolic influences and not with the conduction of nerve impulses. These influences start in the region of the neuronic nucleus and spread over the lines of neurofibrils throughout the whole neurone." Parker speculated no further on the nature of the materials transported: "What the metabolic influences are, it is impossible at present to say. It seems hardly reasonable to think of them as streams of material in the nature of a hormone, emanating from the region of the nucleus and percolating throughout the neurone."

Gerard (1932), summarizing the then new evidence that the nerve impulse depends on oxidative metabolism, proposed that the enzymes necessary to maintain their metabolism are being continually transported down within the nerve fibers from their source in the nerve cell bodies. On cutting the nerve, the loss of those components would lead to inexcitability and the form changes characteristic of Wallerian degeneration. A somewhat similar view had earlier been expressed by Goldscheider, who suggested that a transport of "ferment-like" substances from the cell body moved down along the whole course of the axon to their extremities is required for the nutrition of the axons (Barker, 1899).

Struck by the similarity of the staining of "neurosomes" in neurons to zyomogen granules in the secretory cells of the pancreas and fundus glands of the stomach, Scott (1905) hypothesized that "nerve cells are true secretory cells," and "in the body of the nerve cell a substance is formed from the nucleus and Nissl bodies which gradually passes into the nerve fibres; and also that stimulation of other cells by a nerve fibre is brought about by the passage of some of this substance into the cells on which the fibre acts" (Scott, 1906).

Torrey (1934), a student of Parker, proposed that the substance supplied by the cell body to its nerve fibers was a "hormone-like" material necessary for the maintenance of the nerve fiber. He considered that "the substance involved may be enzymatic in character, produced by the nucleus of the neuron, and transported peripherally by the way of the neurofibrils."

Only within the last quarter century has the fine structure of the neurofilamentous material present in the neuron been resolved by the use of the electron microscope (EM). With advances in EM technique, the filamentous material has been differentiated as microtubules and neurofilaments (Gray and Guillery, 1961; Potter, 1971; cf. Chapter 9). The microtubules in particular have received prominent attention with respect to their role in transport, as will be described in Chapter 10.

5.   Nerve Fiber Regeneration and the Movement of Material

An aspect of nerve closely related to transport is the phenomenon of nerve regeneration. There is evidence indicating that nerve suturing was practiced in the Middle Ages with the aim of restoring nerve function (Ochs, 1980). The practice in any event fell into disuse, so that the observations of Cruikshank made in 1776 were considered to be the first indication that cut nerves could regenerate (Ochs, 1977a). However, evidence of nerve regeneration was not clearly established until much later. Throughout most of the 19th century the dominant belief was that nerves healed by a "reunion" rather than by a regeneration of new fibers. This concept implied the autonomy of nerve fibers and nerve cell bodies, a view based on Valentin's concepts and supported by the work of Schwann and Schiff. Finally, with the acceptance of the neuron doctrine in the early part of the 20th century, the concept of nerve fiber regeneration was accepted as well (Ochs, 1980a).

Ramon y Cajal (1928) in his classical studies on nerve degeneration and regeneration demonstrated that some time after partially compressing nerves with a ligature, a series of bulges and narrowings were seen in the fibers above the region of compression. The beaded form of the fibers was interpreted by him as due to the starting and stopping of regenerative nerve fiber growth. Ramon y Cajal drew attention to the fine fibers growing down through the site of the partial constriction and noted the sprouts (growth cones) at their terminals (Chapter 15), an indication that these fibers retained their capacity for regenerative growth.

More recently Weiss and Hiscoe (1948) studied the effects of partial compressions of nerve trunks and saw similar form changes in the fibers but interpreted them differently. The axonal contents were considered to be flowing or constantly "growing" down within their sheaths at a rate of 1–3 mm/day. The moving column of semi-solid axoplasm on meeting with the obstruction presented by the constriction becomes "dammed up" to cause the beading and tortuosities seen in the fibers. Reasons are given in Chapter 10 that these morphological changes do not represent slow transport and that slow transport has another basis. Weiss' hypothesis had, however, a stimulating influence on modern investigations of axoplasmic transport, as will be discussed in Chapter 10. A somewhat similar concept of a continual outflow of axoplasm within the nerve fibers had been advanced by Young (1944a). He made the observation that isolated portions of nerve fibers cut off from their cell bodies appeared to contract to form folds and spirals and, "the central portion swells as if under the influence of a turgor pressure." Young also pointed out that the oriented molecules in the fiber (the neurofibrils of Parker) are normally under restraint but become disorganized in the isolated nerve stumps. On the second day after cutting, the central ends of the nerve fibers are swollen. And "during the succeeding days the material of the central end flows out from the tube, forming one or several regenerating strands." These phenomena were taken to show that the turgor pressure within the fiber is due to materials supplied by the cell body and moving down within the nerve fibers. Young (1936) had earlier observed an outflow of axoplasm from the cut ends of squid giant axons and from this inferred that the continued production of axoplasm by the cell body is the force for the turgor and downflow of axoplasm in the nerve fibers. The pressure exerted by the downflowing axoplasm was considered to be opposed by the lateral pressure in the walls of the fibers, a pressure due to surface tension (Plateau, 1873). The beaded form of Wallerian degeneration in myelinated fibers distal to a nerve transection was viewed as Plateau figures brought about by the unopposed surface tension (Young, 1949b). Lubińska (1964) showed axoplasm oozing from the cut ends of myelinated nerve fibers, a behavior which seemed to be in accord with Young's concept.

However, Lubińska later found that acetylcholinesterase present in motor fibers has both an anterograde and retrograde movement in nerve (Chapter 4), a result inconsistent with a simple downflow of axoplasm. Analysis based on morphological changes in histological preparations alone cannot characterize transport. The evidence obtained by the use of isotope tracers has clearly shown a much faster rate of transport present in nerve fibers (Chapter 2), and chemical and optical studies (Chapters 3,4) have given us our present understanding of the properties of anterograde and retrograde transport.

B. AN OVERVIEW OF TRANSPORT IN RELATION TO NEURON FUNCTION

The development of the technique and present methodology for the study of transport will be described in the next several chapters. At this point it will be useful to sketch the various aspects of axoplasmic transport dealt with in this book (Fig. 1.2).

As indicated in this figure, the neuron has three well-defined regions: (A) the cell body, the site of the synthesis of materials, (B) the axon (and as well the dendrites) in which materials are moved by axoplasmic transport, and

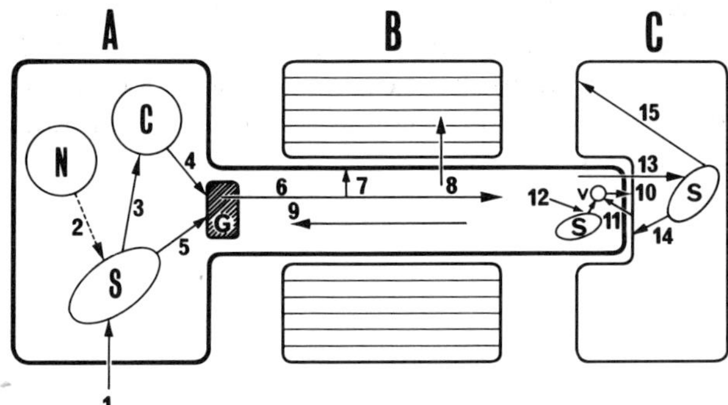

Fig. 1.2 Diagrammatic representation of function of the neuron related to transport. (A) Cell body, (B) nerve fiber, and (C) terminal regions of a schematized motoneuron are shown. Precursors enter (1) the cell with the nucleus (N) controlling synthesis (2). Arrow (3) shows a compartmentalization (C) of synthesized materials moving from the synthesis site and from the compartment into the axon (4,5). (G) represents a gate for some substances, possibly the Golgi apparatus, controlling the egress of materials into the axon. The anterograde transport system moves materials down the nerve fiber (6). Arrows from the transport system represent components supplying the membrane (7) and the Schwann cell (8). The myelin sheath is represented by parallel horizontal lines. Retrograde transport is shown by the arrow directed toward the cell body (9). In the terminal region vesicles (V) participate in neurotransmission (10). Uptake of components (11) in the terminal with a reconstitution of neurotransmitter in the vesicles is indicated (S), with also a contribution of downflowing materials (12). An arrow from the nerve terminal entering the muscle fiber (13) indicates the entry of neurotrophic materials. Some control of synthesis (S) in the muscle fiber of, among other things, receptor proteins, is shown by arrows (14) and (15). These components are inserted into the membrane at the usual site in the end-plate (14) and, after denervation, at new (15) receptor sites appearing in the muscle membrane outside the end-plate region. From Ochs (1974b).

(C) the nerve terminals, either motor as shown, or sensory. In the cell body proteins and polypeptides are synthesized under the direction of the nucleus. These (and other synthesized materials) may be either quickly transported in the axons or be stored in a cell body compartment for later export into the axons (or into dendrites). The egress of various components into the neurites is controlled by a "gate," which could be the Golgi apparatus (Chapter 5). Arrow 6 of Fig. 1.2 directed down the axons indicates the anterograde transport of materials, those required to maintain the structure and function of the fiber including the membrane (arrow 7), and as well some substance required by the Schwann cells (arrow 8). The latter is indicated by the characteristic Wallerian degeneration which occurs when the nerve fiber is separated from its cell body (Chapter 6). Synaptic vesicles must be provided to the nerve terminal, these containing the neurotransmitter required for synaptic transmission and also, in some cases, synthetic enzymes for the local synthesis and turnover of neurotransmitter in the nerve terminals (Chapter 13). Trophic materials are also considered to be carried from the nerve terminal into the post-synaptic cell (arrow 13) to there regulate some of its functions (Chapter 14). Additionally, as shown in the schematization, materials acting to regulate the level of synthesis in the cell body are carried back to it by retrograde transport (arrow 9). Such a regulation is inferred from the chromatolysis seen in the cell bodies following a transection of the nerve fibers (Chapter 6).

2

Properties of Anterograde Transport

As indicated in Chapter 1, morphological methods can give only a limited amount of information concerning transport. Most important in the characterization of transport has been the use of the isotope labeling technique, either by itself or combined with damming, enzyme, or chemical marker studies. In this chapter slow and fast transport as shown by isotope methods will be described. Further analytical methods will be discussed in Chapter 3 and a critical assessment of slow transport taken up in Chapter 10. The relation of slow to fast transport will be further analyzed in Chapter 11, where the concept that slow transport is an aspect of the fast transport mechanism will be advanced.

A. DEVELOPMENT OF ISOTOPE PROCEDURES AND SLOW TRANSPORT CHARACTERISTICS

The technique of isotopic labeling to characterize transport introduced by Samuels *et al.* (1951) was to arrange for an isotopically labeled precursor to be taken up by the neuron cell bodies and then, after its incorporation into various compounds, to allow sufficient time for the labeled synthesized materials to be carried out into the nerve fibers. The nerves were removed to determine the proximo-distal pattern of their content of labeled radioactivity. After intraperitoneal injection of the precursor ^{32}P (as orthophosphate) into guinea pigs, downflow times from days to weeks were allowed before the nerves were removed and cut into proximal, middle, and distal (or into proximal and distal) portions for analysis (Samuels *et al.*, 1951). At earlier times after injection, a greater amount of labeled activity in the form of phosphoprotein was observed in the proximal segments, and, at later times, increasingly more of it was found distally in the nerve. The proximo-distal shift of this labeled component in the nerve with time indicated that it has a slow downflow in the nerve.

From the regression line taken through the analysis points, it was determined that in the first 10-day period, the ratio of radioactivity in the proximal as compared to the distal segments was 1.4, in the next 10-day period 1.05, and in the final 10-day period 0.7. The rate of downflow estimated on this basis was 2.5 mm/day. The points representing the specific activities, however, showed considerable scatter and a correlation coefficient of only 0.645. Some of this variation most likely resulted from the precursor being taken up locally from the circulation by non-neural elements in the nerve trunks. Its local synthesis into various components confounds the subsequent outflow of labeled materials transported within the nerve fibers. As phosphate does not readily pass the blood–brain barrier into the CNS, the amount taken up in the spinal cord to be incorporated by the motoneurons and transported is correspondingly small.

To circumvent the blood–brain barrier and obtain higher levels of uptake of the precursor by the nerve cell bodies, the technique of injecting the isotope directly into the CNS was introduced (Ochs and Burger, 1958). The precursor ^{32}P was injected into the ventral horn of the L7 segment of the cat spinal cord in the vicinity of the motoneuron cell bodies (Fig. 2.1).

The subsequent outflow of incorporated labeled compounds in the motor fibers of the L7 ventral roots showed a proximo-distal gradient with respect to the time elapsing between injection and the taking of the roots for counting, as expected of an outflow in the fibers (Fig. 2.2).

To show that the outflow was not due to a leakage of the isotope between the fibers of the ventral root, compression asphyxiation of the spinal cord was used to destroy the motoneurons before injecting the precursor (cf. Van Harreveld and Marmont, 1939). As the cell bodies were no longer able to synthesize the precursor, a marked reduction in the downflow occurred (cf. Fig. 2.2). This would not have occurred if leakage were responsible for the outflow pattern (Ochs and Burger, 1958). The same result was obtained when sodium cyanide was injected into the ventral horn region of the spinal cord to destroy the motoneuron cell bodies before injecting the precursor (Ochs *et al.*, 1962).

The form of radioactive outflow was assessed by taking a series of small, equal, transverse segments of ventral roots successively along their length (Ochs *et al.*, 1962) and plotting their content of radioactivity against the distance from the ganglion. The curve obtained showed an exponential decline. Plotting the outflow semilogarithmically gives a straight line (Fig. 2.3).

A similar exponentially declining pattern of outflow of labeled activity was also found by Miani (1963) in the hypoglossal and vagus nerves after applying ^{32}P topically to the floor of the 4th ventricle of rabbits for uptake by the cell bodies (Fig. 2.4).

After its uptake by the cell bodies and the incorporation of the precursor, the outflow of labeled radioactivity in the nerves was shown to be present in the form of ^{32}P-labeled phospholipids. There was some variation in the pattern of outflow with time. In the hypoglossal nerve, a declining slope remained present throughout the times studied, while in the vagus nerve there was a later rise and then a fall again of activity with time.

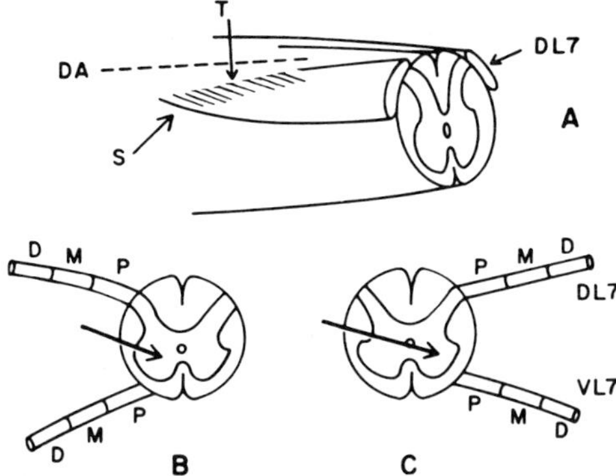

Fig. 2.1 Injection sites and root samples. (A) The "V" formed on the cat spinal dorsal surface of the cat spinal cord indicates the site of origin of the dorsal lumbar 7th roots. The arrow T marks sites in the center of the segment injected from above. S indicates injection from side. DA is the mid-dorsal line. (B) In a cross section of the segment, the arrow indicates injection into the ventral horn region. Root samples were taken after injecting from P (proximal), M (middle), and D (distal) parts of the sciatic nerve on the same side. (C) The arrow indicates injection across the mid-line into the ventral horn region of the opposite side with root samples taken from that side. From Ochs and Burger (1958).

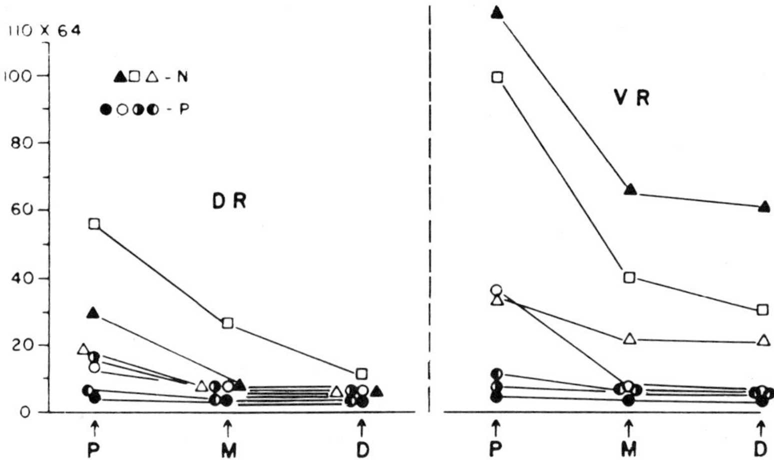

Fig. 2.2 Downflow of radioactivity in roots following a period of pressure-asphyxiation. The cat spinal cord was compressed for over 1 hr and 7 days beforehand to cause asphyxiation and a cell body destruction. Circular symbols represent pressured (P) animals. Angular symbols represent animals not compressed (N). Radioactivities in the ventral roots after ventral horn injection of ^{32}P are represented on the ordinate as counts/15 min/cm of nerve above background multiplied by 64. Abscissa shows proximal (P), middle (M), and distal (D) ventral root segments (cf. Fig. 2.1). From Ochs and Burger (1958).

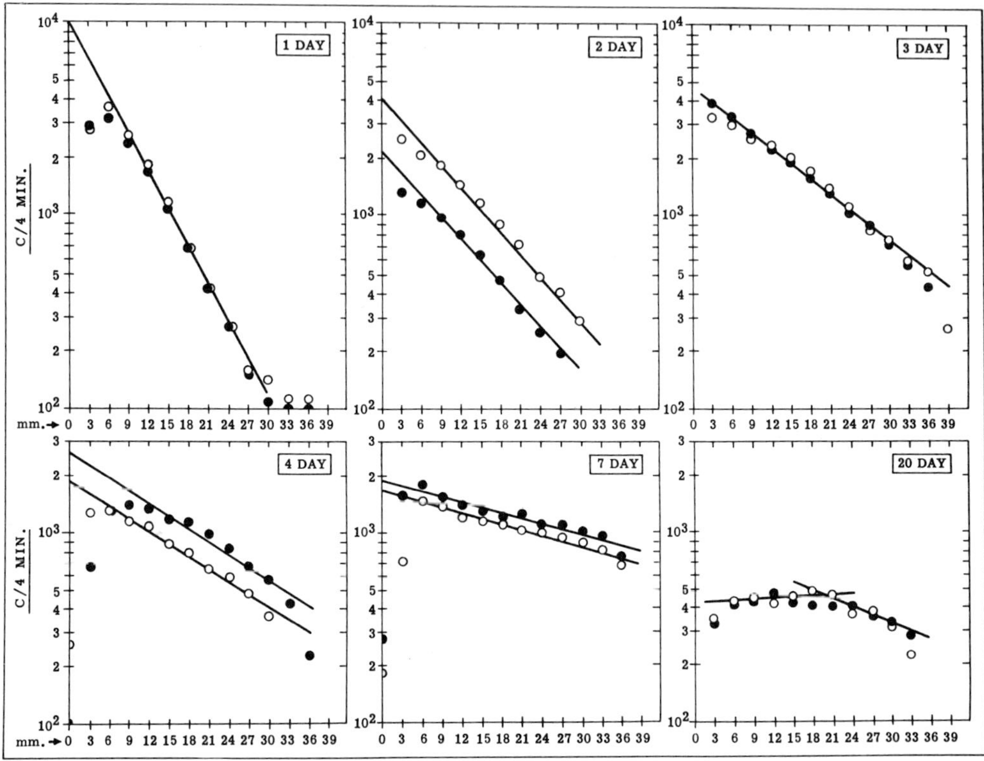

Fig. 2.3 Slow transport in ventral roots. Cord injections of ^{32}P were made in the lumbar L7 segment of the spinal cord of cats as shown in Fig. 2.1 at the days indicated. Open and filled circles represent radioactivity (ordinate) at 3-mm increments (abscissa) in the left and right ventral roots shown in each panel. Origin of abscissa at 0 was taken as the most proximal part of the ventral root. Radioactivity on the ordinates are given on a logarithmic scale with each point representing counts per 4 min. The line fitted to these points is steep early on and more shallow at later times. Irregularities in outflow were found later, e.g., 20-day efflux. From Ochs and Burger (1958).

In an attempt to estimate the rate of transport from such outflow curves, the distance of the half-activity points of the curves was plotted as a function of time, and a rate of 4.5 mm/day was derived on that basis (Ochs *et al.*, 1962). On the other hand, Miani (1963) took the farthest outflow of activity in the nerve as an estimation of rate, one which gave a value of 31 mm/day for transport in the hypoglossal nerve and 70 mm/day for that of the vagus nerve (cf. Lubińska, 1964). Those ''faster'' rates however, only represents a different way of assessing the same slow transport of ^{32}P-labeled components. An explanation of the patterns of slow outflow in relation to fast transport will be given in Chapter 11 and further consideration of phospholipid transport in Chapter 5.

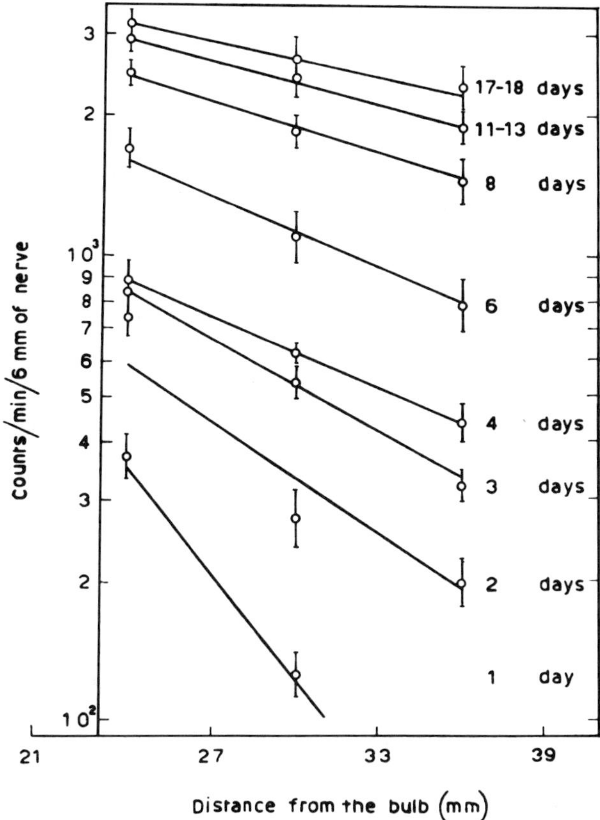

Fig. 2.4 Slow outflow of ^{32}P-labeled lipoprotein. Hypoglossal nerves were taken at different times after topical labeling of cell bodies in the bulb with ^{32}P. Points plotted on a semilogarithmic scale represent the mean value of labeled lipoprotein in ten experiments. From Miani (1963).

Droz and LeBlond (1963) introduced the use of tritium (^{3}H)-labeled amino acids in the study of axoplasmic transport. They injected the precursor ^{3}H-leucine systemically into animals and then at various times afterwards removed the sciatic nerves to prepare sections for histological examination and autoradiography. Nerves, when taken at early times after injection, showed a profuse labeling of Schwann cells in all portions of the nerve due to the local uptake of the labeled precursor from the circulation. However, at a later time, a smaller amount of the ^{3}H-leucine incorporated into proteins was seen to have moved down within the nerve fibers, as indicated by the appearance of grains of radioactivity located over the axons in the autoradiographs.

By counting the grains which appeared within nerve fibers at different times and at successively further distances from the motoneuron cell bodies, a slow transport of the labeled activity at a rate of 1–2 mm/day was estimated by Droz and Leblond. The labeled activity was considered to represent proteins or higher molecular weight (MW) polypeptides because the radioactivity remained present in the tissue sections carried through the usual histological procedures. It could not be due to free ^{3}H-leucine, as the precursor would have been washed from the sections. This inference was established by more direct biochemical evidence of the incorportion of ^{3}H-leucine into polypeptides, as will be described in Chapter 5.

The laborious method of counting grains and the interfering presence of locally incorporated labeled compounds was circumvented by injecting the precursor ^{3}H-leucine directly into the L7 ventral horn region of the cat spinal cord near the motoneuron cell bodies (Ochs, 1965, 1966). The much higher level of uptake of the precursor resulted in a high level of incorporated radioactivity transported within the axons (Fig. 2.5).

The large myelinated fibers had been cut on the bias and thus appear as ovals in the figure. The myelin sheaths of the nerve fibers retain their normal thickness and form when using freeze-substitution (Chapter 9) to prepare the tissue for autoradiography. The axons have large numbers of grains of radioactive-labeled proteins present within them, the grains generally distributed over the whole of the axonal area, as can be seen in the autoradiograph.

The pattern of outflow appearing at longer times may be complex. After injecting cat L7 dorsal root ganglia with ^{3}H-leucine, Lasek (1968) found a similar exponentially declining outflow, but with more irregular, smaller waves appearing at the later times (Fig. 2.6).

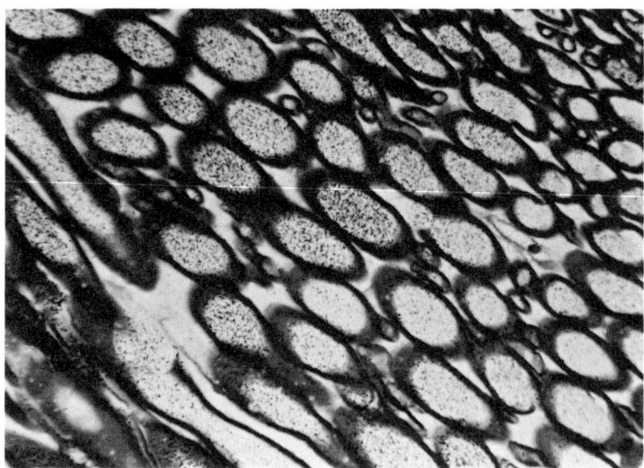

Fig. 2.5 Intra-axonal position of labeled protein. ^{3}H-leucine was injected into the ventral horn of the L7 segment of the cat spinal cord. After three days the ventral root was prepared by freeze-substitution. The plane of section of the ventral root was oblique, the myelin sheaths of the fibers appearing as ovals. Autoradiographic grains are seen over the inside of the axons. From Ochs (1966).

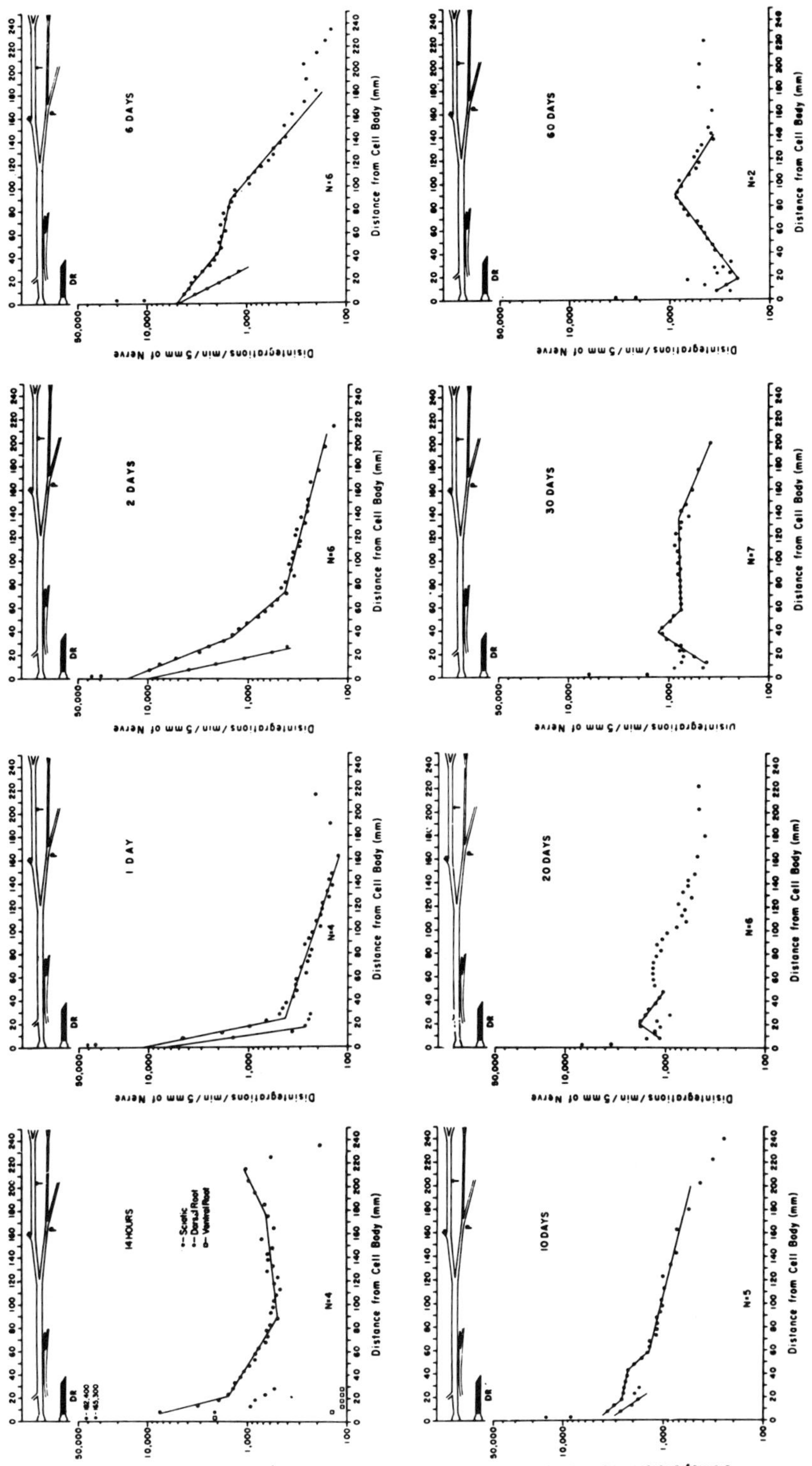

Fig. 2.6 Slow outflow of radioactivity. Five-millimeter segments of combined peroneal (P) and tibial (T) branches of the sciatic nerves (●) and dorsal roots (○) show radioactivity plotted (ordinate) against the distance of the segments from the site of ³H-leucine injection of the dorsal root ganglia (abscissa) at times in hours and days afterwards. The points represent the means of a number of values, denoted by N on each curve. Above each curve is a schematic representation of the segment of nerve used. From Lasek (1968).

The significance of the wave-like outflows seen at the later times in relation to slow transport will be further discussed in Chapter 11.

Another kind of irregularity was seen in the ventral root outflow at still earlier times than shown in Fig. 2.6. After injection of the spinal cord (Ochs, Johnson and Ng, 1967), breaks in the curves of outflow suggested that a much faster outflow was also present (Fig. 2.7).

The fast component in the ventral root shifts rapidly between 1 and 3 hours after the injection of ^{3}H-leucine into the cord. That such an outflow was due to transport and not to a leakage between the fibers was shown by locally freezing the root at dry-ice temperatures with a cooling bar: The freezing causes the nerve fibers to seal without an apparent disruption of other struc-

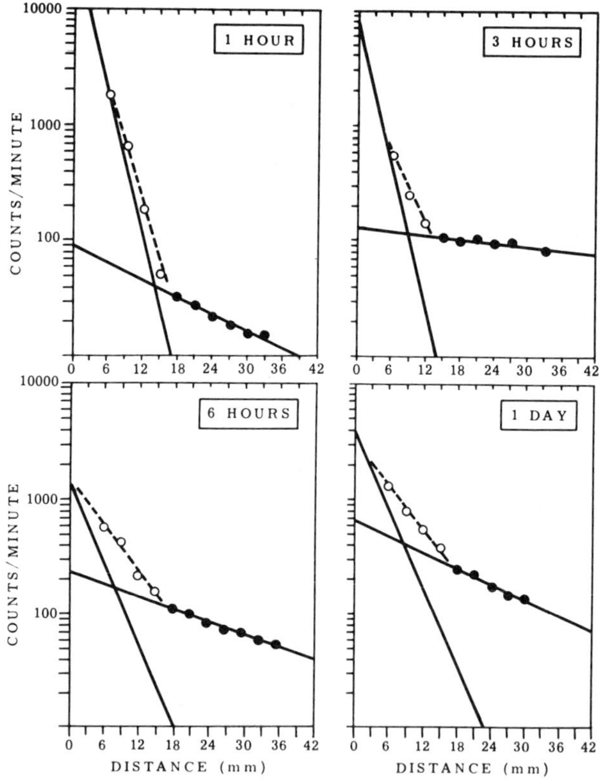

Fig. 2.7 Fast and slow outflow patterns in ventral roots. Following ^{3}H-leucine injections made either into the lumbar 7th or sacral 1st spinal cord segments of cats, the downflow of radioactivity is shown on the ordinates in counts/min plotted along the distance on the abscissae at the different times shown after injection. Two components of outflow are found after subtraction by lines fitted by least mean squares, these representing fast (○) and slow (●) phases. From Ochs and Johnson (1969).

tures in the nerve trunk (Ochs and Johnson, 1969). When this is done several hours before injecting the cord with ^{3}H-leucine, the downflow of labeled radioactivity dams up above the site of the freeze-block region, a result which militates against an interaxonic diffusion and points to a fast axoplasmic transport.

B.   FAST AXOPLASMIC TRANSPORT

Attempts to arrive at a rate of transport on the basis of the furthest extent of outflow of radioactivity in nerves or roots led to erroneous estimations of rate. Rates of transport obtained from the outflow pattern of radioactivity in ventral roots (cf. Fig. 2.7) ranged from somewhere above 100 mm/day (Kidwai and Ochs, 1967) to 800 mm/day (Ochs, 1967). Fast transport was finally accurately assessed when earlier outflow times were systemically explored by the profile of outflow in longer lengths of nerve. The technique and the basic phenomena characterizing fast transport shown by that method are described in the following section.

1.   Methodology to Show the Characteristics of Fast Transport

Because the sciatic nerve preparation has been used in many of the studies of axoplasmic transport to be discussed in this book, it is described here in some detail. The lumbar seventh (L7) dorsal root ganglion, which generally gives rise to a larger proportion of sensory fibers to the sciatic nerve, and the L7 segment of the spinal cord containing the motoneuron cell bodies supplying motor fibers to the sciatic nerve are preferentially selected for injections of labeled precursors. These are shown in relation to the skeleton of the lumbar region of the cat in Fig. 2.8.

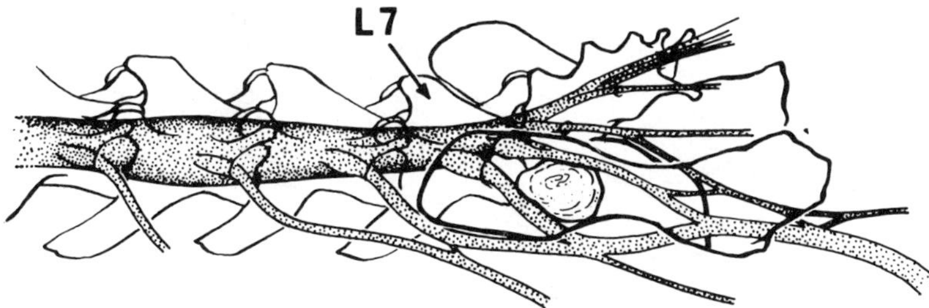

Fig. 2.8 Anatomical relations. The lumbosacral spinal cord, ganglia, roots, and sciatic nerve are shown in relation to the vertebral column. Arrow L7 points to 7th dorsal root ganglion with its nerve branch passing below the sacro-iliac joint to join the sciatic nerve. From Ochs and Ranish (1969).

After exposure of the lumbar region of the cord, small volumes (15–20 μl) of isotonic saline or Ringer solution containing ³H-leucine are injected via a micropipette into the L7 dorsal root ganglion for uptake by the neurons, incorporation and transport in sensory fibers of the sciatic nerve. Alternatively, the isotope solution was injected into the L7 segment of the ventral horn of the spinal cord for uptake by the motoneurons (Fig. 2.9).

The injected precursor spreads throughout the extracellular spaces in the ganglion or ventral horn of the spinal cord and is rapidly taken up by the cells and incorporated into a variety of proteins and polypeptides (cf. Chapter 5). After allowing time for transport into the fibers, the sciatic nerve, dorsal root ganglion, and dorsal root are removed from the animal and cut into 5-mm segments. These are each counted separately as indicated schematically in Fig. 2.10.

The radioactive counts found present in the segments after scintillation counting are plotted typically on a logarithmic scale along the ordinate. The changes in outflow pattern in the nerves of 5 animals taken at different times after L7 dorsal root ganglion injection with ³H-leucine are shown in Fig. 2.11.

In all these cases, a large amount of radioactivity was found remaining in the injected ganglion. In the bottom trace a nerve taken only 2 hrs after injection shows little evidence of outflow. At 4 hr and thereafter, a plateau

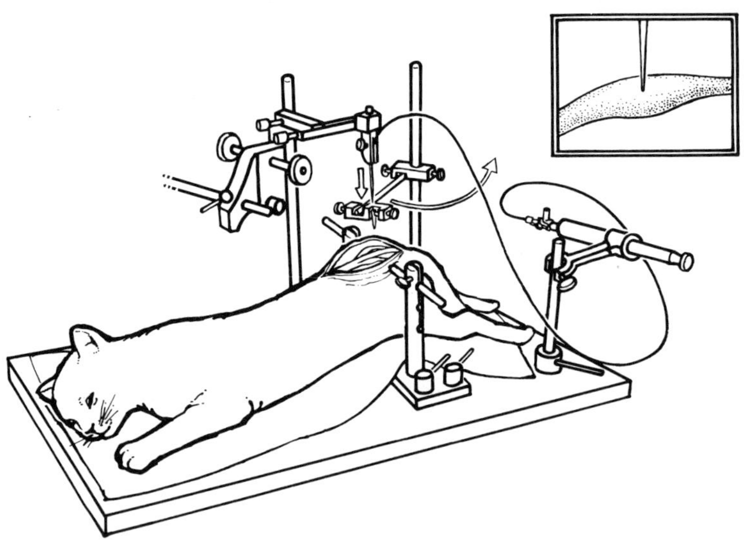

Fig. 2.9 Technique of injection. The injection pipette positioned by a micromanipulator is shown placed over the isotope loading vial. After a volume of radioactive solution is drawn into the pipette, the loading vial is swung away and the pipette is positioned over the ganglion or cord for injection. Also shown is the exposed cord with an arrow pointing to the L7 dorsal root ganglion (cf. inset).

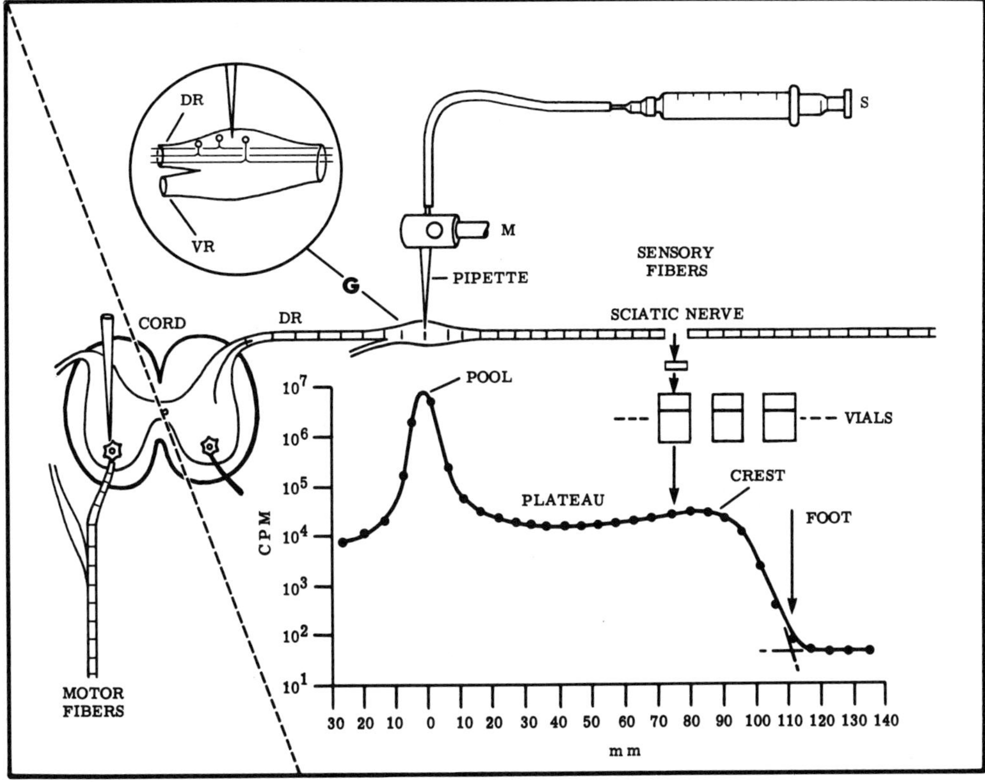

Fig. 2.10 Sampling technique. The L7 dorsal root ganglion is shown above circled in the insert with its T-shaped neurons, one branch ascending in the dorsal root (DR), the other descending in the sciatic nerve. The ventral root (VR) passes below it. A pipette containing ^{3}H-leucine is shown in the cell body region of the ganglion (G). After injection of the precursor, and its incorporation, the subsequent downflow of labeled polypeptides in the fibers is determined by removing the sciatic nerve from animals at various times after injection and sectioning it into equal 3- or 5-mm segments. Each segment is placed in a vial and solubilized, scintillation fluid is added, and radioactivity is measured as shown. The outflow pattern is displayed on the ordinate on a log scale in counts per minute (CPM) with the distance on the abscissa, in mm from the ganglion, taken as 0. Distally in the nerve a plateau of radioactivity is seen to rise to a crest before abruptly falling to baseline levels forming the front of the crest. The transport distance is calculated from the foot (arrow) to the pool of radioactivity in the ganglion. To the left of the dashed line the injection of the precursor into the L7 motoneuron cell body region in the spinal cord is shown. On removal of the ventral root and sciatic nerve at a later time, a similar outflow pattern is seen. From Ochs (1977b).

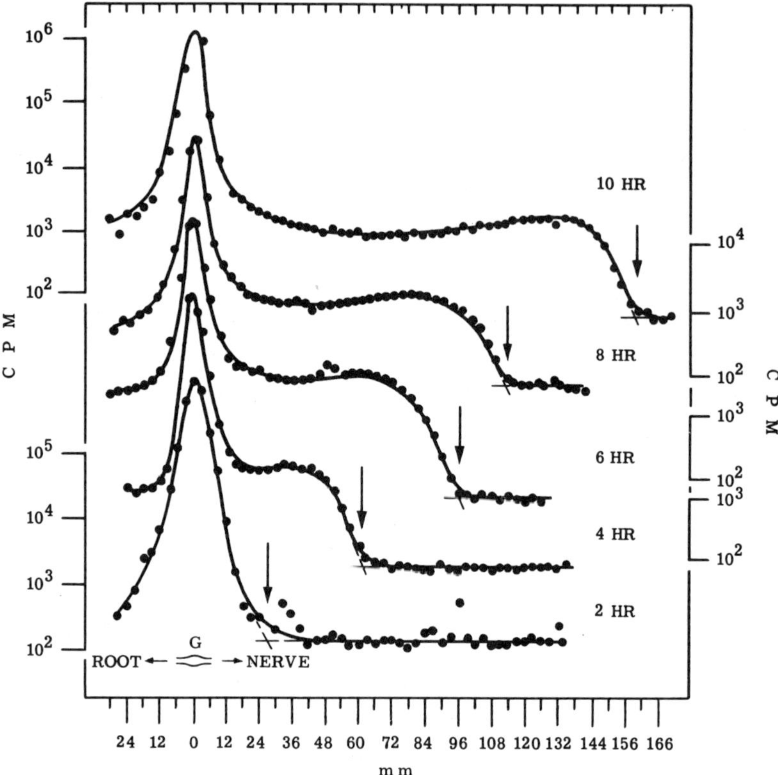

Fig. 2.11 Distribution of radioactivity showing fast transport in sciatic nerves. The nerves of five cats taken 2 to 10 hr after injection of ^{3}H-leucine into the L7 ganglia (G) shows the distribution of radioactivity in 5-mm segments of roots, ganglia, and nerves as described in Fig. 2.10. The nerve taken 2 hr after injection, shown at the bottom with its ordinate on a logarithmic scale at the left, gives no evidence of an outflow; the time allowed for outflow was too short. All the other nerves show the characteristic crests and fronts which advance with time. At the top left a scale is given for the nerve taken 10 hr after injection. Only partial scales are shown at the right for the nerves taken 4, 6, and 8 hr after injection. The abscissa gives distance in mm from the ganglion taken as 0. From Ochs (1972c).

of radioactivity was seen to extend distally into the nerves, characteristically rising to a crest before rapidly dropping to the baseline level of activity. The drop represents the front of advancing radioactivity transported down within the fibers. The fronts show, by the arrows, a progressive distal shift in the nerves with time after injection. Taking the distance from the foot at the front of the crest with the baseline back to the zero point at the center of high activity in the ganglia, and knowing the time allowed for downflow, the rate of transport determined from a large number of such experiments was shown to be constant at 410 ± 50 (S.D.) mm/day (Ochs, 1972a).

The background level of radioactivity in the nerve distal to the front of transported radioactivity represents precursor which had escaped into the circulation to become incorporated by the Schwann cells and other non-axonal elements of the nerve. This was demonstrated by injecting ^{3}H-leucine into the L7 dorsal root ganglion on only one side. The nerve on that side showed the usual outflow, whereas the nerve on the uninjected side showed only the baseline level of activity.

When lower levels of precursor were injected, a lesser amount of incorporated radioactivity was transported in the crest, but the rate of transport remains unchanged. The relation between the radioactivity present in the crest and the pool of radioactivity remaining in the ganglion may not be linear. Over the lower ranges of precursor injected, the crest amplitude increases linearly with respect to the pool of radioactivity in the ganglion. It leveled off when higher amounts of precursor were injected (Ochs, 1976). A number of factors may cause deviations from linearity. The concentration of precursor in the ganglion falls off non-linearly with distance from the site of injection, the uptake of precursor by the cells depending on the size of the ganglion, the injection volume, and the concentration of the precursor. Muñoz-Martínez, Núñez, and Sanderson (1981) found a linear relation between the radioactivity remaining in the ganglion pool and that which appears in the plateau (Fig. 2.12).

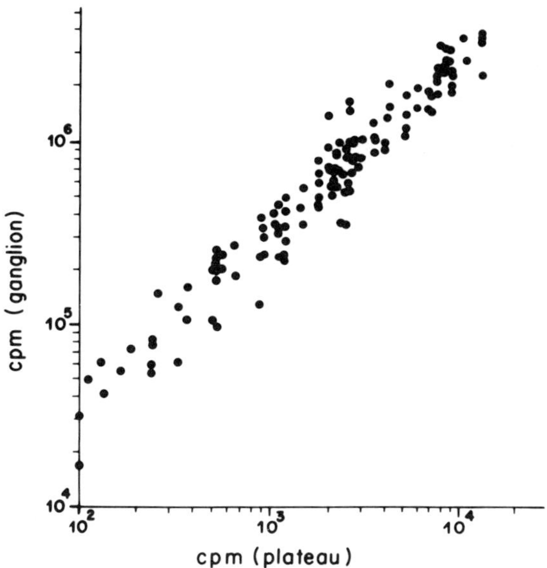

Fig. 2.12 Relation of radioactivity in the plateau to the ganglion. ^{3}H-leucine was injected into the L7 dorsal root ganglia of cats, and the varying amounts of radioactivity present in the plateau shows a regular relation to the height of ganglion peak. From Muñoz-Martínez *et al.* (1981).

Another factor to be considered is that different precursors are not taken up by the cell bodies to the same extent. Amino acids are actively carried into the cells by membrane transport mechanisms (Lajtha, 1964). Other substances, such as Ca^{2+} (Chapter 8), enter neuron cell bodies to a much smaller extent. Some precursors, such as glucose, which are readily taken up by the cells, are rapidly metabolized leaving only a relatively small portion available for incorporation into components transported down the axons (cf. Chapter 5).

2. Transport in Dorsal Root Ganglion Fibers Ascending in the Dorsal Columns

The neurons of the dorsal root ganglion are T-shaped. The initial segment extends from the cell body and branches into two daughter fibers, the dorsal root branch ascending to enter the spinal cord, the other, the sensory nerve branch, descending in the peripheral nerve (cf. routing, Chapter 11). The dorsal roots of the L7 ganglia of monkeys have a relatively long length of 10 cm or more. The rate of transport in them was found to be the same as that in the sensory fibers after injecting their L7 dorsal root ganglia with ^{3}H-leucine (cf. Fig. 11.12).

The dorsal root fibers entering the spinal cord undergo a three-fold division. The long branch, which ascends in the dorsal columns of the spinal cord, is of particular interest in assessing the pattern and rate of axoplasmic transport in a central tract in comparison to the peripheral nerve. The dorsal root L7 ganglia are injected with ^{3}H-leucine, and after a suitably long period of time allowed for transport, the dorsal columns are removed and sectioned in the same manner as sciatic nerves (Ochs, 1972a). The outflow pattern and the front of transport in the dorsal columns were found to be similar to those in the sciatic nerve (Fig. 2.13).

The similarity of the front positions indicates that the rate of fast transport in those CNS fibers is the same as in the peripheral fibers.

The increased level of radioactivity at the entry of the dorsal root into the spinal cord represents an artifact of tissue sampling, namely, the inclusion of collateral fibers dividing locally in the spinal cord. Their radioactivity is adventitiously added to segments taken at the entry of the dorsal root fibers to the cord with which they are enmeshed.

3. Comparison of Transport in Motor and Sensory Nerve Fibers

To assess the transport rate in motor fibers and compare it with sensory fibers, ^{3}H-leucine was injected into the L7 segment of the ventral horn region of the cat spinal cord on one side, and into the L7 dorsal root ganglion on the other side. The crest patterns subsequently found in the sciatic nerves on the two sides were the same; however, the positions of the fronts in the sensory and motor nerve outflows were different (Fig. 2.14).

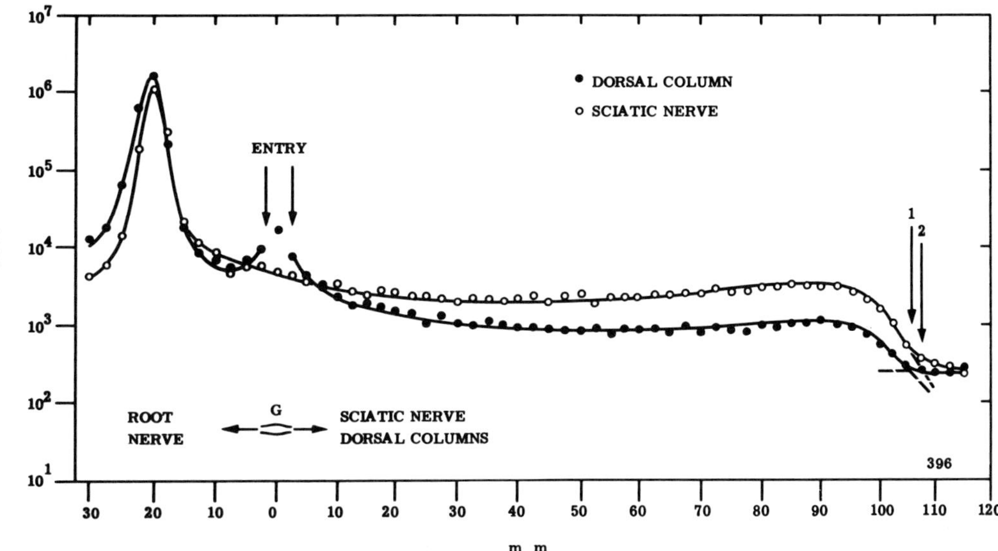

Fig. 2.13 Transport in dorsal columns. Following injection of ^{3}H-leucine into the L7 dorsal ganglia of a large cat, the outflow in the dorsal columns (●) and nerve (○) are seen to be similar. The peroneal and tibial nerve branches of the sciatic nerve were taken down to the ankle, allowing the full pattern of radioactivity to be seen over a long length of nerve. At the entry zone of the dorsal roots into the dorsal column an adventitious increased peak of radioactivity is seen. From Ochs (1972a).

The relative displacement of the two fronts is fully accounted for by the distance between the L7 motoneurons anteriorly and the dorsal root ganglion cell bodies posteriorly. This explanation is supported by comparable studies made in monkeys which have a greater antero-posterior displacement of their L7 motoneurons and L7 dorsal root ganglia. The greater discrepancy in the fronts corresponded directly to the antero-posterior distance between the two sets of cell bodies (Fig. 2.15).

We can conclude from these results that the rates of axoplasmic transport in the sensory and motor fibers are the same.

The rates of fast axoplasmic transport found for various other mammalian species in sensory and motor fibers were closely similar, with an average value close to 410 mm/day, in spite of the marked variations in the lengths of nerves taken from animals of various sizes (Table 2.1).

The relatively small differences in the rates given in Table 2.1 are likely to be due to other factors, mainly to variations in body temperature. As will be described in Section C2 below, even as little a variation as 0.5°C in body temperature would amount to a 15 mm/day difference in the rate of fast axoplasmic transport. Nerves removed from the animal are placed on a piece of stiff cardboard for sectioning. Discrepancies in tension and somewhat different degrees of stretch placed on the nerves due, for example, to differ-

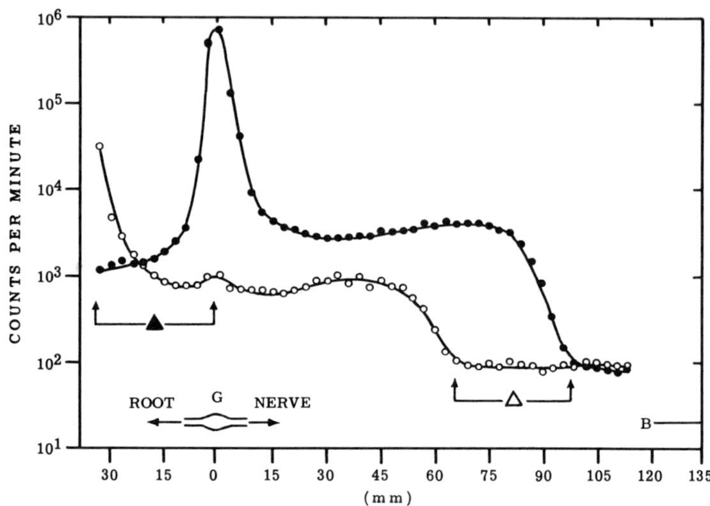

Fig. 2.14 Fast transport in sensory and motor fibers. The L7 motoneuron region of the cat was injected with ^{3}H-leucine on the left side of the cord (○), the L7 ganglion on the right side (●). Six hours later the nerves were removed and similar crests of outflow were seen. The anteroposterior displacement of the fronts of the outflow of radioactivity (△) on the two sides are identical to the anatomical anteroposterior displacement of the cord motoneurons and ganglion cell bodies (▲). From Ochs and Ranish (1969).

ences in their weight, might cause the larger nerves to be somewhat more extended and alter the calculated rate. Taking these factors into account, the rate of fast transport appears to be remarkably constant in the sensory and motor nerves of a wide range of animals ranging from the rat to the goat.

4. Rate of Transport with Respect to Fiber Diameter

The possibility that the rate of axoplasmic transport might vary with the diameter of the nerve fiber was assessed using autoradiography (Ochs, 1972a). After L7 dorsal root ganglion injection with ^{3}H-leucine, small pieces were taken from the nerves at the position where the front of fast advancing radioactivity was estimated to be present. The regularity of the rate made this possible, as shown by the outflow in the remaining portions of the nerve (Fig. 2.16).

The pieces of nerves taken from the front were freeze-substituted for histological examination, sectioned, and coated with photographic emulsion for autoradiography. Myelinated nerve fibers ranging in diameter from 20–22 μm down to 3–4 μm, were seen to contain grains due to radioactivity located over their axons (Fig. 2.17).

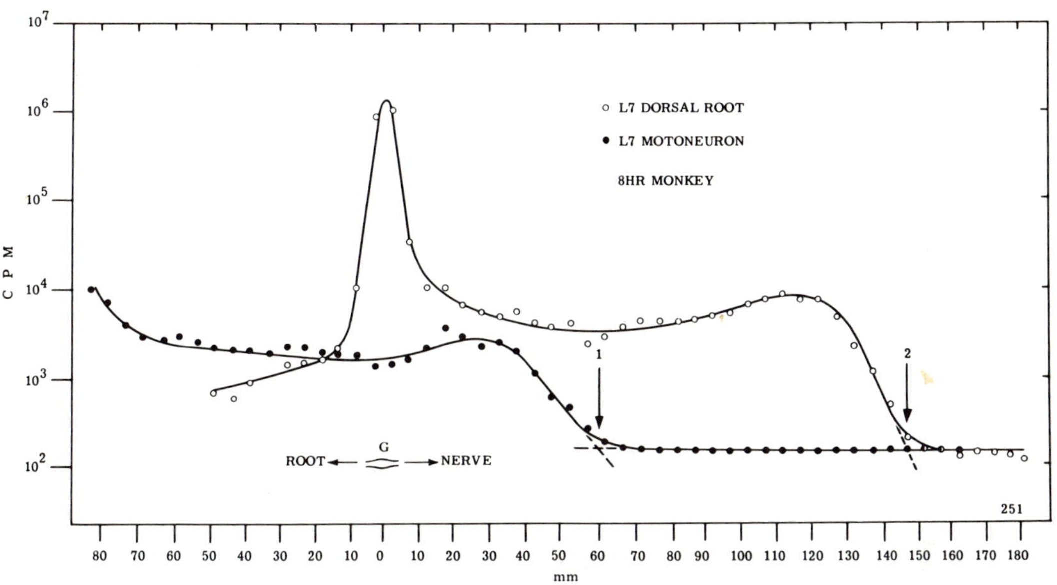

Fig. 2.15 Comparison of downflow in sensory fibers and in motor fibers of the monkey. The L7 dorsal root ganglion (G) was injected on the right side (○) and the L7 motoneurones in the ventral horn of the spinal cord on the left side (●) with ^{3}H-leucine (cf. Fig. 2.14). After 8 hr of outflow, the front on the dorsal root ganglion injected side is more distal (arrow 2) than on the motoneuron injected side (arrow 1). The difference, 85 mm, was identical to the anteroposterior displacement of the cells in the ganglion and the motoneurones in the cord, a distance which is greater in this species than in the cat. From Ochs (1972a).

TABLE 2.1   Rates of Fast Axoplasmic

Nerve type	No. of nerves	Species	Rate[a]
Motor (sciatic)	18	Rat	411±50
	5	Monkey	400±35
Sensory (sciatic)	14	Monkey	416±30
	26	Cat	409±50
	2	Rabbit	394 (360, 415)
	2	Goat	389 (382, 396)
	4	Dog	423±15
Dorsal column spinal cord	3	Monkey	397±57
	4	Cat	391±59

Rate values given as means and S.D. Sensory nerves, after L7 dorsal root ganglia injection with ^{3}H-leucine, motor nerves after cord injection of ^{3}H-leucine for uptake by L7 motoneurons.

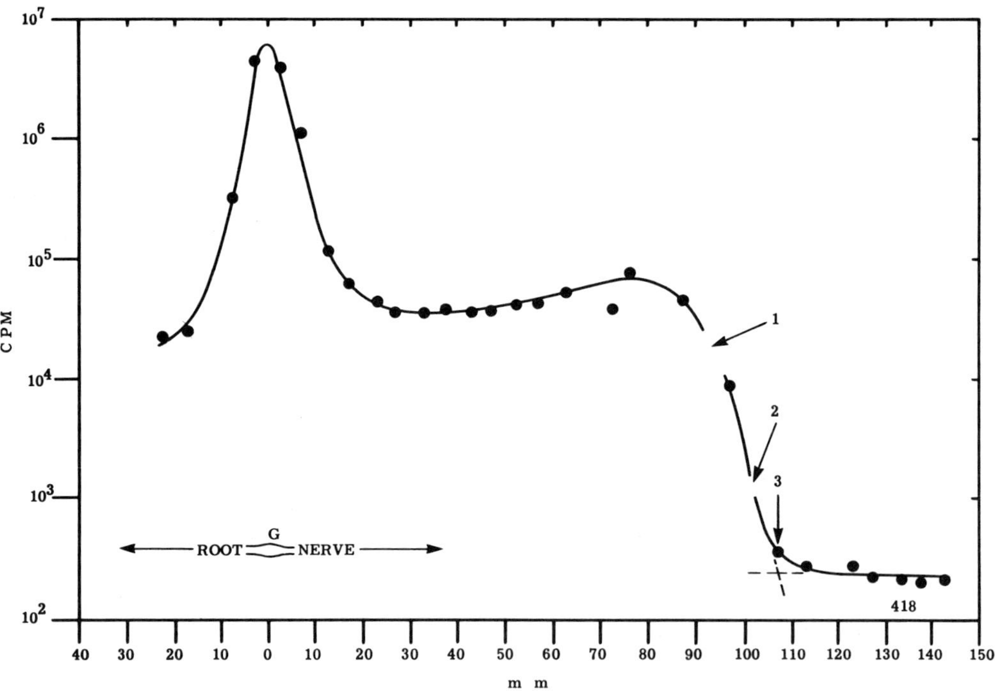

Fig. 2.16 Sections taken from the front for autoradiography. The sciatic nerve was removed 6 hr after L7 dorsal root ganglion (G) injection with ^{3}H-leucine. Arrows 1 and 2 represent segments of nerve taken from the front for histological preparation and electron microscopic autoradiography, and arrow 3 represents the foot of the advancing crest of fast transported activity. The curve determined from the remaining parts of the nerve shows a typical transport outflow. From Ochs and Jersild (1974).

The fact that the grains were seen within myelinated nerve fibers of all diameters at the front of downflow of radioactivity indicates that they all have the same rate of fast transport. If the rate, for example, were slower in the small-diametered fibers, no grains would have been present in them whereas grains would be contained in the larger fibers. Similarly, if the rate of transport were faster in the small-diametered fibers, only they would be seen to contain grains and not the large-diametered fibers.

A fast rate was reported in unmyelinated fibers by Byers *et al.* (1973). Using the procedure described above for evaluating the effect of fiber diameter on transport rate (Ochs and Jersild, 1974), grains were found located inside the unmyelinated fibers as well as the myelinated fibers (Fig. 2.18).

Finding grains of radioactivity in the unmyelinated fibers at the front

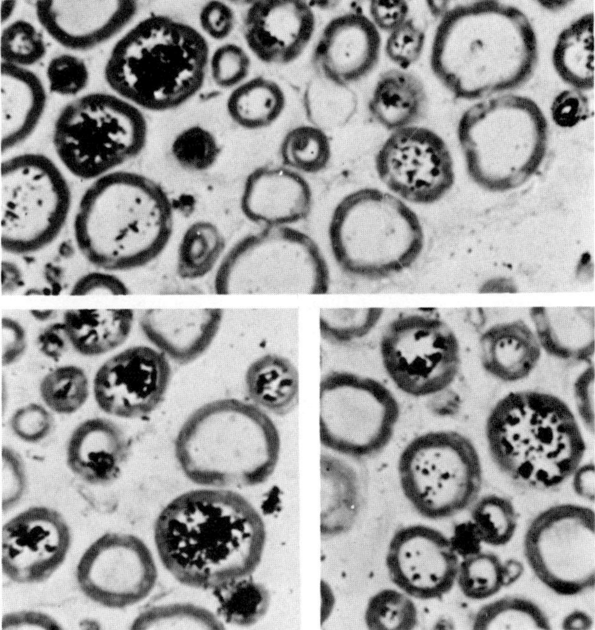

Fig. 2.17 Autoradiographs of nerve segments taken from the front of outflow. After injection of the cat L7 ganglion with ^{3}H-leucine, the nerve was removed 5 hr later and a short nerve segment was taken from the front as in Fig. 2.16 and prepared for autoradiography by freeze-substitution. Myelinated nerve fibers are shown as nearly circular profiles, dark osmium-stained compact myelin sheaths. The fibers ranged in diameter from 3 to 23 μm, with fibers of all diameters labeled but to varying degrees. Large nerve fibers show fairly uniform scattering of labeled grains of radioactivity over the axon. Some fibers show a grouping of grains with a clear space in the center. Small fibers in the size range of about 4 μm also show grains present inside them. From Ochs (1972a).

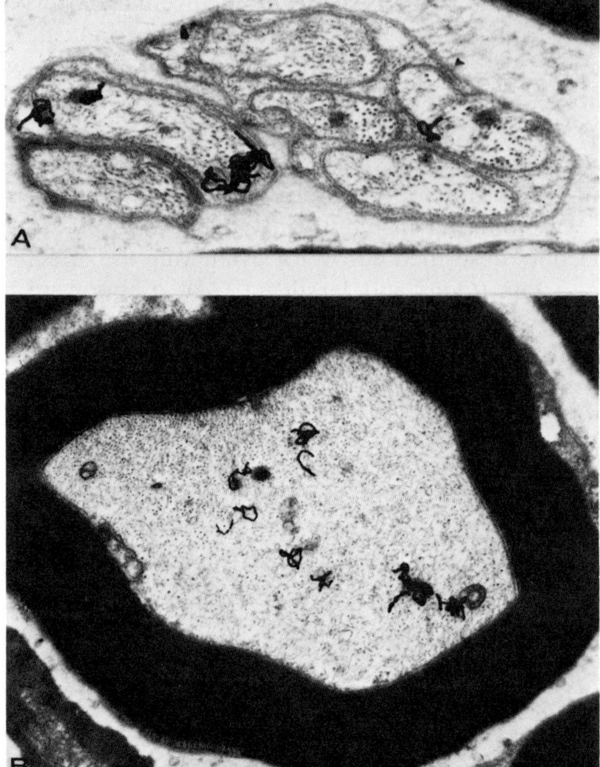

Fig. 2.18 Electron microscope autoradiographs of fibers. Unmyelinated (A) and myelinated (B) nerve fibers taken from nerve segments at the front of the crest of fast
transported activity are shown (see Fig. 2.16). Grains are seen present over two of
the unmyelinated axons and within the myelinated axon. Magnification: (A) 14,400×,
(B) 8,600×. From Ochs and Jersild (1974).

indicates that they also have the same fast transport rate close to 410 mm/
day as myelinated fibers.

In both the light microscopy and EM autoradiographs, it was noted that
some fibers were filled with grains while others showed few or no grains.
Obviously, no radioactivity would be expected in the motor fibers of a mixed
nerve after injecting the L7 dorsal root ganglion with labeled precursor, or in
the sensory fibers after injecting the L7 ventral horn region for uptake by
motoneurons. However, when autoradiographs of the dorsal root fibers were
examined after L7 dorsal root ganglion injection, or ventral roots after L7
ventral horn injection, a number of fibers were found to have little or no
grains of radioactivity. This variability may be related to geometrical differences in the uptake by the cell bodies of the precursor. These geometrical

differences depend on the pattern of spread of the precursor to cells from the injected site, and thus the concentration of the precursor reaching the cells. Perhaps there are also cyclic variations in the synthetic capabilities of the individual cells, with some temporarily low compared to others at the time of injection.

5.   Transport in Nerve Fibers Independent of the Cell Body

In some earlier theories of the mechanism of transport, a force exerted by the cell body was invoked to cause materials to flow down the fibers (Chapters 1 and 10). We now know that the transport mechanism is locally present all along the fibers (cf. Ochs, 1971a; Chapter 10) and that it is independent of a direct effect exerted from the nerve cell body (Ochs and Ranish, 1969). This was shown by injecting the L7 dorsal root ganglia with ^{3}H-leucine and allowing a transport in the fibers for several hours before ligating or otherwise interrupting a further outflow from the cell bodies into their fibers. Those labeled materials which had gained entry to the nerve fibers below the ligation continue to be transported at the same fast rate. As an example, 2 hr after injection of ^{3}H-leucine into the L7 dorsal root ganglia, a ligation was made just distal to the ganglion on one side and a further time of 7 hr for transport allowed. The front of labeled activity in the ligated nerve is seen to have moved down to the same distance as the front in the unligated nerve (Fig. 2.19).

An obvious difference between the two outflow patterns is the much lower level of radioactivity in the plateau behind the crest in the ligated nerve. This is due to the lack of materials which normally would continue to have been contributed by the cell bodies to the fibers for transport at the later times but are prevented from doing so by the ligation. The radioactivity which is found present in the plateau represents labeled components which have dropped off locally in the axons from the advancing crest of radioactivity. Local depositions of labeled components will be further assessed in relation to slow transport in Chapter 11. We note here that by allowing more time to pass before making ligations after an injection of the precursor, the level of the plateau increases. This is the case because more labeled components are contributed to the crest and thus more radioactivity can be locally deposited.

6.   Is There a "Superfast" Wave?

A fairly large number of studies have now shown agreement with a rate of 410 mm/day for the fast axoplasmic transport in peripheral nerves. Those studies will be cited in this chapter and elsewhere in this book. Komiya and Austin (1974) and Tang, Komiya, and Austin (1974), however, described an additional superfast wave of downflow in the peripheral nerve of mice, one

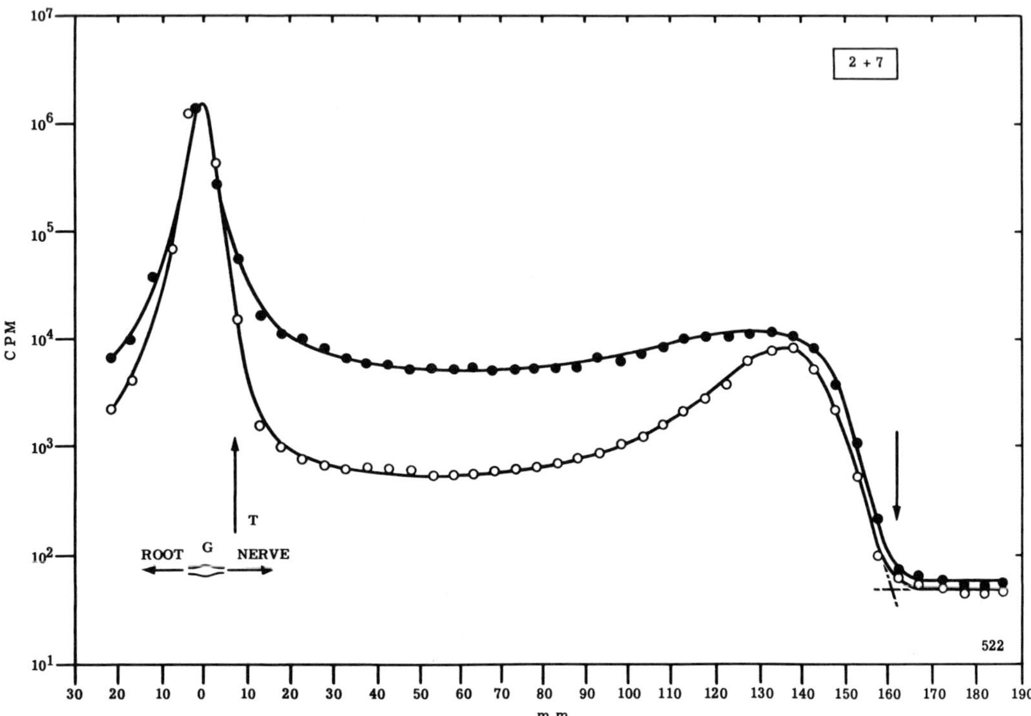

Fig. 2.19 Transport following a ligation. The distribution of radioactivity in an unligated nerve 9 hr after injection of ^{3}H-leucine into the cat L7 dorsal root ganglion shows the characteristic plateau with a rise to a crest before an abrupt fall at its front (●). The arrow shows the intersection of the front with the base line. A ligation (T) was made 2 hr after injection in the nerve on the other (○) just below the ganglion with a further 7 hr of downflow time allowed. The plateau level is much lower than the crest which now appears as a peak. The front has moved, however, to the same position as in the case of the unligated nerve. In each nerve the root, ganglion, and sciatic nerves are sectioned at 5-mm intervals and the activity given in counts per minute (cpm) on the ordinate on a logarithmic scale. From Ochs (1975c).

with an apparent rate of 2000 mm/day. Bisby (1975) failed to find evidence for a superfast wave. Repeated attempts to find it in the cat were also unsuccessful (Ochs, unpublished experiments). The study of axoplasmic transport employing the technique of isotope labeling in mice poses special difficulties. The position of the injection tips in the spinal cord is quite critical and there is increased danger in such a small animal of a diffusive spread of radioactivity into the endoneural space of the nerve. Diffusion is often observed when spinal cord injections are made in the rat where a spread as long as 45 mm in the nerve was seen (Ochs, 1973).

Bradley and Jaros (1973), Boegman, Wood, and Pinaud (1975), and Watson *et al.* (1980) saw what appeared to be superfast waves of outflow in the sciatic nerves of rats treated with pargyline. This agent produces what is considered to be a model of dystrophy (Chapter 12). In contrast to the rate of 390 mm/day measured in control rats, waves with rates of 1,230–1,996 mm/day were reported for the pargyline-treated rats. On the other hand, only the usual fast front of movement was found in pargyline-treated animals by Ranish and Dettbarn (1977), and a failure to find superfast waves was also reported by McLane and McClure (1977). The explanation for the discordant findings with pargyline remains unknown.

C. *IN VITRO* TRANSPORT

1. Transport Rate and Characteristic Pattern of Outflow

As will be described in Chapter 7, transport is closely related to metabolism. This indicates that temperature should have a strong effect on transport. To assess this factor, and, as well, the effects of other physical agents and the actions of various pharmacological agents (Chapter 12), an *in vitro* preparation is required. As shown in the preceding section, axoplasmic transport can continue in nerve fibers isolated from their cell bodies for many hours, as long as an adequate blood supply is maintained. Interfering with the vascular supply of nerve *in vivo* to cause an anoxia results in a rapid block of transport (Chapter 7). This suggested that transport can be maintained *in vitro* if an adequate supply of oxygen is provided, and this was shown to be the case (Ochs and Ranish, 1970). Ganglia were injected with ^{3}H-leucine and after several hours of transport *in vivo,* the nerves were removed from the animals and placed in chambers supplied with 95% O_2 and 5% CO_2 while kept moist at a temperature of 38°C. Alternatively, nerves removed from animals after several hours of transport were placed in 25-ml Erlenmyer flasks containing 15–20 ml of Ringer solution, the medium continuously bubbled with 95% O_2 + 5% CO_2, and the flasks kept at 38°C in a water bath for the full period of *in vitro* transport allowed. In any case, the pattern of outflow of activity in the nerves after several hours of *in vivo* downflow plus additional time of *in vitro* downflow was similar to that obtained after a comparable time of *in vivo* transport (Fig. 2.20).

The pattern of outflow remained the same regardless of how the proportion of time allowed for *in vivo* transport to *in vitro* transport was varied. In all cases the same fast rate was found for *in vitro* transport as *in vivo* transport. The *in vitro* component assessed by subtracting the time of *in vivo* transport shows that it has the same rate of close to 410 mm/day.

Fink and his colleagues (Fink *et al.*, 1972; Kennedy *et al.*, 1972) used the vagus nerve to study the effects of anesthetics and other agents on axoplasmic transport *in vitro*. The cell bodies in the floor of the 4th ventricle of rabbits were first exposed to topically applied ^{3}H-leucine. Then, after

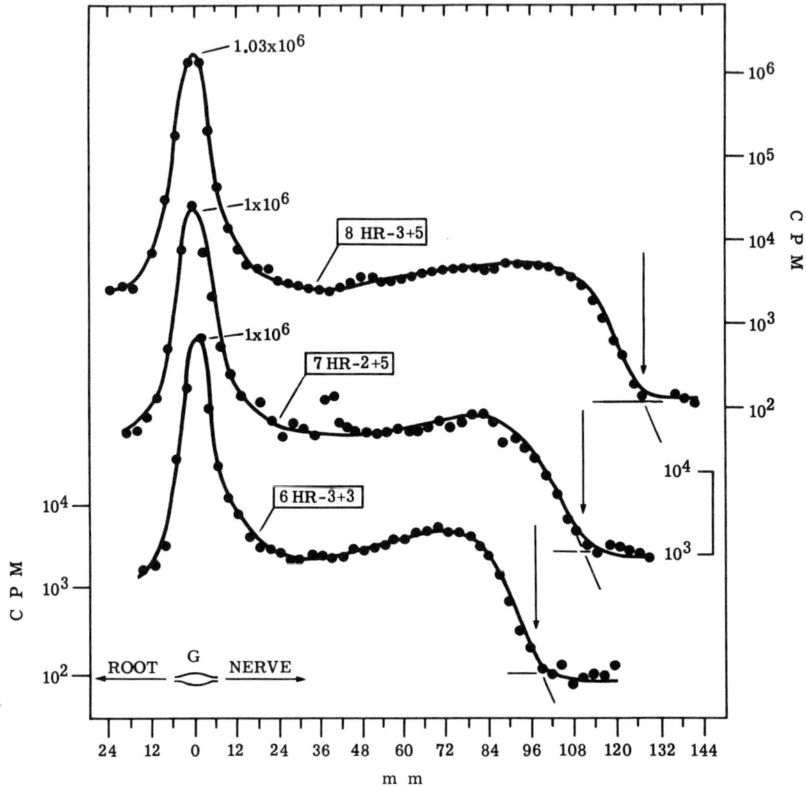

Fig. 2.20 Characteristics of *in vitro* transport. The upper curve shows a 3-hr *in vivo* downflow plus 5-hr *in vitro* downflow for a total transport time of 8 hr. The middle curve shows a 2-hr *in vivo* downflow plus 5 hr of *in vitro* downflow for a total transport time of 7 hr. The bottom curve shows a 3-hr *in vivo* downflow plus a 3-hr *in vitro* downflow for a total transport time of 6 hr. In all cases the same characteristic outflow pattern and fast rate of transport were seen. The ordinate scale on the upper right is for the uppermost curve, the short scale in the middle is for the middle curve, and the lower left-hand scale is for the lowermost curve. Arrows at the front of the descending crests of activity show the intersection of the fronts with the base line. From Ochs and Smith (1975).

allowing several hours of downflow *in vivo,* the vagus nerves were removed, transferred to Petri dishes containing a lactate–Ringer medium, and oxygenated for a further period of *in vitro* downflow. The peaks of the labeled materials were seen to move distally in the nerves at near the usual fast axoplasmic transport rate (Fig. 2.21).

In this preparation, the radioactivity in the plateau behind the peaks was greatly reduced because the vagus nerves were cut off from their cell bodies so that the later outflow from the cells could no longer contribute to the plateau (cf. Fig. 2.19 above). The distal shift in the peaks with time shown in Figure 2.21 gave an apparent rate of 15 mm/hr or 360 mm/day for transport. This is reasonably close to the fast transport rate of 410 mm/day determined for sciatic nerves, and the rate of 380 to 410 mm/day found *in vivo* for the rabbit vagus by Sjöstrand (1969). It is likely that the somewhat smaller rate was due to measuring the peak positions of outflow rather than the fronts. The broadening of the peaks with time and distance acts to lower the computed rate of transport as is suggested by comparable studies made using neurotransmitter and transmitter-related enzymes in adrenergic nerve fibers (Chapter 4).

2. Effect of Temperature on the Rate of Transport

The effect on transport of varying temperature over a wide range on the rate of transport in mammalian nerves was determined using the *in vitro* preparation (Ochs and Smith, 1971a, 1975). A 2-hr period of downflow of labeled components *in vivo* was allowed before removing the nerves for a further

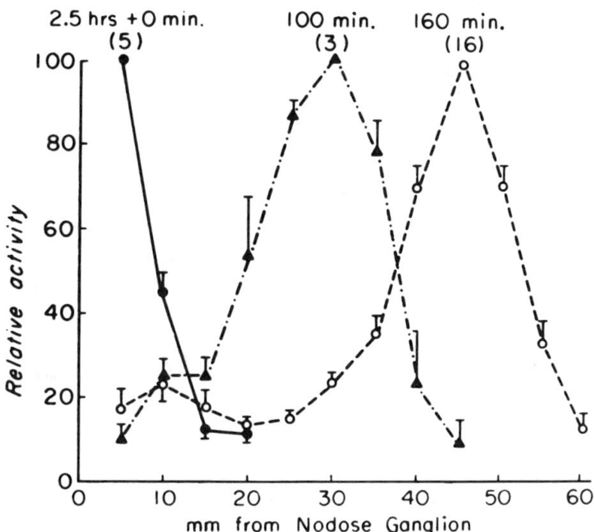

Fig. 2.21 *In vitro* transport in vagus nerve. After topical application of labeled precursor to the floor of the 4th ventricle containing hypoglossal cell bodies, the hypoglossal nerves were taken 2.5 hr later and incubated for a further time of 100 and 160 min. The resulting peaks of labeled radioactivity show an *in vitro* transport at close to the expected fast rate. The numbers of nerves averaged at these times are shown in brackets. From Kennedy *et al.* (1972).

period of *in vitro* transport at a specified temperature. An example of transport *in vitro* at a temperature of 28°C is shown in Fig. 2.22.

The *in vitro* rate of transport at 28°C was determined by subtracting the *in vivo* rate of transport of 410 mm/day. The *in vitro* transport at 28°C so determined had a rate of 212 mm/day, about half the usual rate at 38°C. From a number of such studies carried out at different temperatures, the Q_{10}, the change in the rate of transport caused by a 10°C change of temperature, was calculated:

$$Q_{10} = \frac{k_1}{k_2} \cdot \frac{10}{t_1 - t_2}$$

where k_1 is the rate at the higher temperature t_1 and k_2 is the rate at the lower temperature t_2.

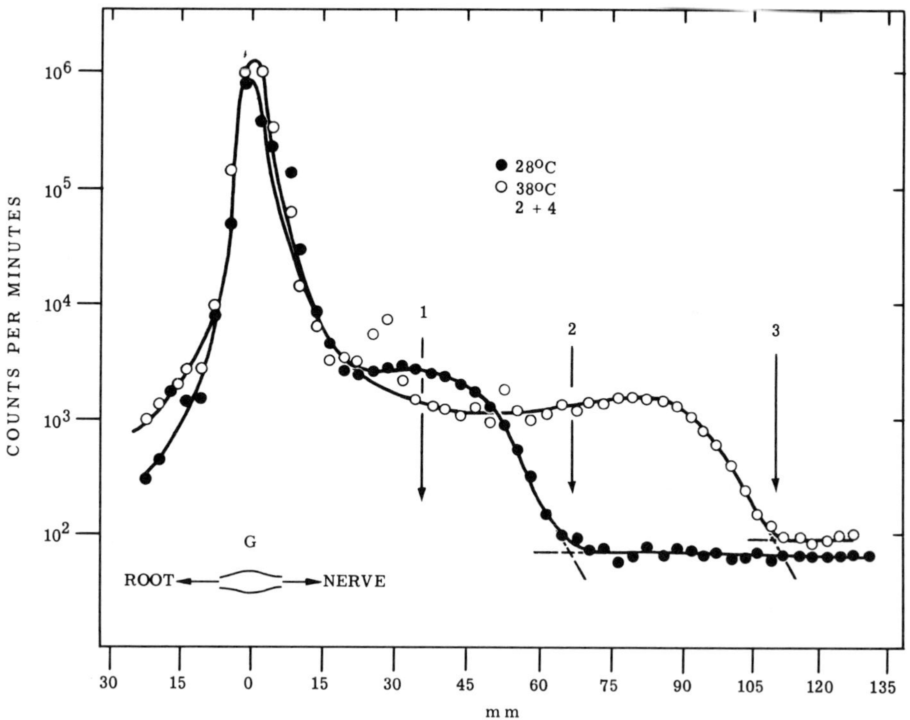

Fig. 2.22 Effect of lower temperature on the rate of transport *in vitro*. Nerves taken 2 hr after injection of the ganglia with ³H-leucine were placed in a chamber for a further 4 hr of *in vitro* downflow. A control nerve (○) was kept at 38°C, the nerve from the opposite side (●) at 28°C. Arrow 1 indicates downflow to the position expected of 2-hr downflow *in vivo*. Arrow 2 shows the extent of additional transport at 28°C for 4 hr *in vitro*. Arrow 3 shows transport in a control nerve at 38°C for 4 hr *in vitro*. From Ochs and Smith (1975).

Over a wide range of temperatures the Q_{10} was found to be approximately 2.0–2.3 (Table 2.2).

Other estimations of Q_{10} have ranged from 2 to 3.5 (cf. Chapter 3). The significance of a Q_{10} in this range is that these are values typical of a chemical reaction rather than of diffusion. The higher Q_{10} finds a rationale in the enzymatic utilization of ATP as the source of energy required to drive the transport mechanism (Chapters 7 and 10).

In addition to the slowing of the rate of transport at lower temperatures, a decrease in the slope of the front of outflow was also seen at lower temperatures (Fig. 2.23).

In this example the nerve was kept at 18°C for 22 hr of *in vitro* downflow and the front had moved to a distance of 115 mm. After subtracting the *in vivo* downflow as described above, the rate of transport *in vitro* at 18°C was determined to be 96–111 mm/day with a Q_{10} close to 2.0. At still lower temperatures the slope was even shallower.

It is interesting that the Q_{10} for transport in the ground squirrel, a hibernating animal, does not change when it is maintained at a low temperature to induce hibernation (Bisby and Jones, 1978). A Q_{10} of 2.51 was found for the sensory fibers of their sciatic nerves as compared to 2.44 for the non-hibernating squirrel and a Q_{10} of 2.62 found in the rat. Similar values were reported by Cosens *et al.* (1976).

Another phenomenon appears at a temperature of close to 11°C, namely, a maintained arrest of fast axoplasmic transport which was termed a "cold-block" (Ochs and Smith, 1975). At all temperatures below the cold-block temperature of 11°C down to 0°C, no transport at all occurred even though the nerves were kept for long periods of time at the low temperature (Fig. 2.24).

TABLE 2.2　Rate at Lower Temperatures

No. of nerves	Temp. (°C)	Rate *in vitro* (mm/hr ± S.D.)	Rate (mm/day)	Q_{10}	Duration *in vitro* (hr)
11	38	17.1 ± 1.2	410 ± 29		6
10	28	8.83 ± 1.6	212 ± 38	1.9	3–5
11	18	4.63 ± 1.4	111 ± 35	2.0	2.6–5
10	18	4.0 ± 0.34	96 ± 8.2	2.1	17–22
4	13	2.18 ± 0.0+	52 ± 0.06	2.3	20.5–22
4	11	—	—	—	22
10	10	0	0	—	16–22
22	8	0	0	—	16–46
2	0	0	0	—	24

[a] The *in vivo* distance was subtracted from the total *in vitro* time to give the rate of transport at various temperatures. Means are given ± S.D. Q_{10}'s were determined at the various temperatures for *in vitro* transport in comparison to the rate determined at 38°C.

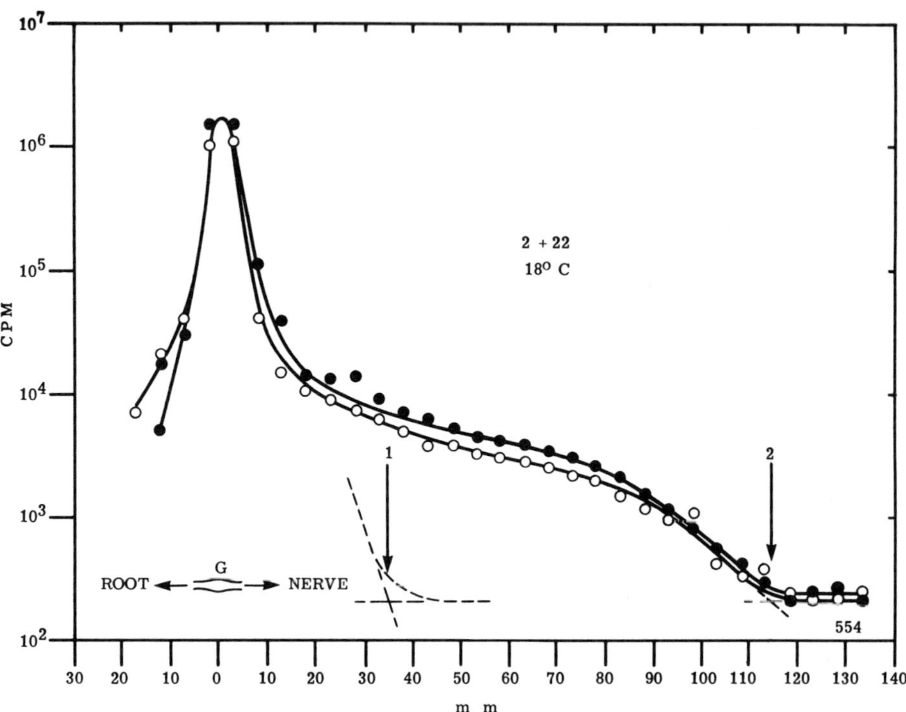

Fig. 2.23 Transport *in vitro* at 18°C. After 2 hr of *in vivo* downflow, nerves were removed from the cat and maintained *in vitro* at 18°C for a total downflow period of 24 hr. Arrow 1 shows the point to which transport would have moved *in vivo* after 2 hr. Arrow 2 shows the furthest downflow in the nerves seen after an additional 22 hr of *in vitro* transport at this low temperature. The slope of the front is reduced, a characteristic effect of transport at this low temperature. From Ochs and Smith (1975).

Bisby and Jones (1978) also found a cold-block of transport below 13°C, as did Brimijoin (Chapter 7). Cold-block is not an irreversible process. When the nerves are warmed back to 38°C, axoplasmic transport resumes (Table 2.3). A small change in rate was seen following recovery from long-lasting periods of cold-block, as will be further described in Chapter 12. Brimijoin (1975) made use of this phenomenon to assess the rate of transport of endogenous components, the adrenergic transmitter noradrenaline, the adrenergic enzyme dopamine-β-hydroxylase and acetylcholinesterase (Chapter 4).

D.   FAST TRANSPORT IN NON-MAMMALIAN NERVES

Outflow characteristics similar to those seen in mammal nerves are found in non-mammalian nerves. Abe *et al.* (1973) injected the 9th dorsal root ganglion of the frog with ³H-leucine and, after allowing time for *in vitro* trans-

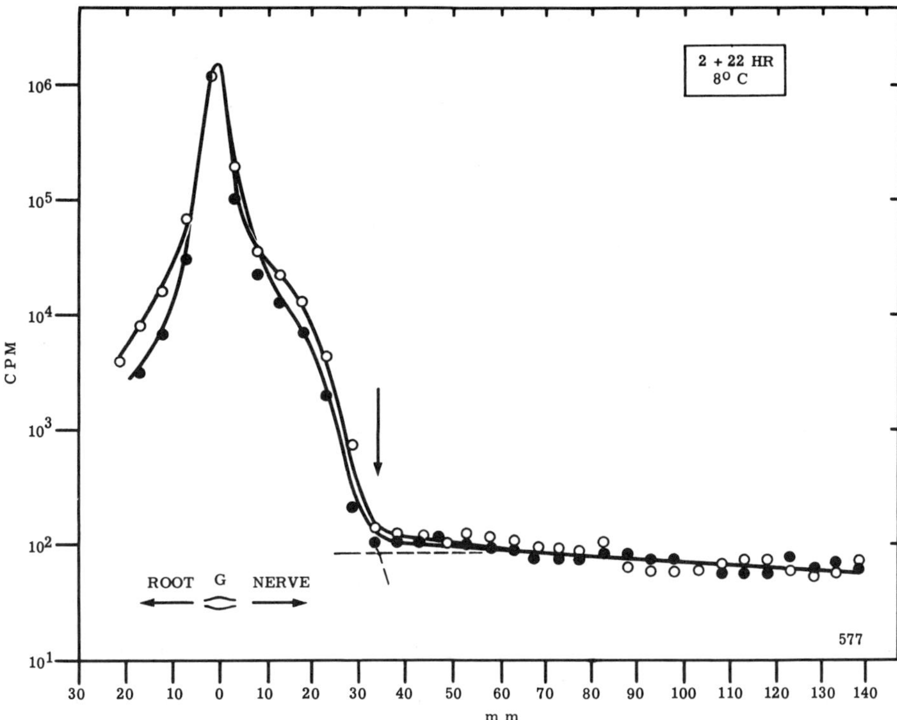

Fig. 2.24 Cold-block of *in vitro* transport. In experiments similar to those of Fig. 2.23, nerves were taken after 2 hr of *in vivo* transport and then incubated at 8°C for a period of 22 hr. Cold-block is shown by the lack of movement beyond that of the transport which had occurred *in vivo* indicated by the arrow. From Ochs and Smith (1975).

port at a temperature of 25°C, obtained an outflow pattern characteristic of mammalian nerves (Fig. 2.25).

Two outflow curves are shown in this figure for two different durations of *in vitro* downflow. From such studies, the fronts were found to advance linearly with time at a fast rate (Fig. 2.26).

When those rates of transport were scaled to a temperature of 38°C and the Q_{10} of 2.6 established for the frog nerve taken into account, the rate of axoplasmic transport was determined to be 405 mm/day, one quite close to that obtained in the mammalian preparation. The somewhat different shape of outflow seen in *b* of Fig. 2.25 when using labeled choline as a precursor will be discussed in Chapter 5.

In another variation of the *in vitro* technique, Edström and Hanson (1973b) employed a special compartmented chamber which allowed ^{3}H-leucine to be applied to the ganglia for its incorporation while transport in the nerve fibers could be assessed in another compartment of the chamber. Leakage of the

TABLE 2.3   Recovery Rates After Cold-Block

Expt.[a]	Temp. (°C)	Time *in vivo* (hr)	Time at low temp. *in vitro* (hr)	Recovery time *in vitro* (hr)	Total distance (mm)	Total distance *in vitro* (mm)	Apparent recovery rate (mm/hr)
A	10	2	22	5	105	71	14.2
	10	2	22	5	105	71	14.2
	10	2	22	5	112	78	15.6
	10	2	22	5	110	76	15.2
B	8	2	17	5	92	58	11.6
	8	2	18	5	105	71	14.2
C	0	2	13	4	95	61	15.2
	0	2	13	4	95	61	15.2
	0	2	16.5	5	100	66	13.2
	0	2	16.5	4	100	66	13.2
	0	2	17	4	112	78	15.6
	0	2	17	4	110	76	15.2
						Mean =	14.38 ± 1.23
						=	318 ± 29 mm/day

[a]A shows a group of cold-blocked nerves kept at 10°C for a period of 22 hr, then transferred to flasks at 38°C for 5 hr of recovery. The movement of the crest during the 5-hr recovery period was determined after subtracting the *in vivo* distance. B shows a similar group of experiments at a cold-block temperature of 8°C and recovery time of 5 hr. C shows a group of nerves at a cold-block temperature of 0°C where recovery times of 4 and 5 hr were used. The overall rate for all groups during the recovery period was determined as 13.2 mm/hr or 318 mm/day.

precursor from the ganglion compartment into the sciatic nerve compartment was prevented by petrolatum jelly seals placed between the partitions. After allowing time for transport of the labeled proteins into the fibers, peaks of radioactivity moved distally in the nerve at a rate characteristic of fast transport (Fig. 2.27).

The amplitude of the most distal peak in the example on the right shows a smaller amplitude than the one on the left. Such variations in the amplitude of radioactivity in the peaks have no effect on their rate. The rate of transport determined at different temperatures gave a Q_{10} which ranged from 2.3 at the lower temperatures to 3.4 at the higher temperatures. The rate of transport calculated by extrapolating to 37°C, and taking the Q_{10} into account was 400 mm/day, again a rate closely comparable to that found for fast transport in mammalian nerves.

The long olfactory nerves of the garfish, with its uniform population of small unmyelinated fibers, has advantages for the study of transport (Gross and Beidler, 1973). The precursor ^{3}H-leucine applied to the olfactory mucosa is taken up by the nerve terminals and incorporated by the olfactory cells. This is followed by a transport of labeled proteins in the olfactory nerves

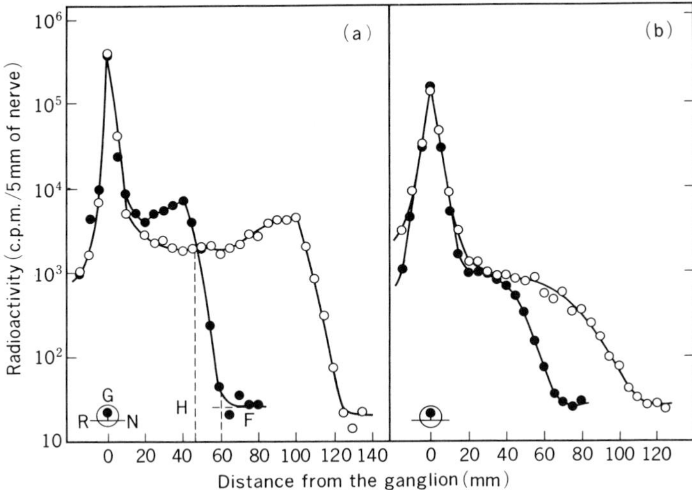

Fig. 2.25 Outflow patterns in frog nerves *in vitro*. (a) The 9th dorsal root ganglia were injected with 1-[4,5-³H]-leucine, and outflow of labeled proteins in the sciatic nerves after 9 hr (●) and 18 hr (○) *in vitro* is shown. A typical fast transport pattern is present with a rate when scaled to 37°C close to 410 mm/day. (b) Outflow of phospholipids after injecting Me³H-choline as the precursor is shown. The rate of the latter is similar but with a greater slope of the phospholipid fronts compared to the protein. From Abe *et al.* (1973).

toward the brain. Throughout the time of transport, the fish are kept at a uniform temperature and oxygenated in a special apparatus (Fig. 2.28).

After various periods of time allowed for transport, the nerves were removed to assess its rate by counting radioactivity in successive sections of nerve, with the characteristic outflow pattern seen in Fig. 2.29.

The rate of transport was determined as usual from the distance to which

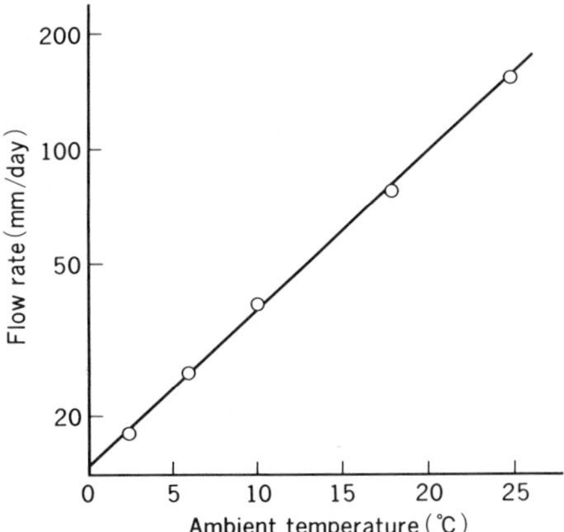

Fig. 2.26 Relation of temperature to the rate of transport. The temperature dependence of fast transport in the frog sciatic nerve (cf. Fig. 2.25) is seen to be linear over the range of temperature from 2.5 to 25°C. From Abe *et al.* (1973).

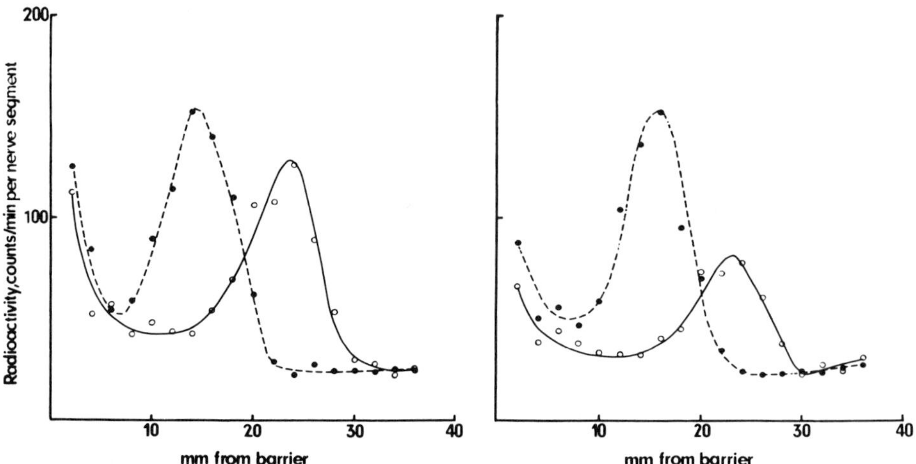

Fig. 2.27 The transport of labeled proteins in frog nerves *in vitro*. Nerves were analyzed for their content by TCA-insoluble radioactivity after a downflow period of 4 hr (broken line) and 5.5 hr (solid line) at 18°C following ^{3}H-leucine uptake and incorporation. Two similar experiments are illustrated. The peaks show variation in the amount transported distally, with *in vitro* rates similar and close to 400 mm/day when scaled to 37°C. From Edström and Hanson (1973b).

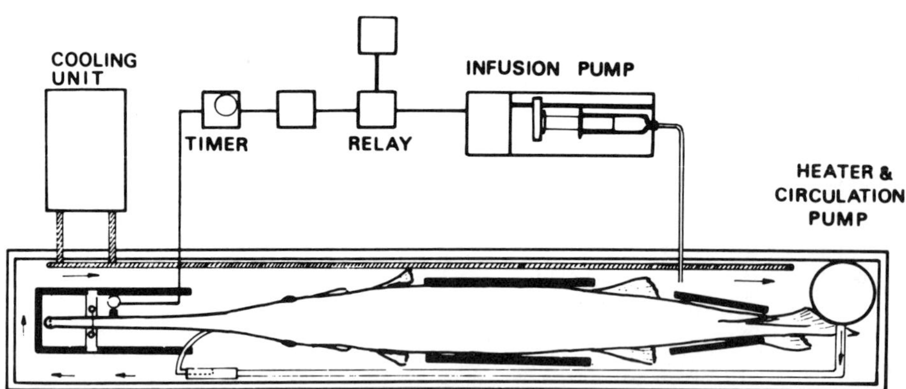

Fig. 2.28 Study of axoplasmic transport in the garfish. The tank shown accommodating the garfish has a circulation pump to pass oxygenated water through the gills at a rate of 2 liter/min. The water temperature is controlled by cooling and heating units. The respiratory reflex of the lower jaw triggers the automatic anesthesia system to supply the anesthetic administered by the infusion pump. From Gross and Beidler (1973).

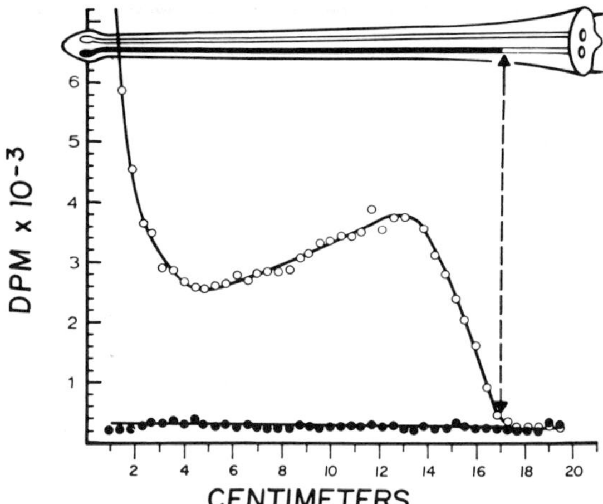

Fig. 2.29 Fast transport in the garfish olfactory nerve: 21.3 hr after the application of ³H-leucine to the left olfactory cavity the nerve is sectioned and radioactivity displayed (○). The center of the olfactory mucosa is taken as the zero point on the abscissa. A crest of outflow of radioactivity shows the characteristic form of fast transport with a rate, when scaled to 37°C, of close to 410 mm/day. Only uniform, low-level activity, presumably caused by systemic labeling, is discernable in the right olfactory nerve (●) whose cell bodies were not directly exposed to the isotope. From Gross and Beidler (1973).

the foot at the front of the crest had advanced during the time allowed for transport. The radioactivity in the nerves was plotted on a linear scale on the ordinate, making the crest somewhat more prominent than in the studies described above wherein a logarithmic scale was used. Gross and Biedler noted a small lag before the labeled components moved down the fibers, a phenomenon not observed in mammalian or frog nerves. The lag may be due to a slower uptake of the precursor from the nasal mucosa in the garfish, perhaps related to its anatomical structure, or to some difference in the transfer of incorporated materials from the cell to the fiber (cf. Chapter 5).

As in mammalian nerve, the rate of transport in the garfish olfactory nerve fibers was reduced at lower temperatures. By extrapolation to a temperature of 37°C, the rate of fast transport was determined to be 403 mm/day, a rate close to the 410 mm/day rate found for mammalian nerves. Another feature of interest was that the cold-block occurred at temperatures between 5 and 10°C, a level somewhat lower than that in mammalian nerve.

The crest of labeled proteins shows a relatively greater fall-off of amplitude with distance in the garfish olfactory nerve than was usually seen in the mammalian nerve (Fig. 2.30).

In part, the fall-off is exaggerated by the use of a linear scale to represent the crest radioactivity. There does, however, appear to be a relatively greater fall-off in the crest in the garfish nerve, particularly at earlier times. As will

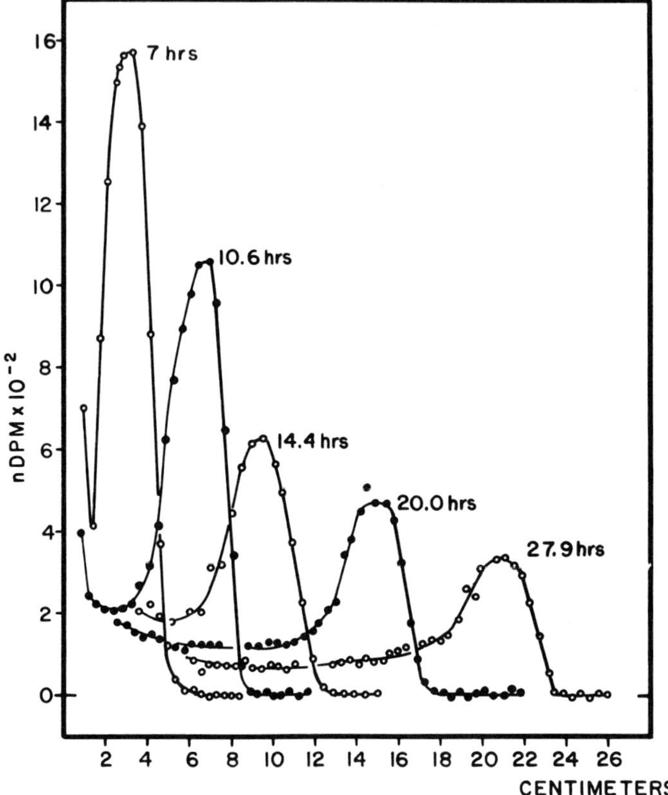

Fig. 2.30 Crest amplitudes during fast transport in garfish olfactory nerves. The five nerves shown were taken at the different times indicated after exposure to the precursor. The peak amplitudes of labeled outflow are seen to decrease with distance, the slope at the front also decreases somewhat. From Gross and Beidler (1975).

be described in Chapter 11, differences in the pattern of fall-off are likely to be related to differences in the binding of components carried by the transport mechanism down the length of the fibers with those materials dropped off early on contributing to slow transport.

In the swan mussel, *Anodonta cygnea,* the CNS consists of three pairs of ganglia joined by connectives, some having nerve fibers lengths as long as 8–10 cm. On injecting ^{3}H-labeled leucine, valine, or tryptophane into the cerebral ganglia for uptake by the cells, Heslop and Howes (1972) found labeled proteins to be transported into the nerve connectives with a pattern similar to that seen in the nerves of higher species.

At reduced temperatures down to approximately 12–15°C, transport rates were found to decrease as it does for the nerves of other species. However, instead of a cold-block appearing at low temperatures, transport leveled off and remained present at temperatures down to at least 4°C. This suggests a special adaptation in the nerves of this poikilotherm to enable them to function at low environmental temperatures.

In the swan mussel nerves, a Q_{10} of 2 was measured for temperatures above 12°C. Using this Q_{10} to calculate the rate of transport extrapolated to 38°C, a value of about 240 mm/day was determined. This rate is substantially lower than similarly extrapolated rates in garfish olfactory and frog nerves, and it may represent a species difference.

Crayfish nerves also appear to have a slow rate of 0.56 mm/day and a fast rate of only about 10 mm/day at a temperature of 9.5–11.5°C (Fernandez *et al.*, 1970). Using a Q_{10} of 2 yields a fast rate of only about 80 mm/day and selecting a Q_{10} of 3, a rate of about 270 mm/day. The uncertainty as to the Q_{10} in these nerves makes it difficult to assess the rate on the basis of the results obtained so far in this species. Evidence of both a fast transport as well as a slow transport was reported for axons in the CNS of the cockroach *Periplanata americana* by Smith (1971).

A low rate is not invariably present in the nerves of invertebrates. Goldberg *et al.* (1978) found the transport of serotonin-containing vesicles in the serotonergic axon of *Aplysia* to have a rate close to the 400 mm/day when the outflow was scaled to a temperature of 37°C and the Q_{10} taken into account (Chapter 4).

Further study of the rates of fast transport in a wider range of lower vetebrates and invertebrate species is called for to determine if the transport mechanism in various species is a well-preserved feature of nerve or if there are well-defined differences which may perhaps be related to molecular differences in their transport mechanism (cf. Chapter 10).

E.   CONTINUOUS AND *IN SITU* MEASURES OF TRANSPORT

Two relatively new techniques have been introduced which have interesting possibilities in that transport may be measured continuously in the same nerve over a period of time.

By using isotopes with sufficiently high particle energy, e.g., ^{32}P, ^{35}S, or ^{14}C, the particles radiated from the isotope can be collimated at sucessive sections along the length of a nerve in a multiwire chamber (Snyder *et al.*, 1976). Taking such measurements at different times gives a measure of the rate of fast transport and allows evaluation of changes in the rate occurring over a period of time following an experimental intervention.

In another approach to dynamic measurement of transport, a miniature scintillation detector was developed by Takenaka, Horie, and Sugita (1978). When placed against a nerve containing isotopes with a high energy particle, the passage of the front and crest can be determined at selected points. Such a technique allows successive measurements to be made over a period of several days thus showing possible changes in slow transport (Takenaka and Ochs, 1980).

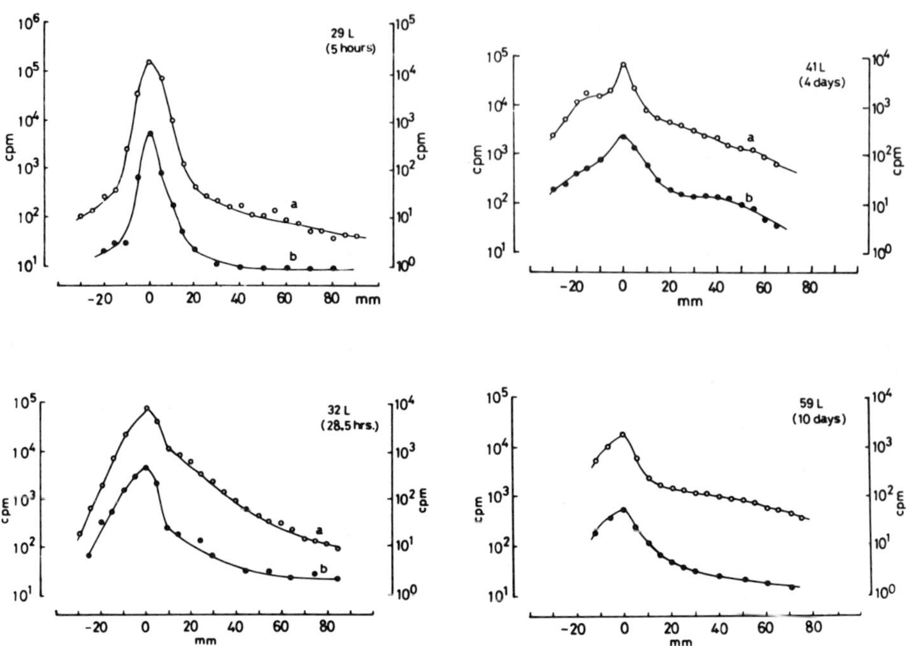

Fig. 2.31 External detection of transport. Radioactivities determined by scintillation counting (a) and by the use of an external semiconductor detector (b) are shown in a series of determinations at the time given in hours and days after injection of the precursor ^{32}P-ATP into the rat dorsal root ganglia. From Takenaka and Ochs (1980).

Although the counting efficiency of these methods is low in comparison to regular scintillation counting (Fig. 2.31) and the detection of the radioactivity is mainly from fibers near the surface, the possibility of making measurements of transport over a period of time allows studies not possible with the usual method, an advantage which will undoubtedly be put to further use in the future.

3

Further Characterization of Anterograde Transport, Dendritic Transport, and Retrograde Transport

A. THE VISUAL SYSTEM

The accumulation of isotope labeled components at ligations or at nerve terminals is a consequence of axoplasmic transport. Weiss and Holland (1967) injected ^{3}H-leucine into goldfish eyes and then in autoradiographs found at a later time after injection, grains of radioactivity located in the optic tecta. This technique was considerably enlarged on and developed by Grafstein and her colleagues using autoradiography and as well scintillation counting of radioactivity in the tecta of goldfish or the lateral geniculate of mice after eye injection (Grafstein, 1967, 1977; Grafstein and Forman, 1980).

Because this system has been widely used to study characteristics of axoplasmic transport, it will be of value to describe it in some detail. A fine hypodermic needle or pipette is used to pierce the eyeball in order to inject a small volume of ^{3}H-leucine or other labeled solutions into the posterior chamber of the eye. The precursor is taken up by the large ganglion cells of the retina, incorporated into proteins, and the labeled proteins transported in their fibers. In fish and in rodents the optic nerves and tracts are almost completely crossed, and as a result the labeled proteins are transported to the tectum in the fish, and to the lateral geniculate ganglia in the mouse and rat, to the side opposite the injected eye. A diagram of the goldfish visual system is shown in Fig. 3.1.

A technical difficulty which must be taken into account in using this system is that the circulation within the retina is extensive and as a result, a considerable amount of the labeled precursor enters the circulation to be carried to the cells of the tectum or to the lateral geniculate to there become locally incorporated into labeled polypeptides and proteins. In a schematization given by Rahmann (1971), injection of eyes with various precursors shows the different patterns expected of axoplasmic transport, local incor-

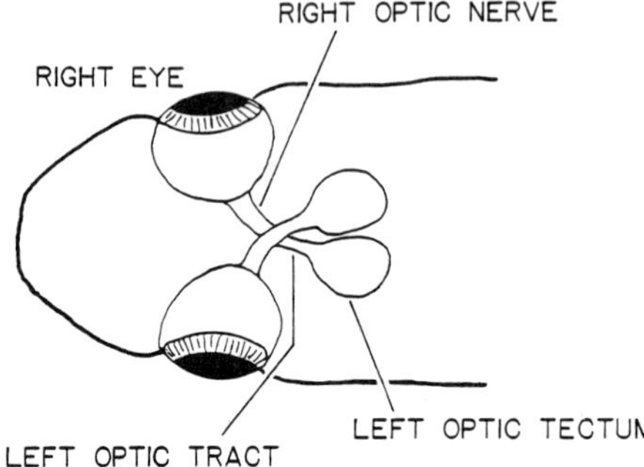

Fig. 3.1 The goldfish visual system. In this species a complete crossing of optic nerve and tracts from the eyes to the tecta of the two sides occurs as shown. From McEwen *et al.* (1971), modified by B. Grafstein.

poration in the tectum following an uptake of precursor from the circulation, and diffuse leakage from the retina (Fig. 3.2).

In this idealized representation, the tectum on the side opposite the eye injected with ³H-leucine shows transport of labeled proteins by the dense distribution of grains in the layers of the striatum where the optic fibers terminate. Actually, the amount of ³H-leucine leaking into the circulation from the injected eye which is picked up and incorporated by the cells of the tectum, gives rise to a much greater amount of non-transported labeling than is indicated in this diagram (see below). Glucose is an example of a precursor which gives rise to a much greater amount of leakage, with radioactivity found distributed in the tecta of both sides. The case of a labeled precursor which is incorporated but not transported and retained in the retina is also indicated. Actually, in those cases a somewhat greater diffusive spread from the retina out into the optic nerve occurs than is shown in this figure (Di Giamberardino, 1971).

To differentiate the radioactivity which is transported in the fibers as opposed to precursor which has leaked into the circulation to become locally incorporated in the tecta, the radioactivity in the tectum (or in mammals the lateral geniculate) ipsilateral to the side injected, which represents locally incorporated precursor, is subtracted from the activity on the opposite side which includes both transported and locally incorporated radioactivity (Fig. 3.3).

When the precursor is injected intraperitoneally the tecta on both sides are equally labeled, as can be seen in autoradiographs where the grains are distributed generally in all layers (cf. Fig. 3.2). On the other hand, the radioactivity which has been axonally transported to the tectum on the side opposite to the injected eye is found mainly located in the 2nd layer of the tectum. From neuroanatomical studies it is known that this layer receives the retinal afferent fibers from the eye of the opposite side.

In studies of transport in the visual system ³H-proline and ³H-asparagine are better precursors than ³H-leucine because the proline and asparagine which leaks into the circulation does not as readily pass the blood–brain barrier of the CNS to become locally incorporated in the tectum (Elam and

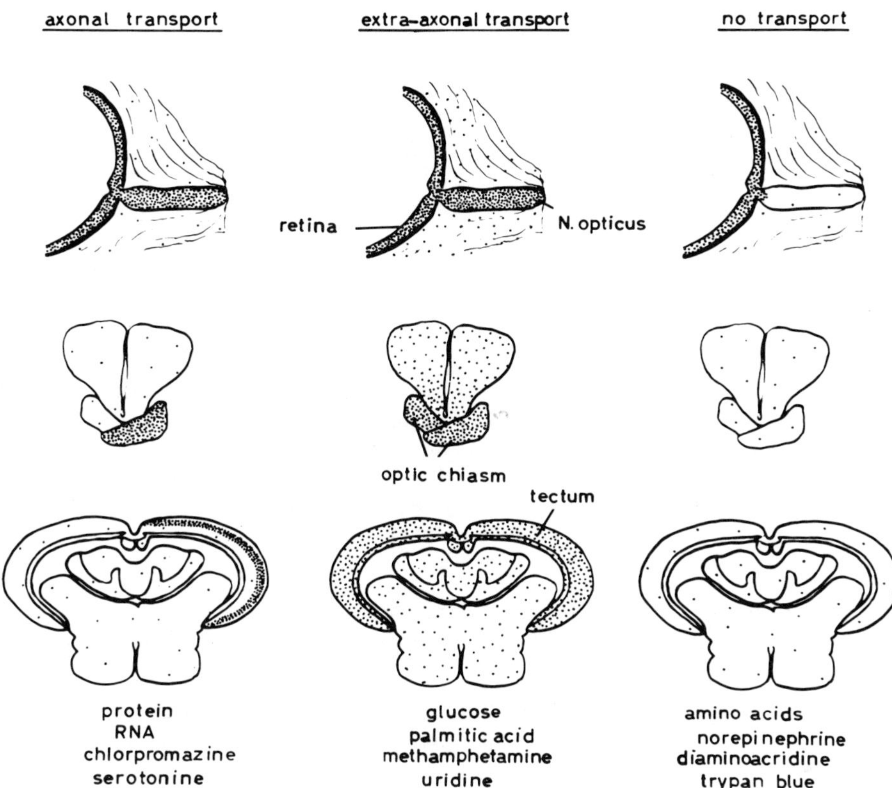

Fig. 3.2 Schematic diagram showing transport or failure of transport. The stippling represents the presence of radioactivity seen in autoradiographs after different labeled precursors were injected unilaterally into the avian eye (cf. Fig. 3.1). Transport is indicated in the left column by labeling of tectum on the opposite side, leakage is indicated in the middle column by labeling of both tecta following uptake from the circulation, and failure of transport and egress into the optic nerve is shown on the right. From Rahmann (1971).

Agranoff, 1971a). Fully half the radioactivity in the ipsilateral tectum represents locally incorporated radioactivity when using ^{3}H-leucine (cf. McEwen and Grafstein, 1968), while with ^{3}H-proline or ^{3}H-asparagine, the background activity is almost negligible (Fig. 3.4).

This results in a much greater differential between transported and locally incorporated radioactivity. This is seen dramatically in autoradiograms taken after unilateral injection of one eye with ^{3}H-proline (Fig. 3.5).

No such advantage accrues when using ^{3}H-proline as a precursor in the cat L7 dorsal root ganglion or spinal cord (Chapter 2). In this system there is relatively less leakage of precursor into the circulation and the amount that does leak out is more greatly diluted in the body. Also, ^{3}H-leucine appears

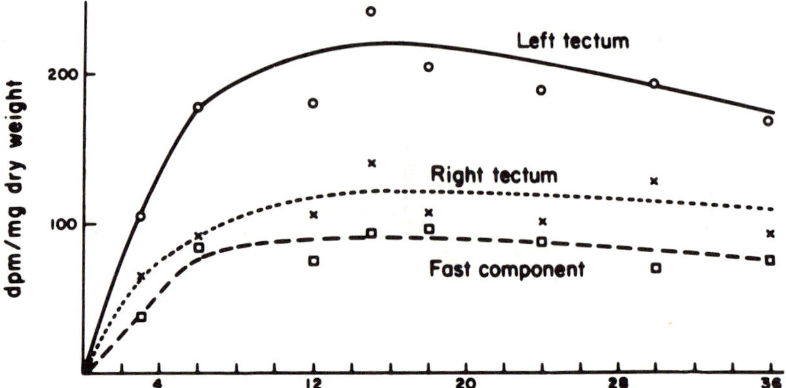

Fig. 3.3 Fast transport shown by accumulation. The time-course of appearance of labeled leucine in the left tectum of TCA-precipitable material after injection of ^{3}H-leucine in the right eye is shown by its rapidity of accumulation of a fast-transported component. This is measured as the difference between the radioactivity in the right tectum due to leakage and local uptake from the left which represents both transport and local uptake. From McEwen and Grafstein (1968).

to be superior to ^{3}H-proline as a precursor in the dorsal root ganglion or spinal cord preparation because of the relatively high number of leucine residues in the proteins labeled and transported in the nerves.

The rate of transport in the visual system may be determined by the time of the earliest accumulation of transported labeled proteins appearing in the tectum (Fig. 3.6).

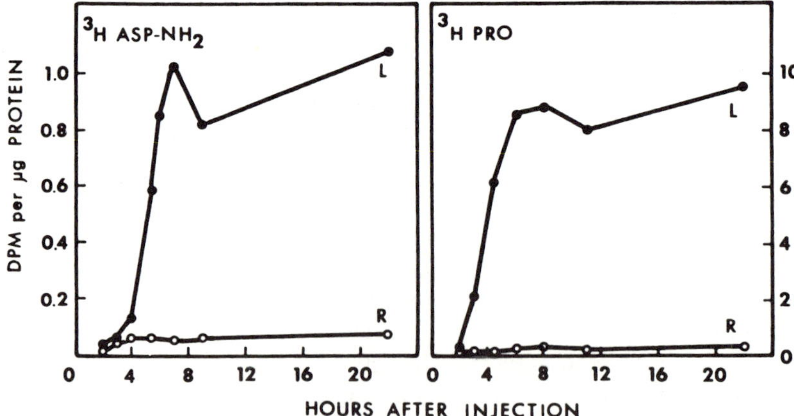

Fig. 3.4 Fast-transport of aspargine and proline. Labeling of the left (L) and right (R) optic tecta at various times after intraocular injection of either ^{3}H-aspargine or ^{3}H-proline into the right eye of goldfish is shown. A much greater asymmetry of labeled proteins transported into the left tectum (L) as compared to adventitious leakage and uptake by the right (R) tectum is seen than when using ^{3}H-leucine (cf. Fig. 3.3). From Elam and Agranoff (1971a).

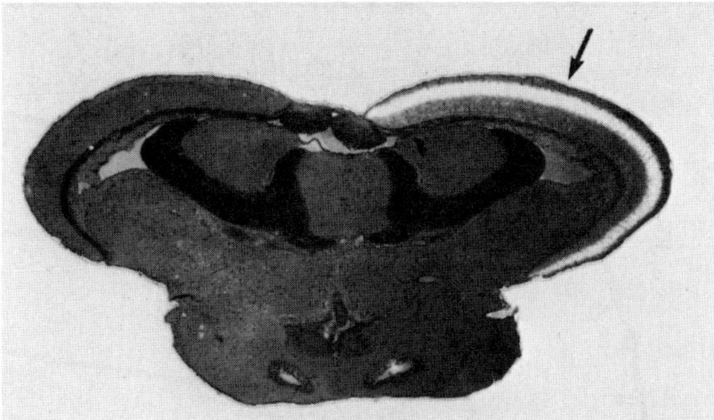

Fig. 3.5 Autoradiography of the tectum following ³H-proline injection. After injection of ³H-proline in one eye, localization of radioactive-labeled protein in the contralateral tectum is shown by a bright band on that side (arrow) in the layers of the cortex on which the entering fibers terminate. From Springer *et al.* (1977).

Both with ³H-leucine and ³H-glucosamine as precursors (the latter incorporated into glycoproteins; Chapter 5), a lag is seen before the rise in accumulated radioactivity due to transport. This lag is to be contrasted to the almost immediate rise of radioactivity which represents uptake of leaked precursor from the circulation. When using ³H-proline as the precursor the uptake of leaked precursor was decreased, allowing the lag due to transport to be more clearly seen (Fig. 3.7).

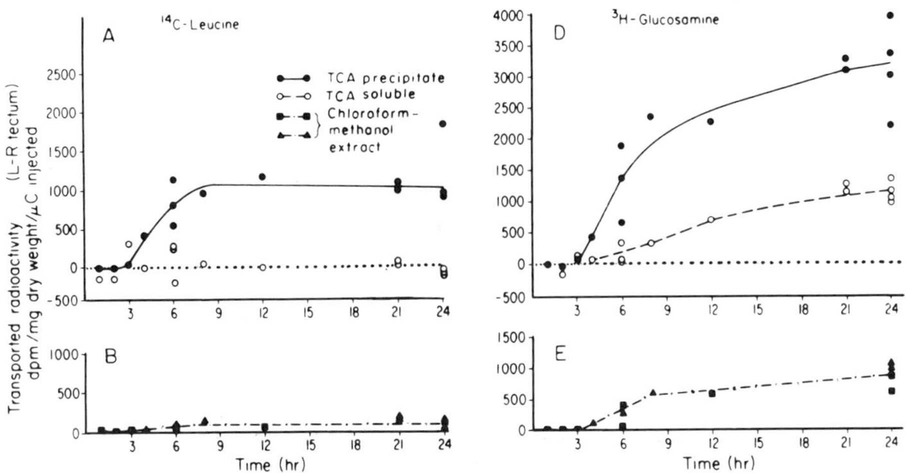

Fig. 3.6 Accumulation curves using glucosamine. (A,B) show fast arrival of labeled TCA-precipitable and -soluble components in the tecta, respectively, after injection of ³H-leucine into the eye of the opposite site. (D,E) show arrival of fast-transported glycoproteins and TCA-soluble components, respectively, after injection of ³H-glucosamine. Both labeled components transported show a latency before their rise. From McEwen *et al.* (1971).

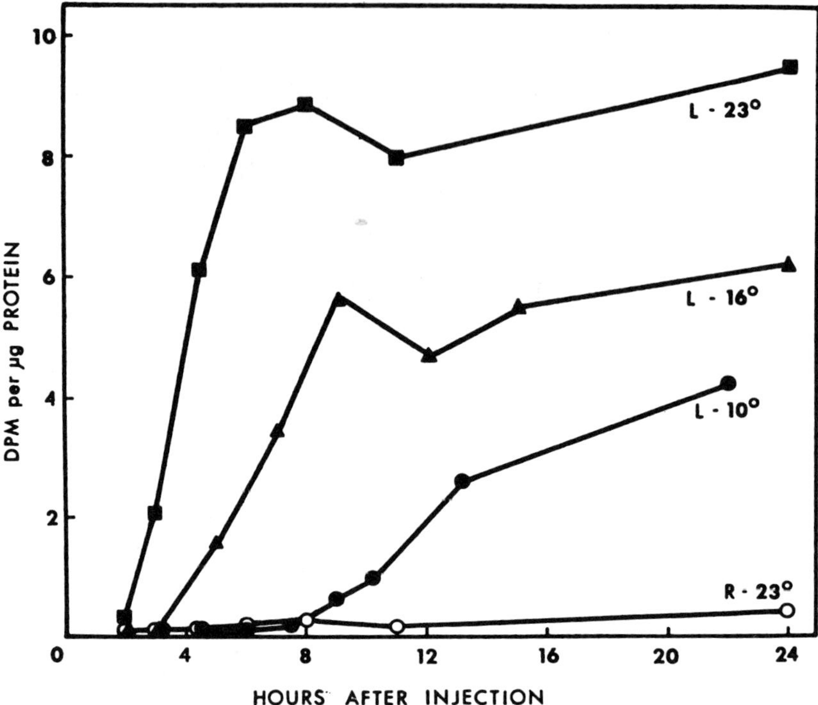

Fig. 3.7 Variation of accumulation at lower temperatures. Fish were stored at the temperatures shown for 2 days prior to injection of one eye with ^{3}H-proline. The latency of fast accumulation becomes longer at the lower temperatures. Groups of four fish were pooled for each point. L, left tectum; R, right tectum. From Elam and Agranoff (1971a).

The rate of fast transport is calculated from the lag time of the earliest appearing rise in accumulated radioactivity in the tectum above background and the measured distance of the cell bodies in the retina to their terminals in the tectum. The lag time, as is evident in this example, increases at the lower temperatures, as expected from the relatively high Q_{10} of transport (cf. Chapter 2). There are some uncertainties inherent in this procedure when applied to the visual system. The time for diffusion of the precursor injected into the vitreous substance to the ganglion cells in the retina can vary, depending on the point of injection. In addition, the differing lengths of fibers arising from the ganglion cells spread out over the surface of the retina, as well as the differing lengths of their nerve fiber terminations over the surface of the tectum, could result in as much as a 3-fold difference in the estimated length of optic nerve and in turn the calculated rate of transport. With these qualifications taken into account, the rate of transport estimated from the earliest rise of accumulated radioactivity when scaled to a temperature of 37°C using a Q_{10} of 3 was given as 190 mm/day (Grafstein, 1977).

Following the first wave of accumulated radioactivity, a later wave of accumulation was seen and interpreted (Grafstein, 1977) as the arrival of labeled materials transported at a slower rate (Fig. 3.8.).

Early and late waves of accumulation were also seen in the lateral geniculate of the mouse on which the retinal ganglion neurons terminate in the mammal as can be seen in *B* of Figure 3.8. Such observations have suggested the concept that two separate transport mechanisms are operating, one subserving fast, the other slow outflow. Other studies indicated more variation in the pattern of accumulation of radioactive labeled components than can be accounted for by two transport systems. Karlsson and Sjöstrand (1971c), for example, have described two fast waves of accumulation (Fig. 3.9).

And, on a slower time scale, two slow waves of accumulation were found as indicated in Fig. 3.10.

Waves of accumulation at an "intermediate" rate have often been re-

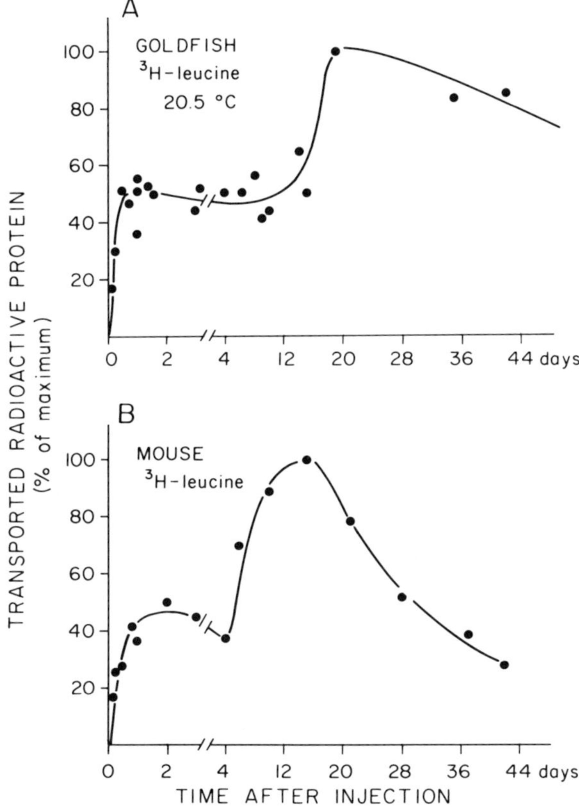

Fig. 3.8 Fast and slow components of accumulation. Comparison of accumulation in the tectum and the lateral geniculate of the goldfish and mouse, respectively, after injection of the eye on the opposite side shows fast and slow phases of accumulation in the two species. From Grafstein (1977).

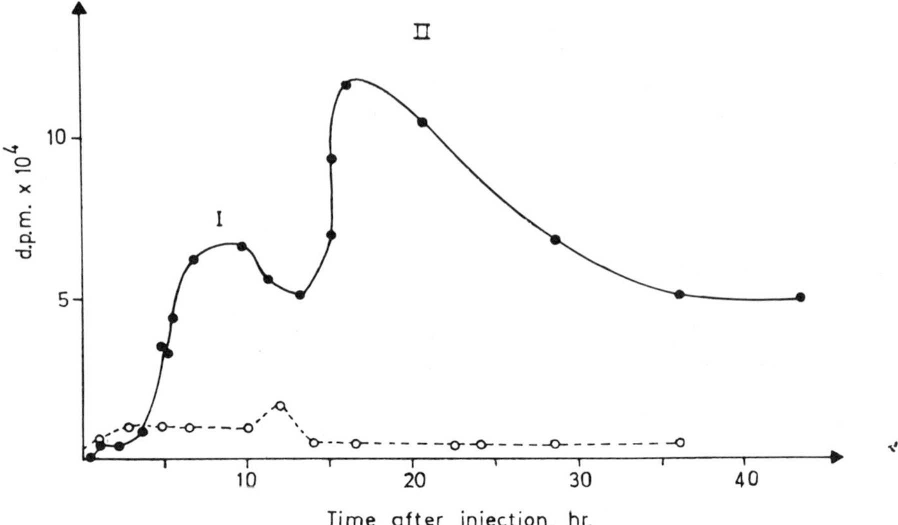

Fig. 3.9 Two phases of fast accumulation. Accumulation of radioactivity in the lateral geniculate bodies at different intervals following intraocular injections of ^{3}H-leucine shows two rapid phases of transport of labeled protein (●) arriving at the lateral geniculate body, curves I and II; (○) indicates TCA-soluble material in the supernatant. Each symbol represents the average of two to four animals. From Karlsson and Sjöstrand (1971c).

ported. Schonbach and Cuénod (1971) reported a fast rate in the avian retinotectal path calculated to be 100–500 mm/day, and a later appearing wave with a calculated rate of 20–60 mm/day. A rate of 40 mm/day and a slower one of 1 mm/day was given by Cuénod and Schonbach (1971). The variations in rate reported in different neural systems and in different species makes their direct comparison difficult. The correlation of multiple rates is further complicated by the different rates reported when SDS and polyacrylamide gel electrophoresis (SDS–PAGE) is used to assess the outflow of labeled polypeptides. This technique will be more fully discussed in Chapter 5, but we note here that five rates of polypeptide outflows have been determined by this method; group I with a rate greater than 200 mm/day, group II 34–68 mm/day, group III 4–8 mm/day, group IV 2–4 mm/day, and group V 0.7–1.1 mm/day (Willard *et al.*, 1974; Willard and Hulebak, 1977). All these multiple rates may be explained by the operation of a number of different individual transport systems. Alternatively, on the basis of the unitary hypothesis, the same transport system responsible for fast transport can, with some additional considerations, account also for slow transport, as will be discussed in Chapter 11.

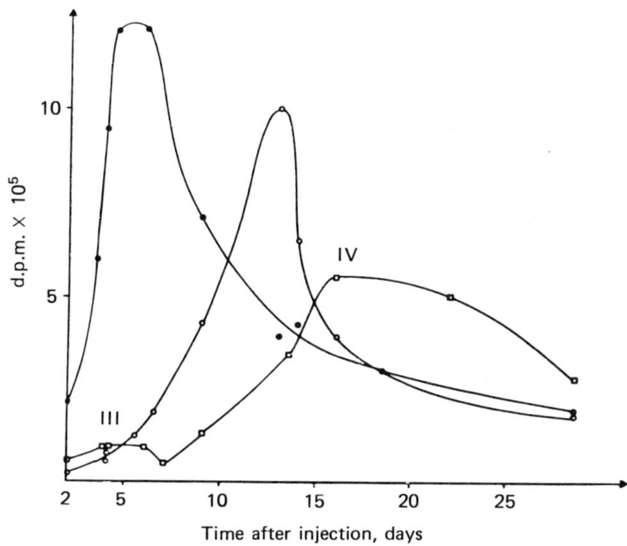

Fig. 3.10 Two phases of slow accumulation. Accumulation, at later times, of radioactive protein in the lateral geniculate following intraocular injections of ^{3}H-leucine is shown. Each symbol represents the average of two animals. The curve at the left represents fast transport and curves III and IV represent the two phases of slow transport reaching their maxima at about 12 and 16–18 days, respectively. From Karlsson and Sjöstrand (1971c).

B.   TRANSPORT INTO DENDRITES

Single motoneurons of the cat spinal cord injected iontophoretically with the amino acid precursor ^{3}H-glycine show, following its incorporation, an outflow of labeled proteins into the dendrites in autoradiographs (Schubert *et al.*, 1971). For the injection procedure, multibarreled microelectrodes were employed. The precursor is loaded in one barrel and after the cell is entered, the ^{3}H-glycine is injected by micro-iontophoresis by passing current from the tip containing the precursor to the tip of another barrel, a method which prevents large currents from crossing the cell membrane and damaging it (Fig. 3.11).

After iontophoretic injection of the precursor, the cell's viability was shown by its ability to give rise to the antidromically evoked action potential responses typical of spinal cord motoneurons. At timed intervals after microiontophoresis, portions of the spinal cord were taken and sections prepared for autoradiography. Grains were located over the injected cell bodies in sections prepared from tissue taken early after injection and in the dendrites in sections taken somewhat later, as expected of a fast transport of labeled proteins from the cell body into the dendrites.

The extent of the passage of radioactivity into the far reaches of the

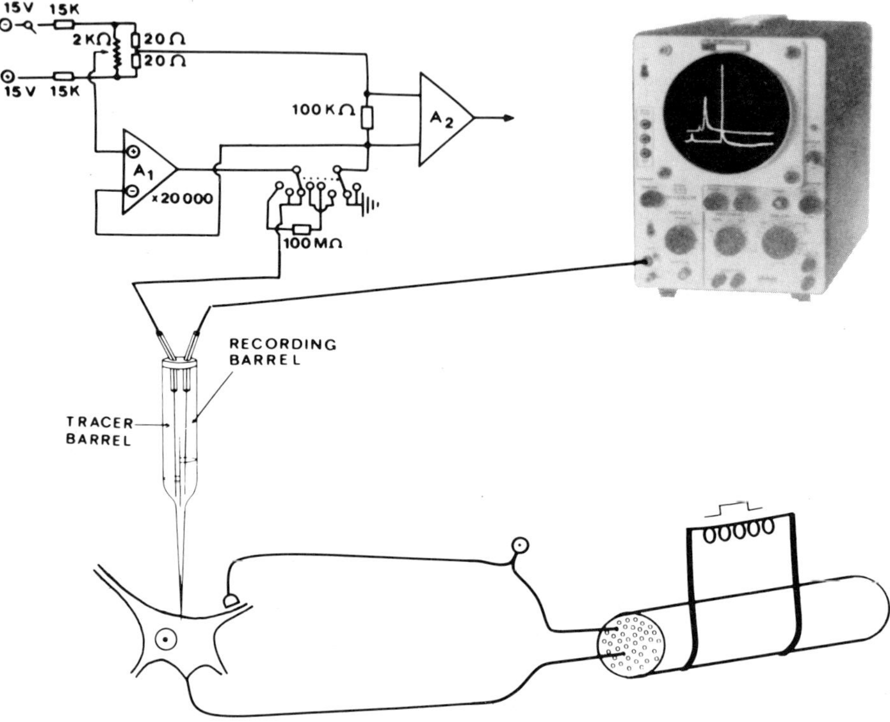

Fig. 3.11 Single-cell injection of labeled precursor. A three-barreled micro-electrode is shown with its tip inserted into the soma of a motoneuron in the spinal cord. The recording electrode barrel is connected through an electrometer stage to an oscilloscope. Iontophoresis current in the range of 10^{-7} to 10^{-8} Å is applied from the barrel containing the precursor to another to eject the precursor into the cell from the tip. The ejection current is measured by a differential amplifier. Also shown in this modified schematization are the stimulation sites on nerve and the recording of responses on an oscilloscope. From Schubert, Lux, and Kreutzberg (1971).

dendrites may be seen from a reconstruction of autoradiographs made of serial sections of the cord containing an injected motoneuron (Fig. 3.12).

Ribosomes are present in the dendrites, at least within the larger branches near the cell body, and presumably protein synthesis can take place there. Yet, as appears from the spread of labeled proteins outward into the dendrites, the major site of synthesis is in the cell bodies. The contribution of materials which may be synthesized in the dendrites remains at present unknown.

A precise determination of the rate of axoplasmic transport into the dendrites is not easily accomplished from such autoradiographs, but it appears to be fast, as judged from the rapid appearance of grains in the far reaches of the dendrites.

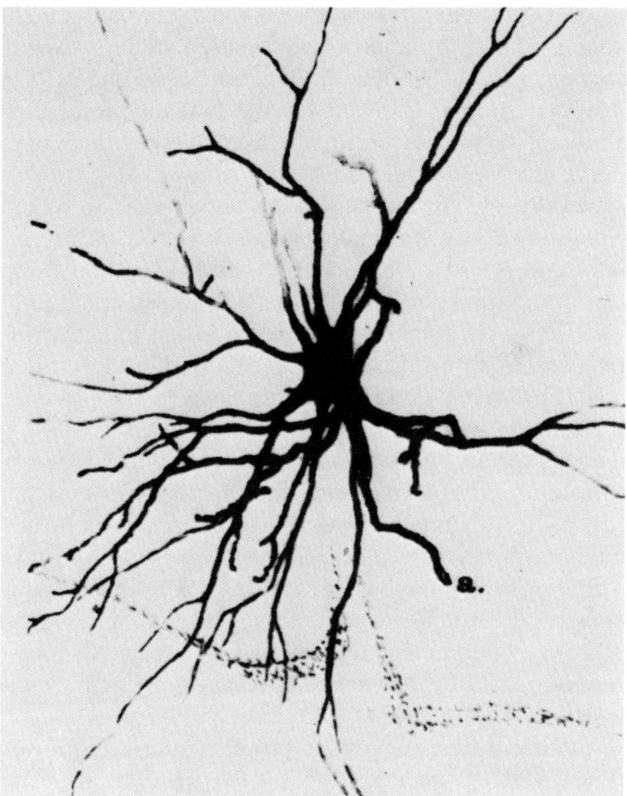

Fig. 3.12 Transport of incorporated protein into dendrites. Reconstruction of the soma and dendritic tree of a motoneuron injected with ^{3}H-glycine showing transport of the incorporated precursor into the dendrites. Successive tissue sections prepared by autoradiography reveals the outflow of labeled proteins into the dendritic arborations. From Schubert, Kreutzberg, and Lux (1972).

The injection of dyes such as cobalt blue or procion yellow into cell bodies, followed by their subsequent spread throughout the dendrites and into axons (Kater and Nicholson, 1973), most likely takes place by their binding to proteins or other components which are normally carried into the dendrites and axons by axoplasmic transport.

C. RETROGRADE TRANSPORT

Retrograde transport is defined as the movement of materials in the axons (and dendrites) toward the cell body. Exposure of nerve terminals to ^{3}H-leucine and other amino acids is not followed by a transport of labeled proteins towards the cell bodies because of the lack of protein synthesis in the nerve terminals (Chapter 5). However, a retrograde movement of proteins within the axons is shown by appropriate ligation studies. After injection of a labeled precursor, such as ^{3}H-leucine, for uptake by the cells and incorporation into proteins, time is allowed for the labeled proteins to be carried down to their distal terminals or to a ligation. Then, an upper ligation is made in the nerve and another period of time allowed for retrograde transport to carry the labeled proteins back up to the upper ligation where it accumulates as shown in such studies made by Lasek (1967), Bray *et al.* (1971), Frizell and Sjöstrand (1974), Ochs (1975c), and Bisby and Bulger (1977).

Rates of retrograde transport for various substances collated by Forman *et al.* (1977a) and Bisby (1980b) range between a rate half that of fast anterograde transport to rates almost as fast as were found in the studies of Edström and Hanson (1973b), Abe, Haga, and Kurokawa (1974), and Sjöstrand and Frizell (1975), among others.

In addition to the turnaround of unspecified phospholipids or of ^{3}H-amino-acid-labeled proteins, the turnaround and retrograde transport of endo-plasmic-reticulum-like material, lysosomes, and mitochondria at a site of injury was indicated by the use of the markers acetylcholinesterase (Chapter 4), acid phosphatase, and monoamine oxidase (Section E below; Schmidt, Yu, and McDougal, 1980). Retrograde transport, like anterograde transport, is blocked by cold (Chapter 2), interruption of oxidative metabolism with sodium cyanide (Chapter 7), and by mitotic blocking agents such as colchicine and vinblastine (Chapter 12).

Convincing evidence that exogenous proteins such as albumin and horse-radish peroxidase (HRP) are taken up by the nerve terminals and carried back up the axons to the cell body by retrograde transport was first given by Kristensson (1970) and Kristensson and Olsson (1971). When injected into a muscle, albumin tagged with a flourescent label, and HRP as shown by its enzymatic reaction, were later identified within the cell bodies of the moto-neurons innervating those muscles. The rate of retrograde transport of HRP in nerve fibers was assessed by injecting the protein into the optic tectum of chicks and then at various times thereafter determining its first appearance in the optic nerves and the retina. By this method its rate of retrograde transport in the fibers was determined to be at least 84 mm/day (LaVail and LaVail, 1974). A somewhat faster rate of 120 mm/day for the retrograde transport of HRP in the visual system of the rat was reported by Hansson (1973). Some of the characteristics of uptake of HRP and other exogenous materials by the terminals for their retrograde transport will be discussed in Chapter 13 and its use in anatomical tracing in part F below.

A protein naturally present in the organism, nerve growth factor (NGF), is retrogradely transported in nerve fibers. The role of NGF in the development and maintenance of the peripheral adrenergic and sensory neurons will be discussed more fully in Chapter 15. Here it is noted that NGF can serve as a marker of retrograde transport in adrenergic (Hendry *et al.*, 1974) and sensory neurons (Levi-Montalcini and Angeletti, 1968). The retrograde transport of ^{125}I-labeled NGF is shown after injection into the forepaws of rats by its appearance in the C6–C9 dorsal root ganglia taken at various times thereafter and prepared for autoradiography (Stoeckel *et al.*, 1975a,b). Such uptake is indicated in Fig. 3.13.

The accumulation of ^{131}I-NGF in the dorsal root ganglia was blocked when the nerve between the paw and ganglia was cut or when colchicine was placed on the nerve to block transport (Chapter 12), evidence that the accumulation of radioactivity is in fact due to a retrograde transport of the labeled NGF within the nerve fibers. The rate of retrograde transport is determined from the earliest indication of NGF accumulation seen 6 hr after exposure of the nerve terminals to the labeled NGF. This time and the distance between the injected paw and the ganglia, estimated at 80 mm, gives an estimated rate of retrograde transport of 13 mm/hr or 320 mm/day. There is some uncertainty, however, as to the exact time at which the accumulation begins. If we consider that the small rise seen at the foot of the accumulation curve at 6 hr is not significant, and that a significant accumulation actually begins after 7 hr, the rate would be 270 mm/day. In any case, the rate so determined is faster than some of the other measures of retrograde transport rate discussed below in Section D.

An old controversy as to whether tetanus toxin ascends into the CNS by transport within the axons or by some extraneuronal route seemed to have

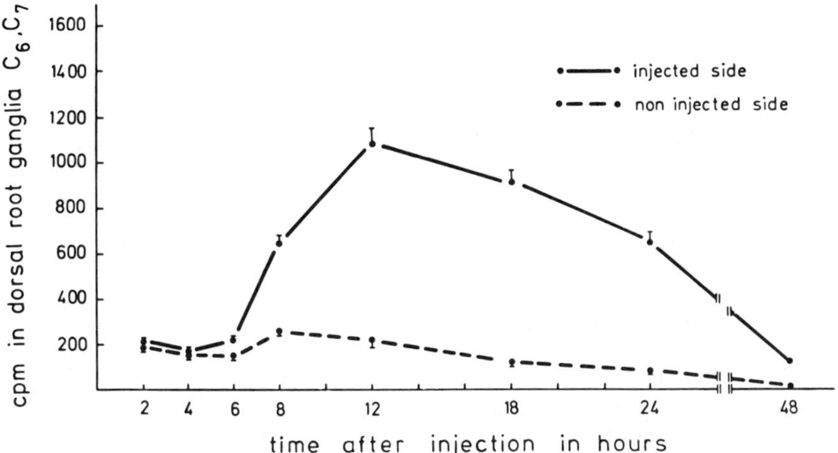

Fig. 3.13 Retrograde transport of labeled tetanus toxin. The time course of accumulation of radioactivity in the dorsal root (C_6, C_7) ganglia is shown after unilateral injection of ^{125}I-NGF into the forepaw of the rat. After its uptake by the sensory nerve fibers, the ganglia of the injected and noninjected sides were removed at times from 2 to 48 hr afterwards and counted. A latency of 6 hr was seen before an accumulation of radioactivity on the injected side, becoming maximal after 12 hr. Each point represents the means ± S.E. of 7–9 animals. From Stoeckel, Schwab, and Thoenen (1975a).

been settled by the experiments of Wright (1955) in favor of an upward spread in the endoneurium between the fibers of a nerve trunk (cf. Chapter 8 on the relation of the nerve fibers to the endoneurium). Some of the earlier evidence which strongly implicated an ascent within the nerve fibers was little attended to; for example, the block of ascent of tetanus toxin by the use of dry ice to freeze the nerve fibers and cause them to seal off which had been reported by Roofe (1947). Direct evidence for an intra-axonal ascent of tetanus toxin was given by Price *et al.* (1975) and by Erdmann *et al.* (1975) using ^{131}I-labeled tetanus toxin. When injected into a muscle, the ascent of the toxin was blocked by crushing the nerve supplying the muscle. In sections of the nerve taken just below the crush an accumulation of labeled toxin within the axons was seen in autoradiographs (Fig. 3.14).

In the studies of Erdmann *et al.* (1975) which independently demonstrated the intra-axonal ascent of tetanus toxin, labeled radioactivity in the ventral root fibers and in motoneuron cell bodies was seen in autoradiographs following injection of the labeled toxin into muscle. The ascent of the toxin was blocked by exposing the nerve to colchicine or to vinblastine, agents effective in blocking axoplasmic transport (Chapter 12).

When the toxin gains entry to the spinal cord it produces tetanus by interfering with synaptic inhibition, suggesting that the toxin acts on the inhibitory neuron boutons terminating on the motoneuron cell bodies and dendrites (Brooks *et al.*, 1957). Evidence for a passage of the toxin from the motoneuron cell body to the terminals of the inhibitory neurons synapsing on the motoneurons has in fact been found (Schwab and Thoenen, 1976).

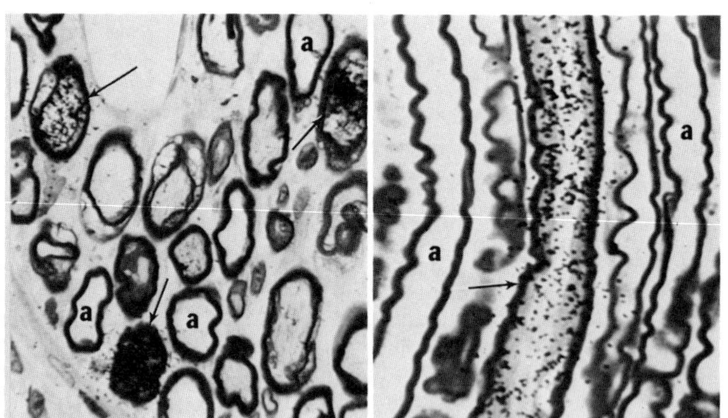

Fig. 3.14 Intra-axonic location of transported tetanus toxin. Autoradiographs of sciatic nerves distal to crushes were made to show the location of labeled tetanus toxin retrogradely transported in the nerve. At the left, axons and myelin sheaths are seen in a cross-section taken just below the crush with silver grains seen present within four axons. Arrows point to three labeled axons with three unlabeled axons indicated (a). On the right, axons were cut in longitudinal section and silver grains are seen present within the axon in the center of the field. From Price *et al.* (1975).

This and other examples of a transfer of substances across synaptic junctions will be described in Chapter 13.

Surprisingly, the retrograde movement of tetanus toxin is not restricted to the motor fibers. An ascent of the toxin within sensory nerve fibers was shown by its subsequent appearance in dorsal root ganglion cells following injections of ^{131}I-labeled tetanus toxin into muscle (Stoeckel *et al.*, 1975a,b). The pathogenicity of tetanus toxin is not related to its presence in the sensory fibers of muscle and this finding remains an interesting and as yet unexplained observation.

A retrograde movement of herpes virus (Goodpasture, 1925) and polio virus (Bodian and Howe, 1941; Howe and Bodian, 1942) in nerves had earlier been inferred from indirect studies. Evidence of the ascent of polio virus in nerve fibers was obtained by briefly freezing nerves with dry ice, a procedure which seals off the nerve fibers with apparently little effect on extraneuronal tissues in the nerve trunk. Several hours later the nerves so treated were exposed to a fluid suspension of polio virus below that site, and the animals did not develop paralysis as did control animals (Howe and Bodian, 1942). If a period of 2–6 weeks was allowed to lapse before the application of the virus, it did not ascend in the newly regenerated fibers or in the endoneural spaces between the fibers. However, an ascent did occur after 10 weeks, when the fibers had sufficiently regenerated (cf. Chapter 15).

Further evidence for a retrograde intra-axonal transport of herpes simplex virus (HSV) in nerve (Johnson, 1974; Wildy, 1967) was provided by the studies of Kristensson, Lycke and Sjöstrand (1971). When injected into the foot-pad, the virus was found within 3 days in the dorsal root ganglia, its ascent blocked by procedures interfering with axonal transport in the nerve. Viral particles were seen in EM localized within vesicular or cisternal structures of the axon (Hill, Field, and Roome, 1972; Kristensson, Ghetti, and Wiśniewski, 1974). This has suggested their movement in ER channels, although this location does not necessarily prove that it does move by this route (Chapter 10), the virus can be translocated from the transport system into the ER at a later time.

The pathogenicity of the virus is explained by its presence in the sensory neuron cell bodies in the dorsal root ganglion to which it is carried by retrograde transport. However, herpes virus is not restricted to sensory nerve fibers. Following the injection of herpes simplex virus (type 2) into the eyes of rabbits, it is taken up by the retinal ganglion cells and transported in its axons to the contralateral superior colliculus (Kristensson *et al.*, 1974). Presumably, the membrane of the retinal ganglion cells and the terminals of peripheral sensory nerve fibers have similar recognition sites for the uptake of the virus. Until recently, there was no way to adequately label components directly to study slow retrograde transport.

A new technique which allows a labeling of proteins within the axons for investigation of retrograde transport was introduced by Fink and Gainer (1980a,b). Radioactively labeled N-succinimidyl-2, 3-^{3}H-proprionate (^{3}H-N-

SP) was injected into the nerve trunk, permitting the alkylating reagent to enter the axons and proprionate proteins. These labeled proteins are then carried by anterograde and a retrograde transport to give rise to the spread of labeled radioactivity from the injection site shown in Fig. 3.15.

The change in the slope of the outflow with time after injection gives a measured rate of anterograde movement of 3–6 mm/day and that of a retrograde movement of 1–2 mm/day.

As will be described in Chapter 7, anterograde transport is dependent on oxidative phosphorylation and it is sensitive to agents which disassemble microtubules (Chapters 9 and 12). Retrograde transport shares the same properties (Chapters 7, 9, and 12).

D. MICROSCOPIC OBSERVATION OF PARTICULATE MOVEMENT AND TRANSPORT RATES

The microscopic visualization of particulate movement within living cells has a long history, starting with the observation of particle movement in amoeba and cyclosis in plant cells. The movement of particles in nerve fibers has been described by Hughes (1953), Burdwood (1965), Pomerat *et al.* (1967), and Rebhun (1972), among others. Dark field and, more recently, Normarski optics microscopy have been used to visualize particle movement by time-lapse cinematography. Single frames of the nerve fibers are taken on movie film at timed intervals and then, by running the film at a higher rate, the character and the velocity of particle movement can be determined. Kirkpatrick *et al.* (1972) and Cooper and Smith (1974) have shown spherical particles, roughly 0.2–0.3 μm in diameter, to have a movement which may be saltatory, the particles stopping and starting at irregular intervals. Rod-shaped bodies, which most certainly by their size and shape are mitochondria, were also seen at times to move at a fast rate. Most of the time they did not move at all. The characteristics of mitochondrial transport will be discussed in more detail in Section E below.

The majority of the smaller particles are seen to move in the retrograde direction, in a ratio of 10:1 with respect to anterograde movement (Cooper and Smith, 1974). As pointed out by Cooper and Smith, the threshold for optical detection of a particle requires that they be at least 0.2 μm in diameter. Actually, at these dimensions a diffraction pattern rather than the particle itself is seen. As a result of this size limitation, there is a systematic bias toward a visualization of the movement of the larger particles with the smaller ones not seen at all. Thus, the larger number of particles observed moving retrogradely does not mean that more material is moving in this direction. On the contrary, as shown by other methods, the bulk of transported labeled materials is known to move in the anterograde direction (Chapter 5). Evidence for different-sized particles accumulated just above and below a block of transport has been obtained by Smith (1980) and by Tsukita and Ishikawa

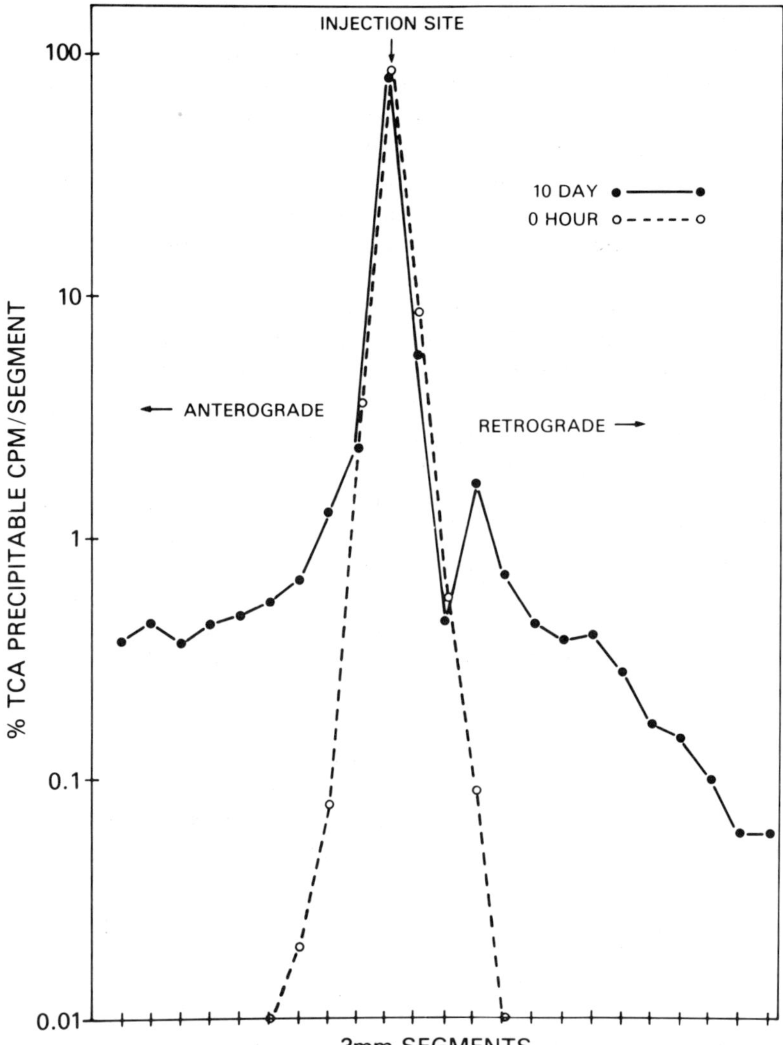

Fig. 3.15 Local labeling of intra-axonal proteins and transport. The distribution of TCA-precipitable proteins in 3-mm segments of nerve is shown after injection of ^{3}H-N-SP superineurally into the sciatic nerve. The nerve, when taken soon afterwards, shows a narrow band of labeling (○), with virtually all the labeled protein confined to segments near the injection site. After 5 days, labeled proteins were found in segments at distances greater than 20 mm proximal to and distal from the injection site (cf. Fink and Gainer, 1980a), and this was spread still further as shown here after 10 days (●). More material can be seen spread in the anterograde direction than in the retrograde direction. Ordinate: counts per min (CPM) of TCA-precipitable material. From Fink and Gainer (1980b).

(1980). Relatively large multilamellated particles were seen in EM to accumulate distal to a cold-block and vesiculotubular structures just above the block by Tsukita and Ishikawa (1980). A fine nylon fiber was used by Smith (1980) to compress individual nerve fibers and here also the retrogradely transported particles were seen to be multilamellated bodies. The bias toward a visualization by light microscopy of a retrograde movement of particles can thus be accounted for by the retrograde movement of the larger size of the multilamellated bodies compared to the anterogradely transported vesiculotubular particles.

The rates given for the movement of the particles visualized in chicken nerve fibers averaged 4.4 mm/hr at 31°C and in *Xenopus* average approximately 1 μm/sec at a temperature of 20–22°C (Kirkpatrick *et al.*, 1972). A relatively high Q_{10} was found for the movement. Leestma (1976) determined a range of velocities of 0.3–0.8 μm/sec for particle movement in the neurites of cultured mouse spinal cord. Much variability in the rate of movement for a given particle was seen but the mean rate can be uniform (Smith and Koles, 1976; Smith, 1979). This can be appreciated from the positions of a single moving particle plotted with respect to time (Fig. 3.16).

The particle shown exhibited a more or less regular progression over a long time period (*A* of Fig. 3.16), although at smaller time increments its velocity underwent marked variations as can be seen in *B–D* of this figure.

Using a computer-based method of collecting data of particulate movement, Forman *et al.* (1977a,b) found the larger particles moving in the retrograde direction in bullfrog nerves to have a wide range of rates. Taking into account the high Q_{10} of 2.5–3.5 of the movement, the collation of retrograde rates determined by various means when extrapolated to a temperature of 37°C were seen to be generally fast (Fig. 3.17, Table 3.1).

The mean retrograde rate of close to 220 mm/day given by the solid line in this figure represents the microscopic studies of Forman *et al.* (1977a,b), a rate somewhat higher than the mean found by other investigators as given by the dashed line. This rate is roughly half that of anterograde transport symbolized by the dotted line which, at 37°C, is 4.5 μm/sec or 410 mm/day.

The variations reported for the rate of retrograde transport are at present not completely understood. Some of the values appear to represent differences in technique, others to a species difference. Smith (1978) found the vesicles in crab nerves to be retrogradely transported at times at rates much faster than had heretofore been reported, approaching that of fast anterograde transport when extrapolated to a temperature of 38°C and the Q_{10} was taken into account. This can be seen also for anterograde transport of some of the particles observed, e.g., by Forman *et al.* (1977a,b) or Cooper and Smith (1974), when applying a temperature correction to scale the rates to 37°C. While fast rates of transport of some components, either in the anterograde or retrograde direction, may occur for a brief time, the mean velocity appears to be that calculated in Fig. 3.17.

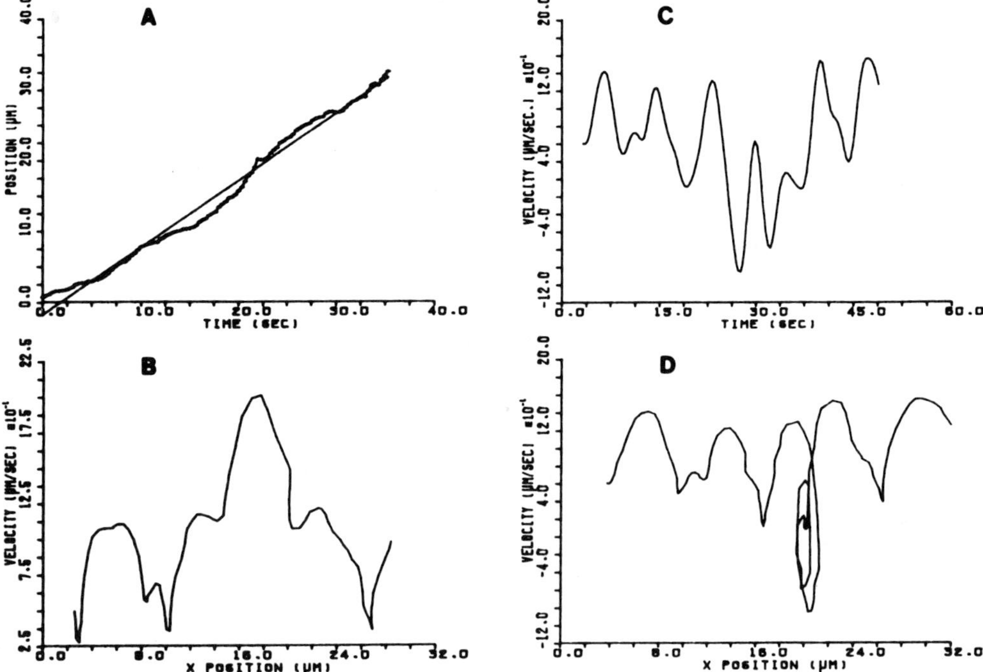

Fig. 3.16 Particle movement shown by microscopy. (A) The axial position of a retrogradely moving particle is plotted with respect to time. The straight line describes the average velocity of the particle. (B) Using the data of A, the change in the velocity of the particle with respect to the distance traveled is plotted. (C) For another particle, its velocity is plotted with respect to time. (D) For the same particle, the velocity is plotted with respect to its axial distance in the fiber. Abscissa: time in sec or distance in μm. Ordinates: distance in μm or velocity in μm/sec $\times$ 10^{-1}. Fiber from *Xenopus laevis* at room temperature. From Smith (1979).

E. TRANSPORT OF MITOCHONDRIA

The rod-shaped bodies approximately 0.12–0.20 μm in diameter and 1–4 μm in length seen in nerves when using dark-field and Nomarski microscopy are, by their size and shape, considered to be mitochondria. Their movement, though infrequent, is not random; they appear to be guided axially in the fiber along some underlying track (Cooper and Smith, 1974). This can be seen most readily where the nerve has been bent, and the rod-shaped particles conform in shape to the underlying path taken by microtubules and neurofilaments (Fig. 3.18).

When the mitochondria move, they do so at a rate approaching that of labeled proteins. However, because the mitochondria only move infrequently and they also move in both anterograde and retrograde directions, their net

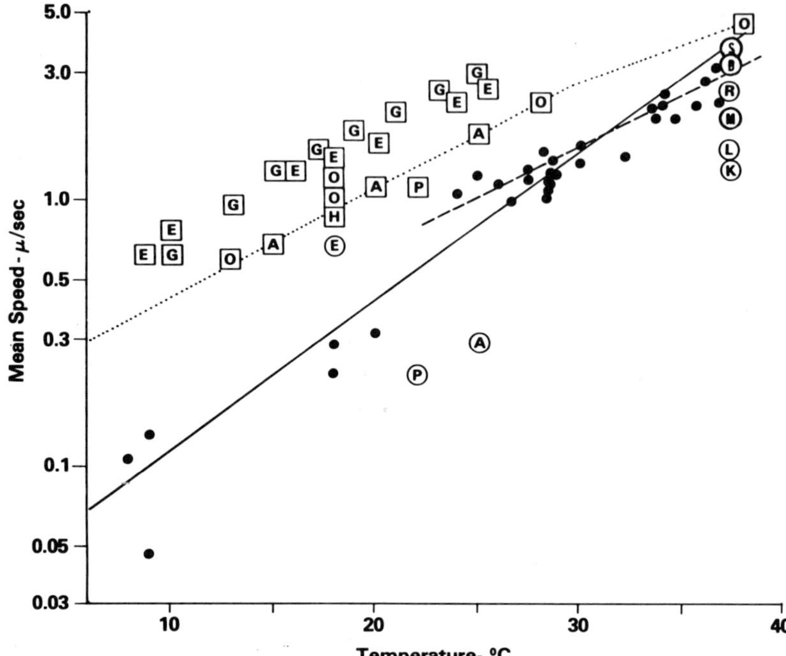

Fig. 3.17 Graphic comparison of average particle speeds with rates of anterograde and retrograde transport. Average mean particle speeds in bullfrog myelinated axons (small filled circles). Rates of axonal transport measured by other methods are given by letters surrounded by circles indicating retrograde transport and squares indicating anterograde transport. Studies in sciatic nerves or branches of the sciatic have, except for ⒢ from the garfish olfactory nerve, and Ⓚ, data from rabbit hypoglossal nerve. The dotted line is the rate of anterograde rapid axonal transport of labeled protein in bullfrog sciatic nerves. Values above 25°C have been extended and these reach 4.74 µm/sec (410 mm/day) at 37°C. From Forman, Padjen, and Siggins (1977b).

anterograde movement is at a slow rate. The first evidence for their transport using a marker of mitochondria, NAD-diaphorase, was given by Friede (1959). He found this mitochondrial enzyme to accumulate above nerve crushes, indicating that the mitochondria are transported to and remain at this site. A large accumulation of mitochondria at a nerve interruption was subsequently seen in EM preparations (Weiss and Pillai, 1965).

Zelená, Lubińska, and Gutmann (1968) have further described the dense accumulation of mitochondria along with vesicles and tubules forming pellets above nerve crushes. These pellets increase in length over the first 18 hr in both the proximal and distal nerve stumps and thereafter were seen only in the proximal stump. The mitochondria concentrated in the pellets are clearly transported to this site from other parts of the axon (Zelená, 1968; Zelená and Gutmann, 1968; Friede and Ho, 1977). This was shown by the drop in

Symbol	Reference	Species	Method[a]
Ⓐ (boxed)	Abe *et al.* (1973)	Bullfrog	LP
Ⓐ	Abe *et al.* (1974)	Bullfrog	LP
Ⓑ	Brimijoin and Helland (1976)	Rabbit	SF
Ⓔ (boxed)	Edström and Hanson (1973a)	Frog	LP
Ⓔ	Edström and Hanson (1973b)	Frog	LP
Ⓖ (boxed)	Gross (1973)	Garfish	LP
Ⓚ	Kristensson *et al.* (1971)	Rabbit	HRP
Ⓛ	Lubińska and Niemierko (1971)	Dog	AChE
Ⓜ	Stöckel *et al.* (1975b)	Rat (M)	LP
Ⓞ (boxed)	Ochs and Smith (1975)	Cat	LP
Ⓟ (boxed), Ⓟ	Partlow *et al.* (1972)	Frog	AChE
Ⓡ	Ranish and Ochs (1972)	Cat	AChE
Ⓢ	Stöckel *et al.* (1975b)	Rat (S)	LP

[a]Methods: LP = movement of radiolabeled protein; AChE = accumulation of acetylcholinesterase at crushes, ligations, or at the ends of isolated segments; HRP = transport of exogenous horseradish peroxidase; SF = stop-flow, M = motor, S = senory nerve. From Forman, Padjen, and Siggins (1977b).

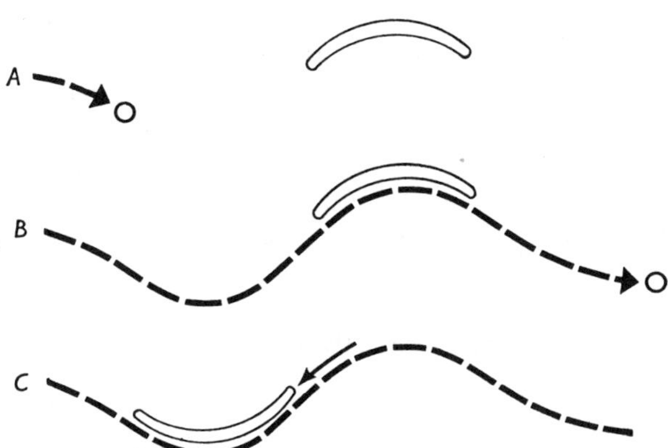

Fig. 3.18 Guided movement of particles. Diagram of a sequence is shown in which a somatopetally traveling particle was seen in a nerve examined microscopically to enter an area of folded axoplasm (A) where it proceeded to travel along a sinuous path along with a rod-shaped organelle. (B) A few seconds after the passage of the particle the rod moved in the reverse direction along the path previously taken by the particle to a new stationary position (C). From Cooper and Smith (1974).

the density of mitochondria counted in the midportion of doubly-ligated pieces of nerve which corresponded to the increases of mitochondria accumulated at the ligations at either end of the nerve pieces.

The origin of the mitochondria from the cell body was indicated by the studies of Barondes (1966). He used systemic injections of β-phenylisopropylhydrazine (PIH) into the brain to irreversibly block the mitochondrial enzyme monoamine oxidase (MAO) in the nerve fibers. At various times thereafter, synaptosomes were isolated from the brain to determine their content of MAO. The PIH injections caused a great reduction of MAO, followed by its slow return over a time period of 4–5 weeks. The slow return was consistent with the slow net transport of a new supply of mitochondria transported in the fibers down to their terminals.

Another technique used to show the slow net anterograde transport of mitochondria was the injection of ^{59}Fe into the ventral horn for uptake by the motoneurons (Jeffrey *et al.*, 1972). Mitochondria marked with ^{59}Fe were subsequently seen to have a slow transport in the motor fibers. Additional support for a slow net anterograde movement of mitochondria was gained using the pattern of accumulation of MAO seen in doubly-ligated nerves (Khan and Ochs, 1975). The accumulation of MAO above distal ligations did not show a "departure curve" as expected of a fast transport (cf. Chapter 4), but instead an accumulation characteristic of slow transport. The accumulation of MAO above the upper ligation of doubly-ligated nerves was only slightly in excess of the accumulation below the lower ligations, a pattern consistent with a slow net anterograde movement of the mitochondria.

Thus, while the mitochondria may temporarily have a fast movement anterogradely and retrogradely, they are for the most part sessile, and their net outflow is characterized as a slow transport. The mitochondria supply the ATP needed for axoplasmic transport (Chapter 7), and presumably some signal related to a need for ATP appearing in the axons could cause the mitochondria to move to a new position. Such a signal could perhaps arise in relation to another function of mitochondria, their participation in Ca-regulation (Chapter 8). The signal in this case could be a local change in the level of free Ca^{2+} in the axon requiring its regulation.

F.   THE USE OF TRANSPORT TO DETERMINE PATHWAYS IN THE CNS

Classically, nerve pathways in the CNS have been determined by making lesions and then following the course of degenerating fibers as revealed by various staining methods. This method, however, results in an unintended injury of fibers lying near those selected for destruction, complicating the interpretation of the data. Also, the sensitivity of the methods used to stain degenerated fibers may be low and the staining method precarious. Even the best silver method, the Nauta technique, may not reveal all the fibers present in a given group. This was seen by comparing it with the use of axoplasmic transport and autoradiography to show the path of fibers with standard methods. O'Steen and Vaughan (1968) injected ^{3}H-hydroxytryptophane into the retina of rats and with autoradiography found fibers projecting to the hypothalamus, a termination not observed by degeneration and silver staining. Goldberg and Kotani (1967) injected ^{3}H-leucine into the eye of bullfrog tadpoles and also found fibers containing grains in autoradiographs terminating as an additional system projecting to the thalamus not seen with the classical technique of degeneration and silver staining.

The use of axoplasmic transport to reveal nerve paths has other advantages. Cowan *et al.* (1972), using transport-autoradiography to study the course of optic fibers in the monkey, pointed out that when times are kept brief, labeled radioactivity may be found localized in the bouton termineaux (cf. Droz and Barondes, 1969). When longer times of downflow were allowed, the fibers were more heavily labeled. Thus, by a choice of times it is possible to determine the pattern of bouton terminations on their target neurons, whether axosomatic or axodendritic, as well as to assess the course of the fibers in the CNS within a given system.

The use of HRP has become of prime importance as a means to determine pathways in the CNS. After injecting HRP for uptake by nerve terminals and its retrograde transport, the cells of origin can be determined by the accumulation of the protein in them. This technique, when used in conjunction with the injection of ^{3}H-leucine to show anterograde transport in the same pathway, has proved to be particularly valuable in such anatomical studies (Cowan and Cuénod, 1975; LaVail, 1975; Livett, 1976). Recent improvements of the HRP method has increased its sensitivity (Mesulam and Rosene, 1979) and extended its range of usefulness. Using the more sensitive method, it was possible to better assess the anterograde transport of HRP. It was shown to have a rate between 288 and 432 mm/day (Mesulam and Rosene, 1979). It is likely that the anterograde transport is carried down at the same rate as proteins and by the same fast transport mechanism (Mesulam, 1982). The rate of the retrograde transport of HRP also approaches that of other retrogradely transported components and it is also likely carried by the same transport mechanism (cf. Chapter 10).

Another method useful in tracing adrenergic, cholinergic, or serotinergic fiber tracts in the CNS is to make a series of lesions in the CNS at various positions in the brain in different animals. The interruption of the transport of the endogenous transmitter or transmitter-related components contained in those fibers (Chapter 4) is followed by the damming up of those substances at the cut ends of the fibers connected to the cell bodies, with their depletion in the fibers amputated from their cells. By means of a number of such transections systematically made in the brain and charting the accumulation and depletion of their component materials, maps of adrenergic, dopaminergic, cholinergic, and serotonergic fiber pathways in the brain and spinal cord have been constructed (Cowan and Cuénod, 1975).

4

Transport of Transmitter and Transmitter-Related Components

With an understanding of the action of neurotransmitters at autonomic and neuromuscular junctions gained through intensive study of this subject carried out for the most part of this century, attention has turned to the transport of neurotransmitters and neurotransmitter-related components in the nerve fibers to their terminals, the subject of this chapter. The relation of transport to the release of transmitters and the resynthesis of transmitter in the nerve terminals will be taken up in Chapter 13. Here we will deal with the intra-axonic transport in the cholinergic nerves of the transmitter acetylcholine (ACh) and its related components, transport in adrenergic nerves of the transmitter noradrenaline (NA) and dopamine (DA) along with their related enzymes, and transport in the serotonergic neuron of the neurotransmitter 5-hydroxytryptamine (5-HT) and its related compounds. Brief mention of several other neurons characterized by their putative transmitters will be made.

A. CHOLINERGIC NEURONS

The establishment of ACh as the neurotransmitter at the neuromuscular junction (Katz, 1966), in autonomic ganglia and in certain CNS neurons (Eccles, 1964; Phillis, 1970), raises the question as to the form in which the neurotransmitter is carried down to the nerve endings. The cholinergic nerve fibers contain, in addition to ACh, the hydrolyzing enzyme acetylcholinesterase (AChE) and the synthesizing enzyme choline acetyltransferase (ChAc):

$$\text{Ch} + \text{Ac} \; \underset{\text{AChE}}{\overset{\text{ChAc}}{\rightleftharpoons}} \; \text{ACh}$$

where *Ch* represents choline, *Ac* acetyl, *ChAc* choline acetyltransferase, and *AChE* acetylcholine esterase.

70

The presence of the enzymes AChE and ChAc as well as ACh in motor nerve fibers led to the suggestion that they might be carried from their site of synthesis in the nerve cell bodies down to the nerve terminals (Feldberg and Vogt, 1948). This would require, however, that ACh and its related enzymes be transported in the axons in a compartmented form. Otherwise, the ACh would be hydrolyzed by the AChE. When the uptake of choline and its synthesis into ACh in the nerve terminals was discovered, the relative proportions of ACh transported to that locally synthesized became an issue, a topic reserved for discussion in Chapter 13.

1.   Acetylcholinesterase (AChE)

Sawyer (1946) found AChE accumulating just above the cut ends of nerves without, however, relating this phenomenon to axoplasmic transport. Later, when evidence for slow transport had appeared, Clouet and Waelsch (1961) and Koenig and Koelle (1961) reasoned that by treating animals with an agent which irreversibly blocks AChE, e.g., difluorophosphate (DFP), the supply of newly synthesized enzyme transported down from the cell bodies should appear in the nerves with a proximo-distal gradient. Believing that AChE was carried down the fibers at the slow rate of several mm/day, they expected more of the newly formed enzyme to appear proximally and then over a period of days and weeks, increasingly greater amounts more distally in the nerve in conformity with a slow spread (cf. Fig. 2.3). In both studies the anticipated proximo-distal gradient was not found. Instead, the enzyme reappeared uniformally along the whole length of the nerve, even as early as a few days after introducing the blocking agent. At the time, this tended to cast doubt on the concept of axonal transport altogether. However, as was subsequently found to be the case, AChE is fast transported, and the times taken to assess the spatial distribution of AChE were already too late to observe a gradient. For this the double-crush technique and a histochemical method to measure AChE in the nerves and show its redistribution was of key importance, a technique first introduced by Lubińska and her colleagues (Lubińska, 1964). Peripheral nerves were crushed and then at different times thereafter, an accumulation of AChE above and below the crushes was found, indicative not only of its anterograde but its retrograde transport as well (Fig. 4.1).

The enzyme accumulated linearly for 26 hr and then over the next several days the increase of enzyme tapered off. When two crushes were made closer together in the nerve, the increase of AChE above the distal crush diminished, the amount of AChE accumulating depending on the length of the nerve segment isolated by the two crushes. This suggested that some portion of the enzyme present within the isolated segment is free to move and accumulate at the crush, as is indicated graphically in Figure 4.2.

To be noted in this figure is the accumulation of AChE both above and below the crushes and the decrease of the enzyme in the middle of the

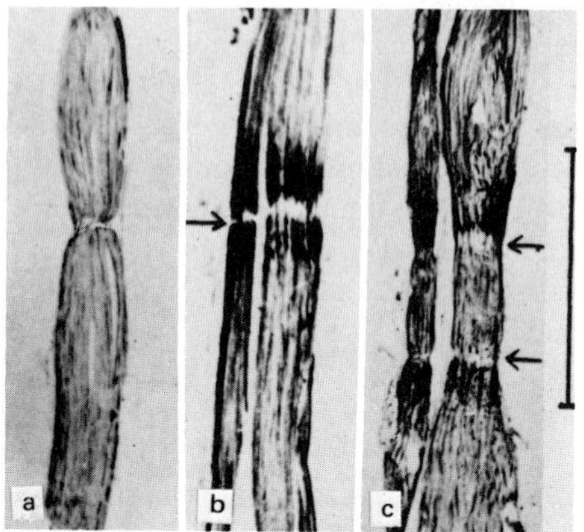

Fig. 4.1 Accumulation of AChE at nerve crushes. Longitudinal section of a crushed sciatic nerve of the rabbit shows the accumulation of AChE near the sites of the crushes (arrows) previously made. No great accumulation of AChE appears soon after a crush (a). The nerve removed 26 hr after crushing, however, shows an increased activity of AChE on both sides of the lesion (b). The sciatic nerve crushed at two places with the distance between crushes 3.4 mm shows a marked increase of AChE activity just above the proximal and below the distal lesion with no increase in the total AChE activity within the short segment isolated by the crushes (c). Scale bar = 3 mm. From Lubińska (1964).

segment isolated by the crushes. The total amount of AChE present within the nerve segment remains unchanged; the AChE becomes redistributed in the fibers isolated by the crushes (cf. Chapter 11) with the enzyme carried by axoplasmic transport in both the anterograde and retrograde directions.

By measuring the accumulations of AChE at double crushes spaced to give different lengths of isolated nerve and over different periods of time, Lubińska and Niemierko (1971) arrived at a rate of anterograde transport of AChE of 220 mm/day and a retrograde transport rate approximately half as fast, 100 mm/day. While this rate of anterograde transport is faster than that previously estimated, it is lower than the rate of labeled proteins (Chapter 2). Evidence for a higher rate of anterograde transport of AChE, one corresponding more closely to that of labeled proteins, was found using the long uniform lengths of sciatic nerve available in the cat and making double ligations at a fixed distance apart (Ranish and Ochs, 1972). The enzyme was measured in 5-mm segments of nerve taken at various times above and below the ligations at the sites shown in Figure 4.3.

The accumulation of AChE above the upper ligation of the double-ligated nerves was linear over a period of 20 hr, while the accumulation above the lower ligation leveled off after 4.4 hr (Fig. 4.4).

The interpretation was that the enzyme within the double-ligated nerve segment free to move accumulates above the distal ligation until, with no more of the freely mobile enzyme available to accumulate at that site,

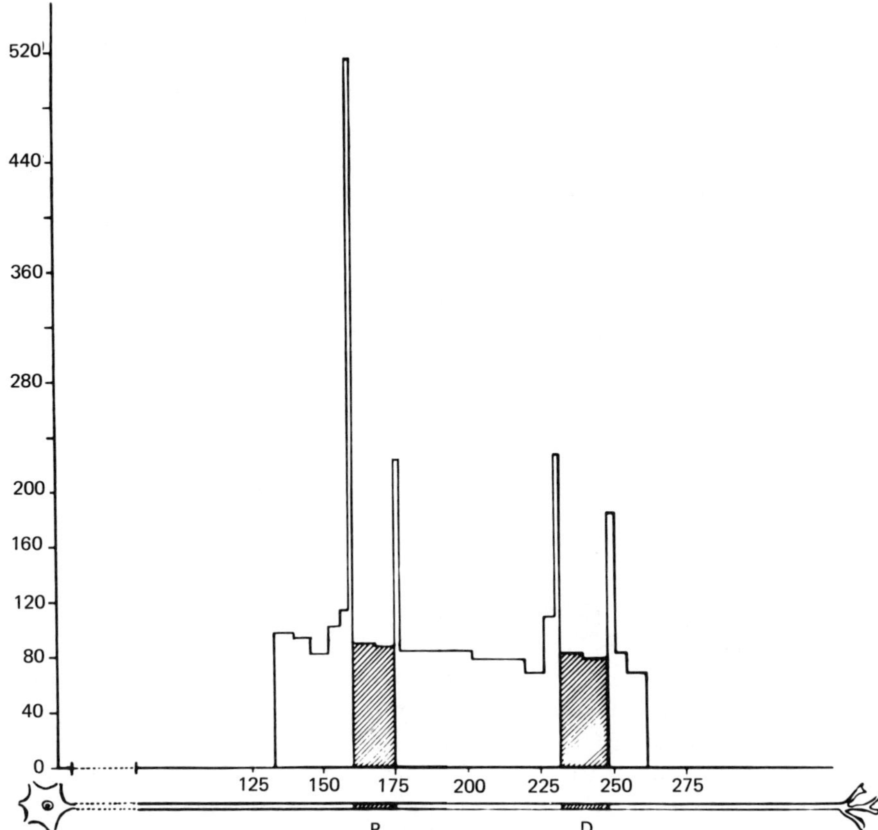

Fig. 4.2 Graphical display of AChE accumulation at a double crush. Crushes were made similar to that shown in Fig. 4.1. A greater amount of AChE activity was seen to accumulate above the proximal (P)) than at the distal (D) crush. Accumulation is also seen at the crushes within the piece of nerve isolated by the crushes. The crushed regions are shown by the hatching. From Lubińska (1964).

it then remains at that level. The "departure" of the curve thus represents a "clearance" by fast axoplasmic transport of the enzyme from within the nerve segment isolated by the two ligations. From the departure time of 4.4 hr and the fixed length of nerve segments of 83.5 mm, the rate of transport of AChE was calculated to be 431 mm/day, a value not significantly different from the rate of fast transport assessed by the use of labeled proteins (Chapter 2). About 10% of the total AChE present in the nerve fibers was calculated to be free to move in the anterograde direction by taking into account the rate of transport in mm/hr, the amount of enzyme accumulated in one hour, and the sample size. Fonnum (1973) also concluded from his studies that 5–20% of the AChE present in the vagus nerve was free to move at a rate of 425 mm/day. Using another method, the stop-flow technique, which will be more fully described in the following section, Brimijoin and Wiermaa (1978)

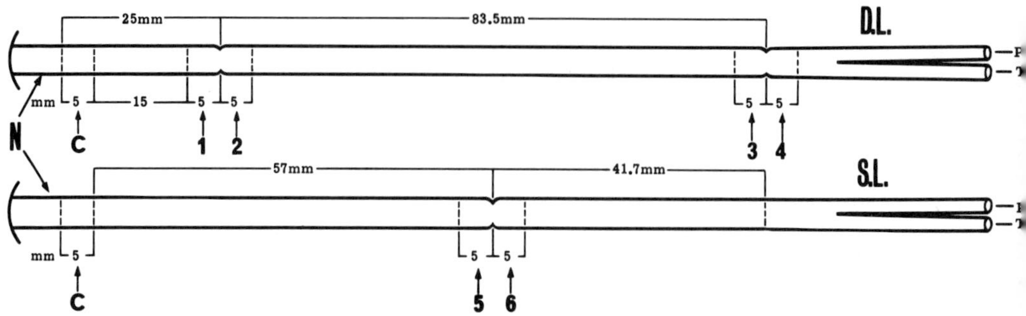

Fig. 4.3 Schematic showing ligations for double-ligation studies. Sites of ligations and sample designations of segments taken for study of AChE accumulation and rate of transport. Segments 1, 3, and 5 are proximal, segments 2, 4, and 6 distal to ligation sites (C) control segments taken away from ligation sites. All samples are 5 mm in length. P and T signify peroneal and tibial branches of the sciatic nerve (N) and D.L. and S.L. signify the double- and single-ligated nerves. From Ranish and Ochs (1972).

found some 10% of the AChE present in the nerve free to move at a rate of 400 ± 35 mm/day. On balance, the various approaches used to assess transport of AChE are in relatively good agreement with an anterograde rate close to 410 mm/day and with some 10–15% of the enzyme free to move.

The bulk of the AChE present in nerve fibers is in a relatively stationary form which, as indicated by EM histochemistry studies, appears to be located close to or in the axolemma, in the ER, and in association with other axon organelles (Brzin *et al.*, 1966; Kasa, 1968). When nerve is homogenized and the subcellular fractions separated by differential centrifugation, most of the enzyme is found in the particulate (microsomal) fraction, with some of the enzyme associated with the mitochondrial fraction (Holmstedt and Toschi, 1959). Nearly all the AChE fast-transported in the nerve and accummulated at ligations was also found in the particulate fraction (Ranish and Ochs, 1972).

The AChE which is free to move in the axons might have a relation to the AChE found in the synaptic junction, i.e., in the synaptic cleft (Ochs, 1974b). This possibility will be discussed further in Chapter 13 with reference to the different species of AChE present in the nerve and nerve terminals identified by their sedimentation rates in Svedberg numbers (Hall, 1973). The 16S species appears to be mostly fast-transported in the nerve fibers (Di-Giamberardino and Couraud, 1978; Fernandez, Duell, and Festoff, 1979; Brimijoin, 1979). Using a double-ligation technique, Couraud and Di-Giamberardino (1980) found the fast rate of the highest MW species (20S, which is likely to be similar to the 16S reported by other workers) to be 408 mm/day in the chicken sciatic nerve, a rate close to that found for undifferentiated AChE. Some 14% of the 20S species was found to be mobile.

A retrograde transport of AChE was also determined from the double-

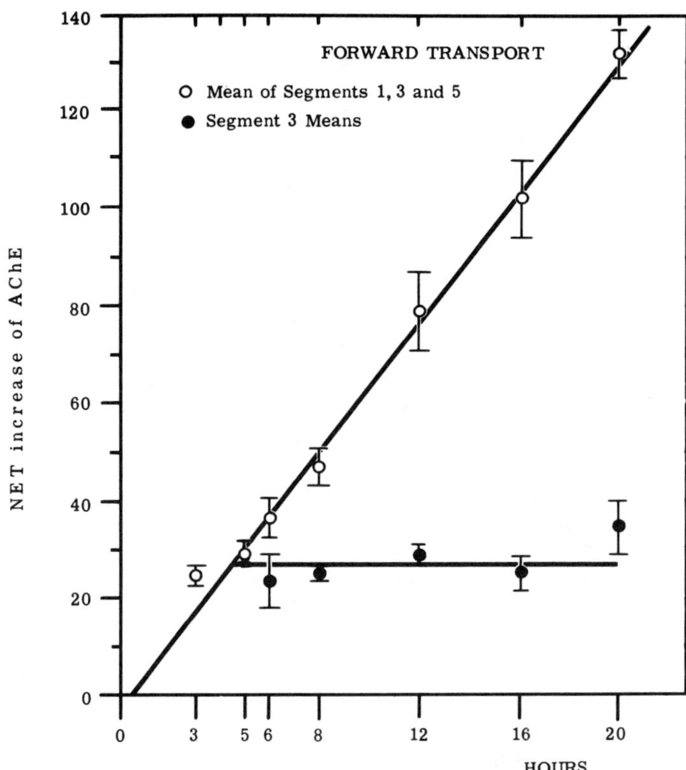

Fig. 4.4 Accumulation of AChE in double-ligated nerves. Using the ligation scheme of Fig. 4.3, the net accumulations of AChE in nerve segments proximal to the ligations at 3–20 hr is shown; the regression line representing the combined values from segments 1, 3, and 5 (O) is described by the equation: $\bar{y} = 6.575x - 1.321$. The line drawn for the means of segment 3 (●) within the isolated portion of nerve taken just above the distal ligation is parallel to the time axis. This represents a depletion of AChE from within the isolated segment as it accumulates at segment 3. The ordinate scale represents net increase of enzyme activity. From Ranish and Ochs (1972).

ligation studies by its accumulation in 5-mm pieces of nerve taken just below the upper and lower ligations at different times. The accumulation of AChE below the two ligations of doubly ligated nerves exhibited a pattern similar to that of anterograde transport (Fig. 4.5).

As can be seen in this figure, the accumulation of enzyme below the lowermost ligation increases linearly with time, but at a lower rate than the accumulation above ligations contributed by anterograde transport. The leveling off of AChE below the uppermost ligation within the doubly-ligated nerve segments, its "departure," occurred at 8.6 hr. Taking that longer time and the length of the isolated nerve segment into account, the rate of retrograde transport of AChE was calculated to be 220 mm/day, i.e., a rate

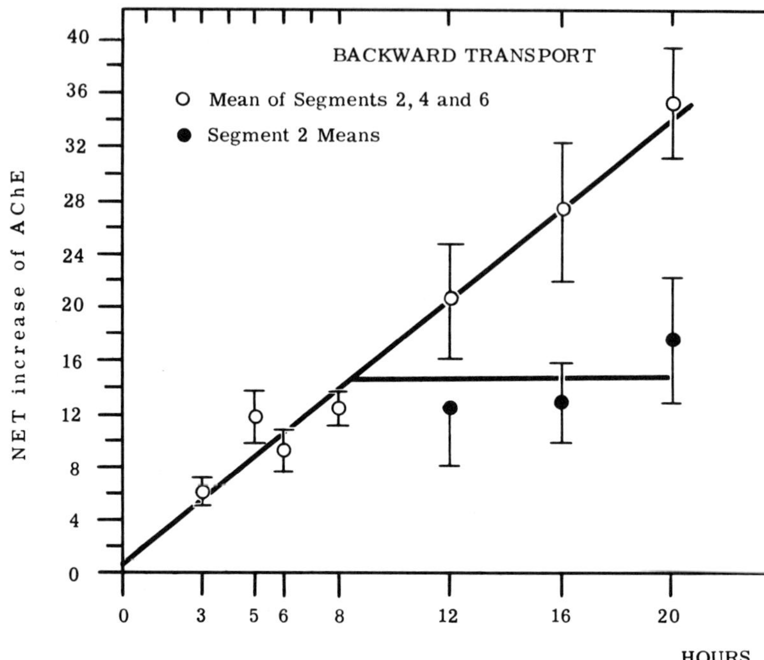

Fig. 4.5 Retrograde transport of AChE in double-ligated nerves. The net accumulation of AChE in nerve segments distal to the ligations at 3–20 hr is shown after
placing the ligations (cf. Fig. 4.3). The regression line shown for the combined
values from segments 2, 4, and 6 (O) is described by the equation: $\bar{y} = 1.69x +
0.48$. The ordinate scale represents accumulation of enzyme activity. The line drawn
for the means of segment 2 (●) at 12 hr and longer after ligation denotes the overall
mean parallel to the abscissa as in Fig. 4.4. The accumulation reflects retrograde
transport of less material at a slower rate. From Ranish and Ochs (1972).

roughly half that of anterograde transport (Ranish and Ochs, 1972). The
percentage of the enzyme free to move in the retrograde direction was also
calculated to be less than that in the anterograde direction, some 5% of the
total AChE. Couraud and DiGiamberardino (1980) in their double-ligation
studies of AChE species in the chicken nerve found the 20S form (i.e. the
16S form of other workers) to be retrogradely transported at a rate of 145
mm/day with 9% of it mobile. Those results again are close to those obtained
for retrograde transport of AChE in the cat.

Other estimates made of the rate of retrograde transport show variations
(cf. Fig. 3.17, Chapter 3). While the rate of anterograde transport of AChE
found by Fonnum (1973) was in good agreement with other measurements
of its fast transport at close to 410 mm/day, he gave a much lower rate of 73
mm/day for its retrograde transport.

In the use of double-ligation to estimate transport in the study of AChE, there must not be an embarrassment of the blood supply within the isolated segment or else the normal course of transport will be impeded (cf. Chapter 7). The ligations need to be made far enough apart so as to include a sufficient blood supply via the nutrient blood vessels (Chapter 7). Such precautions do not apply to the determination of the rate of AChE transport in the sciatic nerves of *Rana pipiens* made *in vitro* by Partlow *et al.* (1972). The velocity of anterograde AChE transport calculated on the basis of the accumulation of enzyme at nerve ligations was 90 mm/day at 22°C with an estimation of the fraction of the enzyme free to move of 12.5%. Considering the high Q_{10} which has been determined for *in vitro* transport in frog nerves (Chapter 2), the rate of AChE transport when scaled to a temperature of 38°C was very close to 410 mm/day. The retrograde rate similarly determined was, however, approximately 1/5 that of the anterograde rate. The reason for the lower rate of retrograde transport found in this preparation in comparison to that of the cat or chicken sciatic nerve remains unknown.

2. Choline Acetyltransferase (ChAc)

Hebb and Silver (1961) in their pioneer studies reported ChAc to be transported in nerve at a slow rate. The enzyme is present in the nerve in soluble form and appears to fit with the general proposition that soluble components are carried at a slow rate in contrast to a fast transport of particulates (Tuček, 1975, 1978). There are now, however, enough instances of a fast transport of soluble components now known to make this view untenable (cf. Chapter 5). Specifically, ChAc, in addition to its slow rate, has been reported to have a fast rate of transport as well (Dahlström *et al.*, 1974; Frizell *et al.*, 1970). Fonnum *et al.* (1973) reported a fast rate of transport of ChAc of 191 mm/ hr in the vagus and 47 mm/day in the hypoglossal nerve, rates based on the estimation that only 6.7% and 5.8%, respectively, of the enzyme present in those nerves are free to move. Davies, Whittaker, and Zimmermann (1977) considered that some 15% of the ChAc is mobile and estimated its rate of axoplasmic transport in *Torpedo marmorata* nerves at 50–140 mm/day. Estimations of rate based on the assumption that the accumulation of the enzyme at ligations is due to a proximo-distal movement of all the enzyme present in the nerve are thus too low.

The possibility of a retrograde transport of ChAc has been suggested by its accumulation just below a ligation made in vagus nerves (Fonnum *et al.*, 1973, 1976). The increase found, however, was small and further studies are required to substantiate this point.

3. Acetylcholine (ACh)

A transport of ACh in mammalian nerve fibers was indicated by its accumulation at the cut ends of ventral root fibers (Evans and Saunders, 1974). However, this technique cannot be used as a direct measure of its rate of axoplasmic transport. The cut ends of fibers seal off in the course of time and local synthesis or hydrolysis of ACh at the cut ends are also factors which could affect an estimation of rate on that basis (Dahlström *et al.*, 1974; but cf. Evans and Saunders, 1974, and O'Brien, 1978).

Nevertheless, a portion of the ACh accumulating at nerve crushes indicates that it is carried in the nerve fibers at a fast rate (Fig. 4.6).

Dahlström *et al.* (1974) estimated the rate to be 5 mm/hr or 120 mm/day.

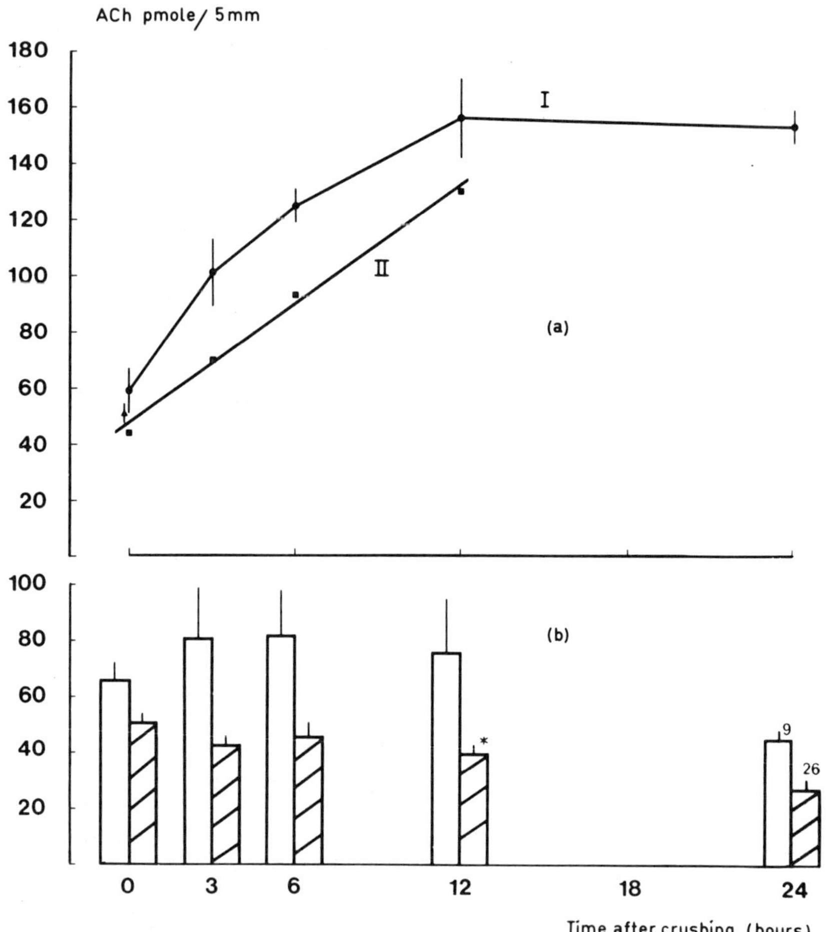

Fig. 4.6 Accumulation of ACh. (a) Change in ACh content in 5-mm nerve segments proximal to crushes from 0 to 24 hr after operation. (I) Original experimental values of uncrushed nerve. (II) Values corrected for local synthesis after subtracting "extra" ACh accumulated above this level in uncrushed nerve. Vertical bars are means ± S.E.M. (b) Fall in ACh content in segments distal to the crushes 0–24 hr after operation. Open bars give values in segments 5 mm immediately distal to crush. Hatched bars give values in nerve segments 10–25 mm distal to the crushes. Vertical lines are ± S.E.M. From Dahlström *et al.* (1974).

Some features of this figure require comment. As shown in the upper curve, a 40% increase in ACh content was found very soon after making a crush, the very fast increase likely due to a local synthesis. Thereafter, the further increase of ACh shows by its linear accumulation a rate consistent with fast anterograde transport. Additionally, smaller increases of ACh seen by Dahlström in the nerve segments just distal to nerve interruptions suggest that some ACh is also transported retrogradely. Accumulations due to retrograde transport must be taken into account when determining the decrease of ACh in the nerve segments taken just distal to a crushed region and ascribed to anterograde transport as shown in part *b* of Fig. 4.6. The decreases of ACh amounted to 15–20%, indicating that this may be the portion of ACh free to move in the anterograde direction. The rate of anterograde transport when estimated on that basis was 5 mm/hr, or 120 mm/day. However, it may be that the amount of ACh free to move is even smaller if, in addition, ACh is brought to this region by retrograde transport. It is important to determine if this is the case because relatively small variations in the amount of ACh at that region can make a considerable difference in the estimation of the rate of anterograde transport.

A fast transport of ACh was also shown using labeled choline as a precursor. There is some limitation, however, in the use of labeled choline to study the transport of ACh. Choline has a low affinity for the uptake mechanism in the membrane of the neuron cell bodies of *Aplysia* as compared to the high affinity for ACh uptake present in the nerve terminals (Suszkiw and Pilar, 1976). However, the enzyme required to acetylate choline is present in the cell bodies and when ^{3}H-choline was injected directly into the cell bodies of the large right R2 cholinergic giant cells, Koike *et al.* (1972) found approximately 50% of the precursor converted to ACh with the remainder incorporated into phosphorylcholine, a component related to membrane turnover (Chapter 5). The subsequent outflow of radioactive-labeled material into the right connective nerve fiber of the R2 neuron, mainly in the form of ACh, was seen as a crest moving at a rate of 17 mm/day at 15°C. A minor component was also seen moving at a faster rate of 55 mm/day. An estimation of its rate of transport at 38°C is likely to have been too low because Koike *et al.* (1972) used a relatively low Q_{10} of 1.6 to scale the rate to the higher temperature of 38°C. Using the higher Q_{10} found for transport in other studies carried out in this laboratory (Chapter 2), the rate of the minor component when scaled to a temperature of 38°C approaches that of the fast transport of labeled proteins in mammalian nerves (cf. Goldberg *et al.*, 1978, and Section C below). The bulk of the ^{3}H-ATP injected which is not fast-transported spreads by diffusion (Koike and Nagata, 1979).

A transport of ACh in the long antennule nerve fibers of the roach was reported by Schafer (1973). The antennules contain unmyelinated axons with diameters less than 0.5 μm, each fiber containing 2–5 microtubules with no neurofilaments present. The cell bodies are located near the antennule surface with a short dendritic tip exposed to the surface through a small pore in the cuticle. After dipping the dendritic tips into a solution of methyl^{3}H-choline and allowing time for uptake and incorporation of the precursor, the rate of transport of labeled ACh in the axons was estimated to be 120–130 mm/day. Those studies were done at room temperature and the rate of ACh, when scaled to a temperature of 38°C and the Q_{10} taken into account, would likely be as fast as that of labeled proteins in mammalian nerve fibers.

The form in which ACh is transported in the fibers is not clearly known. There is evidence that ACh is present both in soluble and particulate form (Hebb, 1963; O'Brien, 1978). A sizeable amount (31–46%) of the ACh present in *Aplysia* nerve fibers was found by Koike *et al.* (1972) in the particulate fraction and a similarly relatively high proportion was found present in the granules isolated from the cholinergic fibers of *Torpedo* (Whittaker *et al.*, 1975).

B.   ADRENERGIC NEURONS

Peripheral adrenergic nerves contain noradrenaline (NA), the form in which the adrenergic neurotransmitter is released from its nerve terminals (von Euler, 1956). The biochemical path of its synthesis is shown in Fig. 4.7 where the equivalent term norepinephrine is used for NA.

As will be described, not only is NA fast-transported in peripheral adrenergic nerve fibers, but dopamine (DA) as well. The last step in the pathway shown in Fig. 4.8, the conversion of NA to adrenaline (A), takes place in the adrenal gland. There, the enzyme phenylethanolamine N-methyl transferase (PNMT) converts NA to adrenaline (A) which is released into the blood stream as a neurohormone. A small number of adrenergic neurons containing PNMT are found in the CNS, with none in peripheral nerve (cf. Cooper *et al.*, 1978).

1.   Transport of Noradrenaline (NA)

In electron micrographs (EM) adrenergic nerves are seen to contain vesicles with an electron-dense granule inside it, the dense core vesicles (DCVs). The DCVs isolated by differential centrifugation from peripheral nerve were shown to contain NA (von Euler and Hillarp, 1956; Geffen and Livett, 1971). Both small and large DCVs are seen in EM sections in adrenergic nerves, those in the size ranges of 400–600 Å and 750–1200 Å, respectively (Hökfelt, 1969; Stjärne, 1966; Banks and Mayor, 1972; Thureson-Klein *et al.*, 1973). The large DCVs are considered to be relatively immature and converted in the terminal part of the axons to the smaller vesicles (Lagercrantz, 1976). However, this is not indisputedly established. While a greater number of the smaller DCVs are usually seen present in the terminal varicosities where transmitter release takes place, both small and large DCVs are found in the cell bodies.

The DCVs are often seen lined up along microtubules (Thureson-Klein *et al.*, 1973), an arrangement suggestive of their transport along the microtu-

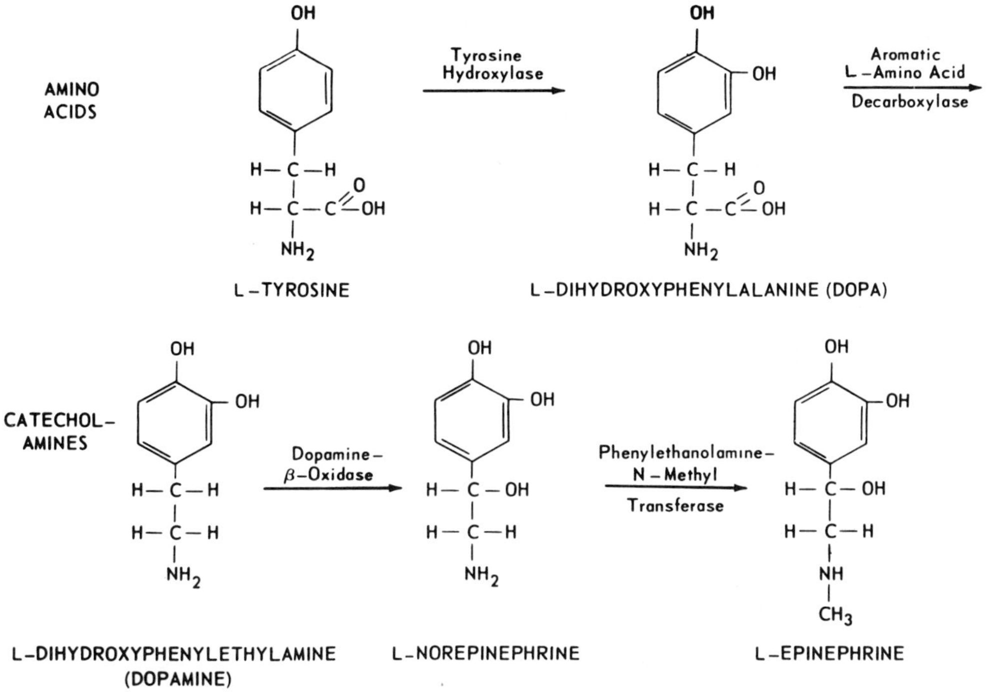

Fig. **4.**7 Biochemical path of noradrenaline (norepinephrine) synthesis. All but the last step occurs in peripheral nerve. Adrenaline (epinephrine) is produced in the adrenals which contains the transferase enzyme. From Geffen (1975).

bules (cf. Fig. 13.2). The large DCVs isolated from homogenized nerve by differential centrifugation are found, after release of their contents by osmotic shock or detergent action, to contain in addition to NA, dopamine-β-hydroxylase (DBH), chromogranin, ATP, Mg^{2+}-activated ATPase, a cytochrome, and phospholipid (Geffen and Livett, 1971). Most likely, DA is also present (Section 2 below). How some of these vesicle components are involved in transport, or in the repackaging of the neurotransmitter NA in the nerve terminals, will be dealt with in part in this chapter and in more detail in Chapter 13.

The fluorescent method of Falck *et al.* (1962) was of considerable importance in allowing for the visualization and the assay of the relatively small amount of NA present in adrenergic nerve fibers. The increased fluorescence due to the accumulation of NA in fibers just above a nerve crush was used to study the accumulation of NA above nerve crushes or ligations and by this means its axoplasmic transport in the fibers (Dahlström and Fuxe, 1964a). A diagrammatic picture of the fluorescent increases is given in Fig. 4.8.

A small amount of fluorescence is also seen just below the ligations suggesting a possible retrograde transport, although corroborating biochemical evidence for a retrograde transport of NA has not been obtained (see below).

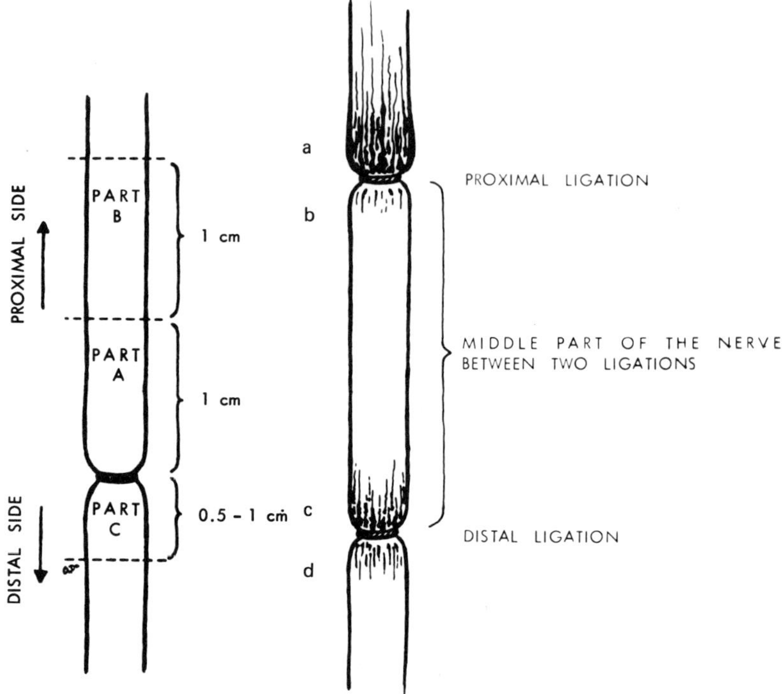

Fig. 4.8 Accumulation of noradrenaline in ligated preparations. Diagrammatic representation of ligations made in sciatic nerve of rat with fluorescent material accumulated just above the sites of constriction made 1 day beforehand. From Dahlström and Häggendal (1966).

The accumulation of NA above nerve crushes or ligations increases linearly with time, and then tapers off at the longer times (Fig. 4.9).

The curves differ somewhat in cat and rat sciatic nerves, but in either case a sufficiently long period of linear increase allows a determination of the rate of fast transport by means of double-crushs or double-ligations. The accumulation of fluoresence above the lower ligation in doubly-ligated nerves falls off quickly compared to its continued increase above the upper ligation. The time at which fall-off above the lower ligation begins gave an estimated rate of fast transport of NA of 5–6 mm/hr (120–144 mm/day) in rat nerves and 9–10 mm/hr (216–240 mm/day) in cat nerves (Dahlström and Häggendal, 1966). For rabbit nerves the rate was estimated as 3 mm/hr or 72 mm/day (Dahlström and Häggendal, 1967). In those assessments of rate, the assumption was made that all the NA present in the nerve fibers was free to move and to accumulate above the ligations. A more recent examination of this point by Dahlström *et al.* (1975) showed that only a part of the NA was free to move, resulting in an upward reassessment of the rate of transport in the rat to 9 mm/hr or 216 mm/day.

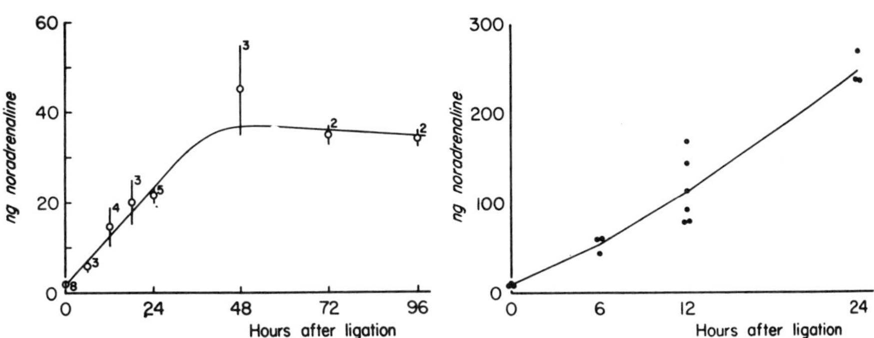

Fig. 4.9 Accumulation of noradrenaline. The curves of accumulation in rat (left ○) and cat (right ●) sciatic nerves above single ligations. From Dahlström and Häggendal (1966).

Evidence for a still faster rate of NA was found by Brimijoin using a cold-block stop-flow technique. A nerve, e.g., the sciatic nerve of the rabbit, is removed from the animal and placed in a chamber with part of the nerve kept at a temperature of 37°C, another part brought to a low temperature of 2°C (Fig. 4.10).

Transport *in vitro* in the part of the nerve kept at 37°C continues as usual with the NA moving down to the cold region where, as a result of the cold-block (Chapter 2) further transport is blocked and the NA dams up at the cold front (Fig. 4.11).

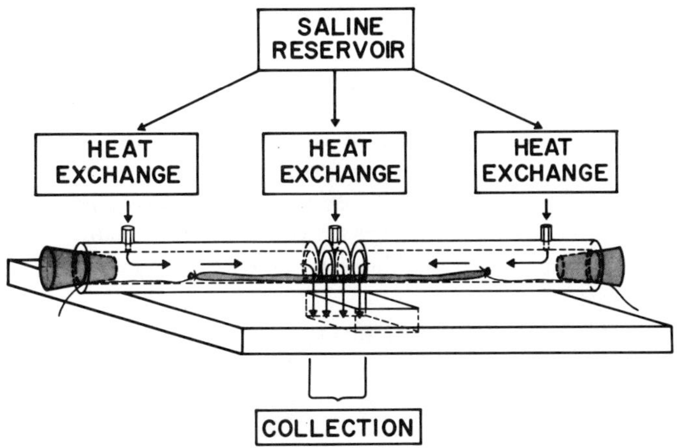

Fig. 4.10 Apparatus for stop-flow analysis. Arrows indicate the direction of fluid flow in a chamber used to bring about a stop-flow of transport by producing a low temperature in a part of the nerve. A nerve is indicated in approximately the position it would occupy during the cooling phase of a stop-flow experiment. The block of transport is produced by low temperature (2°C) in the middle chamber. From Brimijoin (1975).

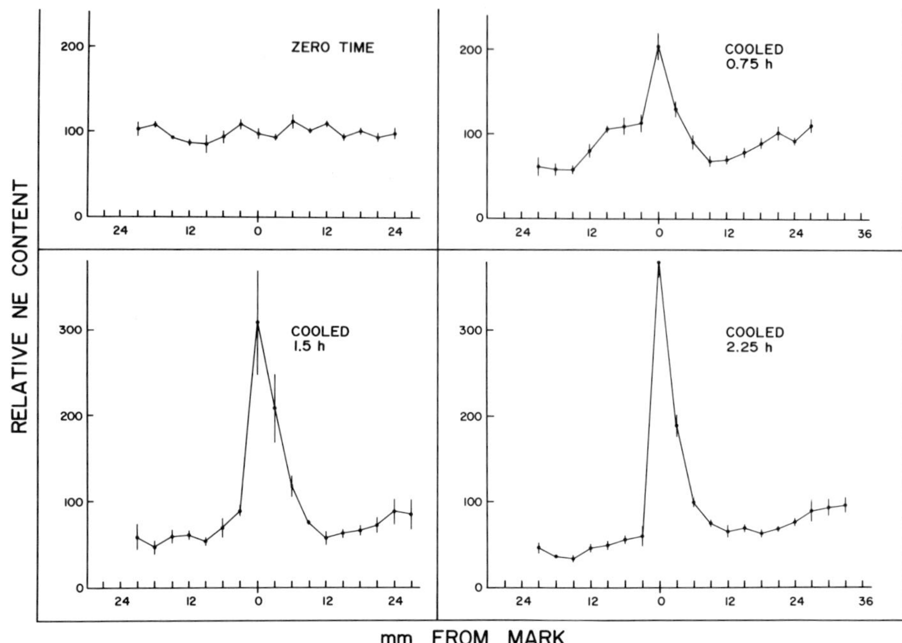

Fig. 4.11 Distribution of noradrenaline (NA) at a cold block. Distribution of NA [given as norepinephrine (NE) on the ordinate] along sciatic nerves locally cooled *in vitro* for various periods of time using the apparatus shown in Fig. 4.10. The amine is seen to increase over a time from 0.75 to 2.25 hr. Means and S.E.M. are given in relative units. On the abscissa the portion of nerve proximal to the cell bodies is to the left and distal to the right. Four to six nerves were used for each group shown. From Brimijoin and Wiermaa (1977b).

After a suitable period of accumulation, e.g., 2 hr, the cooled region is rewarmed to 37°C allowing the dammed up NA to move out into the re-warmed portion of nerve as a wave. A series of such waves is obtained by allowing different periods of time for transport after rewarming; the different positions of the waves allowing the rate of NA transport to be determined (Fig. 4.12).

From the distal displacement of the fronts of the waves, an estimation of transport rate of 430 mm/day was arrived at, while a rate estimation made from the spatial displacements of the wave peaks gave a rate of 293 mm/day (Brimijoin and Wiermaa, 1977b). Taking the peaks as a measure of rate is likely to be misleading because of the broadening of the waves as they are transported in the nerves (Brimijoin, 1977), this process introducing an error in the matching of equivalent positions of the moving waves.

The rate of 430 mm/day calculated from the fronts of the waves is closer to the rate found for labeled proteins, AChE, and to the rate of NA obtained using a double-ligation technique and a sensitive radioenzymatic method to assay NA at the femtomole level (Ben-Jonathon, Maxson, and Ochs, 1978).

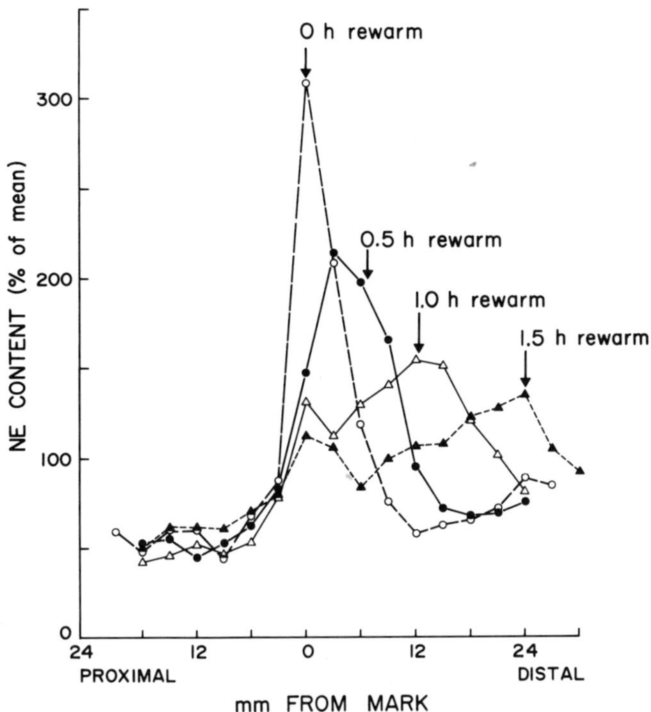

Fig. 4.12 Outflow of noradrenaline after rewarming. After rewarming the nerves which had been locally cooled for 1.5 hr (cf. Fig. 4.11), the transport of accumulated NA, given as norepinephrine (NE), is shown as a series of waves. Each curve represents the means of 6–9 nerves showing the waves of outflow after rewarming for times from 0.5 to 1.5 hr. Proximal portion of the nerve is to the right on the abscissa. From Brimijoin and Wiermaa (1977b).

The peroneal branch of the sciatic nerve in the cat was selected as having a relatively long length of uniform nerve without branches, allowing double ligations to be made over a relatively long length (Fig. 4.13).

As shown in this figure, the level of NA is relatively uniformly distributed over the length of the peroneal nerve. Double ligations were made at a fixed distance of 53 mm. The accumulation of NA in the nerve segments above the uppermost ligation was linear over the 7-hr period investigated, while the accumulation of NA above the distal ligation within the doubly-ligated portion of nerve showed a similar increase for several hours and then a sharp departure (Fig. 4.14).

The pattern of departure is analogous to that seen for AChE in double-ligated nerves (cf. Fig. 4.4) and a similar conclusion was drawn, namely, that the portion of nerve contained within the two ligations had been cleared by fast transport of the NA free to move. Taking the length of the nerve segment and the time of departure of NA above the lower ligature, a fast transport rate of 392 mm/day was found, a value reasonably close to the 410

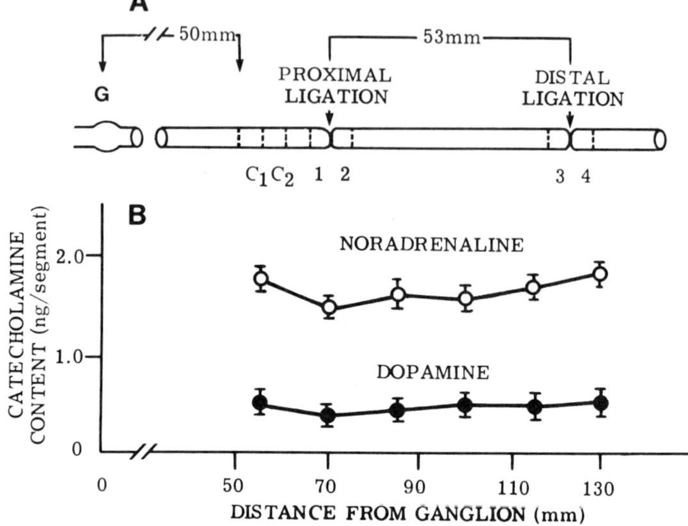

Fig. 4.13 Ligation sites for study of noradrenaline and dopamine (DA) transport. (A) Schematic representation of the cat peroneal nerve showing sites at which ligation sites are made and samples taken. Segments C1 and C2 represent control portions, segments 1 and 3 are nerve pieces taken above the ligation sites and 2 and 4 segments below the ligations. All segments are 5 mm in length. G denotes the L7 dorsal root ganglion. (B) Levels of NA and DA along the nonligated nerves are shown with each point representing the mean ± S.E. of five nerve segments. From Ben-Jonathan *et al.* (1978).

mm/day figure obtained for transport of labeled proteins. One difference noted in the pattern of NA accumulation as compared to AChE was that the level of NA which accumulated above the distal ligation did not remain constant, but showed a decline with time. This phenomenon was related to the constant synthesis and turnover of NA within the DCVs in the course of their transport, a topic to be further discussed in Section 6.

No evidence for an accumulation of NA in nerve segments taken just below the ligations was found, as would be expected of its retrograde transport. A similar lack of a retrograde transport of NA had been reported by Geffen, Hunter, and Rush (1969), Banks, Mangnall, and Mayor (1969), and Brimijoin and Wiermaa (1977b), among others. This indicates that the DCVs are depleted of their contents of NA before they are carried by retrograde transport back to the cell body. As we shall see in Section 3 below, DCVs are in fact being carried back by retrograde transport as shown by the retrograde movement of the dopamine-β-hydroxylase (DBH) contained in them (cf. Section 3).

2. Fast Transport of Dopamine (DA)

The radioenzyme assay used to determine NA was sensitive enough to measure DA at the same time (Ben-Jonathan and Porter, 1976); the DA amount-

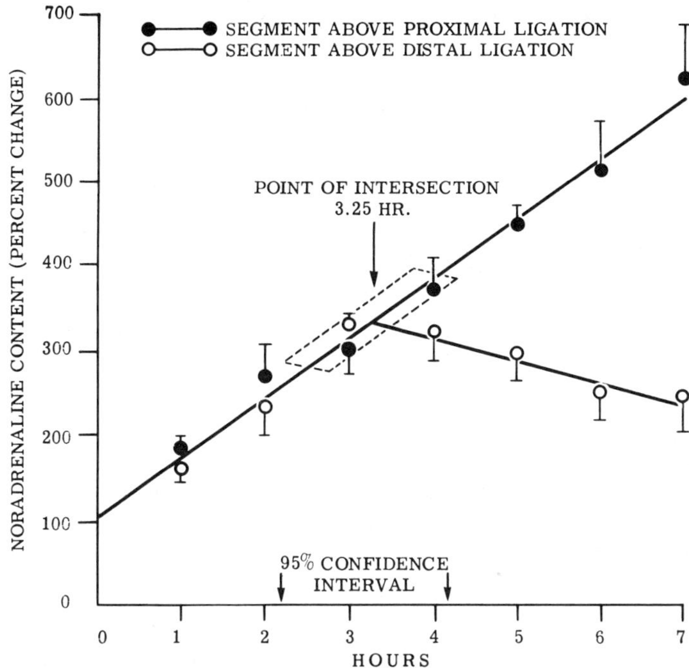

Fig. 4.14 Noradrenaline accumulations in double-ligated nerves. Net accumulation of NA in segments above the proximal and distal ligations placed 53 mm apart in the peroneal nerve as in Fig. 4.13 at times of 1–7 hr. Net accumulation was calculated by dividing the content of NA in the segments above the proximal ligation (segment 1, Fig. 4.13) indicated by (●), and the segments above the distal ligation (segment 3 in Fig. 1) indicated by (○), by the NA content in the control segments (average of C1 and C2 in Fig. 4.13). Each point represents the mean ± S.E. of five to seven nerve segments. The point of intersection at 3.25 hr shown by the arrow was calculated by analysis of variance and polynomial regression. The two arrows above the horizontal axis designate the 95% confidence limits. From Ben-Jonathan *et al.* (1978).

ing to 1/5th that of NA in nerve. In doubly-ligated nerves the pattern of accumulation of DA was found to be similar to that of NA, but with an apparently lower rate of transport of 347 mm/day (Ben-Jonathan *et al.*, 1978), as can be seen in Fig. 4.15.

This could be due to a turnover and loss of DA from the DCVs accumulated at the lower ligation. The turnover of the amines can be prevented by treating animals with disulfiram, an agent which blocks DBH and raises the level of DA in the DCVs. When, over a period of time, the level of DA in the DCVs was raised with disulfiram and the nerves double-ligated to assess the rate of transport of DA by its accumulation at the lower ligations, the rate of anterograde transport was seen to more closely approach a rate of 400 mm/day (Starcher, Ben-Jonathan, and Ochs, unpublished experiments; cf. Section 6 below).

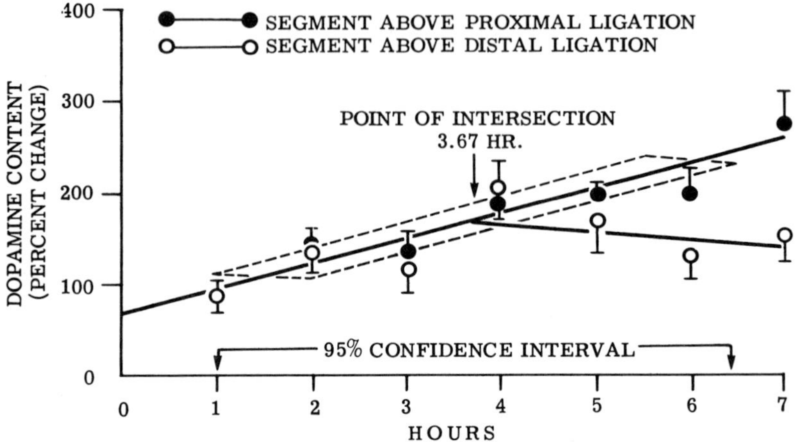

Fig. 4.15 Accumulation of dopamine in segments of double-ligated nerves. Accumulations of DA above proximal and distal ligations placed in the peroneal nerve for 1–7 hr (Fig. 4.13) were determined in the samples that are used to determine NA (Fig. 4.14). The same analysis and significance of symbols applies. From Ben-Jonathan *et al.* (1978).

3.   Dopamine-β-Hydroxylase (DBH)

The enzyme synthesizing NA from DA, dopamine-β-hydroxylase (DBH), is present within the DCVs. While early studies had suggested a lower rate of axonal transport for DBH than that of labeled proteins (Coyle and Wooten, 1972; Thoenen *et al.*, 1973), Brimijoin (1975) using the stop-flow technique found DBH to be transported at the same fast rate as that of NA. The temporal accumulation of the enzyme above a cold-block region is shown in Fig. 4.16.

Using a fixed period of 1.5 hr for the accumulation of DBH, the cooled portion of nerve was then rewarmed to 37°C, and for the different periods of transport allowed into the rewarmed portion of nerve, the waves of outflow of DBH determined (Fig. 4.17).

From the displacement of the peaks of the waves of DBH, an estimation for the rate of anterograde transport of 360 mm/day was derived. While this figure is not too far from the 410 mm/day found using labeling with ^{3}H-leucine (Chapter 2) or the double-ligation studies of NA, it is still significantly lower than the 410-mm/day rate. The difference is likely due to the broadening of the waves as the DCVs are moved out into the rewarmed portion of the nerve, as has been described above for NA. When the fronts of waves rather than their peaks were taken as the measure, a higher rate of 432 mm/day was found, one closely corresponding with the rate assessed for NA (Brimijoin and Wiermaa, 1977b). The close correspondence of the rates

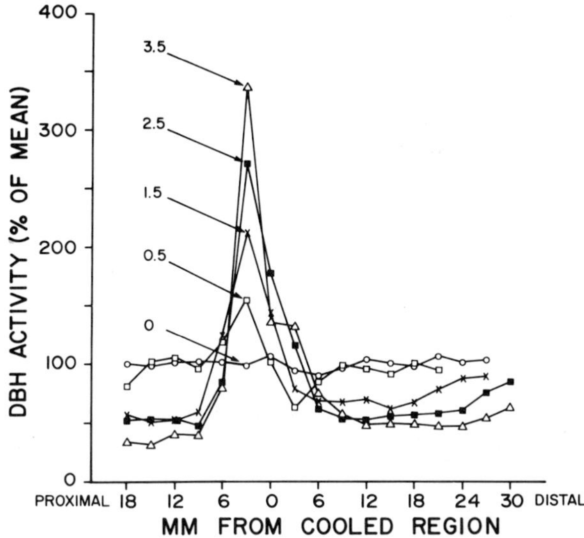

Fig. 4.16 Accumulations of DBH activity above a cold-block. A region of rabbit sciatic nerve was cold-blocked using the technique described in Fig. 4.10 at temperatures between 1 and 3°C for the times in hours indicated by the arrows. Each point in the waves of accumulation represented is the mean of determinations made on 3-mm segments from 4 separate nerves. From Brimijoin (1975).

of transport of NA and DBH is what would be expected insofar as both components are present within the DCVs (Section 6 below).

4. Tyrosine Hydroxylase (TH)

Tyrosine hydroxylase is transported in the nerve as a soluble enzyme at a slow rate of transport (Smith *et al.*, 1969). The early estimations of its rate did not, as Dahlström (1973) pointed out, take into account the relatively small proportion of the enzyme free to move, with most of the enzyme present in the fibers remaining relatively immobile. Evidence for a faster rate of TH was given by Wooten and Coyle (1973) and by Jarrott and Geffen (1972). More recently, Brimijoin and Wiermaa (1977a) have shown both slow and fast waves of TH outflow when using the stop-flow technique (Fig. 4.18).

The unitary hypothesis of transport (Chapter 11) shows how a component such as TH can have both a fast and slow rate of transport.

5. DOPA Decarboxylase (DDC)

Another enzyme in the sequence of NA synthesis (Fig. 4.7) is DOPA decarboxylase (DDC). This enzyme is present in the soluble phase and it appears

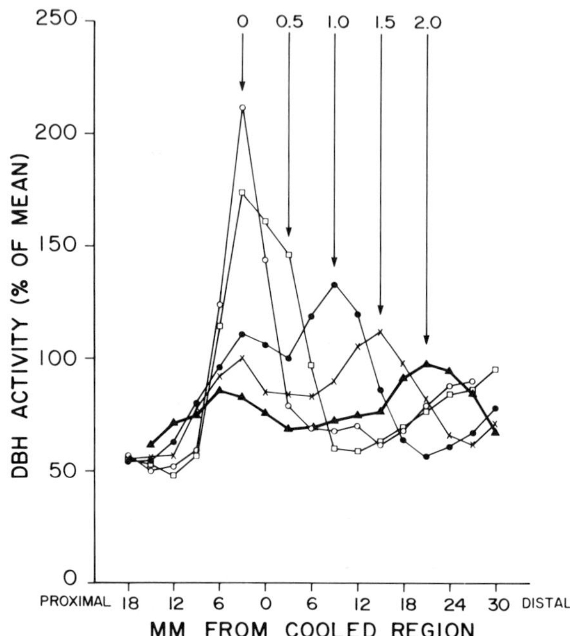

Fig. 4.17 Outflow of accumulated DBH activity after rewarming. The numbers above the arrows indicate duration of outflow (in hours) after rewarming following a period of cold block (cf. Fig. 4.16). Each point represents a means of 4–8 determinations at the time (in hours) given. From Brimijoin (1975).

to only have a slow rate of transport. On comparison to NA its accumulation above nerve ligations shows a considerable lag (Fig. 4.19).

The relatively long delay indicating its slow rate of transport was used by Dahlström (1971) as evidence to show that the early appearing increased levels of NA in the nerve segments above ligatures are due to accumulation and not to an increased local synthesis. This probably holds for the bulk of DDC shown in Fig. 4.19. However, an early rise may be made out in the figure, suggesting a faster component. Using the stop-flow method, Starkey and Brimijoin (1979) found a faster rate of transport of DDC, at a rate of 150 mm/day for the front of the wave. This faster wave constituted about 9% of the enzyme present. The interpretation of an intermediate rate of transport will be taken up in Chapter 11 in a discussion of the unitary hypothesis.

6. Catecholamine Turnover in the Dense-Core Vesicles (DCVs)

McLean and Keen (1972) studied the NA accumulated at ligations in nerves superfused with adrenergic blocking agents. The tyrosine hydroxylase inhibitor H22/54 caused a decrease of NA in the nerve and pargyline, a MAO inhibitor, an increase in its level. When both agents were used to superfuse

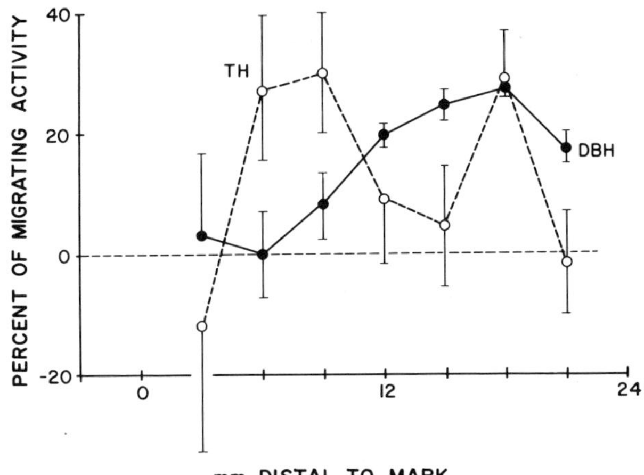

Fig. 4.18 Outflow curves of tyrosine hydroxylase after rewarming. Comparative outflow of accumulated tyrosine hydroxylase (TH) (O) and dopamine-β-hydroxylase (DHB) activity (●) is shown after rewarming following a cold block. Nerves were subjected to local cooling for 3 hr and rewarmed for 1.25 hr. For each enzyme, the activity in the segments distal to the reference mark is given after subtraction of their baseline levels of activity expressed as a percentage of the total activity of enzyme. Note fast and slow TH waves. From Brimijoin and Wiermaa (1977a).

the nerve above ligations simultaneously, the level of NA remained close to that found in untreated nerves.

The authors concluded from this result that a local synthesis of NA in the DCVs accumulated above ligations does not account for its increase. The relative constancy of NA in the DCVs in the unligated nerve, with evidence of an entry of dopamine into the DCVs, suggests a constant turnover of NA in the DCVs. Animals treated with disulfiram to block DBH activity (Ben-Jonathan et al., 1978), showed an increase in the level of DA in their nerves along with a decreased content of NA within doubly-ligated nerve segments. This indicates that disulfiram has a direct effect on the DCVs in the nerve fibers other than its block of synthesis of NA in the cell bodies. The results are in accord with the concept that as the DCVs are transported down within the nerve fibers, they undergo at the same time a steady-state turnover of amines. Dopamine continually enters the DCVs to be synthesized into NA which then leaves the DCVs to be degraded by the monamine oxide (MAO) of the mitochondria. This concept is schematized in Fig. 4.20.

Further details of the transport mechanism indicated in this figure will be discussed in Chapter 10. The dynamic turnover of amines in the DCVs pictured could explain the decline in the level of NA in the DCVs accumulated above the distal ligation of double-ligated nerves seen after the departure time (cf. Fig. 4.14). This could be due to a greater diffusion of NA from the DCVs as a result of the loss of some factor required by the DCVs

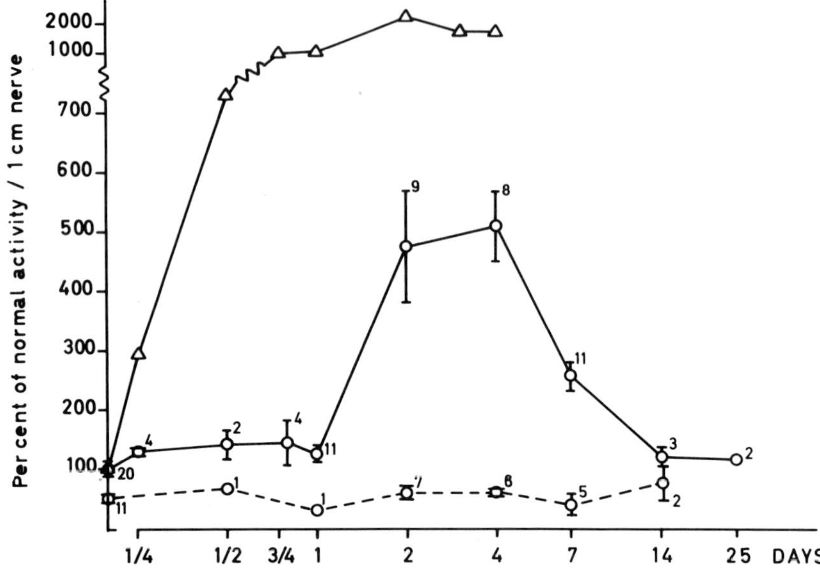

Fig. 4.19 Accumulation of dopa decarboxylase activity and noradrenaline levels above a constriction. Enzyme activities in 1 cm of rat sciatic nerve are given at different times after constriction of nerve expressed as percentages of their normal value (mean ± S.E.M.). The small figures represent the number of experiments performed. (O——O) Constricted sciatic nerve of normally innervated rats. (O – – – O) Constricted sciatic nerves of lumbar sympathectomized rats. (△) Accumulation curve for noradrenaline. From Dahlström and Jonason (1968).

when accumulated at the ligated site, or to a more ready oxidization of the amine by the monoamine oxidase of the mitochondria which is also being accumulated at that site.

The rate of 400 mm/day found for anterograde transport of NA, while only slightly less than the 410 mm/day rate of labeled proteins and not statistically significant, may be due to the dynamic turnover of amines in the DCVs, a situation to be contrasted with the higher rates measured when using stable markers such as radioactively labeled proteins or an enzyme like AChE. The still lower apparent rate of DA of 360 mm/day described above (Fig. 4.15) appears also be related to the continual turnover of amines in the DCVs and the greater lability of its contained DA and perhaps its greater rate of oxidation.

The relatively large immobile fraction of DCVs in the axons could represent DCVs which have dropped off from the transport mechanism (cf. Chapter 11), those DCVs acting as a temporary storage of neurotransmitter, the dropped-off DCVs are later redistributed and carried down the fibers to the terminals when an additional supply of transmitter is needed. The great number of DCVs present in the nerve terminals may also serve such a storage function. Their relatively high concentration in the nerve terminals in relation

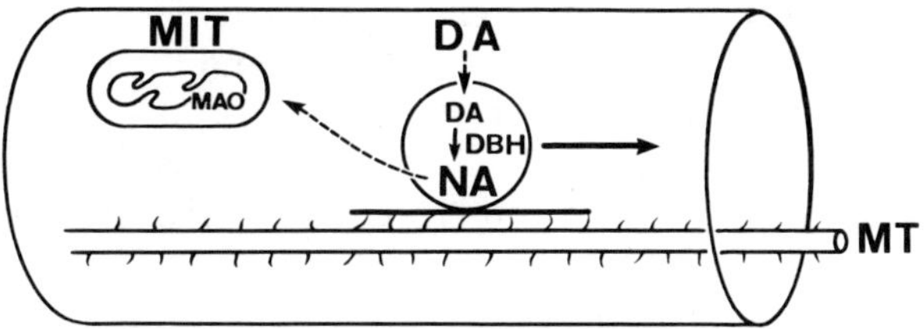

Fig. 4.20 Schematization of amine turnover within the dense-cored vesicles (DCVs). Turnover of amines in the vesicle represented by a circle in the course of its transport down within the adrenergic nerve fiber is shown. As the vesicles are carried along the microtubules (MT) in accordance with the transport filament hypothesis (Chapter 10), the dopamine (DA) enters the DCV, and is converted by dopamine-β-hydroxylase (DBH) to NA, which leaves the DCV to be oxidized by the monoamine oxidase (MAO) of the mitochondrion (MIT).

to the high level of turnover of the neurotransmitter at this site is a topic to be more fully discussed in Chapter 13.

C.   SEROTONERGIC NEURONS IN *APLYSIA* NEURONS

Serotonin (5-hydroxytryptamine, 5-HT) is a neurotransmitter found in some neurons in the CNS of vertebrates and ganglia of invertebrates. Its metabolic pathway has some similarity to that of the catecholamines (Fig. 4.21).

The precursor ^{3}H-5-hydroxytryptophane (^{3}H-5-HTP) when taken up after topical application, or injected into the cell body of the giant metacerebral neurons of the cerebral ganglia (GCN) of *Aplysia,* is converted by the enzymatic action of amino acid decarboxylase to ^{3}H-5-HT: The labeled serotonin is then transported into the two fiber branches of those neurons (Goldberg *et*

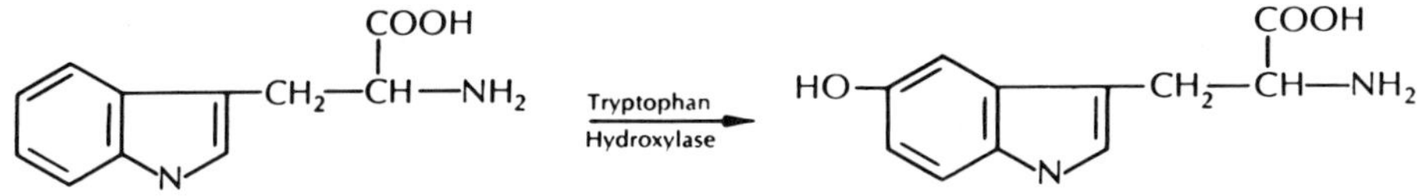

Fig. 4.21 Metabolic pathway of serotonin. The synthetic pathway of serotonin from tryptophane has some similarity to that of NA (cf. Fig. 4.7). From Cooper, Bloom, and Roth (1978).

al., 1976). This serotonergic neuron has some advantage for single-cell studies because of the large size of its cell body and the length of its axons. The morphology of the neuron with its giant axon dividing into a branch leading to the lip, the lip nerve, and a branch which enters the cerebro-buccal connective and its ganglion, the cerebro-buccal nerve branch, is shown in Fig. 4.22.

The ^{3}H-5-HT is transported in these nerve fibers within DCVs which have been identified biochemically and by EM as 5-HT-containing vesicles (Schwartz *et al.*, 1975; Shkolnik and Schwartz, 1980). To obtain a pattern of outflow of ^{3}H-5-HT in the fibers of the giant cell, the cell body is injected with the precursor ^{3}H-5-HTP, a period of time allowed for downflow of labeled serotonin before making a ligation just below the cell body to prevent further downflow from the cells, and an additional period of outflow allowed (Goldberg *et al.*, 1978). The labeled serotonin which had entered the fibers before making the ligation is seen to continue to move in the axons as a peak of labeled radioactivity (Figure 4.23).

This is to be contrasted to the small diffusive spread of the labeled precursor from the cell bodies into its fibers which typically shows a decrement with distance. The actively transported ^{3}H-5-HT moves as a peak of radioactivity at a fast linear rate in each of the two nerve branches. The rate of movement of the peak when scaled to a higher temperature of 37°C with the Q_{10} taken into account, was calculated to be close to 410 mm/day (Goldberg *et al.*, 1978).

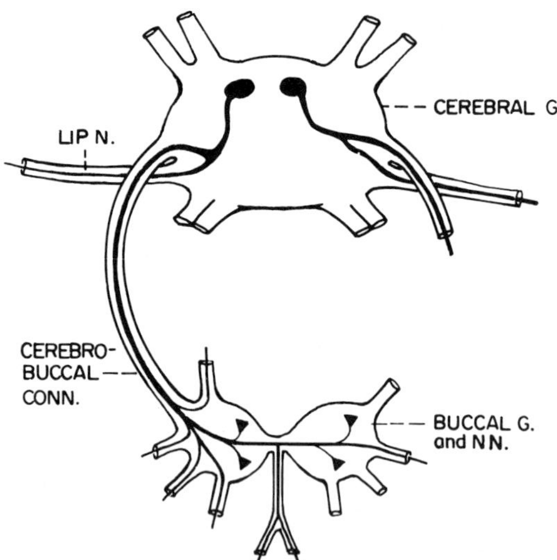

Fig. 4.22 Topography of giant cell cerebral ganglion of *Aplysia*. The serotonergic giant cell has a branched axon with one of its nerve branches leading to the lip nerve and the other branch entering the cerebro-buccal connective passing to the buccal ganglia. From Schwartz *et al.* (1975).

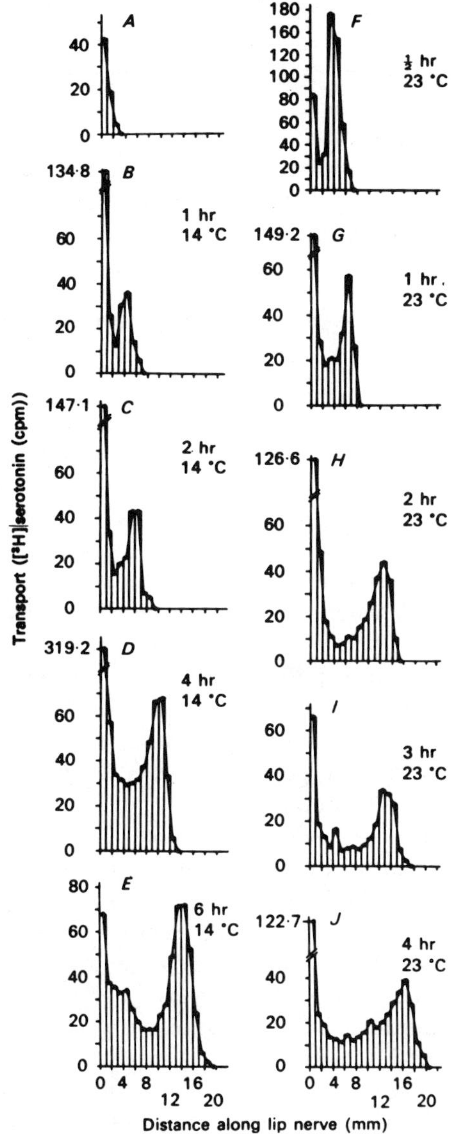

Fig. 4.23 Outflow of labeled serotonin. After injection of the giant cell with ³H-serotonin (cf. Fig. 4.22) and allowing a period of outflow, a ligature block was made near the cell body and further outflow allowed at 14°C (A,C,E) and at 23°C (F,H,J). This technique results in a peak of labeled serotonin moving down the fiber at a fast rate which, when scaled to 38°C, is close to 410 mm/day, the fast rate of transport in the mammal. From Goldberg *et al.* (1976).

Goldberg *et al.* (1976) reported an increase in rate when an increased amount of labeled serotonin is forced into a nerve branch. Ordinarily the cerebrobuccal nerve branch transports about twice the amount of ^{3}H-5-HT than the lip nerve. Cutting the cerebrobuccal branch close to the ganglion diverts more labeled material into the lip nerve with an increase in rate. Discussion of this experiment will be deferred to Chapter 11 in relation to the phenomenon of "routing." Here we point out that such increased rates of transport may last only a short time.

D. OTHER PUTATIVE TRANSMITTERS AND THEIR TRANSPORT

Transport in the fibers of the R3-R14 *Aplysia* having their cell bodies in the parieto-visceral ganglion may be assessed after incubating the ganglia in media containing ^{3}H-labeled amino acids while superfusing the nerve with a non-labeled media (Price *et al.*, 1979). After loading with ^{3}H-glycine, the amino acid was found to be transported as such at a rate of 70 mm/day. Transport was blocked by vinblastine (Chapter 12) and by low Ca^{2+} (Chapter 8), findings in accord with an axoplasmic transport of ^{3}H-glycine in these axons rather than its diffusion. Grains of labeled activity were found present over the R3–R14 axons and not in the surrounding glia in autoradiographs, as expected of its active transport. Additionally, a retrograde transport of ^{3}H-glycine was observed in these neurons.

Both the sympathetic and parasympathetic ganglia of mammals contain peptide neurons of various types and a variety of substances likely to be transmitters are found in autonomic nerves, but little information is available as to the characteristics of transport in them. Evidence for the transport of substance P neuropeptides in the vagus nerve has been obtained (Lundberg *et al.*, 1980). A transport of opioid receptors in the vagus nerve (Young *et al.*, 1980) has also been reported. Still less is known of the transmitters in the large number of visceral neurons which are not adrenergic or cholinergic (Burnstock, 1979). These neurons include the possible purogenic neuron containing the transmitter ATP (Burnstock, 1972) and neurons containing serotonin, GABA, and a wide variety of peptides including enkephalin, substance P, and somatostatin. Perhaps the somatostatin transported in sensory rat peripheral nerve (Rasool *et al.*, 1981) has some role related to transmitter action, at least the high molecular weight form present in the dorsal roots. It is of interest that the rate they calculated for somatostatin transport in the sensory fibers was 415 mm/day.

A high concentration of glutamate has been found in the dorsal root as compared to the ventral root, a distribution in accord with a suspected role of glutamate as the excitatory transmitter within the spinal cord (Graham *et al.*, 1967). Roberts *et al.* (1973) corroborated the higher level of glutamate, with a lesser abundance of the amino acids threonine and arginine in the dorsal roots as compared to the ventral roots. On injection of ^{14}C-glucose into rat dorsal root ganglia, some of it was incorporated into glutamate with its subsequent slow passage out into the nerve, a pattern, however, which could be due to diffusion rather than transport. Johnson (1974) injected ^{14}C-glutamine into the L7 dorsal root ganglia of cats and found the labeled glutamate to be carried distally down into the sensory nerve fibers at a calculated rate of 400 mm/day with the characteristic pattern of fast transport (Fig. 4.24).

The significance of a fast axoplasmic transport of glutamate distally in the sensory fiber is, however, not clear with respect to its proposed role in the CNS as a neurotransmitter. Similar questions arise with regard to the fast anterograde transport of substance P in the sensory fibers of nerve (Lundberg *et al.*, 1980; Brimijoin *et al.*, 1980).

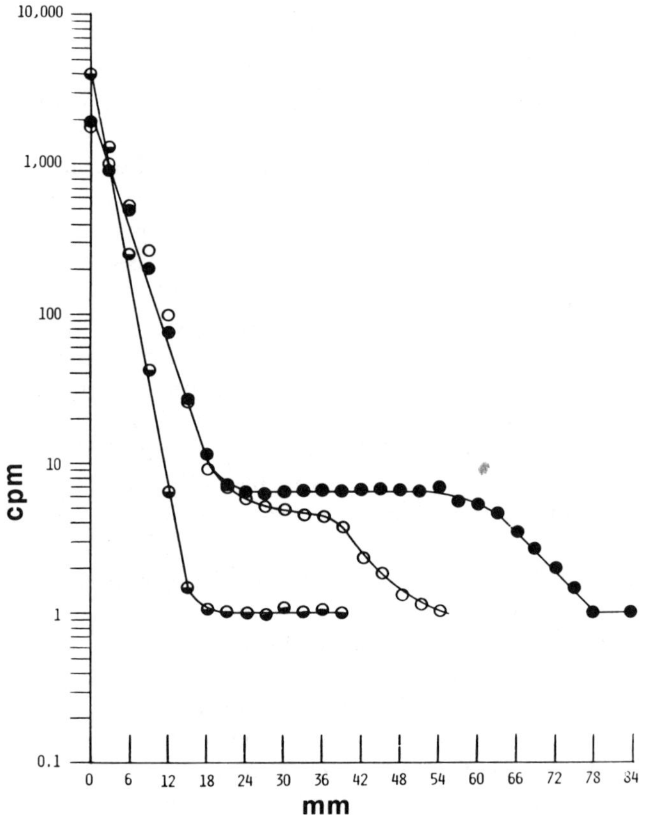

Fig. 4.24 Outflow of labeled glutamine. The proximo-distal transport of labeled glutamine is shown 1, 3, and 5 hr following injection of U-^{14}C-glutamine into the dorsal root ganglia. The times at which nerves were taken to assess outflow were 1 hr (◑), 3 hr (○), and 5 hr (●). From Johnson (1974b).

E.   TRANSPORT OF TRANSMITTER AND TRANSMITTER-RELATED COMPONENTS IN THE MAMMALIAN CNS

The neuronal systems in the CNS identified by their content of neurotransmitter and related enzymes can only be briefly touched on here as they relate to axoplasmic transport.

Histochemical staining for AChE reveals a number of cholinergic fiber tracts in the brain which characteristically contain ACh and ChAC in addition to AChE. The septal-hippocampal projections are rich in AChE (Shute and Lewis, 1963; Lewis and Shute, 1967). Transport of AChE in those fibers was indicated by making lesions in the fimbria between the cell bodies in the septal nucleus and their terminals and finding a dramatic fall of AChE in the hippocampus. The changes found were those expected of a transport of this transmitter-related component from its site of synthesis in the cell bodies out within its fibers. Additionally, ACh was found in the interpenduncular nuclei, the site on which those fibers terminate (Kataoka *et al.*, 1973). We may presume a transport of ACh in those fibers as well.

Dahlström and Fuxe (1964b,c) have, by means of transections, shown the presence of catecholamine-containing tracts in various parts of the CNS. The fluorescence technique was used to show changes in catecholamines in sections cut from freeze-dried blocks along the course of the nerve tracts. After making transections in the fiber paths, fluorescence was seen in the NA-containing neurons to increase in the fibers remaining connected to the cell bodies and to be depleted in the portions of nerve fibers separated from their cell bodies. Separate pathways of DA-containing fibers could also be traced by similar methods. A major system of DA fibers was found to have their cell bodies in the substantia nigra with their fibers projecting to the striatum. A similar approach based on an immunocytochemical method has also been used to trace the path of adrenergic neurons (Reis *et al.*, 1978). Axoplasmic transport of amines in the CNS is also studied by the injection of suitable labeled precursors. These are taken up by the cell bodies with an outflow of labeled amines in the fibers identifying the paths of aminergic fibers (Bobillier *et al.*, 1975).

The Hillarp–Falck fluorescence technique also reveals the presence of 5-HT by its somewhat different color. By making transections at different positions within the CNS, the anatomical paths of serotonergic fiber tracts in the CNS have been revealed.

Histaminergic fibers in the CNS were suggested by changes in the level of 1-histidine decarboxylase, an enzyme required for synthesis of histamine, on making nerve transections. The level of the enzyme in the cortex and hippocampus was seen to fall after cutting the medial forebrain bundle to separate the fibers from their nerve cell bodies (Garbarg *et al.*, 1974).

Most of the work dealing with the various fiber types in the CNS indicates that they have a slow rate of transport. Howevever, recent studies using labeled amino acids has revealed the presence of faster rates of transport. It is likely that more such cases will be found of fast as well as slow transport similar to the recent findings of fast and slow rates of components in the peripheral nervous system (cf. Chapter 11).

In addition to the above-named neurotransmitters and their fiber systems, there are now a number of neurons in the CNS which contain substances considered to be putative neurotransmitters (Phillis, 1970; Cooper, Bloom, and Roth, 1978; McGeer, Eccles, and McGeer, 1978). These include, in addition to amines and amino acids, a wide range of peptides (Emson, 1979). It is to be expected that these will be examined before long for their transport properties.

Streit *et al.* (1979) reasoned that in those neurons where an identified neurotransmitter is released from its nerve terminals the neurotransmitter or its precursor should also be selectively taken up by the nerve terminals and carried back to their cell bodies by retrograde transport. When the pigeon optic tectum, to which fibers of the ICP nucleus project, was exposed to ^{3}H-glycine, a dense labeling of the cell bodies in the ICP nucleus was seen in autoradiographs. In those studies the glycine appeared to be retrogradely transported in a bound form. This was indicated by its presence after fixing the tissues for histological preparation with either formalin or glutaraldehyde. Retention of labeling with formalin is suggestive of its presence in bound form in that formalin generally will fix only the larger molecular species and allow smaller molecular components, including free glycine, to diffuse away. These techniques were used by Cuenod and his colleagues (Streit *et al.*, 1979) to examine the striatum, substantia nigra, and raphe systems of the rat for retrograde uptake of various labeled precursors and neurotransmitters. By this means, GABA, serotonin, and dopamine were seen to be transported. Evidence for a retrograde transport of NA was also found. A retrograde transport of NA would appear to be in confict with the negative findings for the retrograde transport of NA in peripheral nerve (Geffen and Rush, 1968). This would suggest either that the concentration of ^{3}H-adrenaline used was low in the peripheral nerve studies carried out by Geffen and Rush, or that there is a fundamental difference with respect to the retrograde transport of NA in peripheral and central neurons.

The properties of transport in CNS neurons have not been as well studied as in peripheral neurons. We would expect however, that the same mechanism underlies transport in the neurons of the CNS as in the peripheral nervous system. This would indicate the presence of fast transport. This is the case in one CNS system which has been fairly well characterized, the hypothalamo-neurohypophysial system (HNS) to be described in Chapter 13.

5

Somal Synthesis of Transported Components

A.   THE SITE AND PROCESS OF SYNTHESIS

The biochemical steps by which protein synthesis is carried out in cells, and the organelles identified with this process are schematized in Fig. 5.1.

Ribosomes are key organelles in protein synthesis. In association with the sheetlike cavities of the endoplasmic reticulum (ER), they are referred to as the rough endoplasmic reticulum (RER). These constitute the Nissl particles stained by analine dyes (Chapter 6).

The general sequence by which proteins are synthesized is that the deoxyribonucleic acid (DNA) forming the double helical chains in the nucleus specifies by its nucleotide sequence the composition of another polynucleotide chain, messenger ribonucleic acid (mRNA). The latter is carried from the nucleus to the ribosomes in the cell cytoplasm where, as the mRNA moves along the ribosomes it designates the step-by-step addition of specific amino acids brought to it by transfer RNA (tRNA) to form a polypeptide chain. These in the case of membrane components may be "exported" into the axon in vesicles via the Golgi apparatus, as indicated in Fig. 5.1. In addition, proteins formed on free ribosomes not associated with ER may be exported by a path bypassing the Golgi apparatus.

1.   Specification of the Path of Synthesis in the Cell and Export into the Axons

The amino acids within the cell body may be synthesized into polypeptides or enter a path leading to energy metabolism or other metabolic processes. As noted in Chapters 2 and 3, there are specific carrier mechanisms present in the cell membrane for the uptake of various classes of amino acids into the cell body (Blasberg and Lajtha, 1966; Lajtha, 1975). If these membrane carriers for uptake into the cell are bypassed by injecting amino acids directly into the nerve cell body (Chapter 3), the injected amino acids may not become incorporated into polypeptides but instead become metabolized or diffuse from the cell (Eisenstadt *et al.*, 1973; Gainer *et al.*, 1975). These

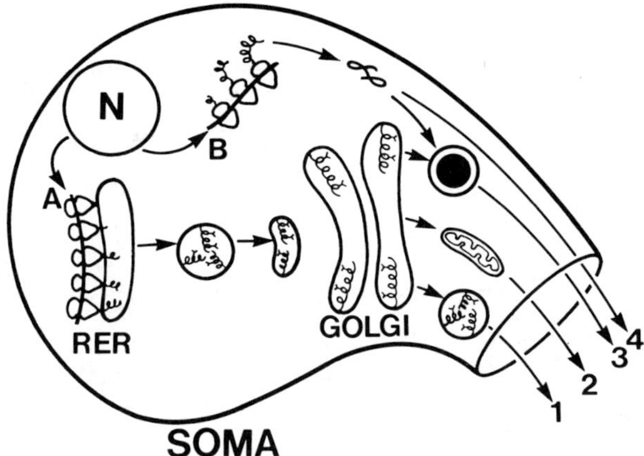

Fig. 5.1 Schematization of cell body protein components synthesized in the cell body. (A) mRNA from the nucleus (N) is shown moving along the ribosomes of the rough endoplasmic reticulum (RER) to synthesize proteins in the vesicles. The vesicles formed containing the new proteins pass to the Golgi apparatus, glycosylated, and then exported into the axon as vesicles (1). (B) Other proteins are synthesized on free ribosomes to follow a path bypassing the Golgi apparatus. Mitochondria (MIT) and other particulates, e.g., DCVs (cf. Fig. 4.20), may receive part of their structure via the Golgi apparatus, from path (A) and part from path (B) before they are exported into the axon (3).

results suggest not only that amino acids are compartmented in the cell (Balázs and Cremer, 1972), but that their movements within the cell are directed by translocation mechanisms rather than by free diffusion. Such translocations may be carried out along microtubules by the same mechanism underlying axoplasmic transport in the fibers (cf. Chapter 10).

An understanding of the steps by which the proteins are synthesized has been gained from studies of exocrine cells, those of the pancreas and liver (Palade, 1975; Blobel and Dobberstein, 1975), and of bacteria. Exocrine cells can serve as a model for the neuron in that both exocrine cells and neurons have a high level of protein synthesis; the difference is that the exocrine cell excretes its protein extracellularly while the neuron "exports" it into its own axons and dendrites.

The first part of polypeptide chain formation consists of an initiating sequence of amino acids which acts as a special code or "signal" sequence. This signal sequence enables the newly formed polypeptide to enter the cisternal space of the rough ER through a special channel in its membrane (Fig. 5.2).

After a length of the polypeptide chain has entered the cisternal space, the signal portion is enzymatically removed. As the lengthening polypeptide chain continues to enter the cisterna, interacting sulfhydril groups, hydrogen bonds, and other bonding forces cause the polypeptide chain to fold. In

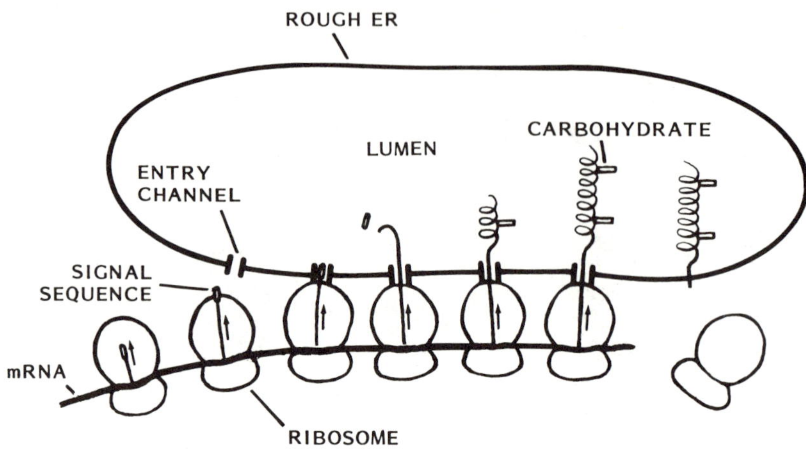

Fig. 5.2 Synthesis of protein in the RER. Illustrated are the essential features of the signal hypothesis whereby the growing protein chains are transferred across the membranes of the RER (cf. Fig. 5.1). Signal codons in the mRNA are shown by a short open bar. The signal codons enter through pores in the RER membrane and then by endoproteolytic action are removed from the polypeptide chain which is completed and folded within the intracisternal space of the RER. Modified from Palade (1975).

effect it is trapped within the cisterna as the folded protein becomes too large to pass back through the ER membrane channel. The vesicles with their contained folded proteins move to the Golgi apparatus where they coalesce with its inner disc portions. Additions of carbohydrate to polypeptide chains in the formation of glycoproteins takes place in the Golgi apparatus. Vesicles then bud from the Golgi apparatus to enter the axon for transport down its length. Additionally, soluble proteins synthesized on free ribosomes and not present in vesicles are moved down within the fibers after bypassing the Golgi apparatus. Details of these processes and the synthesis of vesicles and mitochondria are as yet little understood.

2.   Synthesis in the Cell Body vs. the Fiber

The presence in the cell body of the synthetic machinery necessary for protein synthesis, and its lack in the axons, would in itself suggest that all synthesis is carried out in the cell body. Singer and Salpeter (1966) had presented autoradiographic evidence suggesting a local entry of labeled amino acids into the axons, these then synthesized into proteins. However, the amount of such local entry and incorporation is small, and the functional importance of this finding remains uncertain. It appears that materials gain entry into the Schmitt–Lanterman clefts of the myelin sheaths, most likely components which undergo turnover in the myelin sheath. It is conceivable that a small amount of such substances could pass via this route into the axons, but this remains problematical in the nerves of the vertebratae.

That synthesis of polypeptides and proteins takes place in the cell bodies rather than in axons was indicated by making double ligations in nerves in order to trap labeled proteins, polypeptides, and some free labeled leucine within a length of nerve (Ochs *et al.*, 1970). Then, after 16 or 21 hr, the isolated pieces of nerve were taken for analysis of their contents. No evidence for a shift from lower- to higher-MW polypeptides was found, as might be expected of a synthesis of proteins occurring in the axons. Such evidence does not, however, exclude the possibility of a low level of synthesis below the resolution of the assay method used.

If there was a significant amount of protein synthesis taking place in mammalian axons, we might expect to find evidence of it in the nerve terminals. Such a synthetic capability had been claimed for synaptosomes prepared from brain (Autilio *et al.*, 1968; Austin and Morgan, 1967). However, a part of the protein synthesis which has been measured in synaptosomes is due to that occurring in the mitochondria within the synaptosomes, and part to the adventitious adherence of ribosomes to the synaposomes in the course of their isolation by differential centrifugation. When these two sources of synthesis are eliminated, the level of polypeptide synthesis in the synaptosomes appears to be quite low or nonexistent (Barondes, 1973; Morgan, 1970; Cotman and Taylor, 1971; Gambetti *et al.*, 1972). There is, however, evidence of ribosomes present in the axons of immature rats (Zelená, 1972). And, there is evidence of some RNA species transported in regenerating axons and to some extent in normal mature axons (Chapter 6). Such evidence gives some ground for the possibility of a local synthesis in axons, at least under certain circumstances.

In contrast to vertebrate nerve, measurable amounts of material can pass from the exterior medium into the invertebrate giant axon (Gainer *et al.*, 1975). When labeled amino acid was present in an incubation medium, labeled polypeptides were subsequently found in the axoplasm expressed from the giant axon (Gainer *et al.*, 1977). It appears that the polypeptides are synthesized by their satellite cells and translocated into the axon (Lasek, Gainer, and Barker, 1977), presumably by endocytosis. A local synthesis of polypeptides in these axons is made unlikely by the absence of ribosomal RNA. Koenig (1978, 1979) on the other hand, considers that a local synthesis can occur in the axon of the Mauthner cell of goldfish, where he gave evidence for the presence of ribosomal RNA.

3. Block of Protein Synthesis

Protein synthesis is blocked by cycloheximide and puromycin. Puromycin acts at the point where tRNA brings the individual amino acids to the ribosomes in the process of polypeptide chain formation (Yarmolinsky and de la Haba, 1959). Puromycin blocks protein synthesis because it has a molecular structure close to that of tRNA, with an amide present instead of an ester. As a result, the peptidyl-puromycin derivative entering the growing polypeptide chain prevents it from lengthening any further. Puromycin injected into the cat L7 lumbar ventral horn region just before an injection of the precursor ^{3}H-leucine blocks the outflow of labeled polypeptides into motor fibers of the ventral roots (Ochs, Johnson, and Ng, 1967; Ochs and Johnson, 1969; Ochs, Sabri, and Ranish, 1970). And, when injected into the eye, it blocks the outflow of labeled proteins to the tectum (Sjöstrand and Karlsson, 1969). The block of outflow of labeled proteins by puromycin injected into the L7 dorsal root ganglion before that of ^{3}H-leucine is shown in Fig. 5.3.

The same block was seen when the time between the injection of puromycin and ^{3}H-leucine was shortened or even when both were given at the same time. When puromycin was injected *after* ^{3}H-leucine, only a partial blocking effect was seen when puromycin was injected only 5 min after ^{3}H-leucine (Fig. 5.4).

When the interval between the injection of the precursor and the blocking agent was lengthened to 15–20 min, no block was seen (Ochs *et al.*, 1970). This indicates that the synthesis of polypeptides and proteins had been completed by that time.

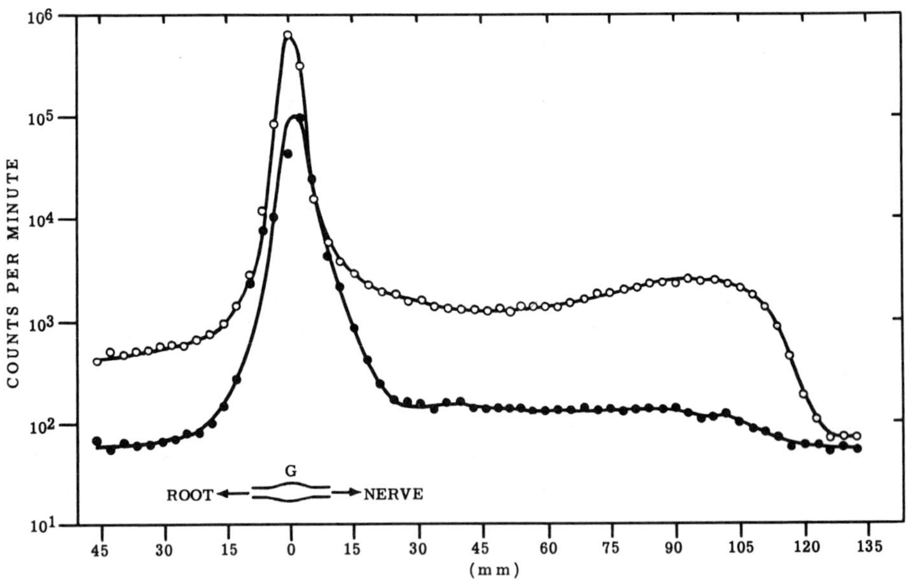

Fig. 5.3 Block of protein synthesis. The effect of puromycin to block protein synthesis and downflow is shown. The left L7 ganglion (●) was injected with puromycin, the right (○) with an equal volume of Ringer-lactate solution. One hour later both ganglia were injected with ^{3}H-leucine and 7 hr of outflow was allowed. A marked reduction of radioactivity is apparent in nerve on the puromycin-injected side. From Ochs, Sabri, and Ranish (1970).

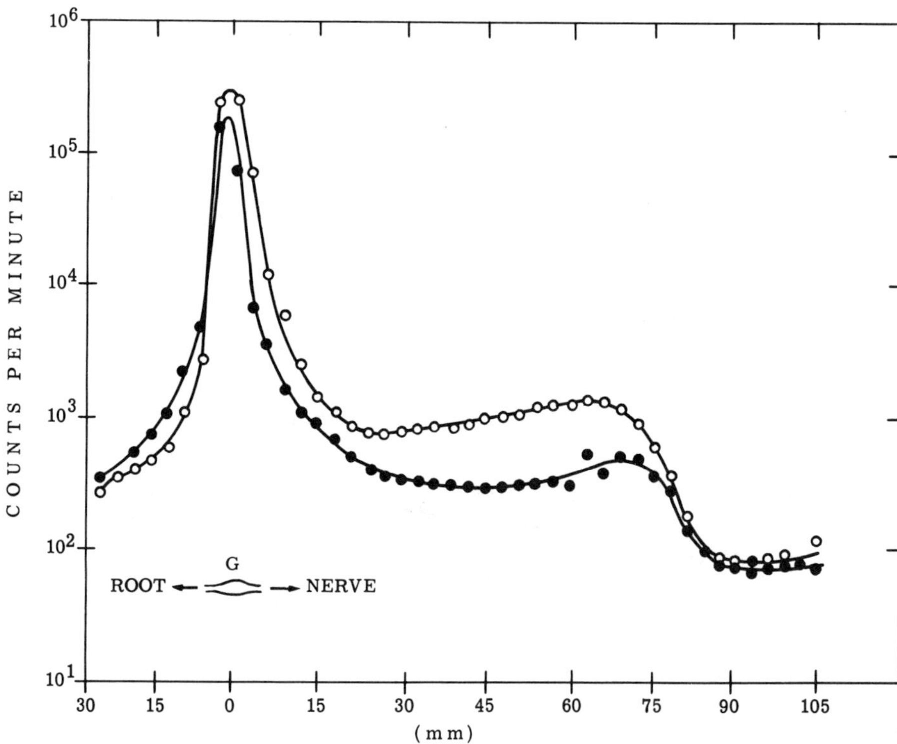

Fig. 5.4 Effect of puromycin injected after the precursor. Puromycin was injected into left L7 ganglion (●) 5 min after injection of ^{3}H-leucine. This results in only a partial block as shown in comparison with the right side (○), which had been similarly injected with Ringer solution. Nerves were taken 5 hr after ^{3}H-leucine injection for determination of outflow. From Ochs, Sabri, and Ranish (1970).

The protein synthesis blocking agent cycloheximide blocks synthesis by binding to ribosomes and preventing the translation of mRNA along the ribosomes (Wettstein *et al.*, 1964). Cyclohexamide, or its analog acetoxy-cycloheximide, injected into the goldfish eye before or at the same time as the precursor ^{3}H-leucine, caused a markedly reduced transport (McEwen and Grafstein, 1968). Edström and Mattsson (1972a) found a block of incorporation and downflow in frog nerves using an acccumulation technique. Cyclohexamine injected into the cat dorsal root ganglion before ^{3}H-leucine causes a full block similar to that of puromycin (Ochs *et al.*, 1970), and as in the case of puromycin, when injected 15–20 min *after* the precursor no block of synthesis and outflow of labeled polypepides occurred (Fig. 5.5).

The similarity of the results obtained with cyclohexamine and puromycin, in spite of the difference in their mode of action, supports the conclusion that protein synthesis is completed in the cell bodies within a time of approximately 15–20 min after uptake of the amino acid precursor. One other

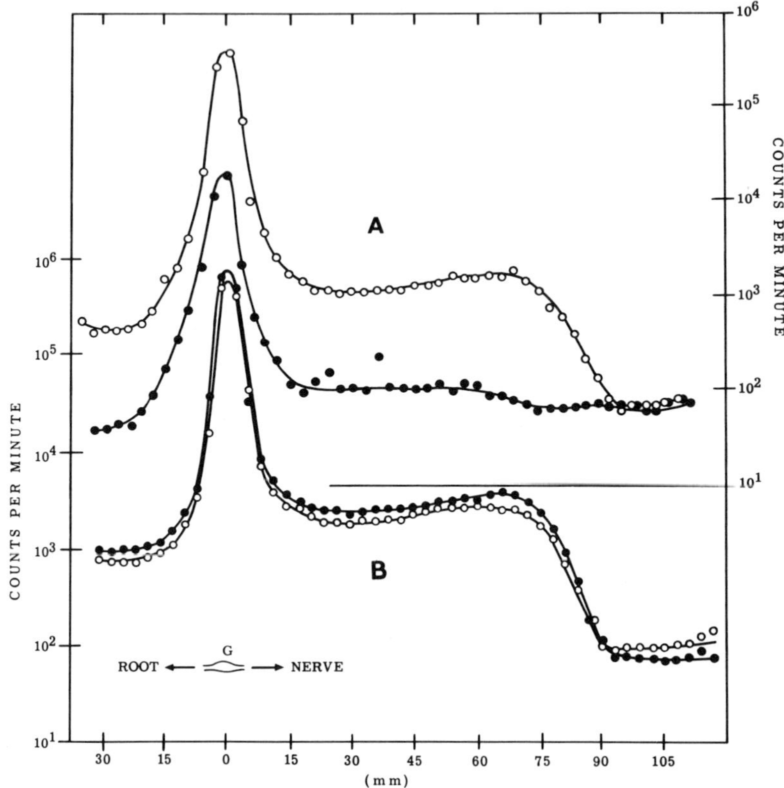

Fig. 5.5 Protein synthesis block with cycloheximide. (A) Cycloheximide was injected into the left L7 dorsal root ganglion (●) and Ringer solution into the right (○) 20 min before the injection of ^{3}H-leucine into each of the ganglia. Nerves taken 4 hr after ^{3}H-leucine injection show a block of protein synthesis and outflow produced by cyclohexamine. (B) Cycloheximide was injected into the right L7 ganglion (○) and Ringer solution into the left (●) 30 min after ^{3}H-leucine was injected into each ganglion. Nerves taken 5 hr after ^{3}H-leucine injection showed no effect on synthesis and outflow. From Ochs, Sabri, and Ranish (1970).

possibility had to be excluded, namely, that the blocking agent moves down into the axons quickly enough to block a possible local synthesis of protein in the nerve fibers. The injection of ^{3}H-puromycin into the L7 dorsal root ganglia gave little support to that possibility in that it moves at a slow rate in the nerve fibers (Ochs *et al.*, 1970).

In similar experiments using the goldfish optic system, McEwen and Grafstein (1968) found that protein synthesis was not blocked by acetocyclohexamide until it was injected 3 hr after the precursor. The longer time required for the completion of protein synthesis in the retinal ganglion neurons of goldfish optic as compared to the mammal could be the result of the lower temperatures at which those studies were done, perhaps to a slower diffusion of the blocking agent to the ganglion cells in the goldfish eye, or a species difference relative to protein synthesis.

Puromycin and cycloheximide block the synthesis and downflow of the cholinergic enzyme AChE (Burešová and Tuček, 1977) and the transmitter NA (Banks *et al.*, 1973a). The block of downflow of NA by these agents is likely to be due to the block of synthesis of the protein membrane of the vesicle containing the NA and/or the protein chromogranin which is contained within the DCV (Chapter 4).

4. Gating

The term "gating" refers to the process by which materials synthesized in the cell body are exported into the axon for their subsequent axoplasmic transport. At least a part of this gating action takes place in the Golgi apparatus. This route was suggested by the different locations of radioactivity in neurons shown in autoradiographs of CNS neurons taken at timed intervals after injecting the precursor ^{3}H-leucine systemically (Droz, 1965b). Grains were seen located first over the Nissl bodies (rough ER) of neuron cell bodies some 15 min after injection, and then found located over the Golgi apparatus after 30 min. Soon thereafter, grains were seen located over the axons.

Materials do not simply diffuse from the cell body into the axon; the gating mechanism selects materials for export into the axons (Ochs *et al.*, 1967). This was suggested by the studies described above where puromycin was used to block protein synthesis. The resulting polypeptide chain fragments containing ^{3}H-leucine were not transported into the nerve fibers, presumably because those materials are not able to pass through the gating mechanism to enter the axons. Another test to determine whether a foreign substance could be exported was to inject ^{3}H-cycloleucine, an amino acid which is taken up by the cells but not normally synthesized into polypeptides, into the L7 ventral horn region for uptake by the motoneurons. Little evidence of its subsequent axoplasmic transport into the motor axons was seen (Ochs *et al.*, 1967).

A specific action of Co^{2+} on the gate mechanism was proposed by Hammerschlag *et al.* (1977). The basis for that hypothesis was their finding that Co^{2+} does not block synthesis in the cell bodies or readily block axoplasmic transport in the axons but does interfere with the transport of labeled proteins when applied to the cell bodies (Hammerschlag *et al.*, 1976). The permeability of the perineurial sheath of nerve (Chapter 8) to Co^{2+} is low. Cobalt does effect a block of transport in the desheathed nerve preparation *in vitro* in concentrations of 5mM (Ochs, unpublished observations). This suggests that Co^{2+} may not be specific for the gating mechanism. However, it is possible that the gating mechanism is more sensitive to the action of Co^{2+} than the transport mechanism in the axon as indicated by the greater effetiveness of Co^{2+} to block at the cell body level compared to the axonal level when using the desheathed preparation (Hammerschlag, 1980).

The anatomical location of the gating mechanism has not been definitively established. Some components, for example, soluble proteins, may not pass through the Golgi apparatus. Further study of Co^{2+}, and other specific agents affecting a block of the gate mechanism, would be useful in assessing the path taken from the cell to the axon for the various molecular species transported.

5.   Compartmentation

The outflow of labeled components which appear in the axons at later times (Chapters 2 and 11), led to the inference that after their synthesis some labeled materials enter a "compartment" in the cell body from which they subsequently exit into the axon over a period of time of days or longer (Ochs *et al.*, 1970). A schematization of compartmentation is shown in Fig. 5.6.

One route from the site of synthesis to the gating mechanism and thence to the axon accounts for fast outflow, another from the site of synthesis to the compartment and thence to the Golgi complex, for a later appearing outflow into the axons. The proteins may undergo some changes in the compartment, presumably to fit them for the role they will assume in the axon when they are transported. Some indications of changes with time of the compartmented proteins are described in the following sections.

B.   PROTEINS SYNTHESIZED FOR TRANSPORT

1.   Isolation of Labeled Proteins by Subcellular Fractionation

The first indication that ^{3}H-leucine had been incorporated into labeled proteins transported into the axons was the presence of radioactivity remaining in autoradiographs of nerve sections carried through the usual preparative

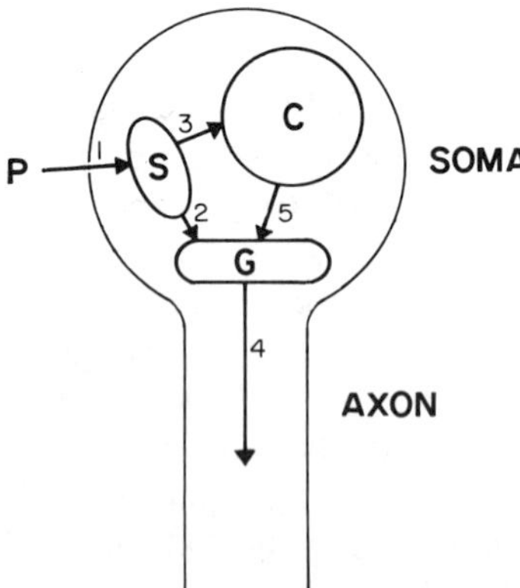

Fig. 5.6 Diagrammatic representation of compartmentation. After entry of the precursor (P) indicated by arrow 1, synthesis (S) causes some material (2) to rapidly enter the Golgi apparatus (G) and the axon for transport (4). Another portion (3) enters a compartment (C) for later exit (5).

steps of histology (Droz and Leblond, 1962). A soluble labeled amino acid such as ^{3}H-leucine would be washed out in the process. Direct biochemical evidence for its incorporation and transport was subsequently obtained (Ochs *et al.*, 1967; Kidwai and Ochs, 1969) using a modification of the technique of subcellular fractionation developed by Whittaker and by de Robertis for brain. The sciatic nerve containing transported labeled radioactivity was homogenized and the homogenate subjected to differential centrifugation. After a first centrifugation at 800g for 20 min, a "nuclear" (N) fraction was pelleted. The supernatant centrifuged at 20,000g/30 min yielded a "mitochondrial" (M) fraction. The supernatant was then subjected to high speed centrifugation at 100,000g/90 min to give a "particulate" (P) fraction and the high-speed supernatant (S) fraction which was found to contain labeled soluble proteins. The labeled protein in the high-speed supernatant fraction was precipitated with TCA. The protein was hydrolyzed, and after separating the amino acids by paper chromatography, the spot which had migrated as leucine was found to contain radioactivity (Ochs *et al.*, 1967). In those studies it was important to show that the incorporated labeled leucine does not readily exchange its tritium label with the hydrogen of water and other amino acids. That this does not occur to any significant degree was indicated by the restriction of the label to the leucine spot in the chromatographs and the similarity of the relative labeling of the various subcellular components when using ^{14}C-leucine as a precursor to that seen with ^{3}H-leucine. A relatively high percentage of the radioactivity is seen to appear in the particulate (P) fraction at the earlier times, i.e., at times related to fast transport (Table 5.1).

In their studies of the optic tectum, McEwen and Grafstein (1968) compared the activity found in the high-speed supernatant after centrifugation at 200,000g for 45 min with that of the pellet, the fraction they referred to as the "particulate fraction." They found at early times a relatively higher level of radioactivity appearing in the particulate fraction in comparison to the high-speed supernatant, the level of radioactivity in the supernatant appearing in greater amounts at later times. The shift in the radioactivity present in the particulate and soluble fractions with time is indicated in Fig. 5.7.

At the earliest time of 16 hr, a time McEwen and Grafstein took to represent fast-transported materials to the tectum, more radioactivity was present in the particulate fraction than in the high-speed supernatant soluble fraction. At 11 days, a time representing slow transport, the particulate fraction still had a relatively high content of radioactivity with respect to the soluble protein, but the level of radioactivity was increased in the soluble fraction. It should be noted, however, that the level of radioactivity in the pellet fracton had greatly increased at 11 days, and even after 44 days the radioactivity in the particulate fraction was high in comparison to earlier times. This demonstrates that the particulate component is not specifically related to fast transport, although it is labeled to a relatively greater degree at earlier rather than later times. The most significant change is the greater

TABLE 5.1 Subcellular Fractionation of Labeled Components[a]

Time	N		M		P		S	
(hr)	SA	Per-cent	SA	Per-cent	SA	Per-cent	SA	Per-cent
3	330	5.1	1,115	29.2	3,341	39.2	921	26.4
5	667	7.1	1,500	31.6	5,056	33.9	1,518	27.3
7	943	3.6	2,209	51.2	4,320	18.9	2,293	26.3
14	2,250	6.8	4,510	29.9	12,647	21.1	11,175	42.1

Subcellular fractionation of sciatic nerve homogenates gives the following fractions: nuclear (N); centrifuged at 800g for 20 min; "mitochondrial" (M), centrifuged at 20,000g for 30 min; particulate (P) and supernatant (S) fractions separated at 100,000g for 90 min. Specific activity (SA) is given in counts per minute per milligram of protein. Percentage of activity is given with respect to the activity of the homogenate. Time is given in hours after injection of ganglia.

proportion of soluble protein labeled at later times in comparison to the particulate fraction, a shift with time shown in cat sciatic nerves in Table 5.1.

In the mammalian preparation, the first 7 hr of outflow can be generally taken to represent fast transport, followed at later times by slow transport (Chapter 2). The proportion of radioactivity in the particulate fraction can be seen to fall off with respect to the soluble protein fraction at the later times, beginning most noticeably at 14 hr. Again, it can be seen that while the specific activity in the soluble protein fraction rises to a greater degree at the later times, the amount of radioactivity in the particulate fraction still remains high, although its rate of rise is not as great. The particulate fraction represents a heterogeneous population. Some of it is in the form of vesicles but it also includes fragments of mitochondria, membranes, and other organelles. This obviates a simple correspondence of fast transport with vesicles destined for the terminals, and slow transport with soluble components supplying the axon proper (cf. Chapters 10 and 11).

2. Soluble Components

The high-speed supernatant fraction when passed through Sephadex G-100 columns for gel filtration showed two major peaks of labeled protein radioactivity, the relative proportion of the peaks changing as a function of the time allowed for transport (Fig. 5.8).

The first emerging outflow from the columns, Peak I, contains labeled protein components with molecular weights (MW) from above 450,000 down to approximately 10,000 daltons. Peak II represents polypeptides from 10,000 down to several hundred daltons. Relatively more radioactivity was found in Peak II at earlier times than that after longer times of transport. After a day, relatively more radioactivity was seen present in Peak I as shown in this

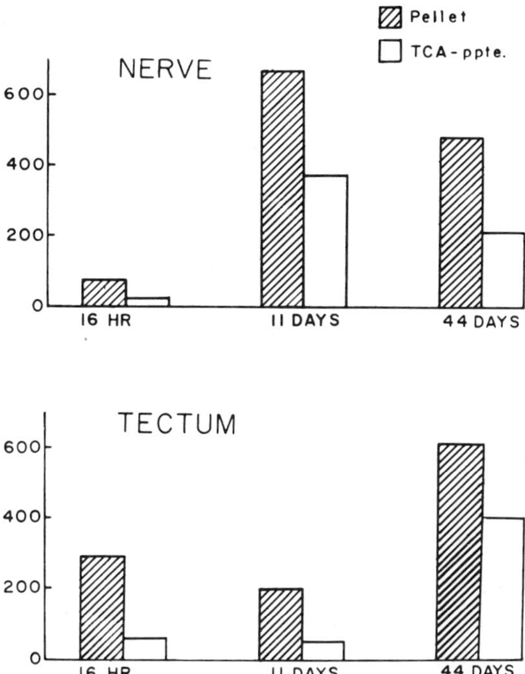

Fig. 5.7 Relative labeling of particulates and soluble proteins. The relative amount of labeling found in the pellet representing particles compared to soluble proteins in the high-speed supernatant (TCA-ppte) is shown at various times (in hours and days) after incorporation of the precursor and transport of labeled components into the optic nerve and the tectum. From McEwen and Grafstein (1968).

figure. Thus, there is a shift with time toward a downflow of more higher-MW proteins during slow transport.

The labeled components of Peak I could be further separated using Sephadex G-200 columns, into those designated as Peak Ia and Peak Ib (Fig. 5.9).

By calibrating the columns with known proteins (Morris and Morris, 1975), Peak Ia was found to contain proteins ranging in MW from above 450,000 down to about 150,000 daltons and Peak Ib proteins in the range of 50,000–150,000 daltons. A prominent component of Peak Ib is tubulin, the subunit of microtubules present mostly as the dimer, which will be further discussed in Chapter 9.

Other types of columns have been used to further subdivide Peak I. Three peaks could be separated with Biogel A5M columns, these designated Peak Ia, Ib, and Ic (Iqbal and Ochs, 1975). The latter contains an important species, a calcium-binding protein (CaBP) of approximately 15,000 daltons, including the regulator protein calmodulin (Chapter 8), which is implicated

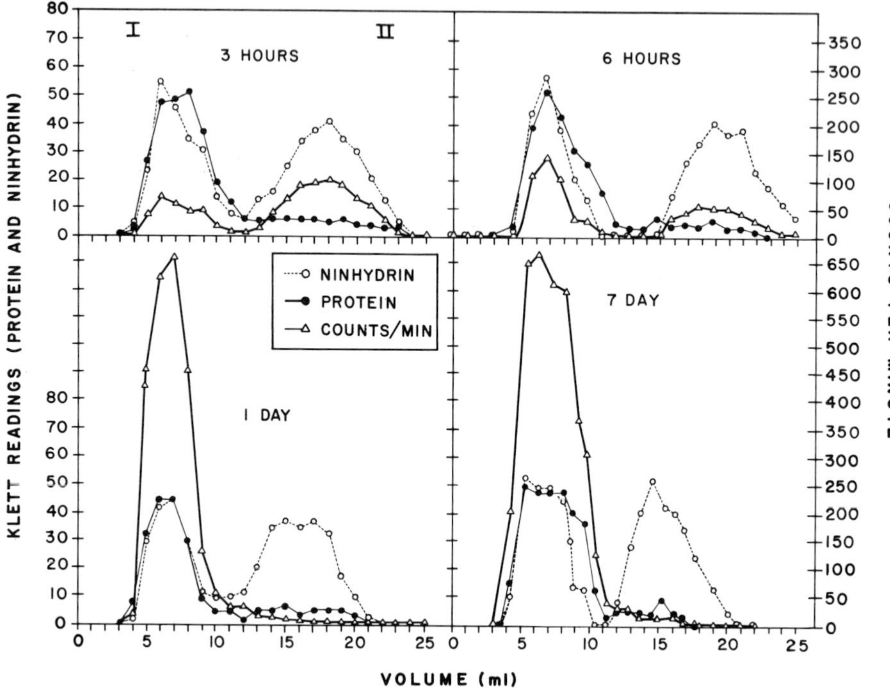

Fig. 5.8 Labeling of high- and low-MW proteins. Labeled proteins in high-speed supernatants of ventral root homogenates were prepared by differential centrifugation of ventral roots at various times after cord injection with ^{3}H-leucine and separated by gel chromatography using Sephadex G-100 columns. The first peak at the left (peak I) after the void volume represents the higher-molecular-weight proteins, the second peak (II) represents lower-molecular-weight polypeptides with some free leucine also present. These were indicated by Lowry (Klett readings) and ninhydrin reaction readings, respectively. Radioactivity given in counts per minute shows a relatively lesser amount present in the higher-MW proteins in the first peak at early times of 3 and 6 hr, with greater amounts appearing after a day. From Kidwai and Ochs (1969).

in the mechanism of transport (Chapter 10). As yet, little is known regarding the transport of small polypeptides. Small molecular weight polypeptides have been reported to be transported at a fast rate (Ochs *et al.*, 1967). What is required is a more specific analysis for a better understanding of their role in the nerve fiber.

One such specific small polypeptide, somatostatin, has recently been found fast-transported in rat sensory fibers of the sciatic nerve using a double-ligation technique (Rasool *et al.*, 1981). The low molecular weight form was found to be anterogradely transported in the sensory fibers at a rate, when the mobile fraction (19.2%) was taken into account, of 415.6 mm ± 15.3 mm/day. This rate is very close to that of labeled proteins and the rates

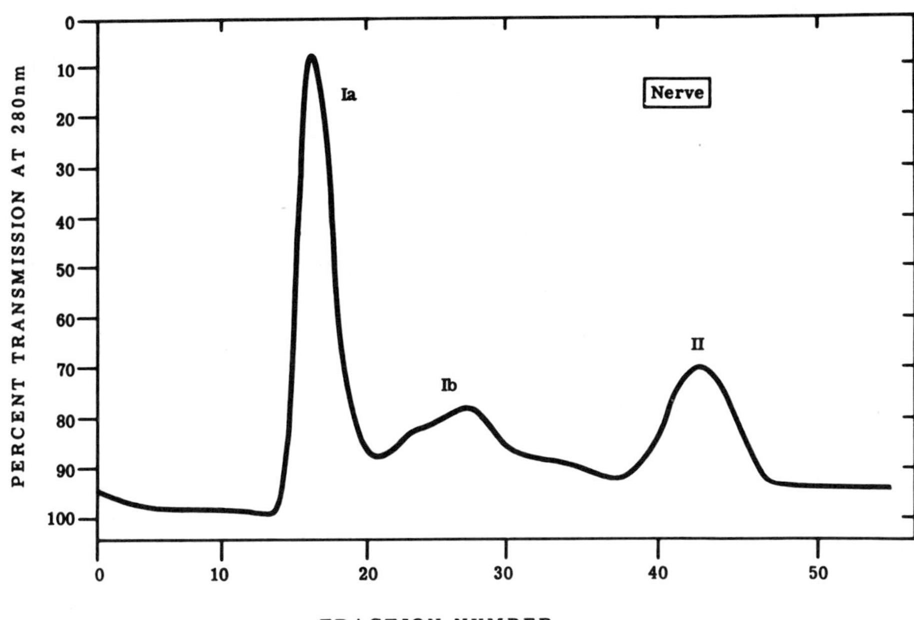

Fig. 5.9 Further separation of peak I proteins. Peak I (cf. Fig. 5.8) could be further separated into a higher-molecular-weight group peak Ia of at least 400,000 and a peak Ib group containing 50,000–120,000 dalton components using Sephadex G-200 columns. Peak II remains a uniform peak. From Sabri and Ochs (1973).

obtained for AChE and NA when using the double-ligation technique (Chapter 4). The significance of the role of somatostatin in peripheral sensory fibers remains to be determined.

Components in the soluble fraction have been characterized on the basis of their isoelectric point (pI) by the method of isoelectric focusing (Kidwai and Ochs, 1969; Sabri and Ochs, 1972; James and Austin, 1970). Either a fluid column or a thin-layer gel is prepared with ampholytes added to it to form a gradient of pH. Then, under the influence of an electric field, the various proteins move until they each reach a region of pH where the charge on them and the electric field are in balance, i.e., where the pI for the individual protein has been attained. The proteins remain in their relative positions when later the column is fractionally eluted or gel slices are taken for identification of the various polypeptide bands which have been separated by this means (Fig. 5.10).

Nerve fibers taken at earlier times after injection of ^{3}H-leucine show labeled proteins present with pI peaks at 5.5 and 4.0. Small shifts in their pIs are seen at the later times, suggesting some differences in their composition or form (Sabri and Ochs, 1973). It is unknown whether shifts in pI can occur as a result of changes in the proteins which take place during their compart-

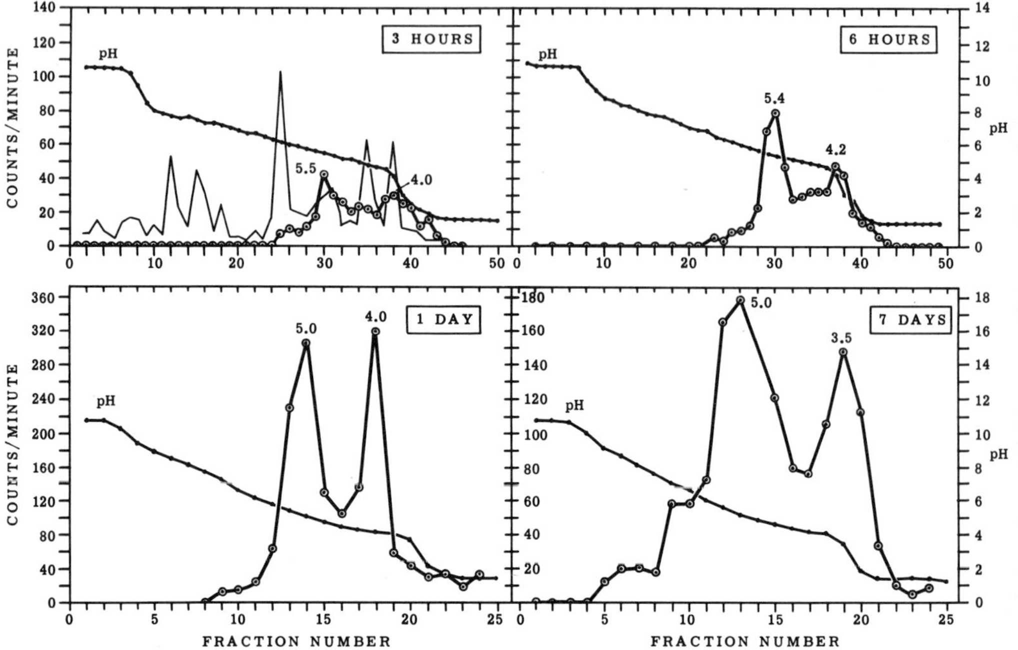

Fig. 5.10 Isoelectric focusing of peak I proteins. The first peak of soluble proteins in the Sephadex G-100 effluents of root homogenates at the times (in hours and days) after injection of the labeled precursor was subjected to isoelectric focusing. The pH (●) and radioactivity (◉) in counts/min are shown within the upper left panel, the latter by the thin line. The concentration of protein is given by the Lowry reaction after TCA precipitation and resolubilization. An increase in acidic proteins with time is indicated. From Kidwai and Ochs (1969).

mentalization in the cell bodies, or after they are dropped off locally in the axons where they undergo a process of turnover (Chapter 11). Radioactively labeled proteins in chicken nerves were found by Austin and his colleagues to have similar pIs (Fig. 5.11).

This technique has been greatly extended to analyze polypeptides in 2-dimensional studies where it is combined with acrylamide gel electrophoresis (Section 5 below).

3. Transport of Amino Acids and Other Low MW Species

A transport of a small amount of free amino acids had been indicated when, after precipitating proteins in the high-speed supernatant of homogenized nerves with TCA, the radioactivity seen remaining in the TCA-soluble supernatant was identified as leucine (Ochs *et al.*, 1967). Free ^{3}H-labeled leucine was also observed in the supernatant when using gel chromatography and electrophoresis (Kidwai and Ochs, 1969; Sabri and Ochs, 1973). The trans-

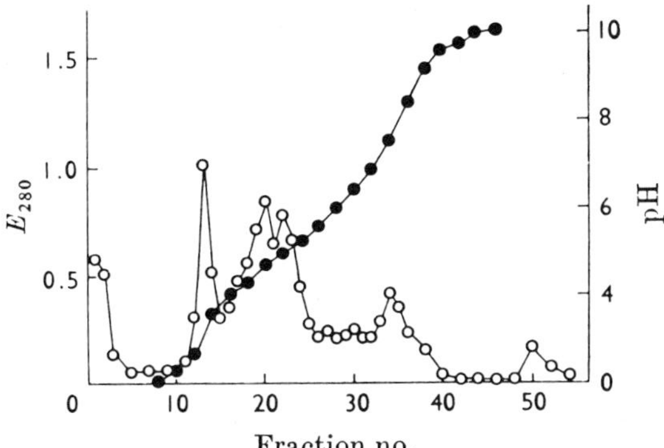

Fig. 5.11 Fractionation of proteins by isoelectric focusing. Soluble axoplasmic protein was fractionated by gel filtration and further fractionated by isoelectric focusing. The pH gradient is given by (●). Four major components with pI values of 3.4, 4.7, 5.0, and 7.55 were resolved (○). From James and Austin (1969).

port of free leucine in nerve has been contested by Karlsson (1977). This point was re-examined (Iqbal and Ochs, unpublished experiments) and a small amount of free leucine was found present in the crest of fast transport in an amount greater than that in the plateau behind it, evidence for its transport rather than its diffusion. It is necessary to establish this point insofar as the exponential decline of leucine with distance from injected eyes indicates that amino acids can diffuse for some distance in the fibers (Di-Giamberardino, 1971). A thorough analysis of the diffusion of ACh and GABA recently carried out by Koike and Nagata (1979) shows an exponentially declining profile of ACh over a length of 10–20 mm, a pattern due to diffusion and one clearly different from that of transport.

Evidence that amino acids can be transported has been given by Csanyi *et al.* (1973). They injected ^{14}C-proline into the eyes of fish and found labeled proline appearing distally in the optic nerves before that of labeled proteins. This was considered to be an argument against a possible action of a proteolytic enzyme in the nerve acting on labeled proteins as the source of the amino acid. Csanyi *et al.* (1973) point to the high-temperature dependence of the outflow they found as further evidence of its transport. Weiss, Schmid, and Wagner (1980) gave additional support for a transport of amino acids.

Transport of small molecular species was reported for D-glucosamine by Forman *et al.* (1971), polyamines by Ingoglia and Sturman (1978), and glutamate and glutamine by Johnson (1974a,b; 1977). Additionally, evidence for the transport of glycine, a presumptive neurotransmitter in neurons R3-R14 of *Aplysia* (Chapter 4) was presented by Price *et al.* (1979). They found ^{3}H-glycine to be transported as free glycine, not only anterogradely but retrogradely as well. And, in a study directed to the use of ^{35}S-sulfate as a precursor of mucopolysaccharides, it was noted that some free ^{35}S-sulfate was also transported as such (Elam and Agranoff, 1971b; Elam *et al.*, 1970).

If, as is indicated by some of these results amino acids are transported, their role in the axon remains unknown. They may possibly act as neurotransmitters, neuromodulators, as neurotrophic substances (Chapter 13), or perhaps serve as precursors for mitochondrial synthesis of certain proteins (Austin and Morgan, 1967; Barondes, 1968).

4. Transport of Membrane Components

An important group of components synthesized in the cell bodies and transported in the fibers are the glycoproteins, an important constituent of membranes, membranous structures in the fiber and the axolemma. A schematization of how the sugar moieties are attached to the glycoproteins in the course of their synthesis in the cell body (Fig. 5.1) and subsequently inserted into the membranes of the axons with their carbohydrate portions exteriorized is shown in Fig. 5.12.

At various points along the axon the vesicles drop off from the transport mechanism to attach to and become inserted into the axonal membrane, a similar process occurring at the nerve terminal (Chapter 13).

Brunngraber (1970) estimates that some 3–10% of the brain proteins are in the form of glycoproteins. Neuronal and glial surfaces contain glycoproteins. The carbohydrates covalently bind to the proteins as single, di, tri, oligosaccharides, or as complex branched heteropolysaccharide chains (Brunngraber, 1972). The predominant sugars are *N*-acetyl-neuraminic acid (NANA), mannose, galactose, *N*-acetylglucosamine, and fucose. The labeled sugars ^{3}H-fucose and ^{3}H-glucosamine which are incorporated into sialic acid serve as markers for glycoproteins (Zatz and Barondes, 1971). When these sugars are taken up by the neuronal cell bodies, they are incorporated and subsequently are fast-transported in the fibers as glycosylated proteins (Forman *et al.*, 1972). ^{3}H-glucosamine was found incorporated into proteins of molecular weight greater than 40,000 daltons, and associated with the membrane fraction (Karlsson and Sjöstrand, 1971a,b). The precursor is converted first to *N*-acetylglucosamine, thence to *N*-acetyl-galactosa-

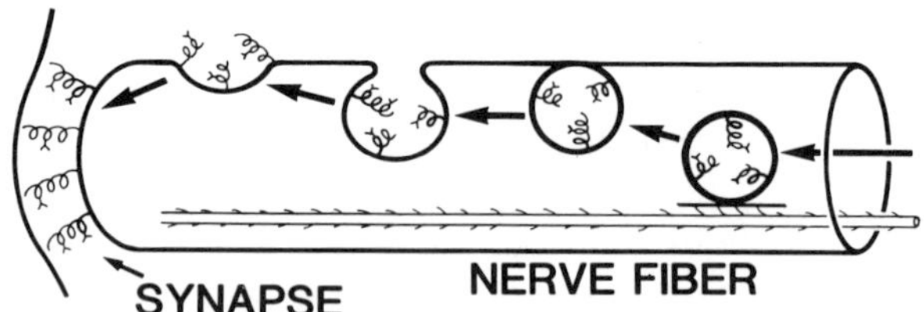

Fig. 5.12 Schematization of vesicle insertion into the axon membrane. The glycoproteins pass through the Golgi apparatus (Fig. 5.1) where sugars (e.g., fucose) are added and the glycoproteins moved down as vesicles to sites where, on merging with the axolemma, the glycoproteins are exteriorized. Transport of the vesicles along microtubules by the transport filament mechanism is indicated (cf. Fig. 10.2A).

mine and sialic acid before it glycosylates the protein. When protein synthesis is blocked, the glycosylated materials remain in the nerve cell bodies, as shown using the markers ^{3}H-fucose, ^{3}H-glucosamine, or ^{35}S-sulfate for glycoproteins (Edström and Mattsson, 1972b). Karlsson (1979) showed that the major glycoproteins fast-transported bind to the lectins wheat germ agglutin and Concanavalin A. When either of these lectins are coupled to Sepharose in affinity chromatography columns, the glycoproteins labeled with ^{3}H-fucose solubilized from the nerve by a non-ionic detergent bind to them, showing that the fast-transported labeled glycoproteins contain exposed *N*-acetyl-D-glucosamine. The possibility earlier advanced that glycosylation occurs externally, on the membrane surfaces of synaptosomes, appears rather to be due to a contamination of the synaptosomes and the participation of the mitochondria in the synaptosomes (Barondes, 1973).

An interesting observation (cf. Fig 2.25) recently re-examined (Stromska and Ochs, 1978), was that the shape of the front of outflow of the labeled glycoproteins seen when using ^{3}H-fucose or ^{3}H-glucosamine as precursors, shows a more shallow sloping front than that of the ^{3}H-leucine-labeled protein outflow (Fig. 5.13).

This suggests some additional time required for the completion of the glycoproteins in the Golgi apparatus. Blocking lipid synthesis with fenfluaramine decreased incorporation of ^{3}H-choline into phosphatidyl choline and

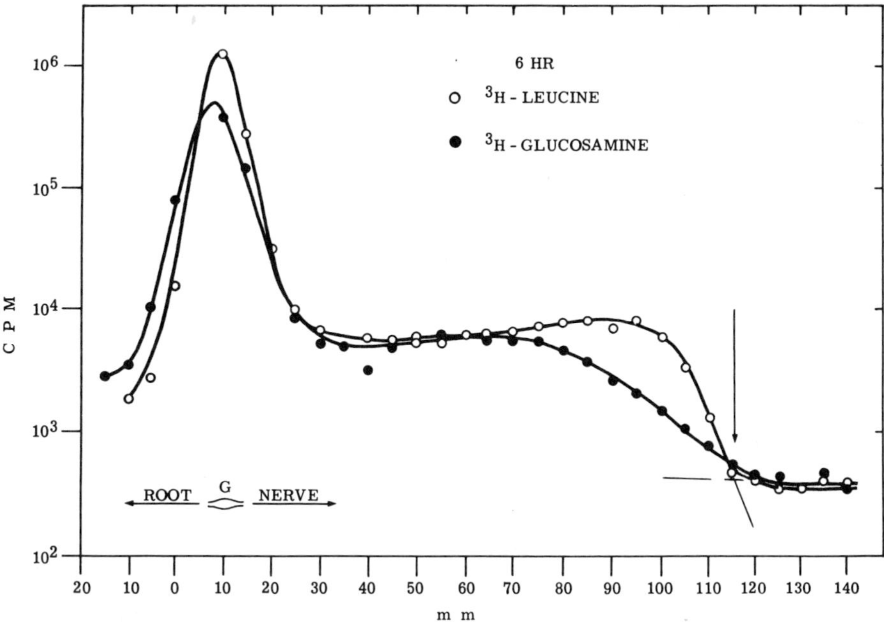

Fig. 5.13 Patterns of glycoprotein transport. After injecting ^{3}H-glucosamine into the ganglion, the glycoprotein radioactivity in the outflow shows that a slope at the front is less steep than the front of ^{3}H-leucine-labeled proteins (cf. Fig. 2.25).

the decrease was proportional to the decrease of ^{3}H-leucine-labeled trans-ported material (Longo and Hammerschlag, 1980). Inhibition of cholesterol synthesis with an analogue also blocked transport of ^{3}H-leucine-incorporated materials. The results indicate an integral participation of lipoprotein synthe-sis at the site of initiation of outflow of membrane components into the axon.

Cholesterol, a component of the membrane bilayer, was also shown to be fast-transported. Labeled acetate injected into rat eyes was incorporated into cholesterol extracted from the lateral geniculate at early times after injection (Blaker, Toews, and Morell, 1980). Me-^{3}H-choline injected into the 9th ganglia of bullfrogs (Abe *et al.*, 1973) became incorporated into a major phospholipid species, phosphatidyl choline, which was fast-transported in the nerve fibers (cf. Part *b* of Fig. 2.25).

The identification of labeled phosphatidyl choline fast-transported in the nerve was accomplished by extracting it with choloroform-methanol from nerve segments taken from the crest region of fast outflow and chromato-graphing the material on silica-gel thin layers. When exposed to iodine vapor, a labeled spot was found located at the same site as a pure sample of phosphatidylcholine similarly chromatographed. On the other hand, Rostas, Austin, and Jeffrey (1979) found a slow outflow using [3]II-cholesterol in the chick eye (Fig. 5.14) .

Dziegielewska, Evans, and Saunders (1980) also found a slow transport of phospholipids after injecting ^{3}H-choline into the rat spinal cord, at a rate of no more than 20 mm/day.

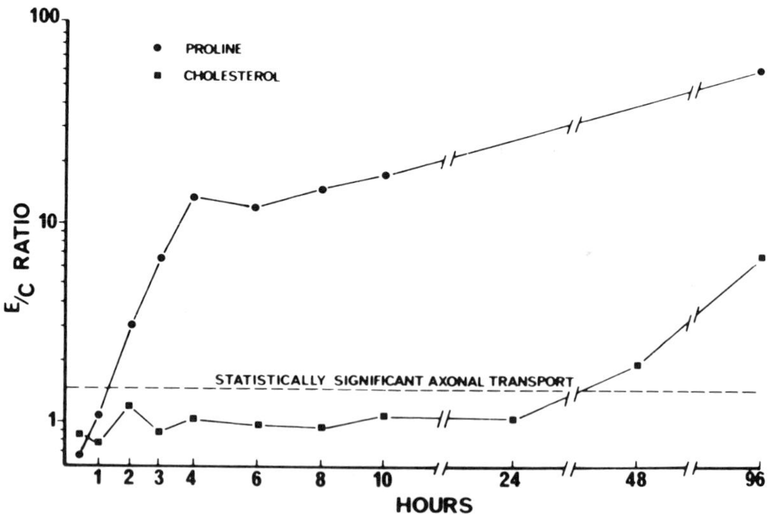

Fig. 5.14 Accumulation of glycoprotein and lipid. Axonally transported radioactivity in the tectum after intraocular injection of ^{14}C-proline (●) or ^{3}H-cholesterol (■). Each point represents the average of two animals. A slow transport of cholinesterol-labeled lipids is indicated. From Rostas, Austin, and Jeffery (1979).

Some of these differences may be due to the precursor used. Rostas *et al.* (1979) found a fast transport only when using a labeled metabolic precursor of cholesterol, ³H-melvonic acid. The point of entry of the various components into the final assembled form may determine whether a labeled component appears as one which is fast or slow-transported. When ³²P in orthophosphate form is used as the precursor, it does not label fast-transported phosphoproteins to a great extent (cf. Chapter 2), as shown in comparison with ³H-leucine-labeled proteins (Fig. 5.15).

Some small proportion of ³²P may possibly be fast transported along with phosphoproteins, but for the most part ³²P does not appear to be in a pool available for rapid phosphorylation of phosphoproteins. The possibility that the phosphorous in ATP may more readily contribute to fast transport was examined (Takenaka and Ochs, 1980). A fast outflow was not, however, found, either by the usual method of assessing outflow or by the method of external detection of ³²P developed by Takenaka, Horie, and Sugita (1978) (Chapter 2).

Using *N*-acetyl-³H-glucosamine as a precursor, a minor portion of the

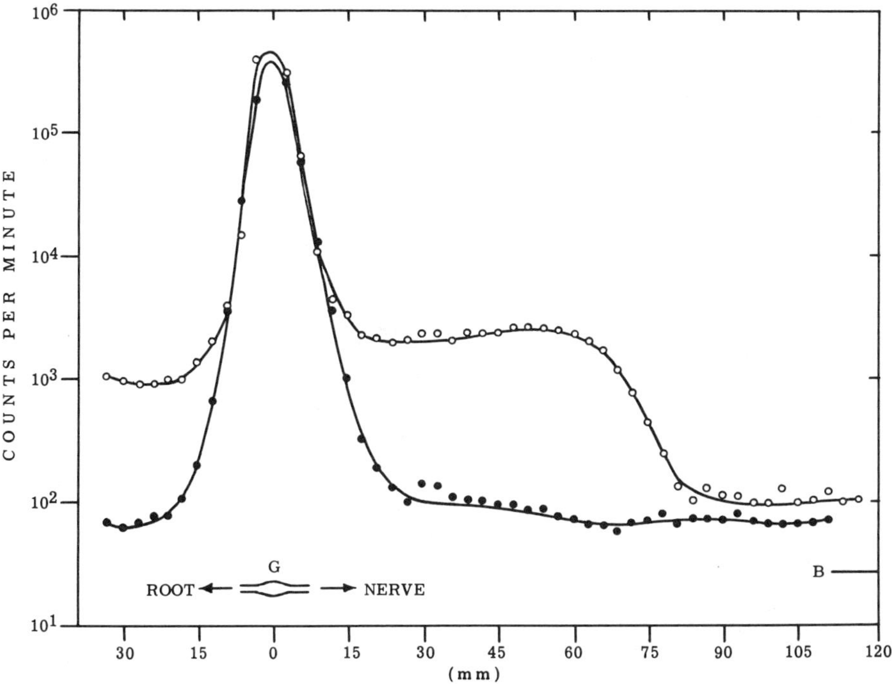

Fig. 5.15 Transport of ³²P phosphorous-labeled components. Comparison of downflow patterns after ³²P and ³H-leucine injections into cat L7 dorsal root ganglia is shown. A fast transport of ³H-leucine-labeled proteins (○) after 5 hr is to be compared to the slow outflow on the ³²P-injected side (●). From Ochs and Ranish (1969).

radioactivity transported was found incorporated into acidic mucopolysaccharides (Held and Young, 1972). The precursor ^{35}S-sulfate also labels transported mucopolysaccharides, those characterized as chondroitin sulfate and
heparan sulfate (Elam and Agranoff, 1971b; Elam *et al.*, 1970).

There is a relatively greater transport of glycoproteins and glycolipids in
the developing organism, presumably to supply the larger amount of membrane formed during the maturation of nerve fibers. This was indicated by
the larger proportion of ^{3}H-fucose-labeled glycoproteins versus ^{14}C-proline-
labeled proteins transported to the optic lobe of the chick embryo as compared to the mature animal (Bondy and Madsen, 1971). A similar observation was made by Gremo and Marchisio (1975). The proportion of membrane
required by regenerating nerves is relatively high and a high level of labeled
glycoproteins is transported in the hypoglossal nerves in the first 1–4 weeks
of its regeneration in compared to normal nerve (Frizell and Sjöstrand, 1974).

Recently, Griffin *et al.* (1981) used autoradiographic EM to show the
location of labeled glycoproteins in normal and in regenerating nerve fibers.
In normal axons some of the glycoproteins fast-transported are seen localized
to the axolemma with a substantially greater amount distributed throughout
the axoplasm (cf. Chapter 11). In the regenerating fibers, even as early as a
day after regeneration has started (Chapter 15), a substantially greater amount
was seen localized to the axolemma, and after 7–14 days almost all the
radioactivity found was restricted to the axolemma. Such evidence is in
accord with a turnover of glycoproteins in the axolemmal as well as at nerve
terminals (Chapters 10 and 15).

Using an immunofluorescent technique, Willard and his associates found
that two polypeptides (fodrin) with molecular weights of 250,000 and 240,000
transported at an intermediate rate of 40 mm/day (Group II in his system of
notation isolated by SDS–PAGE as described in the following section) are
highly concentrated in the axon near the axolemma (Levine, Skene, and
Willard, 1981). Unlike integral proteins of the membranes, fodrin can be
readily extracted from the nerve. It is not present on the external surface, as
shown by lack of staining with the immunofluorescent stain on exposure of
the surface. When the permeability of the membrane is increased, the stain
can enter and bind to it in the axon. The internal location of fodrin near the
axolemma raised the possibility that it may be part of a transport process
viewed as a lining or sleeve moving down from the cell body. The problem
of multiple transport mechanisms, in this case of a special protein moving
down the axon at an intermediary rate, one different from slow or fast rates
(cf. Chapter 2), or of a unitary system of transport with subsequent local
redistribution accounting for apparently different rates, will be discussed in
Chapter 11. Here we note that fodrin would, on the basis of the unitary
hypothesis, be dropped off from the transport mechanism to become positioned peripherally in the axon where it plays some as yet not well-defined
role.

5. Polypeptides Assessed by SDS–PAGE

The various classes of proteins and polypeptides described in the proceeding sections were isolated from nerves by relatively gentle methods whereby their complex forms are retained. Their component proteins can be isolated using the detergent SDS and heating. With this technique the various organelles, membrane complexes, and large proteins are disassembled into their monomeric polypeptide chains. The SDS-dissociated polypeptides are then separated on polyacrylamide gels by electrophoresis (SDS–PAGE), the polypeptides separated on the basis of their molecular weights, their electrical charges neutralized by the SDS bound to them. The polypeptides are seen as individual bands by the use of staining with a dye such as Coomassie Blue. If the polypeptide had incorporated a precursor such as ^{3}H-leucine, its content of radioactivity may be assessed by slicing out the bands and counting the radioactivity in it with a scintillation spectrometer, or by exposing the gels to photographic plates for autoradiography. ^{35}S-methionine is often used as a precursor in order to increase the level of radioactivity and shorten the exposure time for autoradiography. To further increase the counting sensitivity, fluorescent compounds are added, the procedure referred to as fluoroautoradiography. Willard, Cowan, and Vagelos (1974) analyzed portions of the optic nerve, optic tract, lateral geniculate, and superior colliculus taken at different times after injecting ^{35}S-methionine into the rabbit eyes and found by this means a variety of polypeptide bands containing labeled radioactivity. They showed a proximo-distal shift in the nerves with time. Several of the higher MW components appeared at a distal site early, others at later times. This is indicated in a similar study by Levine and Willard (1980) of the labeled particulate fraction carried out in the guinea pig optic system (Fig. 5.16)

From the temporo-spatial shift in the patterns of labeled outflow taken at various times, Willard and his colleagues concluded that there are 5 rates of transport in the optic nerves in all (Willard *et al.*, 1974; Willard and Hulebak, 1977; Levine and Willard, 1980). The significance of multiple rates with regard to mechanisms of transport will be discussed further in Chapters 10 and 11. In a similar SDS–PAGE study made in segments of peripheral rat nerve where all the contents particulate and soluble were assessed, a triplet of polypeptides bands of 200,000, 160,000, and 68,000 daltons was identified as the subunits of the neurofilaments and the 57,000- and 53,000-dalton bands the tubulin subunits of the microtubules. These were slow-moving proteins (Hoffman and Lasek, 1975; Lasek and Hoffman, 1976).

Two groups of slow moving proteins were identified in rat motor fibers; SCa consisting mainly of the tubulin and triplet proteins of microtubules and neurofilaments respectively moving at approximately 1 mm/day, and SCb which consists mainly of actin of microfilaments and a number of other components moving at 3–4 mm/day.

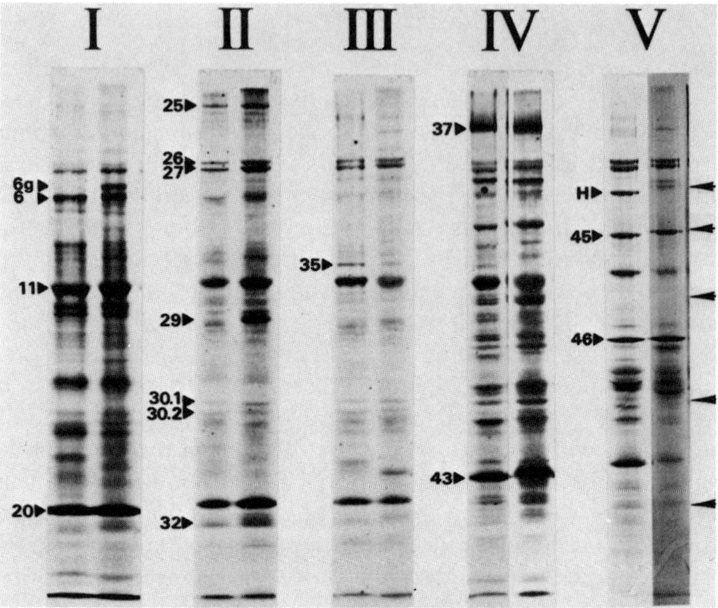

Fig. 5.16 Polypeptides transported in the optic system of guinea pigs at different rates. Autoradiographs (fluorographs) of segments of the visual system containing labeled polypeptides transported in the nerves taken at different times after the precursor was injected for uptake by the retinal ganglion cells. The optic nerve (ON), optic tract (OT), and colliculus were sectioned and each piece subjected to high-speed centrifugation. The resulting pellets were solubilized and then subjected to electrophoresis. As shown in each column at the times indicated, five transport groups are observed in the guinea pig and rabbit (right and left sample of each pair, respectively) for the given segment taken at the post-injection times as follows: I—superior colliculus, 1 day; II—distal OT, 1 day; III—distal OT, 4 days; IV—distal OT, 8 days; V—distal ON, 20 days. The arrows to the right are molecular weight markers. From top to bottom, they correspond to molecular weights of 200,000, 150,000, 100,000, 50,000, and 25,000 daltons. The solid arrowheads and numbers to the left of each column indicate some of the most distinctive polypeptides of each group. Modified from Levine and Willard (1980), courtesy of M. Willard.

In the guinea pig retinal ganglion cell axons, the slower component (SCa) has a rate of 0.25 mm/day and the somewhat faster component (SCb), a rate of 2–3 mm/day (Black and Lasek, 1980). The positions of the polypeptide bands at different distances from the eye at different times after injection of the precursor are shown in Fig. 5.17.

The similar analysis of Mori, Komiya, and Kurokawa (1979) has shown the presence of a number of different rates for these components. The manner in which these proteins are moved down the fiber will be discussed further in Chapters 10 and 11.

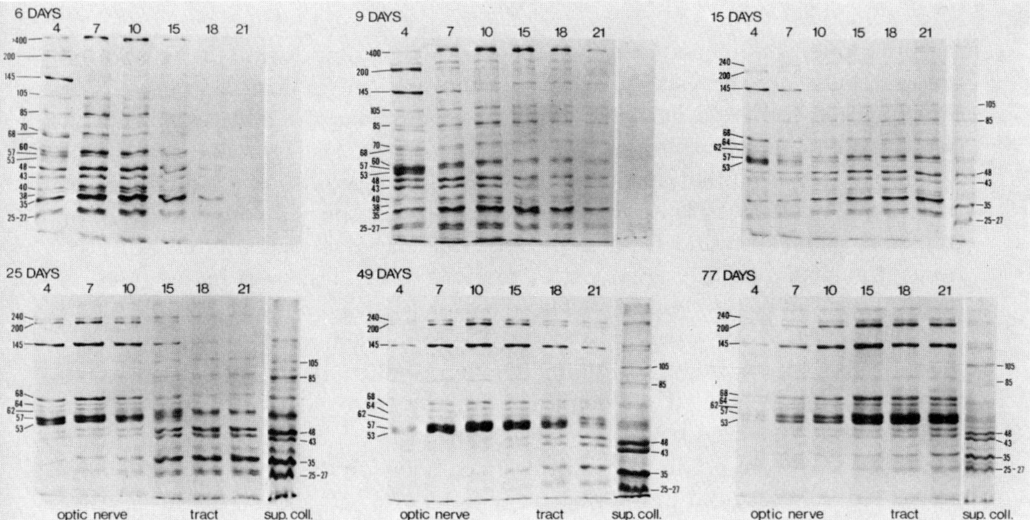

Fig. 5.17 Outflow of labeled proteins associated with slow transport. Fluoroautora-diograph of labeled proteins taken from guinea pig visual system after their eyes were injected with labeled amino acid precursor animals sacrificed at 6 to 77 days later. Segments along the nerve were analyzed by SDS–PAGE to give the distribution of the slowly transported labeled proteins. Each of the columns represents the proteins in consecutive 3 mm segments of the optic nerve, tract and supe colliculus at the distances in mm from the eye give. The SCAa and SCb components are represented by the triplet numbers at the left giving molecular weight x 10^3. From Black and Lasek, 1980.

To be taken into account in the analysis of the composition of transported polypeptides, are the variety of nerve fibers present in a mixed nerve. The olfactory nerve of the garfish, which has a fast axoplasmic transport with outflow characteristics similar to mammalian nerve (Chapter 2), consists of a homogeneous population of unmyelinated fibers with a mean diameter of 0.42 μm (Easton, 1971). This simplifies some aspects of the analysis of transport. Cancalon and Beidler (1975) took the crest and the plateau portions of garfish olfactory nerves after a period of fast transport for an analysis of their content of polypeptides. These two nerve regions were homogenized and after high-speed differential centrifugation, separated on sucrose gradients. The subfractions so obtained when analyzed by SDS–PAGE were found to contain labeled polypeptides ranging from 50 to 150,000 daltons with several peaks evident at 126,000, 54–58,000, and 35,000 daltons. In general, there were no dramatic differences between the labeled components present in the crest and in the plateau regions after 10 and 20 hr, suggesting that there is no large differential in the drop-off of components along the nerve (cf. Chapter 11).

A somewhat similar result was found for the pike olfactory nerve, which also has a uniform content of unmyelinated fibers with a mean diameter of 0.2 μm (Kreutzberg and Gross, 1977). The composition of fast-transported polypeptides in the peak (crest) and saddle (plateau) regions (Weiss *et al.*, 1978) showed three major peaks of 157,000, 137,000, and 118,000 daltons to be labeled. In this preparation the relative increase in the radioactivity present in the high-MW polypeptides with distance suggested more of a drop-off of the lower-MW polypeptides from the advancing crests in these axons as compared to the garfish olfactory nerves (cf. Chapter 11).

SDS–PAGE of frog sensory nerves dorsal root, ganglion, dorsal roots, and ventral roots taken after ^{35}S-methionine injection of the dorsal root ganglia or of the motoneuron region of the spinal cord, showed a large number of different labeled polypeptide species transported (Barker *et al.*, 1976). The composition of labeled polypeptides was generally similar in the motor and sensory nerve fibers. This was shown in segments of nerve taken just above ligations made in the nerves and roots after allowing time for the accumulation of labeled materials. Barker *et al.* (1977) identified 14 polypeptide species transported in those nerves ranging from 10,000–100,000 daltons. It is possible to resolve a much larger number of polypeptide species by using 2-dimensional SDS–PAGE (Stone *et al.*, 1978). In this technique, the polypeptides extracted from nerve are first separated by isoelectric focusing along one dimension of a gel plate and then by SDS–PAGE in the 2nd dimension. Autoradiographs of the polypeptides so separated showed a relatively large number of labeled polypeptides (Fig. 5.18).

The spots indicating the various proteins are separated with respect to their molecular weights along the ordinate and by the isoelectric points (pI) shown on the abscissa. At least 45 polypeptides so determined were seen to be commonly present in the sensory nerve dorsal roots and ventral roots. Further development of this technique has revealed an even greater number

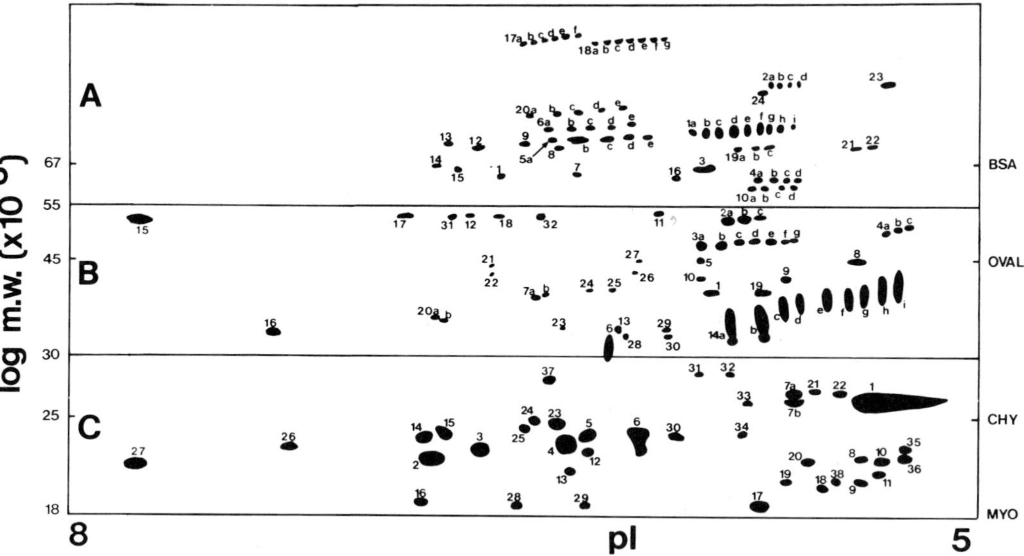

Fig. 5.18 Two-dimensional separation of labeled proteins. Forty-five rapidly transported proteins common to sensory nerve, dorsal root, and ventral roots were separated by isoelectric focusing along the abscissa (pI) and then by SDS–PAGE in the vertical direction. Eighty-four spots have been assigned numbers based on their positions in the autoradiograph. The proteins were arbitrarily divided into three sections on the basis of their molecular weights. Within each section spots were assigned numbers. In cases where a set of spots appeared to be a single species with charge modifications, the spot of greatest intensity was assigned a number, the other charged form designated by lowercase letters. Molecular weight determinations were made in accordance with the following standards: bovine serum albumin (BSA), ovalbumin (OVAL), chymotrypsinogen (CHY), and myoglobin (MYO). From Stone and Wilson (1979).

of individual proteins, again with surprisingly little difference in the poly-peptides present in the dorsal root and sensory nerve. This is of special interest with respect to the phenomenon of routing (Chapter 11), where we expect some differences to occur. Small differences between the sensory and motor nerve and components in different mammalian nerves were reported (Black and Lasek, 1978). A cautionary note was sounded by Neale *et al.* (1980) in their analysis by SDS–PAGE of fast-transported proteins in frog and rat; a number of dissimilarities in the proteins were seen which could be due to species differences.

C.   CHANGES IN TRANSPORT RELATED TO MATURATION

Protein synthesis is generally considered to occur at a higher rate in the immature animal than in the adult, and one might suppose that a higher rate of axoplasmic transport might also be present in the nerve fibers of the immature animal as well. The reverse has, however, been reported. An increase in the rate of fast transport in the course of maturation of the chick visual system was reported by Marchisio and Sjöstrand (1972). This was also seen in the rabbit by Hendrickson and Cowan (1971). Following injection of ^{3}H-leucine into one eye, a fast rate of transport of labeled proteins in the optic nerves was estimated by its earliest accumulation in the superior colliculus to be 120-mm/day 6 days postpartum. This rose to a rate of 150 mm/day at the end of the 3rd week of life to attain at the end of the 4th week, a rate of 200 mm/day. Thereafter, the rate remained fairly level at the adult rate of between 200 and 240 mm/day. Slow transport on the other hand, was reported to decline from 5 mm/day at the end of the 1st week of life to 2 mm/day in animals over 4 weeks of age. Conversely, the rate of slow transport in the immature kitten determined after injecting the L7 dorsal root ganglion with ^{3}H-leucine as determined from the slope of the outflow of radioactivity in the sciatic nerve, was 2–3× greater than in the adult cat (Lasek, 1970). Droz (1965a) also reported that slow transport is faster in the immature rat with a rate of 2–2.5 mm/day, as compared to a rate of 0.6–0.9 mm/day in the adult.

Such reported differences in the rates of fast and slow transport when determined by the accumulation of labeled components at nerve terminals might, however, be due to differences in the amounts of material transported, particularly so with regard to slow transport, rather than to rate changes *per se* (cf. Chapter 11). The rate of fast transport as a function of early maturation can be more precisely assessed from the position of the front of labeled proteins if a sufficiently long length of nerve is available. This is the case in kittens as young as 2 weeks postpartum where the usual outflow pattern was seen (Ochs, 1973). The rate of fast transport determined from the front of the outflow of labeled radioactivity in the nerves of immature kittens was 389 mm/day as compared to the rate of 410 mm/day in the adult animal, a difference which is not, however, statistically significant. These results suggest that the transport mechanism itself remains unchanged although the amounts and perhaps also the kinds of materials transported may change in the course of maturation, differences determined by the genetic control from the nucleus. Judgement on this point must, however, be reserved until younger animals can be examined. An unchanged rate in older animals (cats and dogs) also points to a relative stability of the transport mechanism (Chapter 12). Only in very old rats, 3 years and more, was a small decline in the rate of fast transport indicated (Stromska and Ochs, 1982). This may not be an indication of a basic alteration of the transport mechanism, however, but some associated undetected neuropathological alteration as will be further discussed in Chapter 12.

6

Degeneration and Retrograde Reaction

The classical sequence of changes in the form of myelinated fibers amputated from their cell bodies which characterizes Wallerian degeneration (Chapter 1), includes a separation of myelin from the nodes, ovoid formations of the myelin sheath changing to beads, followed by the resorption of the remaining myelin (Ramon y Cajal, 1928; Young, 1949b). Recently it has been recognized that still earlier morphological signs of degeneration appear after a latency with such changes moving rapidly down the fiber. Those changes and the loss of excitability and conduction of the nerve impulse appearing within several days following nerve fiber interruption will be discussed in Part A below. Additionally, changes which appear in the nerve cell bodies, alterations known classically as chromatolysis, and also referred to as the retrograde reaction, which will be discussed in Part B. This process is related to the regeneration of nerve fibers, a subject to be dealt with in Chapter 15.

A. WALLERIAN DEGENERATION

1. Proximo-Distal Degeneration

The marked changes characteristic of Wallerian degeneration in myelinated fibers observed by light microscopy in appropriate histological preparations are shown diagrammatically in Fig. 6.1.

Not shown in this figure are the later appearing changes seen after several weeks or months when all the myelin has been resorbed leaving the tube-like endoneurial sheaths and the associated cells forming the Bands of Bungner (cf. Section A1 Chapter 15).

An earlier change signalling degeneration was found by Lubińska (1977). In individual fibers teased from the distal amputated stump of rat phrenic nerves soon after making nerve transections, an indentation of the myelin appeared at the center of the internode just under the Schwann cell nucleus. This was soon followed by swellings and indentations appearing to either side of the original indentation (Fig. 6.2).

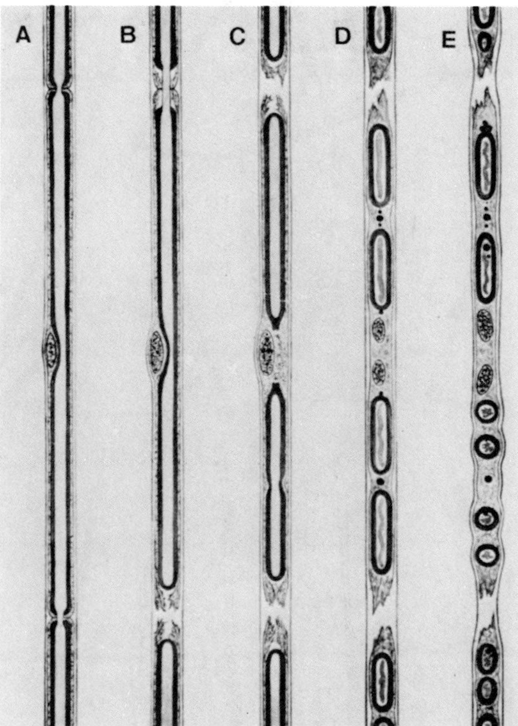

Fig. 6.1 Wallerian degeneration. The classical appearance of form changes of myelinated fibers in the amputated portion of nerve are shown days and weeks following transection. A normal fiber is shown (A). Early in degeneration, after several days, there is a retraction of myelin from the nodes (B). Later (C), long ovoids are formed which in a matter of weeks (D, E) become smaller and round up into vesicular structures which are eventually resorbed. From Young (1949b).

These changes seen in fibers as early as 26 hr after transection were soon followed by a series of segmentations of the fibers, the ovoid formations shown in Fig. 6.1. The time of the appearance of the nuclear indentations and swellings in the center of the internodes was related to the fiber diameter. The thinner fibers were the first to exhibit this change, then the smaller myelinated fibers, the larger ones and finally, the very largest fibers (Lubińska, 1977).

Using the first appearance of indentation and ovoid formation at the center of the internode, Lubińska was able to answer an old problem, namely, whether Wallerian degeneration occurs simultaneously all along the whole length of the transected nerve fiber, or if there is a proximo-distal spread of degeneration moving down from the site of interruption (Joseph, 1973). In most previous studies carried out to decide between these alternatives, the fast rate of change had not been properly taken into account and the earliest times investigated were usually already too late. The proximo-distal gradient

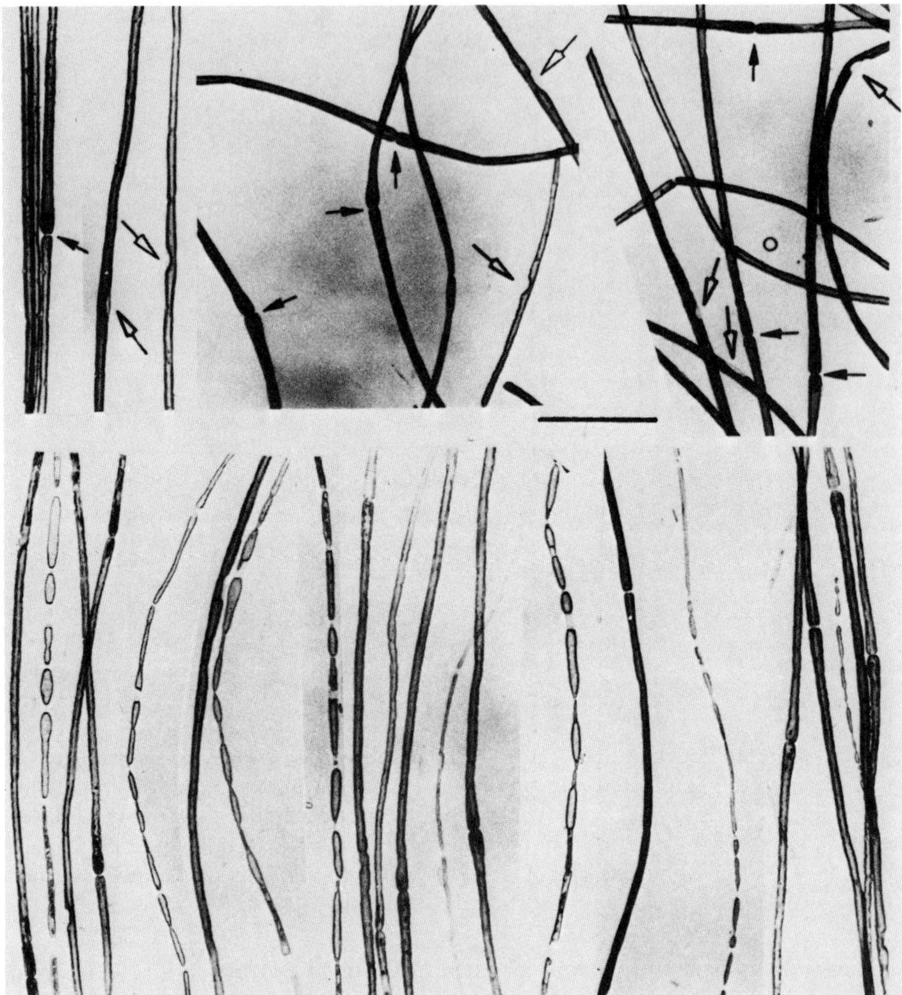

Fig. 6.2 Earliest stages of degeneration. Upper panel: phrenic nerve fibers were taken 26 hr after an upper transection. Normal-appearing nodes of Ranvier are shown by filled arrows. The slight enlargement at either side and depression appearing under the Schwann cell nucleus as the earliest sign of degeneration are indicated by open arrows. Bottom panel: fibers at a slightly later time are seen to be broken into ovoids in the internodal region. Other fibers are continuous. From Lubińska (1977).

of degeneration of the lateral-line fibers of the fish which had been reported by Parker and Paine (1934) was most likely seen because the observations were made at a lower temperature, this resulting in a reduced rate of degeneration (Waller, 1852a). Lubińska (1977) found, after a latency of some 18 hr, degenerative changes appearing first in the thinnest fibers with 50% showing degeneration 25 hr after transection. Then, after 27.5 hr, degeneration was seen in the medium-sized fibers and after 31 hr in the thicker fibers. At 44 hr the thickest group of fibers were still developing their full degree of degeneration (Fig. 6.3).

Taking pieces of nerves at various distances from the site of transection, a Wallerian degeneration was found to spread proximo-distally, first in the smaller-diameter fibers, then in the larger ones (Fig. 6.4).

From the earliest appearance of the indentations in nerve fibers taken at two sites 5 and 15 mm below a transection, Lubińska was able to calculate the apparent rate of the proximo-distal spread of degeneration as 252 mm/hr in the thinnest fibers, 154 mm/hr in the medium-sized fibers, 95.7 mm/hr in the thick, and 46.5 mm/hr in the thickest fibers. The rate in the thinnest fibers, approaching that of fast transport, is striking. From such results the inference drawn was that some factor or signal substance required to maintain the form of the fiber is being supplied by axoplasmic transport and on interruption its loss initiates degeneration. This factor may include a substance required by the Schwann cell, as suggested by the earliest change appearing as an indentation in the myelin sheath in the center of the internode just under the Schwann cell body. The greater latency of onset and the lower apparent rate of the proximo-distal spread of degeneration in the larger-diametered fibers is likely due to their having a greater pool of the required factor present in them. Thus, they can last for a longer time than the smaller-

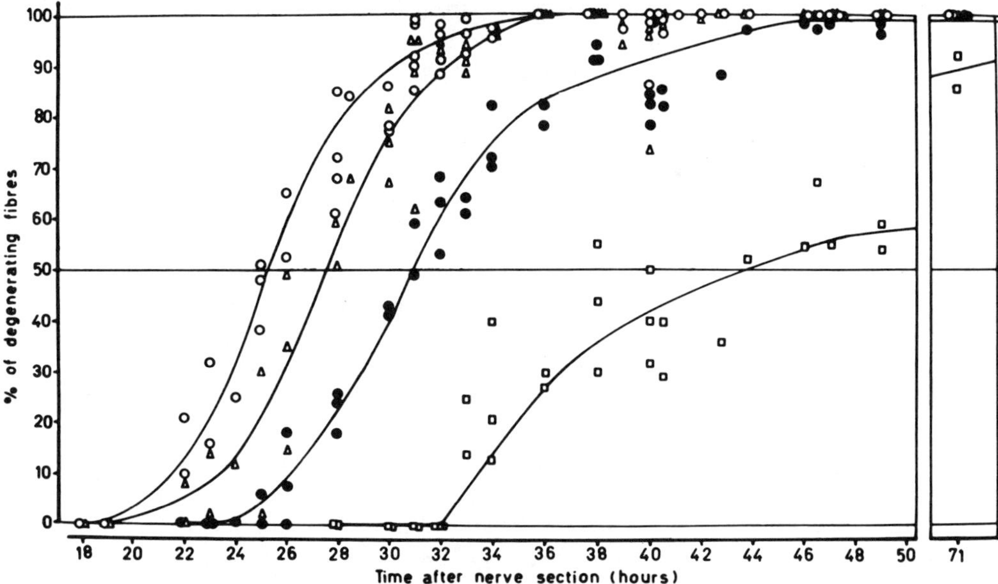

Fig. 6.3 Time course of earliest changes in Wallerian degeneration in relation to fiber diameter. Using the appearance of the earliest form changes in the fiber (cf. Fig. 6.2) in the phrenic nerves of rat at different times after nerve section, the percentage of degenerating fibers at various times (in hours) is seen to appear as a function of fiber diameter; thin (○), medium (△), thick (●), and thickest (□) showing a progressively later appearance of form changes. From Lubińska (1977).

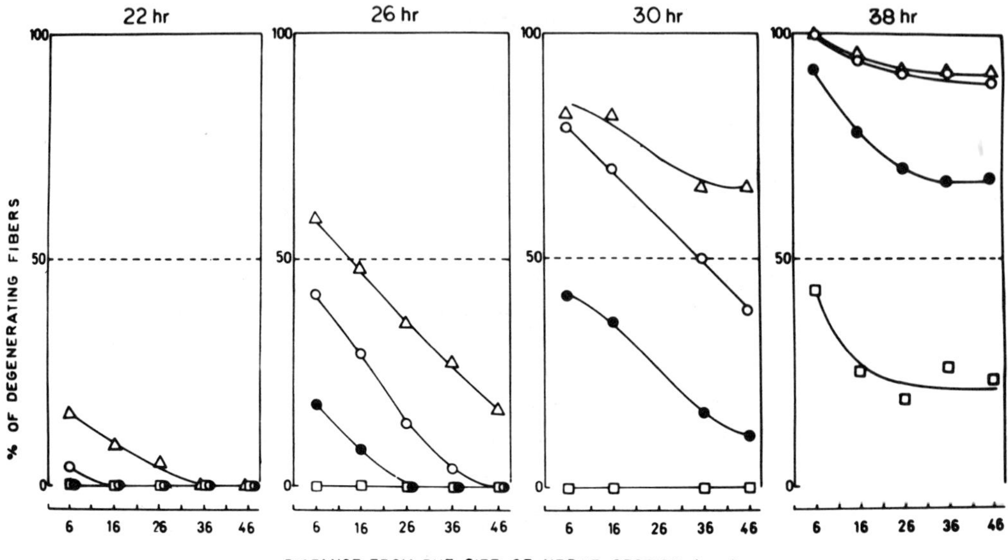

Fig. 6.4 Longitudinal spread of Wallerian degeneration. A spread of degeneration is shown by the appearance of indentation under the Schwann cell as the first sign of Wallerian degeneration (cf. Fig. 6.2) at the indicated distances distal to a transection (in millimeters) on the abscissa at 22, 26, 30, and 38 hr. Fibers from thin to thick diameters (see symbols in Fig. 6.3) show a progressively slower proximo-distal spread of form change in relation to their diameter. From Lubińska (1977).

diametered fibers. In addition, the thinnest fibers also have a smaller volume-to-surface-area relation, and perhaps the greater requirement for the supply of a needed factor to maintain their form and function causes it to be used up faster in the smaller fibers.

The nature of the substance passing from the axon to the Schwann cell suggested by the initiation of Wallerian degeneration remains unknown. It is also possible that some needed structural components of the myelin sheath are supplied by the axon to the Schwann cells. A transfer of phospholipids from the axon to the myelin was suggested by the autoradiograhic studies of Droz (1979) and his colleagues. On comparing the localization of grains when using ^{3}H-glycerol and ^{3}H-choline as phospholipid precursors in the nerve cell bodies (Chapter 5) with ^{3}H-lysine as a precursor for proteins, radioactive grains due to the phospholipid-labeled components were seen located in the myelin sheath. Droz (1979) considers ^{3}H-choline to enter into the sheath via the Schmitt–Lantermann incisures, the ^{3}H-choline then incorporated by the Schwann cell cytoplasm into myelin-forming phospholipids.

Degeneration of the myelin sheath is not invariably seen over all the length of fibers in the mammalian CNS, a peculiarity which may be related to a supply of the myelin sheath of a number of fibers in the CNS by oligodendrocytes (Bunge, Bunge, and Ris, 1961).

The degenerative changes which take place in the axons after they are cut off from the cell bodies were studied by Donat and Wiśniewski (1973). Using EM, they described the destruction of neurofilaments and microtubules which did not show a proximo-distal progression. Analysis of degeneration in unmyelinated fibers has received comparatively little attention. Friede and Martinez (1970) found most unmyelinated axons to have undergone degeneration within 2–3 days after sciatic nerve transection. Thomas *et al.* (1972) reported still earlier appearing degenerative changes, some within 24 hr (cf. also Williams and Hall, 1971), and even earlier degenerative changes were reported by Dyck and Hopkins (1972). In the EM studies of the latter, some 5% of the unmyelinated fibers showed degenerative changes as early as 9–11 hr after crushing the cervical sympathetic trunk in the rat, although such changes were more regularly seen 22 hr after injury. It is significant from the point of view of a possible proximo-distal gradient of degeneration in the unmyelinated fibers that Dyck and Hopkins found empty spaces at sites where unmylinated fiber should be present to decrease in sections examined progressively more distally from the site of injury, an observation which suggests a proximo-distal gradient of degeneration in unmyelinated fibers.

Some invertebrate nerves do not undergo a clear-cut Wallerian degeneration. Hoy *et al.* (1967) found that when the giant axon innervating the claw muscle of the crayfish is cut, the ends can reunite by fusion and allow motor responses to return sooner than could be accounted for by a proximo-distal regeneration of the axon (cf. Chapter 15). In EM studies fine nerve fibers were seen to sprout from the proximal end of the cut nerve fiber and make contact with the distal cut end of the fiber (Nordlander and Singer, 1973). Presumably, some neurotrophic materials (Chapter 14) supplied from the proximal end of the fiber serves to maintain the distal amputated portion of nerve fiber in a viable state.

2. Loss of Excitability During Wallerian Degeneration

The time at which excitability is lost in the course of Wallerian degeneration has been reported to differ in various nerve fiber groups. The visceral afferents of the monkey were found to become inexcitable within 50–60 hr after nerve transection, followed 5–10 hr later by the loss of electrical excitability of the larger somatic nerves (Heinbecker *et al.*, 1932). However, the onset of inexcitability may not be simply related to nerve fiber size in that myelinated autonomic fibers were seen to become inexcitable before the unmyelinated fibers.

The large motor fibers were found to lose their excitability all along their whole length at the same time (Gerard, 1932; Young, 1949b; Gutmann and Holubář, 1950). However, we can now appreciate that those experiments had been carried out at times too long after nerve transection to have revealed a rapid proximo-gradient of loss of excitability. Rosenblueth and Del Pozo (1943) had in fact reported a proximo-distal gradient in the loss of excitability following transection. Several factors may be responsible for the loss of excitability; a separation of myelin from the nodes, the loss of some specific component required by the axonal membrane such as channel proteins or pumps, or, a general change in the membrane.

An increase in nerve activity was said to lead to an earlier loss of excitability in degenerating nerves (Cook and Gerard, 1931). The sciatic nerves of dogs were cut and the distal portion of nerve on one side given a repeated series of electrical stimulations *in vivo,* with contractions of the foot used to indicate adequate nerve excitation. Contractions in response to stimulation were found to decline and then fail some 50 hr after cutting the nerve. This occurred at a somewhat earlier time in nerves repeatedly stimulated than in unstimulated nerves. Those experiments, however, were complicated in that a discrimination between changes in the nerve augmenting Wallerian degeneration and a failure of neurotransmission in the course of degeneration could not be made. In similar experiments, Titeca (1935) recorded the muscle responses, and as well, nerve action potentials in the distal amputated nerves. He found neuromuscular transmission in the cat to fail 2 days after nerve transection, while the nerve remained excitable for an additional day before action potential responses failed. Those experiments showed that neurotransmission was more susceptible to failure than the general loss of nerve excitability due to Wallerian degeneration. This experiment was repeated (Braden and Ochs, unpublished experiments) and the earlier failure of neuromuscular transmission before that of nerve excitability was confirmed. Additionally, neuromuscular transmission was seen to fail sooner when the sciatic nerves were cut closer to the muscle than at a higher level. From the distance between the cut positions and the time of loss of neuromuscular transmission at the two positions, a rate of some 60 mm/day was estimated for the velocity of failure to occur, a rate close to that found for the apparent rate of degeneration of the larger nerve fibers as determined by Lubińska (cf. Section 1). This would suggest the depletion of some material contained in the larger motor fibers as the factor responsible for the loss of neuromuscular responses. The apparent rate of 60 mm/day is lower than the rate of loss of miniature end plate potentials (MEPPS) in rat diaphragm muscle recorded by Miledi and Slater (1970). On comparing the loss of MEPPs in muscles with their motor nerves cut close to and far from the muscle, the higher rate of 360 mm/day found by Miledi and Slater is one close to that of fast transport. Those studies will be described in more detail in Chapter 13 where the relation of transport to events occurring at the nerve terminals in connection with neurotransmission will be taken up.

B.   RETROGRADE REACTION (CHROMATOLYSIS)

1.   Morphological Changes During Chromatolysis

Along with the loss of staining of the Nissl particles in the cytoplasm of the cell, the classical signs of chromatolysis include swelling of the nerve cell bodies and changes in the position of the nucleus and the staining of the nucleolus. Such chromatolytic changes typically appear after a lag period of some days following sectioning of the nerve fiber and gradually increase to a maximum over a period of several weeks, with a slow return to normal appearance over a period of months. All these changes may not be seen. Some nerve cells may show little swelling, or even a shrinkage (LaVelle and LaVelle, 1958), and cell death may occur, as is commonly the case in the neurons of the newborn (Fig. 6.5).

Cell death may be the more common occurance in certain neurons such as those of the trigeminal nucleus (Aldskogius and Arvidsson, 1978). In neuron populations where chromatolysis appears there is usually a marked variation in the degree of chromatolysis seen. Such variation in an apparently homogeneous population of cells suggests the possibility that there are cyclic variations in their metabolism. A factor which determines the degree of chromatolysis seen is the distance from the cell bodies at which the fibers are interrupted. When the fibers are cut closer to the cell this often results in a more profound chromatolysis or in cell death (Van Gehuchten, 1903).

Another variation in the pattern of the retrograde reaction related to cell type is that some cells, for example those of the facial nuclei of mice, show an early increase in the degree of Nissl staining, a basophilia, following transection of their fibers rather than a chromatolysis (Torvik and Skjörten, 1971). An increased basophilia was also seen in goldfish retinal ganglion

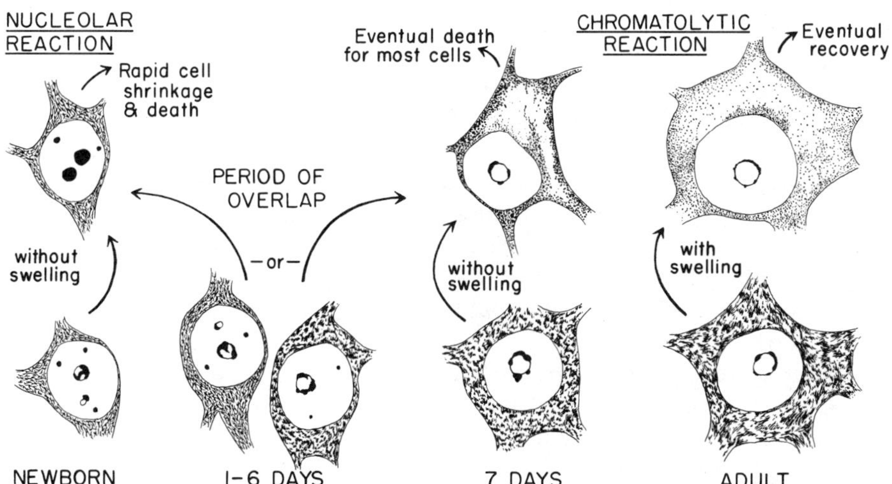

Fig. 6.5 Nucleolar and chromatolytic changes in the nerve cell body. Schematic representation of the types of nerve cell reaction to cutting of its axon are shown at different postnatal ages. The Nissl substance is stained with thionin, the nucleolar apparatus by the Feulgen technique. In the cells of the newborn, 1–7 days old, cell death is common following a transection of the nerve fiber, whereas in the adult, recovery from chromatolysis eventually occurs. From LaVelle and LaVelle (1958).

cells after sectioning optic nerves (Murray and Grafstein, 1969). The fundamental change which occurs during chromatolysis is a dispersion of the ribosomes. Short segments of granular ER are found rather than ribosomes associated along the length of the usually long parallel profiles of the ER cisterns as is characteristic of normal cell bodies. And, an increase in neurofilaments may appear as schematized by Pannese (1963). An increase in microtubules and rough ER also occurs (Price and Porter, 1972).

In neurons with several branching daughter fibers, cutting one of the branches may result in only a mild degree of chromatolysis, or none at all. An interesting example of this is seen in the dorsal root ganglia neurons where little or no chromatolysis appears after cutting the dorsal root, while the usual chromatolysis occurs on transecting the peripheral nerve branch (Hare and Hinsey, 1940; Lieberman, 1976). This could well be the result of the routing phenomenon (Chapter 11). As will be discussed, some 3–5× more material is transported in the sensory fiber branch as compared to the dorsal root branch. If we presume that the amount of material carried back to the cell body by retrograde transport is also proportionally greater in the sensory branch than the dorsal root branch, this may explain the difference in results on cutting the two sets of branches. The cell is supplied with sufficient retrogradely transported material from the sensory branch so that the loss of the dorsal root branch is relatively small. The relation of retrogradely transported materials to chromatolysis will be discussed more fully in Section 4 below.

2. Nucleic Acid and Protein Changes During Chromatolysis

An early hypothesis advanced to explain chromatolysis was that increased amounts of nucleic acids are carried from the cells into the injured fibers for their regeneration with, as a result, their loss from the cells (Hydén, 1943). This concept was based on findings made by Hydén and his colleagues using the micro-spectrophotometric technique of Caspersson to assess the nucleic acid content of neuron cell bodies. Sections of tissue containing the cell bodies are placed under a microscope provided with a small spot of UV light so that nucleic acid absorption could be assessed over each small region of the cell body. Within the first 10 days after transecting hypoglossal nerves, an apparent fall in the level of RNA in the cytoplasm of the motoneurons was seen, presumably RNA needed for new protein production which had passed into the regenerating nerve fibers. In actuality, RNA is not lost from the cell bodies of chromatolytic neurons. The cell bodies take up water and become swollen (Gersch and Bodian, 1943), the swelling causing a decrease in the concentration of nucleic acid with the absolute amount of RNA remaining unchanged (Brattgård, Edström, and Hydén, 1957; Hydén, 1960). The relative changes in cellular volume, protein, and RNA content early in the course of chromatolysis are shown in Fig. 6.6.

During the early or "latent" period occurring soon after transecting the nerve fibers, little apparent change in cell volume or in the level of RNA was

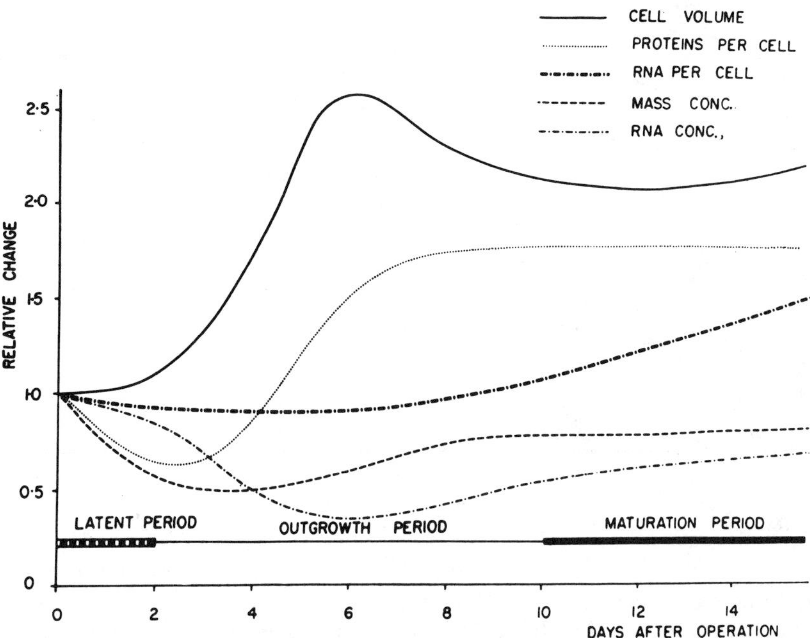

Fig. 6.6 Time course of chromatolysis. Cell constituents in the soma; protein, RNA, and the concentration of RNA and cell materials are shown at different times after section of the axon. A large increase of cell volume occurs during the "outgrowth" period. From Brattgård, Edström, and Hydén (1957).

seen. Only during the "outgrowth" phase, when the fibers are regenerating (Chapter 15), do RNA and protein levels increase. In the following "maturation" period, RNA levels and protein remain high and only later, when the regenerating nerve fibers have reached their target cells and mature, do the protein and RNA levels in the cell revert back toward normal (Fig. 6.7).

Using biochemical methods, Watson (1968) found an earlier increase in nucleic acid in the nucleolus, one beginning about 2 days following nerve interruption (Fig. 6.8).

The earliest and greatest amounts of increased nucleolar nucleic acid were seen when nerve lesions were made close to the cell bodies as compared to those made further distally.

The increase was shown to be due to RNA, as indicated by the effect of ribonuclease to block it. The temporal sequence of a movement of RNA from the nucleolus to the cytoplasm of the perikaryon was further shown by the injection of ^{3}H-orotic acid, a precursor of RNA. In tissues taken at different times after injection of the precursor and prepared for autoradiography, the nerve cells showed grains of radioactivity located first over their nucleus, and then later over the cytoplasm (Fig. 6.9).

The incorporation of ^{3}H-uridine into RNA in the nucleus and cytoplasm changes as a function of time after nerve injury (Fig. 6.10).

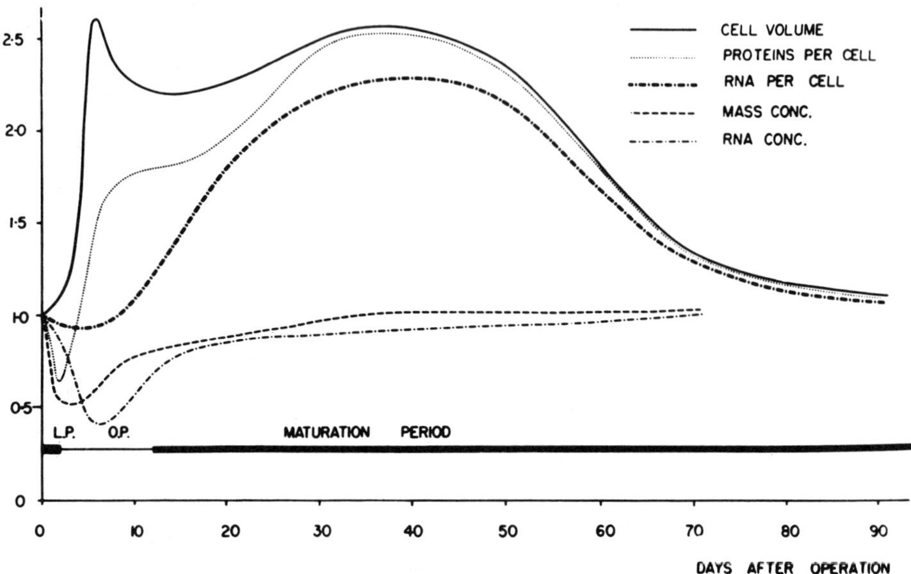

Fig. 6.7 Later time course of chromatolysis. The time scale in this figure is longer than that shown in Fig. 6.6. In the later stages of regeneration (30–60 days) a large increase in RNA and protein of the cell body is observed with a return to normal levels toward the end of the maturation period. From Brattgård, Edström, and Hydén (1957).

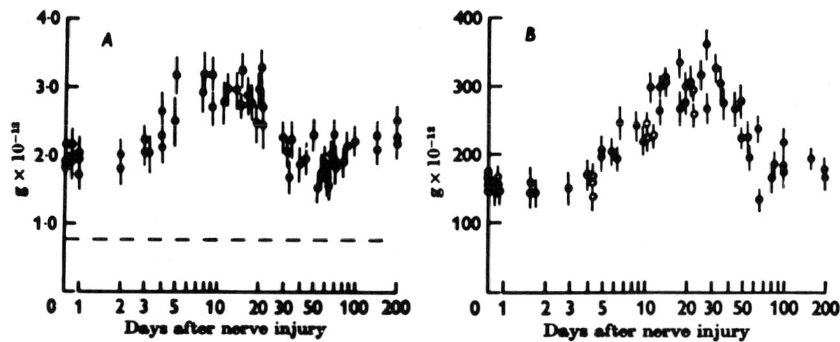

Fig. 6.8 Early nucleolar nucleic acid changes after nerve injury. (A) The nucleolar nucleic acid in the nerve cell body is shown at the time (in days) indicated on the abscissa following nerve injury. The horizontal dashed line represents the nucleic acid content of nucleoli that resisted digestion with ribonuclease. (B) Total cell body nucleic acid (○) and RNA extracted from nerve cell bodies with ribonuclease (●) is shown. The standard error of the mean is indicated by the vertical bars. From Watson (1968).

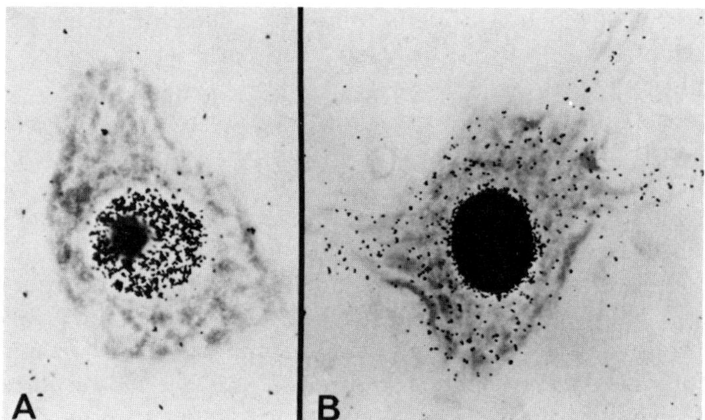

Fig. 6.9 Location of labeled RNA. (A) Incorporation of the RNA precursor ^{3}H-orotic acid is followed 15 min after orotic acid injection by its incorporation in the nucleus and nucleolus as shown by autoradiography. (B) Four hours later a greater amount of labeled radioactive RNA is present in the perikaryon and in the dendrites. After Schubert and Kreutzberg (1975).

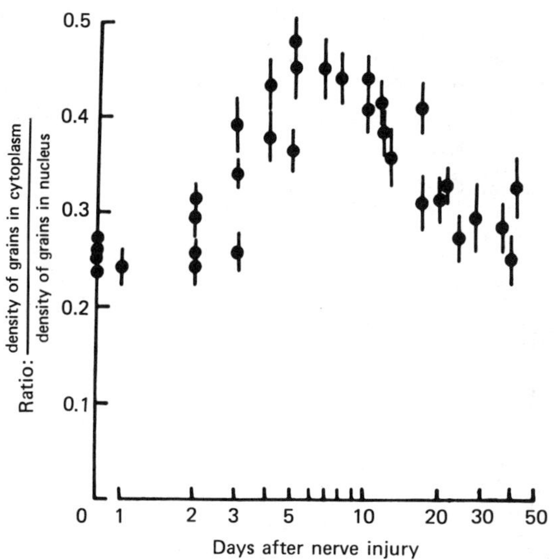

Fig. 6.10 Shifts in RNA after nerve injury. The ratio of radioactive grains in the cytoplasm seen in autoradiographs with respect to the nucleus 8 hr after injecting 5-^{3}H-^{7}uridine is shown over a period of days after the nerve was crushed. The increase in the ratio at times from 3 to 20 days follows an earlier, relatively greater increase in nucleolar labeling (cf. Fig. 6.8). From Watson (1968).

The amount of labeling of cells following uptake of ^{3}H-leucine was seen by Haddad *et al.* (1969) to have an even earlier appearance after nerve section (Fig. 6.11).

These observations are in accord with a shift toward an increased protein synthesis early in the course of chromatolysis. Similar augmented incorporation of ^{3}H-leucine into proteins in the cell bodies was found, for example, in the rabbit dorsal root ganglia (Miani *et al.*, 1961; Francoeur and Olszewski, 1968) and goldfish retinal ganglion cells (Murray and Grafstein, 1969). However, an increase in the labeling following incorporation of labeled amino acids may not always be a good indicator of the level of proteins synthesis following nerve transection. In some cases, an apparently diminished incorporation may occur (Engh *et al.*, 1971), the result of a greater relative degree of proteolysis as compared to synthesis (Watson, 1968).

3. Downflow of RNA and Polyamines in Nerve Fibers

While appreciable amounts of nucleic acids are not normally transported in mature axons, evidence for the transport of some RNA species has been obtained using radioactive-labeled precursors of the nucleic acid. On injection into goldfish eyes, labeled RNA was isolated from the tectum (Ingoglia *et al.*, 1973). Austin *et al.* (1966) and Bray and Austin (1968) had reported a slow outflow of RNA in the acid-soluble fraction of the sciatic nerve following the injection of ^{14}C-orotic acid into the spinal cord of the chicken. However, as Jeffrey and Austin (1975) pointed out, the pattern of RNA outflow differed from that of labeled proteins and could possibly be accounted for by diffusion. They found labeled RNA present between the fibers in autoradiographs of nerves taken after ^{14}C-orotic acid injections in contrast to the intra-axonic location of labeled proteins after injecting ^{3}H-leucine. The diffusive spread of the labeled orotic acid in the endoneurial space of the

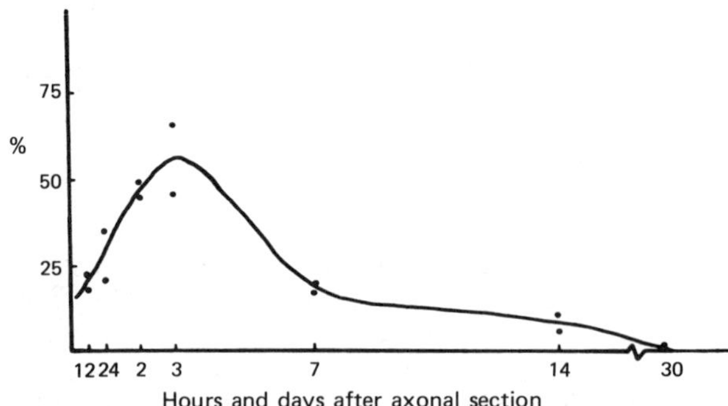

Fig. 6.11 Early increase of protein in cells. The number of silver grains per cell representing labeled protein increases soon after axonal section. From Haddad *et al.* (1969).

nerve trunk, a compartment readily accessible to the spinal cord, would be followed by its local incorporation by the Schwann cells. Yet another route was suggested by Autilio-Gambetti *et al.* (1973) and Gambetti et al. (1973). They considered the RNA precursor to be transported down within the axons and then to move out of the axons to become incorporated by the Schwann cells. This possibility is experimentally difficult to discriminate from an adventitious diffusion of the precursor in the endoneurial space with its subsequent incorporation by the Schwann cells.

The downflow into the optic nerve of ^{3}H-guanosine, ^{3}H-uridine, or ^{3}H-orotic acid injected into goldfish eyes was differentiated from the slower-moving RNA into which these labeled precursors are incorporated (Ingoglia *et al.*, 1973). Evidence that the RNA is in fact transported in the nerve fibers, was gained by making crushes in the optic nerve and finding 5 days later RNA accumulation at the crush with a block of its transport beyond that point. When regeneration commenced, a close relation was seen between the position of labeled RNA carried into the regenerated fibers and that of labeled proteins using injections of ^{3}H-proline. When the regenerating fibers reached the tectum, autoradiographs made after eye injections of ^{3}H-uridine showed grains present within the tectal layers known to receive the terminations of the optic nerve fibers (Ingoglia *et al.*, 1975).

Relatively more RNA is transported in regenerating as compared to normal axons (Ingoglia and Sturman, 1978). It is interesting that the 4S species of RNA, one close to tRNA, is transported in the regenerating fibers and not that of ribosomal RNA (Ingoglia, 1979). This 4S species of RNA is also found in squid and *Myxicola* axons (Lasek *et al.*, 1973), as well as in the rat optic nerve (Brink and Karnovsky, 1973). The polyamines putrescine, spermidine, and spermime are believed to control the growth and poliferation of cells by their action on nucleic acids, particularly on RNA (Cohen, 1971). A fast axoplasmic transport of ^{3}H-putrescine was found in the regenerating optic nerves of goldfish but not in intact mature nerves (Ingoglia *et al.*, 1977). The polyamine ^{3}H-spermidine was transported at an intermediate rate in both intact and regenerating optic nerves and ^{3}H-spermine was found to be transported in both intact and regenerating nerves at a fast rate. The preferential transport of putrescine in regenerating nerves may account for the report that this polyamine is transported in the nerve fibers of the immature Zebra fish (Fischer and Schmatolla, 1972), but that it was not seen to be transported in the fibers of the mature rat (Siegel and McClure, 1975).

At present, little is known regarding the role of polyamines in nerve and the significance of the transport of some polyamine species and not that of others in normal as compared to regenerating nerve fibers remains obscure.

4. The Signal for Chromatolysis and Changes in Transported Species

How does the cell body ''know'' that its axon is cut, in other words, what is the ''signal'' which initiates the sequence of changes occurring in the chromatolytic cell? Nittono (1923) and Abelous and Lassalle (1928) proposed that products derived from the degenerating nerve pass into the circulation to be carried to the spinal cord. There, through the recognition of those substances by the nerve cell body, chromatolysis is initiated. This theory was tested by removing all of the distal part of the transected sciatic nerve along with the tissues of the peripheral hind-leg field it innervates (Ochs *et al.*, 1961). The usual pattern of chromatolysis was found in the nerve cell bodies. Further, pieces of amputated nerves implanted in normal animals did not induce chromatolysis. Those experiments gave no support for the concept that some blood-circulated substance acts as a signal for chromatolysis. Instead, the experiments suggested that some factor normally ascending in the fibers to the cell bodies was missing when the fibers were cut off, this acting as the signal for chromatolysis. If, normally, the rate of protein synthesis in the cell body is repressed by some substance returning to the cells by retrograde transport, its loss would result in a derepression and an increased level of protein production. Alternatively, some ascending signal could initiate an augmented synthesis. Support for that possibility was the finding that actinomycin D, which blocks synthesis by acting at the DNA-RNA step, prevents chromatolysis when given just before nerve transection (Torvik and Heding, 1967).

Derepression of protein synthesis may not be the first step in chromatolysis; the signal might arise from the cytoplasm (Gurdon and Brown, 1965). As another hypothesis, Cragg (1970) suggested that an interruption of the nerve fibers might cause an increase in the rate of fast axoplasmic transport, which in turn could lead to a depletion of proteins from the cell body, the depletion acting as the "signal" to initiate an increased protein synthesis and chromatolysis.

A first attempt to determine if there are changes in the rate of transport during chromatolysis was made using ^{32}P as the precursor to study transport (Ochs *et al.*, 1960). The precursor was injected into the region of the L7 motoneurons of the cat spinal cord and changes in its outflow assessed in the L7 ventral roots at various times after initiating chromatolysis by cutting the sciatic nerve in the thigh. The changes in outflow seen in the fibers of the ventral roots on the transected side as compared to controls were irregular, and gave little support for an increased rate or synthesis in the course of chromatolysis. This was again studied with an improved technique with clearly no change in the rate of axoplasmic transport seen during chromatolysis (Ochs, Dalrymple, and Richards, 1962; Ochs and Johnson, 1969; Johnson, 1970). On the other hand, using the goldfish visual system, and measuring the accumulation of labeled activity in the optic tectum, Grafstein and Murray (1969) reported a two-fold increase in the rate of fast transport and a three-fold increase in the rate of slow axonal transport following nerve transection. Carlsson, Bolander, and Sjöstrand (1971) also reported a significant increase in the rate of fast transport in cat roots 4 weeks after transection and no change in the rate of slow transport. The latter finding was in accord with a subsequent study of Grafstein (1971), who found no difference in the rate of slow transport in chromatolyzed as compared to control goldfish nerves. Kreutzberg and Schubert (1971) reported a significant increase in the amount of labeled proteins carried down in nerves after distal transections by fast transport but could not ascertain any change in the rate.

Apparent changes in the rate based on accumulation studies might reflect differences in the amounts of labeled materials transported rather than rate changes *per se*. In view of that possibility, and to accurately assess any possible small changes in rate resulting from chromatolysis, the advance of the front of fast-transported labeled proteins was used to compare transport in those nerves with normal ones (Ochs, 1976). To initiate chromatolysis, cat sciatic nerves were transected in the popliteal region of the hind limb below the level at which the nerves were subsequently taken to determine fast transport. No effect on the rate of axoplasmic transport was seen in nerves examined over a period of 6–165 days following nerve interruption as compared to the control nerves on the contralateral side. This is indicated by the example shown where outflow was examined 25 days after initiating chromatolysis (Fig. 6.12).

The position of the fronts and shape of the outflow in control and chromatolytic nerves are closely similar. The average rates of 432 ± 34 (S.D.) mm/day determined for the controls and 424 ± 33 (S.D.) mm/day for

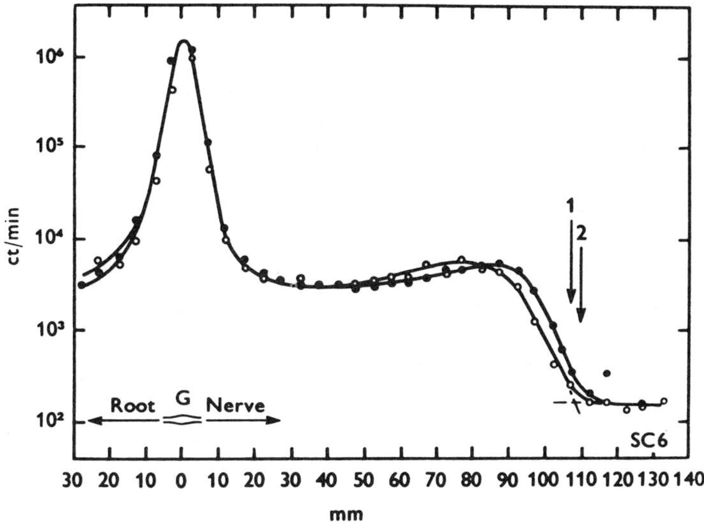

Fig. 6.12 The lack of effect of chromatolysis on rate of fast transport. After section of the sciatic nerve 25 days beforehand (●), the rate and the overall amount of fast-transported labeled proteins in the crest is similar to that of the control (○) as shown following equal injections of their L7 dorsal root ganglia with ³H-leucine. From Ochs (1976).

chromatolytic neurons were not significantly different. The similarity in rates gave no support to the "depletion" hypothesis.

Differences in the total amount of fast-transported components in the chromatolytic neurons as compared to normal neurons were looked for by comparing the amount of labeled activity present in their crests (cf. Fig. 6.12). These were also not found to be significantly different, indicating that the overall amounts of fast-transported proteins are not much changed in chromatolytic neurons. This does not mean, however, that synthesis remains unaffected. The level of AChE for example, falls greatly in the chromatolytic motoneurons of the spinal cord after nerve transection (Schwarzacher, 1958), with a corresponding profound decrease of AChE in the motor fibers of the ventral roots (Johnson, 1970). Recovery of AChE back to control levels occurs only after a prolonged period of time (Ranish and Ochs, 1972). The change in the outflow of AChE indicates a shift in the pattern of synthesis in the chromatolytic motoneurons. However, insofar as AChE represents only a small portion of the total protein synthesized, the overall level of labeled proteins carried down in the crest by fast transport is not appreciably altered. Synthesis and transport of other specific components besides AChE are most likely also altered as the result of changes in the pattern of synthesis in chromatolytic neurons. A differential shift in chromatolytic neurons was indicated by the decrease in AChE and choline acetyltransferase transported with a rise in ACh (O'Brien, 1978). Similar selective changes have also been reported by Reis *et al.* (1974) for CNS neurons. On transecting the adrenergic fibers supplying the hypothalamus, the activities of enzymes related to NA synthesis show changes in their cell bodies located in the locus coeruleus. An early brief rise in DBH and TH activities (Chapter 4) was seen several days after fiber transection followed by a decrease to 50% of normal with a more gradual fall in the level of NA. The levels of all these components returned to normal after several weeks. A similar early brief rise and fall in TH was seen in the dopaminergic cells of the nigrostriatal neurons during chromatolysis with a return to normal levels in several weeks (Reis *et al.*, 1978). These and similar studies indicate a shift in the program of synthesis in cell bodies occurring during chromatolysis.

While little overall difference in the various polypeptide species separated by SDS polyacrylamide electrophoresis was apparent in chromatolytic as compared to normal nerves, changes of a few polypeptide species were reported (Theiler and McClure, 1978). Dorsal root ganglia were exposed to ^{3}H-proline *in vitro* and the levels of labeled proteins in control dorsal root ganglia and ganglia whose fibers were previously cut were compared. The same amount of the precursor ^{3}H-proline appears to be taken up and incorporated by the control and the chromatolytic cell bodies and subsequently transported. Similarly, after labeling with ^{32}S-methionine, nerves analyzed by SDS-PAGE showed that 26 labeled polypeptides identified in control and chromatolyzed sensory nerves remained the same. One small peak representing a 42,000-dalton component appeared 5–15 days after nerve transection and a 19,000-dalton component disappeared. Two other protein peaks were seen to disappear from the chromatolyzed neurons after 50 days.

The initiation of chromatolysis has been explained by the loss of a "signal" substance which normally is carried to the cell by retrograde transport, a substance acting to repress synthesis. Another hypothesis advanced by Bisby and Bulger (1977) is that retrograde transport in the injured nerves carries substances back to the cell bodies at an earlier time and in excess amount to trigger chromatolysis. Nerves were injured to initiate chromatolysis, the cell body region injected with ^{3}H-leucine and labeled proteins were allowed to be transported down to the site of injury before making an upper crush. The anterogradely transported labeled proteins were turned around at the lower injury site and then, following their retrograde transport, labeled materials were collected just below the upper crush. The crush sites are diagrammed in Fig. 6.13.

To study the process in motor fibers, the rat L5 motoneuron region in the cord was injected with ^{3}H-leucine. The labeled material was carried down to the lower injury site in the motor fibers (made at different times beforehand) where it undergoes a turn-around and is then carried back by retrograde transport to accumulate just below the upper crush. The labeled proteins appear in greater amounts at earlier times after initiating chromatolysis some time beforehand as compared to normal nerves not previously injured (Fig. 6.14).

A similar phenomenon was seen in sensory nerves after injecting the rat L5 dorsal root ganglion with ^{3}H-leucine following the same sequence of first making a lower injury followed by a crush made at a higher site to collect the labeled material carried back by retrograde transport (Fig. 6.15).

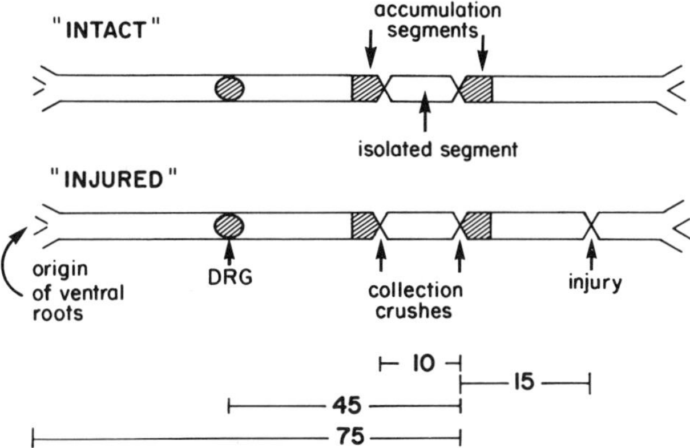

Fig. 6.13 Schematization of nerve crushes used to assess retrograde transport. Rat sciatic nerve is diagrammed to show the locus of nerve crushes made to study retrograde transport by the accumulation of labeled proteins at the crushes. From Bisby and Bulger (1977).

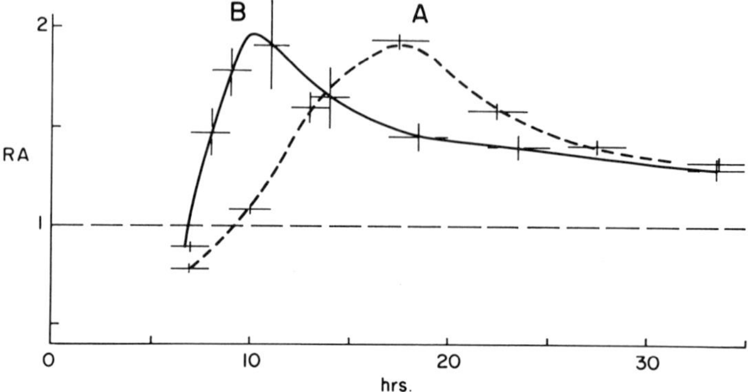

Fig. **6.14** Retrograde transport in crushed motor nerves. Accumulation of labeled proteins at collection sites below crushes (cf. Fig. 6.13) is earlier after a preceding crush injury (B) than in controls (A). Horizontal axis shows hours after ^{3}H-leucine injection into the cord. Each data point covers the 2- or 3-hr period of accumulation, and shows the mean ± S.E.M. of at least five determinations. From Bisby and Bulger (1977).

The total amounts of labeled materials transported in the anterograde direction in the normal and chromatolyzed sensory neurons down to the distal crushes were seen to be closely similar, while the amounts of retrogradely transported radioactivity in motor and sensory nerve fibers were greater in the chromatolyzed as compared to the normal neurons at the upper collection crush.

Chromatolysis is diminished when axons are interrupted at sites further from the cell bodies. This was explained by Bisby as due to more of the materials retrogradely transported undergoing a drop-off locally in the fibers (Chapter 11) in the course of their ascent. Thus, with longer lengths of

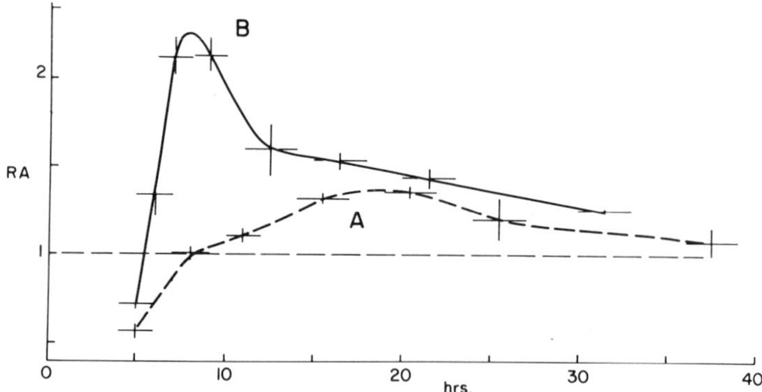

Fig. **6.15** Retrograde transport in injured sensory axons. The earlier return in sensory fibers following a crush injury is seen to occur as in the case of motor fibers (cf. Fig. 6.14). From Bisby and Bulger (1977).

nerve, less material would be returned to the cell bodies to induce chroma-tolysis. In support of this possibility, larger amounts of labeled material were found retrogradely transported when crushes were made closer to the cell bodies.

Thus, two logical possibilities remain to account for chromatolysis, the loss of some signal substance which is normally returned by retrograde transport from the distal length of nerve, or the return to the cell bodies of an excess amount of some material. The recent studies of Dziegielewska *et al.* (1980) on transport of phospholipids indicate that the signal from the site of injury must be carried retrogradely at a rate of at least 140 mm/day. This estimate fits well with a number of studies indicating a retrograde transport rate of approximately 200 mm/day (Chapter 3).

Another finding pointing to a special role of retrograde transport has been obtained with the use of 2-deoxyglucose (2-DG). One of the first signs of chromatolysis, appearing as early as 24 hr after axotomy, is an increased uptake of 2-DG by the cell bodies (Singer and Mehler, 1980; Kreutzberg and Emmert, 1980). Application of colchicine to those nerves blocked the in-creased uptake of 2-DG (Fernandez *et al.*, 1981). This suggests that some special substance from interrupted nerves transported retrogradely may be the signal initiating chromatolysis (Ochs *et al.*, 1961; Cragg, 1970).

7

Metabolism and Energy Supply for Transport

The possibility that fast axoplasmic transport is due to a passive diffusion of materials in the fibers (Magid, Fischer, and Schmattola, 1973) is made improbable by the unchanged shape of the front of labeled proteins as it moves down the nerve over considerable distances at a linear rate (cf. Nadelhaft, 1974; Copeland, 1976a,b). Additionally, transported components are not sorted out on the basis of their size or molecular weight along the length of a long nerve (Chapters 5, 10, and 11) as would occur with diffusion. A further strong argument against diffusion was the finding that transport is closely dependent on oxidative metabolism, the subject of this chapter. The requirement for oxidative metabolism was first indicated by the early failure of transport in nerve occurring within 15–20 min after interrupting its blood supply (Ochs and Ranish, 1970; Ochs and Hollingsworth, 1971). The further analysis of the relation of metabolism to transport was made possible by the study of nerve *in vitro* where the effects of metabolic blocking agents could be determined (Ochs, 1974a; Ochs, 1981a,b). Many of those studies were first carried out using sheathed nerves. While the perineurial sheath acts as a permeability barrier for a number of agents, a subject to be fully discussed in Chapter 8, the presence of the sheath does not modify the results obtained with metabolic blocking agents, as will be described in this chapter.

A. CIRCULATION AND TRANSPORT *IN SITU*

1. Vascular Supply of Nerve

Nerve trunks are highly vascularized by a number of small blood vessels, the *arteriae nervori*, which enter at various points along the nerve trunk to supply it with blood (Fig. 7.1).

The *arteriae nervori* supply a network of anastomosing blood vessels in the epineurial surface of the nerve (Adams, 1942; Lundborg and Brånemark, 1968; Sunderland, 1978). From these surface vessels arteriolar branches spring to form a continuous longitudinally oriented vascular network. Lateral

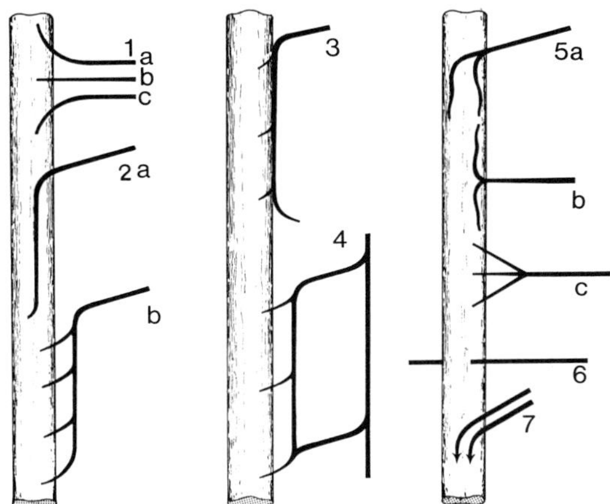

Fig. 7.1 Local arterial supply of nerve. The *arteriae nervori,* which arise from the larger blood vessels which often accompany nerve trunks, have various patterns as shown in these diagrams. The vessels penetrate the sheath or run along the nerve in branches or loops. From Sunderland (1978).

connecting branches extend down to supply a network of vessels on the perineurial sheath covering the nerve bundles or funiculi. In turn, fine pre-capillary arterioles branch from the surface vessels of the perineurium to pass transversely through the sheath. These arterioles terminate in a capillary network inside the funiculus, the endoneurial space contained within the perineurial sheath, as diagrammed in Fig. 7.2.

The capillary endothelium is the main permeability barrier between the circulation and the nerve fibers in the endoneurial space *in vivo* (Chapter 8).

2. Elimination or Occlusion of Blood Supply

The dependence of transport on blood circulation was shown by cutting the aorta to quickly bleed animals. This was done at various times during the course of fast axoplasmic transport initiated by injecting the L7 dorsal root ganglia with ^{3}H-leucine. The sciatic nerves were left *in situ* for a further period of time and then removed to determine the distance to which the front of labeled radioactivity had moved after the nerve had been made anoxic by bleeding (Ochs, Sabri, and Ranish, 1969). The fronts of radioactivity were found to have moved down the nerve for no more than a distance of 5–10 mm at most after bleeding, a distance representing less than 15–30 min of transport during the anoxic state (Fig. 7.3).

The dependence of transport on a maintained blood supply was also shown by devascularizing a portion of the nerve. Simply stripping the *arteriae nervori* from the sciatic nerve over a long length of sciatic nerve has surpris-ingly little effect on transport because of the extensive collateral circulation

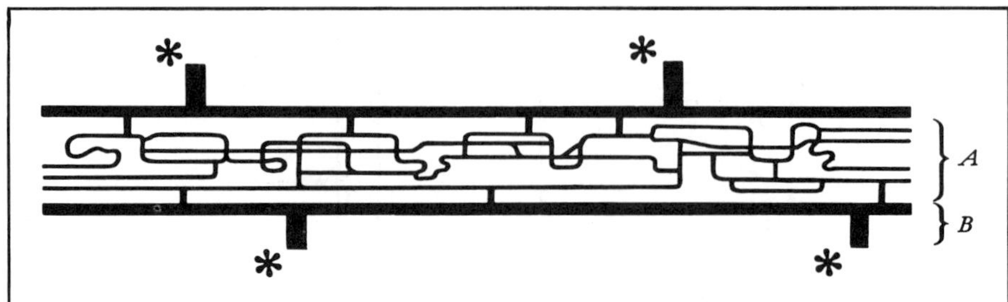

Fig. 7.2 The microvascular architecture within a nerve funiculus. A schematization of peripheral nerve showing the relationship between intrinsic and extrinsic systems. In A, the intrinsic system, the vessels in the funiculus are continuous throughout the entire length of the nerve and consist mainly of capillaries. In B, the extrinsic system the *arteriae nervori* supplying the epineural vessels (✻) have numerous anastomoses along the entire length of the nerve as is also indicated in Fig. 7.1. From Lundborg and Brånemark (1968).

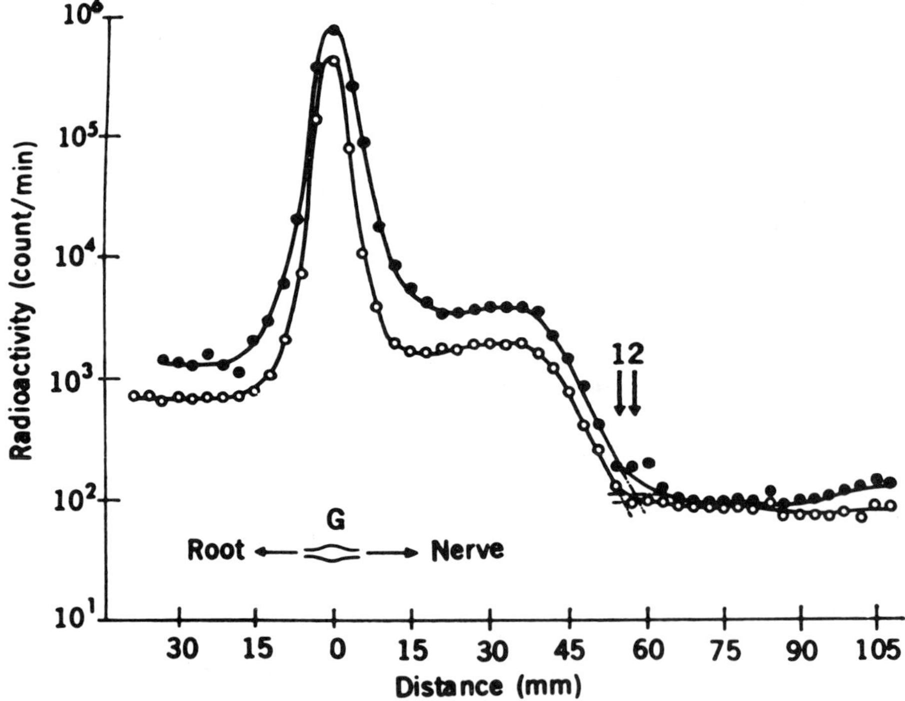

Fig. 7.3 Transport block following anoxia. Three hours after injecting the L7 cat dorsal root ganglia with ^{3}H-leucine the sciatic nerve on one side was removed (○) and its front of outflow is represented by (arrow 1). The nerve on the right side (●) was left *in situ* after circulation was stopped for an additional 3 hr. Little additional transport in the anoxic state occurred as indicated by arrow 2. From Ochs and Ranish (1970).

present within the funiculi of the nerve trunks. To completely interrupt the blood supply, a major arterial vessel supplying the nerve, the inferior gluteal artery, has to be occluded as well (Ochs, 1975d). After nerve devascularization and occlusion of the inferior gluteal artery, labeled proteins were seen dammed up above the devascularized part of the nerve, with a failure of transport into that region (Fig. 7.4).

Another method used to interfere with the blood supply *in vivo* is to compress the hind limbs of cats with blood pressure cuffs raised to pressures above systolic blood pressure. In the human, the ischemia in the arm cuffed at high pressures causes motor and sensory functions to be lost within about 17 min (Lewis, Pickering, and Rothschild, 1931). Compression of the hind limbs of the cat with pressure cuffs is less easily done than in humans

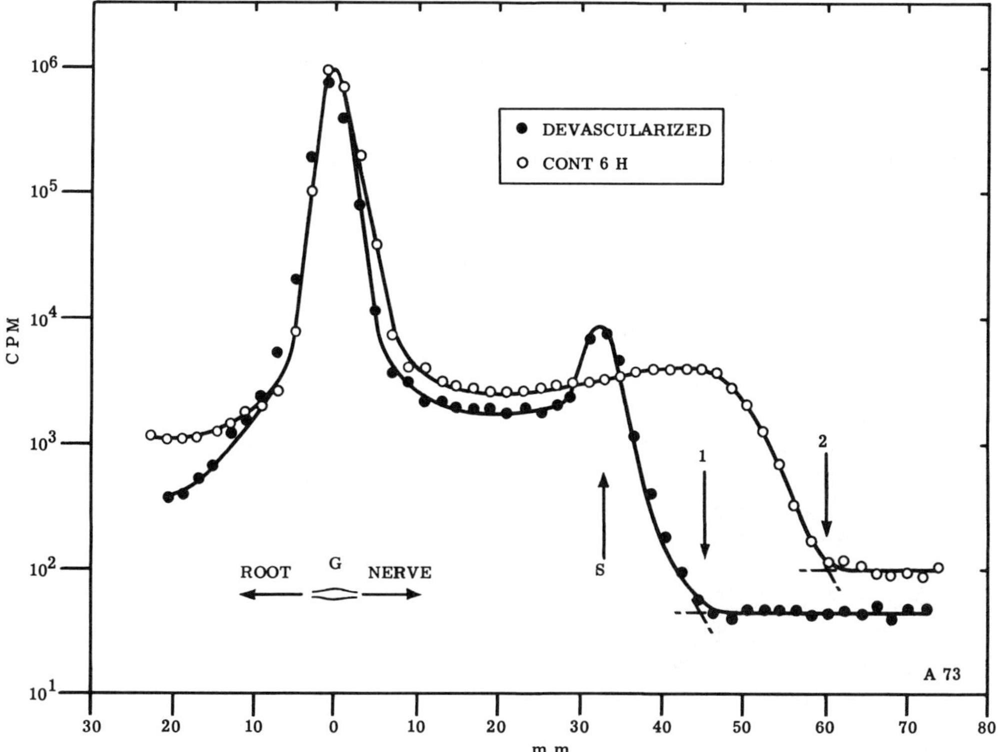

Fig. 7.4. Failure of transport into devascularized sciatic nerve. After stripping nutrient vessels from the sciatic nerve and additionally clamping off the blood supply from the inferior gluteal artery at the sciatic notch to the nerve, transport following injection of the L7 dorsal root ganglia with ³H-leucine is blocked (●). Arrow S shows stripping from the point indicated distally, with a block of downflow appearing in the devascularized region at arrow 1. Arrow 2 indicates normal (○) downflow in the vascularized nerve of the other side for the downflow time of 6 hr allowed. From Ochs (1975d).

because of the tapered shape of the cat hind limb and the protection afforded to the blood vessels by the bones (Denny-Brown and Brenner, 1944). Relatively high cuff pressures of 250 mm/Hg were needed to cause the block of nerve conduction seen after about 30 min (Bentley and Schlapp, 1943). A block of action potential responses using cuff compression occurring in about this same period of time was also found by Frankenhaeuser (1949) in the hind limbs of rabbits.

To produce compression anemia of the hind limbs of cats, blood pressure cuffs were placed high on the thigh and pressures of 300 mm of Hg were applied. This higher pressure was found necessary to consistently cause a block of fast axoplasmic transport (Fig. 7.5.).

A damming of labeled activity was seen above the compressed region with no entry of labeled activity into that region. The block is considered due to anoxia, although there is some question as to whether in addition, a mechanical occlusion of the nerve may be responsible for some of the block of axoplasmic transport seen. A similar consideration applies to the use of an increased intra-ocular pressure to block transport from the retinal ganglion cells (Editorial, 1976; Quigley, Guy, and Anderson, 1979). Some *in vitro* and *in situ* experiments dealing with the reversibility of the block by anoxia and compression bearing on this point will be discussed below in Part E.

B. *IN VITRO* STUDIES OF OXIDATIVE METABOLISM

1. Oxidative Metabolism

To study the connection between oxidative phosphorylation and transport *in vitro*, a standard method was followed. The L7 dorsal root ganglia were injected with ^{3}H-leucine and transport of labeled components into the fibers allowed for several hours before the nerves were removed from the animals and placed into chambers oxygenated with 95% O_2 + 5% CO_2 for a further period of *in vitro* transport. The maintained movement of the labeled materials which had gained entry to the nerve fibers in the animal shows that *in vitro* transport has taken place. The front of labeled radioactivity had its usual pattern and moved at the usual rate of fast transport for times up to 6 hr or more. This occurs without the need of an exogenous supply of glucose added to the medium, indicating that an adequate endogenous source of metabolites is present in the nerve fibers (Ochs and Ranish, 1970). Similar long-lasting transport *in vitro* was also found by Kirkpatrick *et al.* (1973), Banks and Mayor (1972), Edström and Hanson (1973a), and others.

The close dependence of transport on a continued supply of oxygen was shown by comparing downflow in nerves placed in a chamber filled either with 95% O_2 + 5% CO_2 or an atmosphere of N_2. The oxygenated nerve shows the normal pattern of downflow while the front of labeled materials in

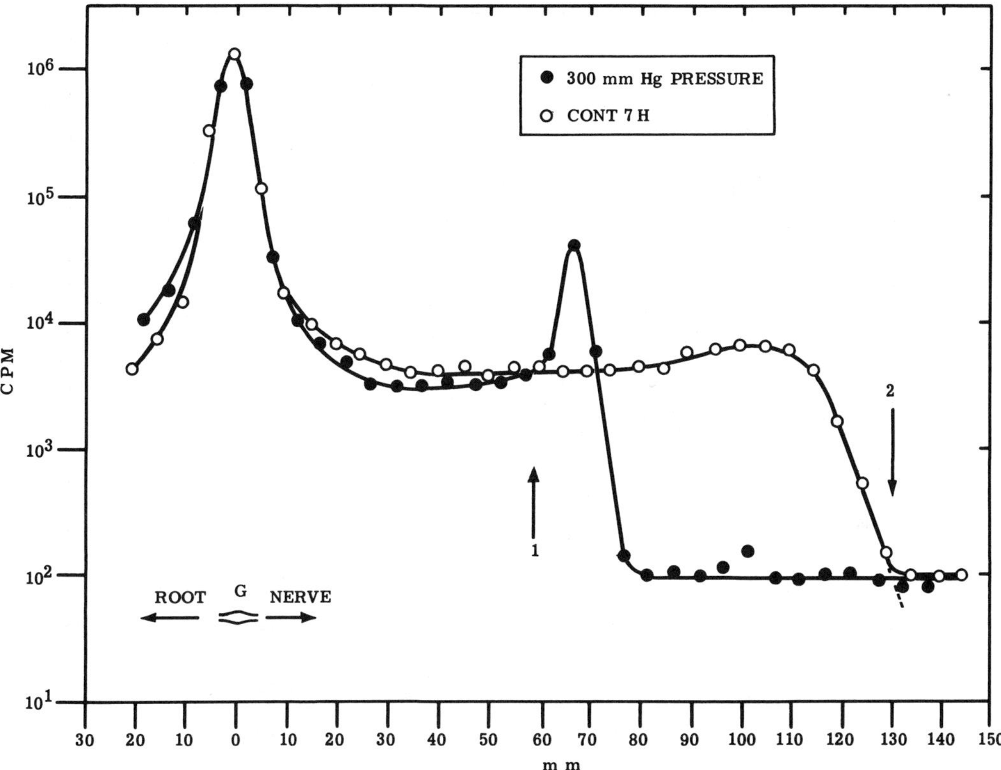

Fig. 7.5 Effect of pressure ischemia to block transport. A cuff with an applied pressure of 300 mm Hg consistently blocked fast axoplasmic transport (●) initiated by injection of the L7 ganglia with ^{3}H-leucine. The upper margin of the cuff indicated by arrow 1 does not show the upper site of the block. The effective zone is seen to be more distal in the nerve. The control nerve (○) on the other side shows transport at its normal fast rate (arrow 2) for the downflow time of 7 hr allowed. From Ochs (1975d).

the N_2 anoxic nerve does not move much beyond the point to which it had been carried in the animal (Fig. 7.6).

The downflow in the nerves made anoxic could have lasted no more than 15–20 min, showing that the reserve in the nerve of metabolically derived energy-rich phosphate compounds ($\sim$P) in the form of adenosine triphosphate (ATP) and creatine phosphate (CP) are in relatively short supply (see below). A similar fast block of transport was seen using various metabolic blocking agents to block oxidative phosphorylation and the production of ATP. Nerves *in vitro* exposed to cyanide (CN) showed the same rapid block of fast transport (Fig. 7.7).

Both N_2 anoxia and CN block at the terminal step of oxidation, by interfering with the action of cytochrome oxidase.

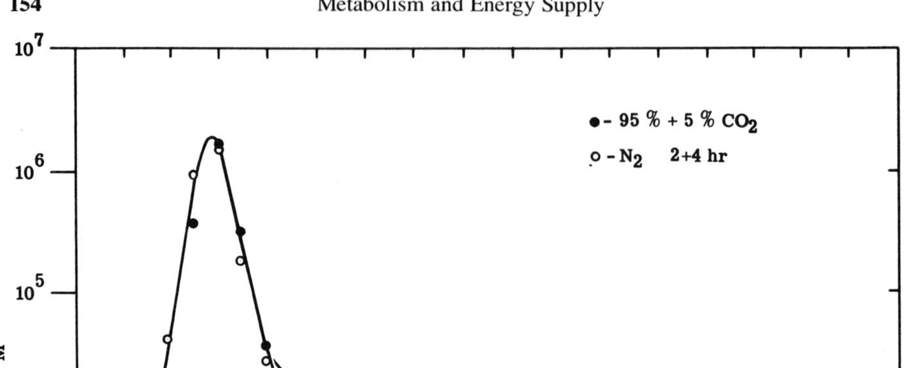

Fig. 7.6 Effect of anoxia to block transport *in vitro*. One nerve removed from a cat 2 hr after injecting the L7 dorsal root ganglion with ^{3}H-leucine was placed in a chamber containing 95% O_2 plus 5% CO_2 and kept moist at 38°C (●) for an additional 4 hr of *in vitro* transport. The rate of transport *in vitro* was that expected of fast transport in the animal (arrow 2). The other nerve (○) was similarly treated except that it was exposed to N_2 while in the chamber for 4 hr. The front advanced no further beyond the 2 hr of downflow than that which had taken place in the animal (arrow 1). From Ochs (1972c).

Another agent blocking oxidative metabolism, azide, also produces a similar rapid block of axoplasmic transport (Ochs, 1972b). This agent combines with the Fe-protoporphyrin of the oxidative chain to block electron transfer and thus prevent oxidative phosphorylation. Similarly, 2,4-dinitrophenol (DNP) also blocked fast axoplasmic transport within approximately 15 min in the *in vitro* preparation (Ochs and Hollingsworth, 1971). This finding is of particular significance in that DNP uncouples oxidative phosphorylation without interfering with electron transfer *per se*. This adds support to the concept that transport requires a continued supply of ATP.

Edström and Hanson (1973a), using an accumulation of labeled proteins

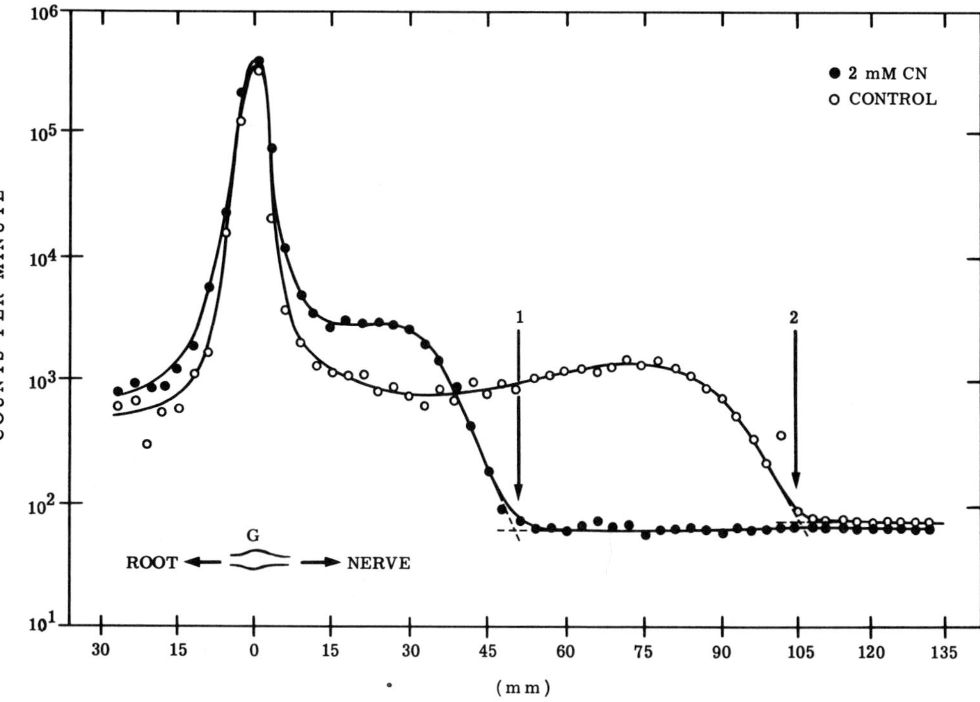

Fig. 7.7 Effect of cyanide to block *in vitro* transport. In an experiment similar to that shown in Fig. 7.6, after 3 hr of downflow in the animal, one nerve was removed from the animal and exposed to cyanide (CN) in a chamber for 3 more hours (●). A block was seen with little more downflow beyond that which had occurred in the animal. The other nerve (○) without CN present showed the downflow normally expected of the total 6 hr of downflow (arrow 2). From Ochs (1971c).

at ligations to assess transport, also found DNP and NaCN to produce a complete block of *in vitro* transport in frog nerves (Fig. 7.8).

Edström and Hanson (1973a) also showed that retrograde transport was blocked by these agents, indicating that its underlying mechanism, like that of anterograde transport, is dependent on oxidative metabolism and a supply of ATP (cf. Chapter 10).

The similar times of 15–20 min within which oxidative metabolism and ATP production were blocked by these various agents supported the inference that the essential factor required to maintain transport is the supply of ATP (Ochs, 1974a). This is indicated in Fig. 7.9 where the sites of block are indicated with respect to the production of ATP by the metabolic machinery.

To determine the relation of ATP to the fast axoplasmic transport mechanism (cf. Chapter 10), ATP and CP levels were measured in nerves after blocking oxidative metabolism (Sabri and Ochs, 1972). It was necessary to measure both ATP and CP to obtain the total high-energy phosphate ($\sim$P)

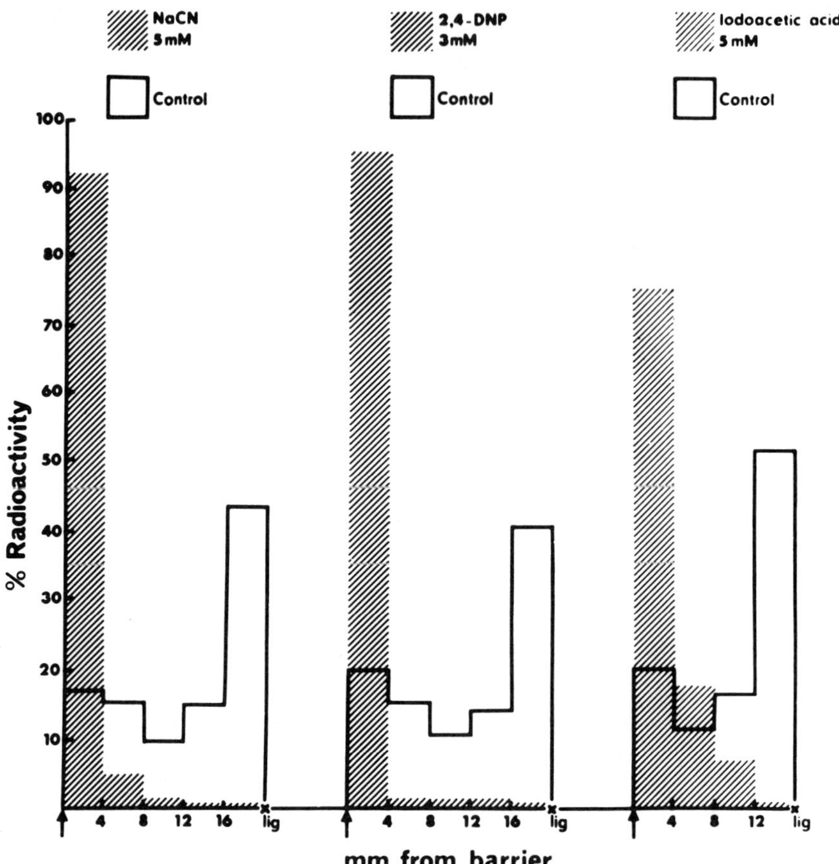

Fig. 7.8 Effect of metabolic blocking agents on transport. Transport is shown by the accumulation of labeled proteins at a distal ligation (lig) after incubating the dorsal ganglia with ^{3}H-leucine. One preparation was used as control (open bars). To the other nerves (hatched bars), either NaCN, 2,4-DNP, or IAA was added. The results, representing typical examples of three experiments, shows a block of transport by the failure of accumulation after exposure of the nerves to these metabolic blocking agents. From Edström and Hanson (1973a).

because, as classically shown for muscle and other tissues, CP acts to quickly restore the level of ATP via the enzyme creatine phosphokinase (Lohmann, 1929; Lehninger, 1975). A combined level of ATP and CP of 1.0–1.4 μM/ g was found in well-oxygenated cat sciatic nerves. After 15 min of anoxia, the ~P level fell to approximately half, to 0.5–0.6 μM/g, at the time when axoplasmic transport was blocked (Ochs, 1974a). This suggests that only half the amount of ~P measured in the nerve is available to the transport mechanism. The remainder of the ~P is compartmented, possibly in Schwann

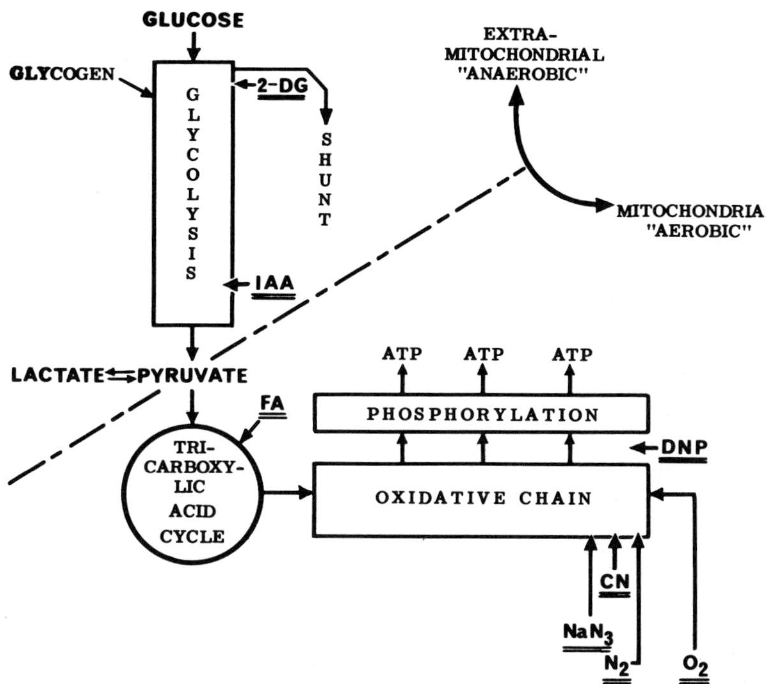

Fig. 7.9 Schematic diagram of metabolism in nerve. Glucose from the outside and glycogen in the nerve are shown entering the glycolytic pathway with 2-DG and IAA acting at steps in this path. Fluoracetate (FA) is converted to fluorocitrate and has a blocking action in the Krebs (tricarboxylic) acid cycle in the mitochondria. In addition to N_2 anoxia, several blocking agents, CN and NaN_3, act on the oxidative chain to block production of ATP by the mitochondria. DNP acts as an uncoupling agent at the sites where ATP is produced. From Ochs (1974a).

cells or in some other pool in the nerve fibers inaccessible to the axoplasmic transport mechanism.

There is some variability in the estimation of $\sim$P in the nerves of different species. Somewhat higher values were found in rabbit sciatic nerve by Stewart *et al.* (1965) who in their studies showed a similar fall during anoxia. Younger cats have higher levels of $\sim$P than adult cats, most likely because of the greater proportion of nerve fiber axoplasm to fibrous and other extra-neural substances present in their nerve trunks. This may be part of the explanation for the difference in $\sim$P noted beween the rabbit and cat nerves. Another factor in assessing the action of metabolic blocking agents is the permeability of their sheaths, in particular the perineurial sheath, to a given agent. That subject will be dealt with in Chapter 8 where the desheathed nerve preparation is described in detail. However, it is important to know if agents used to block oxidative metabolism may, because of differences in their permeability through the perineurial sheath, show a different action in sheathed and desheathed nerve preparations. The time of block of transport produced by N_2, anoxia, CN, and DNP in the desheathed preparation described in Chapter 8 was investigated. These agents were all seen to bring about a block of axoplasmic transport at a time of approximately 15–20 min in both the sheathed and desheathed nerves (Ochs, unpublished experiments). Thus, the similarity in the rapidity with which all these agents act points to an action unrelated to the presence of the perineurial sheath.

2. Glycolysis

When glycolysis was blocked by iodoacetic acid (IAA), the failure of transport had a slower time course (Ochs and Smith, 1971b). This agent inactivates the sulfhydril groups of glyceraldehyde-3-phosphate dehydrogenase (GAPD), the enzyme catalyzing the metabolic step in glycolysis from glyceraldehyde-3-phosphate to 1-3-diphosphoglycerate (Lehninger, 1975). With IAA added to the incubation medium containing nerves in which transport is progressing *in vitro*, the front of labeled activity transported in the nerve shows a gradual declining slope of the front with a complete block of transport after 1.5–2 hr (Fig. 7.10).

This prolonged decline before block occurs does not represent a slow entry of IAA into the nerve fibers. Nerves exposed *in vitro* to blocking concentrations of IAA showed a complete inhibition of GAPD enzyme activity after only 10 min of exposure to the agent (Sabri and Ochs, 1971). Furthermore, IAA blocked axoplasmic transport in the desheathed nerve preparation with the same pattern and at similarly long times as in sheathed nerves (Ochs, unpublished experiments).

Using the accumulation of labeled activity above ligations as a measure of transport, Edström and Mattsson (1976) found a partial block with IAA (Fig. 7.8) in comparison to CN and DNP. The prolonged decline in transport produced by IAA before a complete block at the relatively longer time of 1.5–2 hr can account for the apparent partial block in their studies. The declining transport occurring with IAA would allow some accumulation to occur as compared to the more rapid block of oxidative phosphorylation produced by CN, DNP, or N_2 anoxia.

The longer time of 1.5–2 hr elapsing before transport was blocked by IAA in contrast to the 15–20 min after block of oxidative phosphorylation indicates that metabolites below the level of IAA block of GAPD are able to supply ATP and maintain metabolism for some period of time. Those metabolites could, for example, be α-ketoglutarate or acetyl-CoA entering the tricarboxylic acid cycle (cf. Fig. 7.9). When these metabolites are exhausted, ATP production fails and fast axoplasmic transport ceases. In conformity with that expectation, ~P levels in nerves exposed to IAA *in vitro* showed a fall to half of control levels after 1.5–2 hr, at the time when fast axoplasmic transport was blocked (Sabri and Ochs, 1972). The correspondence of the

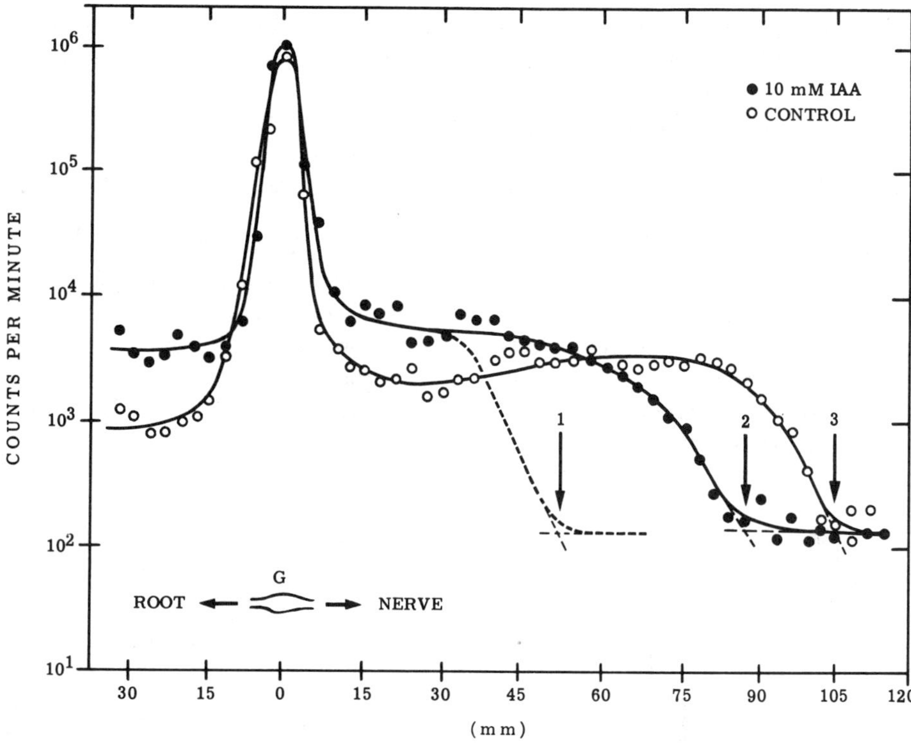

Fig. 7.10 Glycolysis block with IAA and transport. Arrow 1 at the foot of the dashed curve shows the expected downflow of labeled proteins 3 hr after removal of the sciatic nerve from the cat following injection of its L7 dorsal root ganglia with ^{3}H-leucine. Arrow 3 indicates the foot of the crest after an additional 3-hr downflow *in vitro* at the expected distance of fast transport in the control nerve *in vitro* (O). Arrow 2 indicates the foot of outflow in the nerve (●) exposed to 10 mM IAA during the time of *in vitro* transport. From Ochs and Smith (1971b).

fall of ~P levels to the same level when fast axoplasmic transport was stopped by the two different classes of metabolic blocking agents at the two different times is in accord with the concept that the crucial factor maintaining axoplasmic transport is the supply of ATP to the underlying transport mechanism.

While it is known that IAA can block a variety of glycolytic and oxidative enzymes in isolated enzyme systems (Webb, 1966), the relative specificity of IAA in blocking GAPD in peripheral nerve was suggested by its lack of inhibition of lactic acid dehydrogenase (LDH), another sulfhydril enzyme participating in the glycolysis path (Sabri and Ochs, 1971). A similar relative specificity of IAA for GAPD was seen in frog sartorius muscle (Padieu and Mommaerts, 1960) and in degenerating retina (Tieri *et al.*, 1962). Additional evidence for a relative specificity of the action of IAA on GAPD in nerve was obtained by the reversal of the IAA block of transport when pyruvate (Ochs and Smith, 1971a) was added to the *in vitro* medium, as shown in the example of Fig. 7.11.

Pyruvate enters the tricarboxylic acid cycle below the point in glycolysis at which GAPD acts, at the lactic LDH step, allowing oxidative metabolism and ATP production to proceed when glycolysis is blocked by IAA at that point. Similarly, l-lactate is able to reverse the effect of IAA block after its uptake at the LDH step with the formation of pyruvate which then enters the Krebs cycle (Fig. 7.9) . Twenty times more l-lactate is required to attain the same reversal as pyruvate (Ochs and Smith, 1971a). This difference in the effectiveness of lactate and pyruvate depends on the equilibrium constant of LDH to the metabolites, this depending on the isoenzyme composition of LDH (Fondy and Kaplan, 1965). The heart (H) muscle form of LDH more readily utilizes lactate than does the skeletal muscle (M) form. The isoen-

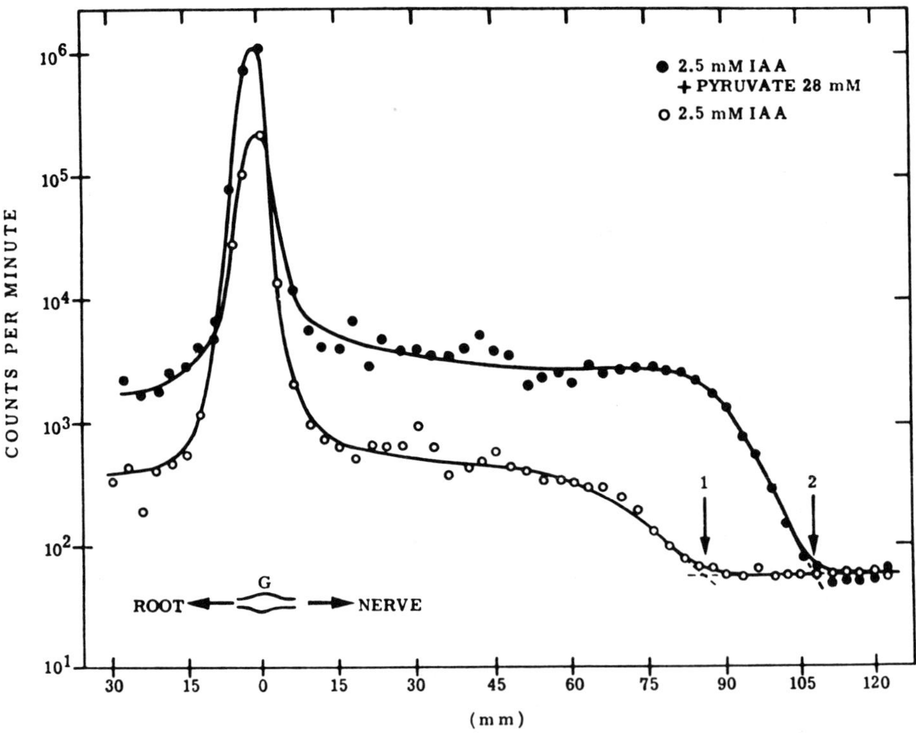

Fig. 7.11 Effect of pyruvate to reverse the IAA block of glycolysis and transport. One nerve *in vitro* treated with 2.5 mM IAA alone (O) (cf. Fig. 7.10) showed the usual declining pattern and transport failure (arrow 1). The other nerve, with 28 mM pyruvate added (●), shows a pattern of downflow *in vitro* close to the normal and to the distance expected of the usual fast rate of transport. From Ochs and Smith (1971b).

zyme pattern determined for the LDH of cat sciatic nerve was seen to be closer to that of the H than that of the M pattern (Khan and Sabri, unpublished observations) in correspondence with the effectiveness of lactate in reversing the block by IAA.

Even with sufficient pyruvate or lactate added to the medium containing IAA, some small effect of IAA on the nerve remains, as seen by the somewhat lower crest compared to that in the untreated nerve (Fig. 7.11). This suggests that IAA might have some other small effect on some enzyme other than that on GAPD. In addition to sulfhydril-containing enzymes, tubulin contains sulfhydril groups. Their oxidation may perhaps cause changes in the microtubules and thus affect transport (Chapter 12). However, the relatively small change in the form of labeled outflow with an adequate level of pyruvate or l-lactate present, indicates that the action of IAA on other sulfhydril sites are not of overriding importance for transport. Other sulfhydril blocking agents than IAA, however, have a less specific action on GAPD and can affect transport differently, agents such as mercurobenzene and certain heavy metals, as will be described in Chapter 12.

Glycolysis is blocked by fluoride at the point where phosphoenolpyruvic acid is formed from 2-phosphoglyceric acid by the enzyme enolase. Fluoride was seen by Edström and Mattsson (1976) to produce a partial block of transport in accumulation studies. Fluoride would be expected to block glycolysis with the same declining slope found with IAA and the partial block with fluoride in the accumulation studies of Edström and Mattsson would have the same explanation. A block of transport by fluoride was also shown by Banks *et al.* (1973a) using as a measure of transport the accumulation of NA-containing DCVs above nerve ligations. Addition of pyruvate to their nerve preparations treated with fluoride allowed axoplasmic transport to continue in normal fashion.

Glycolysis may be blocked at the fructose-1-phosphate step at the onset of glycolysis (Fig. 7.9) by 2-deoxy-D-glucose (2-DG). This agent is phosphorylated in cells, the resulting 2-DG-phosphate competing with endogenous D-glucose-phosphate for phosphoglucose isomerase (Wick *et al.*, 1957). However, 2-DG was found to have relatively little effect in blocking fast transport, even when high concentrations were used (Ochs, 1972d). This was also seen in the desheathed nerve preparation (Ochs, unpublished experiments). Apparently, the endogenous glucose-6-phosphate is favored by the isomerase.

3. Tricarboxylic Acid Cycle (Krebs Cycle)

A block of the tricarboxylic acid cycle (Fig. 7.9) is effected by the use of sodium fluoroacetate (FA). In the operation of the tricarboxylic acid cycle, acetyl-CoA and oxaloacetate condense when catalyzed by citrate synthetase to form citrate, which is converted to *cis*-aconitate and isocitrate by aconitase, these three tricarboxylic acids existing in equilibrium. Citrate synthetase also catalyzes conversion of monofluoroacetyl-CoA (derived from fluoroacetate), to form fluorocitrate, which then inhibits aconitase to block the cycle at this point. When sheathed nerves were exposed *in vitro* to FA in concentrations of 4–10 mM, a block of fast transport was seen to occur after approximately 1.0–1.5 hr (Fig. 7.12).

This delayed action is not due to a low permeability of the perineural sheath. A similar later time of block of transport with FA was seen in the desheathed nerve preparation (Ochs, unpublished experiments). As in the case of IAA, the longer block time seen with FA indicates that some metabolic intermediates can continue to supply reducing equivalents to the mitochondria until the supply is exhausted and ATP production ceases. The level of ~P assayed when fast axoplasmic transport was blocked by FA showed again a fall of ~P to half of control levels at that time (Ochs, 1974a). This

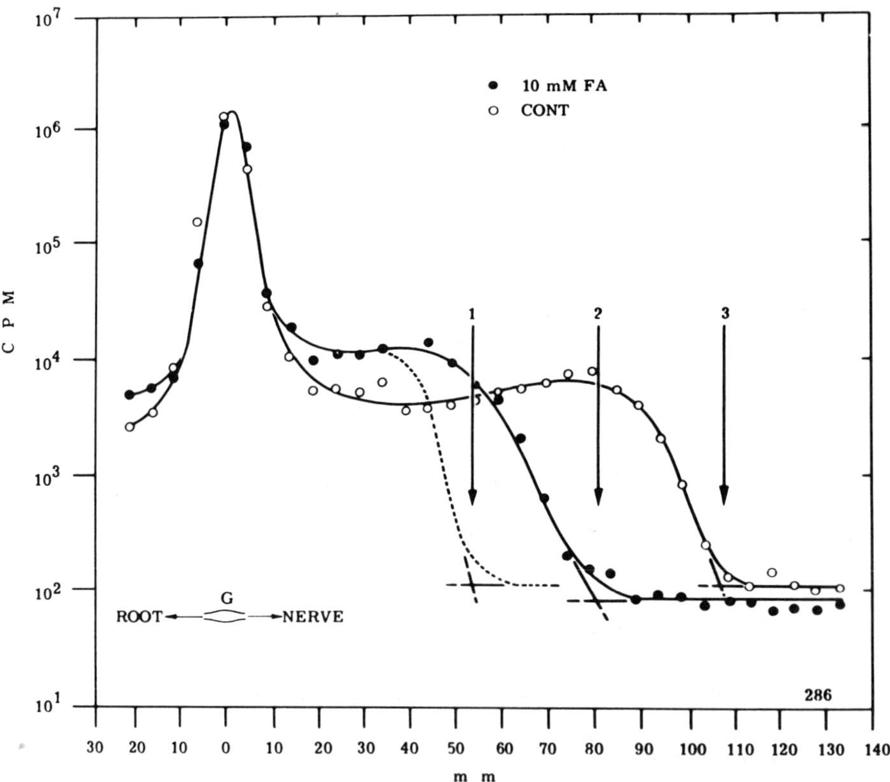

Fig. 7.12 Block by fluoroacetate (FA) of Krebs cycle and transport. The dashed curve indicates the downflow of activity expected in the animal after injection of the ganglia with ³H-leucine and 3 hr of downflow (arrow 1). The control nerve (○) after an additional 3 hr of *in vitro* transport shows a front which has moved to the distance expected of fast transport (arrow 3). The nerve exposed to 10 mM fluoroacetate (FA) *in vitro* for 3 hr (●) shows a block of transport after approximately 1 hr (arrow 2). From Ochs (1974a).

correspondence of the block of transport with the reduction of $\sim$P further supports the proposition that a supply of ATP is required to maintain the operation of the axoplasmic transport mechanism.

C. THE LOCAL SUPPLY OF ENERGY TO THE TRANSPORT MECHANISM

A supply of ATP to the transport mechanism is required along the entire length of the nerve fiber. This was inferred from the block of transport produced by covering nerves with short strips of plastic 1 cm in width coated with petroleum jelly so as to prevent access of oxygen to that part of the nerve (Ochs, 1971a). The strips were placed on nerves removed from animals in which transport of ^{3}H-labeled proteins had been initiated, at a point calculated to be just forward of the advancing front of labeled proteins. The chambers were filled with 95% O_2 + 5% CO_2 and a period of time allowed for transport. A block of transport was seen with a damming of labeled radioactivity just above the anoxic site (Fig. 7.13).

Several interesting features of the local anoxic block may be noted. The damming of labeled components occurred just at the forward edge of the anoxic region, the steeply descending front of radioactivity remaining at the forward edge of the anoxic zone for at least 4 hr without a change, indicating a lack of diffusion into the anoxic region. This was the case even though the fibers are morphologically unaltered and the block was reversible after periods of anoxia of this duration (see below).

The maintained sharp front of radioactivity at the edge of the anoxic region led to the inference that oxygen does not diffuse very far into the anoxic region of the nerve. If it did, ATP would be generated and a transport of labeled materials to that distance into the anoxic zone would have been seen. Oxygen can diffuse radially through most of a nerve 2–3 mm thick. From the equations of Krogh (1919a,b) and Landis and Pappenheimer (1963), the pressure gradient of oxygen through a thickness of 1–2 mm is sufficient to maintain an adequate oxygenation of nerve with its relatively low rate of oxygen utilization. As a rough measure of the limits of oxygen diffusion in the sciatic nerve, covering a length of nerve only 5 mm long to make the region anoxic was still effective in blocking fast axoplasmic transport. This indicates that oxygen is not likely to diffuse for more than 2–3 mm into the covered region from either end. The block of transport maintained at the edge of the anoxic region also indicates that ATP and CP do not diffuse from the oxygenated portion of nerve into the anoxic region. Again, if this were the case we would expect some degree of transport into the anoxic region. From these considerations we can infer that ATP is locally produced all along the nerve fiber to supply energy to the transport mechanism.

A rapid resumption of transport was seen upon reoxygenation of the nerves when the parafilm and petrolatum jelly covering the nerve was removed. An

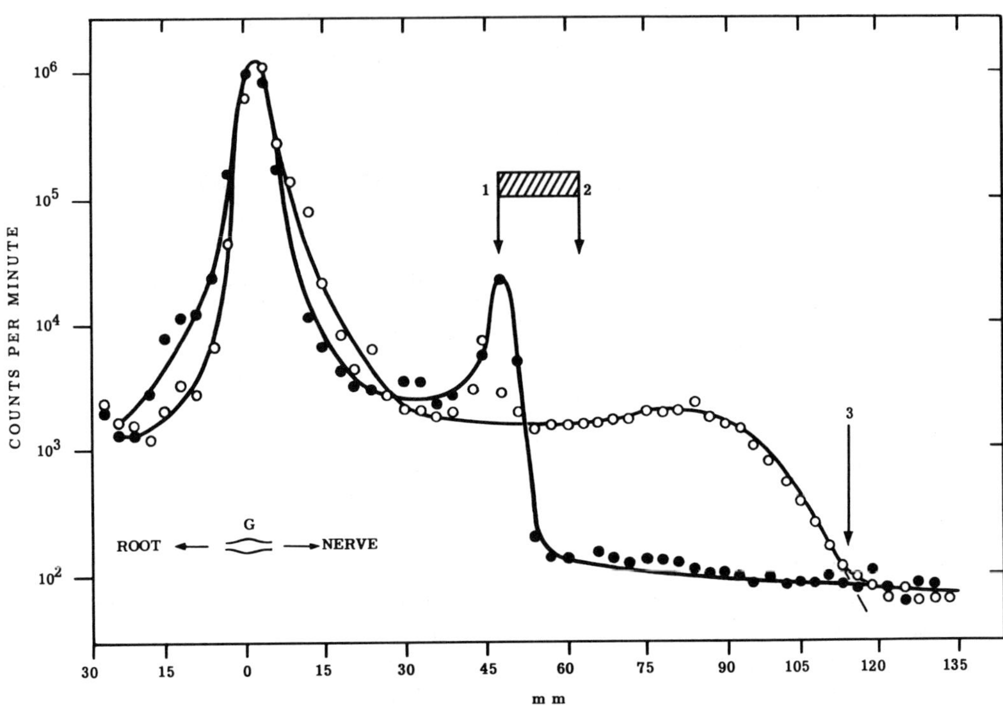

Fig. 7.13 Block of transport by local anoxia. Two hours after injection of the ganglia with ³H-leucine, the nerves were placed in a chamber, flushed with 95% O_2 + 5% CO_2, and kept at 38°C for 4 hr of *in vitro* transport. The front of radioactivity is seen to move down in the control nerve (○) to the distance expected of fast transport, as indicated by arrow 3. The opposite nerve (●) was similarly prepared for *in vitro* transport but covered with 15-mm strips of parafilm coated with petrolatum jelly while in the chamber at a distance of 45 mm from the ganglion at the site indicated by the hatched bar and arrows 1 and 2. Damming is shown by the rise of radioactivity to a peak at the most proximal edge of the anoxic region during the time of *in vitro* transport with a rapid fall to baseline levels near the upper edge of the anoxic region. From Ochs (1971a).

example is shown where the nerve was covered to produce an anoxic block lasting 1 hr and then uncovered for a further period of downflow (Fig. 7.14).

The front of radioactivity can be seen to have moved down to the position expected of the time axoplasmic transport had occurred in the animal plus the time during which the nerve was oxygenated in the chamber. The short-fall of the downflow in comparison to the control is accounted for by the time the nerve had been covered to make it anoxic. There was no evidence of a lag in the resumption of axoplasmic transport on restoring oxygenation. Another phenomenon to be noted is the disappearance of the dammed-up radioactivity during the recovery phase, the excess of labeled materials in that region becoming redistributed in the fibers (Chapter 11). With a shorter

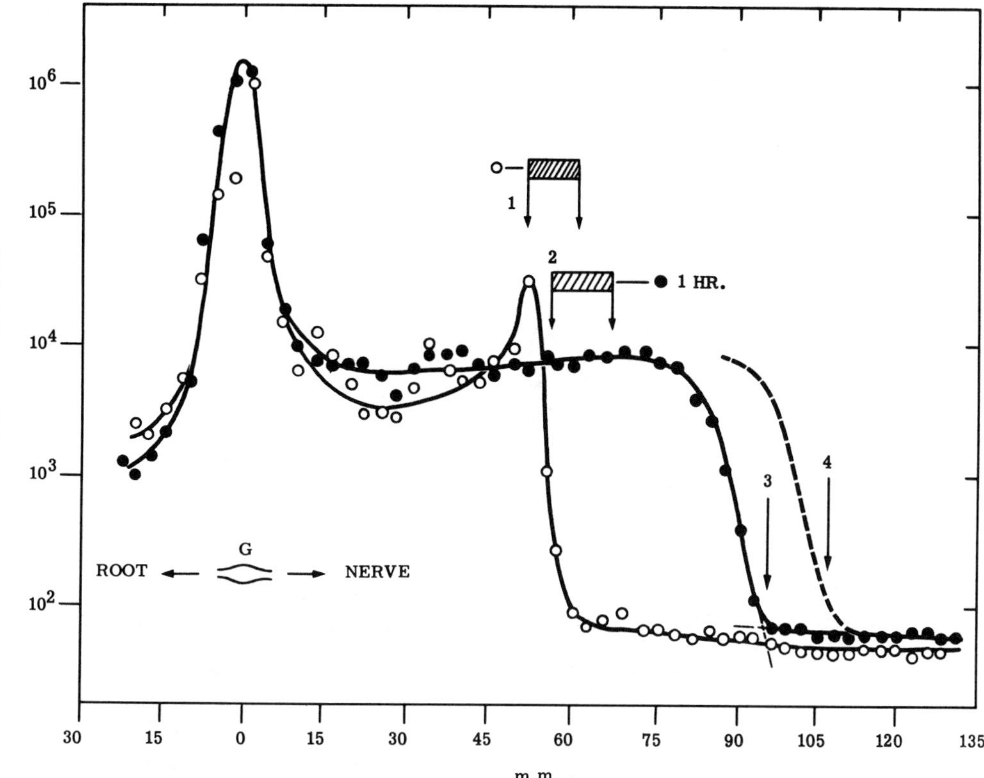

Fig. 7.14 Recovery from local anoxia. Nerves were made locally anoxic in a chamber at the positions indicated by the two bars (cf. Fig. 7.13) following a period of downflow in the animal. One nerve (○) remained locally covered for all of the 3 hr of *in vitro* downflow and a block of transport is seen (arrow 1). The covering over the other nerve (●) was removed after 1 hr (arrow 2) with a subsequent further movement of the front to the position shown by arrow 3. The dashed line and arrow 4 indicates the expected downflow in which nerves had not been made anoxic. From Ochs (1971a).

downflow time, a remnant of the dammed-up radioactivity might be seen (Fig. 7.15).

These observations show that the excess of labeled materials dammed up does not overload the transport mechanism. This point was recently studied by Brimijoin *et al.* (1979) using temperature differences to add an extra load of transported components (Chapter 11).

Local anoxias lasting more than 1.5 hr appeared to be irreversible. However, reversibility from periods of anoxia much longer than this can occur when a sufficient time is allowed for recovery from the anoxia (Leone and Ochs, 1978). Only relatively short recovery times are possible in *in vitro* studies. Longer periods of recovery were made possible using pressure cuffs

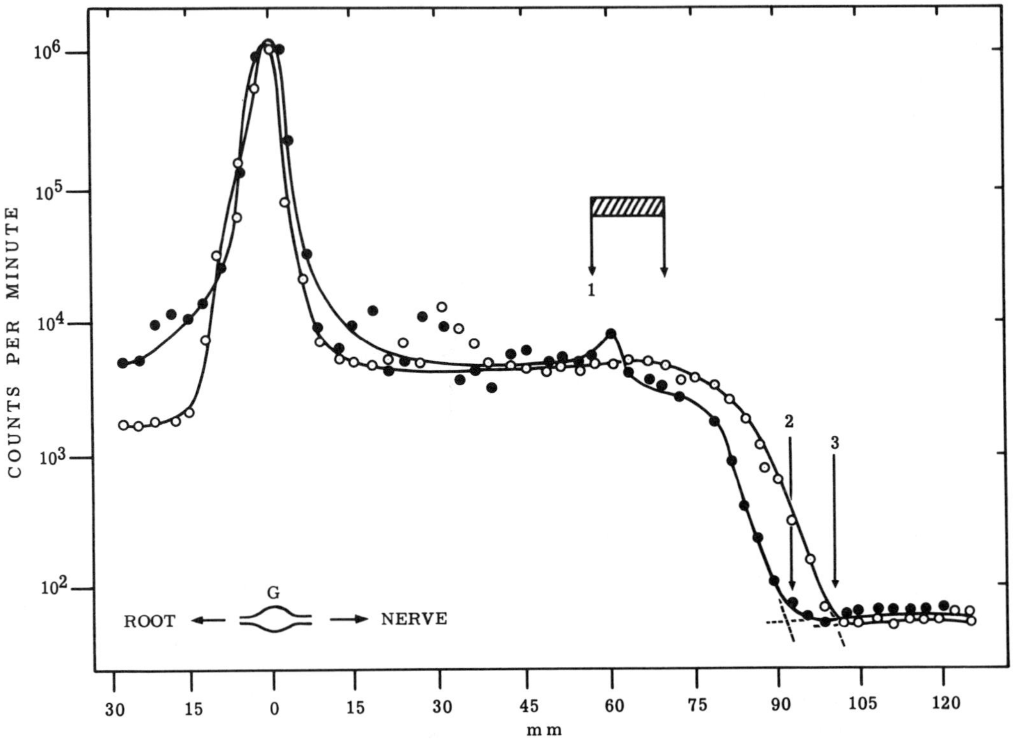

Fig. 7.15 Recovery from local anoxia and dammed proteins. A local block of oxidative metabolism was produced as in Fig. 7.14 by placing parafilm strips and petrolatum jelly on the nerve at the site indicated by the bar and arrow 1, and removed from the nerve (●) 1 hr later. The front of its downflow shown by arrow 2 fell short of the downflow in the control nerve (○) indicated by arrow 3. The shorter time allowed for downflow after removing the local anoxia revealed a small residual amount of radioactivity remaining at the dammed site. From Ochs (1971a).

to produce anoxias *in vivo*, and transport assessed after hours or days of recovery, as will be discussed in the following two parts of this chapter.

D. RELATION OF TRANSPORT TO ELECTRICAL ACTIVITY DURING AND AFTER ANOXIA

Some relationships of membrane properties to axoplasmic transport, the effect of high rates of stimulation and changes in ionic permeability, will be discussed in Chapter 12. Here we deal with the relation of axoplasmic transport to action potentials during and after a determined period of anoxia.

Until recently, the production of ATP in the nerve was considered only with respect to the supply of energy to the sodium pump required to maintain the asymmetry of Na^+ and K^+ across the membrane on which resting membrane and action potentials depend. When oxidative metabolism in the giant nerve axon is blocked by CN, azide or DNP, ATP levels fall and with the failure of the sodium pump there is a gain of Na^+ by the axon and a loss of K^+ resulting in depolarization and a lack of excitability (Hodgkin and Keynes, 1955). This takes a relatively long time in the giant axon because of its large volume. Resting membrane potentials and action potentials remain fairly high for times longer than 90 min after initiating a block of oxidative phosphorylation. Mammalian nerves, on the other hand, show a much quicker failure of excitability with a complete block of action potential responses within approximately 6–30 min (Leone and Ochs, 1978). This is due to the relatively small axonal volume of the myelinated fibers which allows sufficient Na^+ to accumulate and K^+ to leave the axons relatively quickly to thus reduce the resting membrane potential and block the excitation of action potentials.

Considering that both the sodium pump and transport mechanisms require ATP, it was of interest to determine the correlation of the effects of anoxia on axoplasmic transport and the electrical responses of mammalian nerve. A chamber was provided with connections to electrodes led in through the cover so as to permit the stimulation and recording of the nerve to be carried out while *in vitro* transport was going on (Fig. 7.16).

Sciatic nerves were taken from cats 2 or 3 hr following the injection of their L7 dorsal root ganglia with ^{3}H-leucine so as to allow a period of *in vivo* downflow, the nerve placed in a chamber which was then flushed and filled with 95% O_2 & 5% CO_2. In the course of transport *in vitro*, the oxygenated nerves were stimulated at intervals to test their ability to give rise to maximal action potentials. After obtaining control responses, the chambers were flushed and filled with N_2 to initiate a period of anoxia. Action potentials remained present for a time before falling in amplitude and becoming completely blocked within 6–30 min. On flushing and replacing the N_2 in the chamber with 95% O_2 + 5% CO_2, a recovery of action potential amplitudes was seen to quickly return and then, somewhat more slowly, to regain their control amplitudes (Fig. 7.17).

The block of action potential responses occurs at approximately the time when axoplasmic transport is blocked by anoxia, namely within 6–30 min. The block time of electrical responses was found to depend on the history of the nerve. With a first period of anoxia, the block time found for action potential responses averaged 22 min, while a second period of anoxia produced a block at an average time of 11 min (Leone and Ochs, 1978). The reason for the shortening of the time required for block with a second anoxia remains unknown.

In any event, the general correspondence between the time at which a block of axoplasmic transport and excitability occurred after the initiation of anoxia led to the inference that a common pool of ATP supplies both their underlying mechanisms (Ochs, 1974a). The requirement for ATP has been

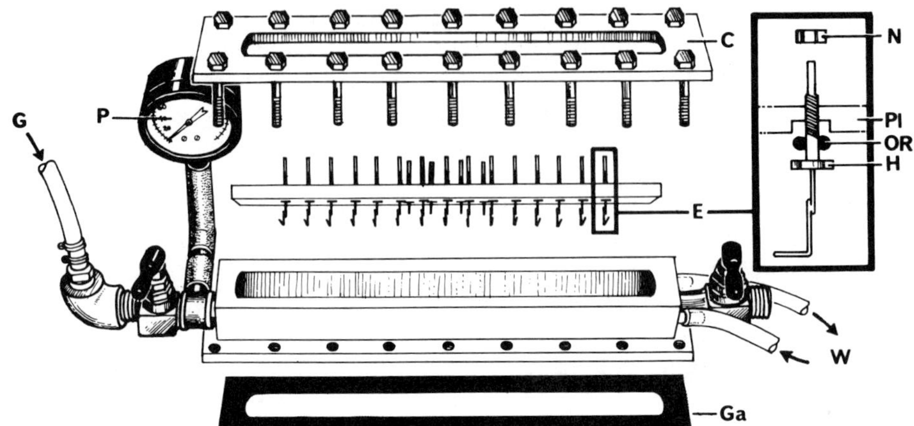

Fig. 7.16 Chamber for stimulating and recording action potentials *in vitro* in nerves transporting labeled proteins. The lid of the chamber shown was made of plastic with pins set in it at 1.5-cm intervals. Angled silver electrodes (E) were soldered to the bottom of the pins on which the nerve was placed and suspended in approximately the center of the chamber when the lid was set over the chamber. A rubber gasket (Ga) was placed between the chamber and lid. Screws around a metal cover (C) press the plastic lid to the chamber to seal it. Valves at either end control the entry and exit of gases (G) which are used to flush and fill the chamber. Typically, the gas used to maintain transport is 95% O_2 plus 5% CO_2. After the chamber was flushed and filled to a small excess of pressure above atmospheric pressure as indicated by the gauge (P), the valves were closed. A small amount of Ringer's solution placed on the bottom of the chamber allows water vapor to come into equilibrium with the gases present in the chamber, preventing the nerve from drying. The walls of the bottom are hollow, and through it a heated water supply (W) regulated at 38°C is circulated by a pump. Connections made to the projecting pins from the top to a stimulator and preamplifier lead to an oscilloscope so that the nerve can be stimulated and recorded from various places along its length. The insert E shows details of the electrode connections. A nut (N) holds the pin to the plastic lid (Pl) with an O ring (OR) between a head piece (H) on the pin to act as a seal. From Ochs (1974c).

integrated into a model for axoplasmic transport, the transport filament mechanism (Ochs, 1971b; 1972c; 1981a,b). A full discussion of that model will be given in Chapter 10. Here we point only to the general relation of the common pool of ATP supplying the sodium pump and the transport mechanisms schematized in Fig. 7.18.

When, after a period of anoxia, ATP supply falls below some critical level needed to drive their underlying mechanisms, both transport and excitability are blocked. On resumption of oxidative metabolism, a new supply of ATP becomes rapidly available as indicated by the rapid return of action potential responses (see Fig. 7.17). And, if the anoxic period is not too long, a rapid return of transport is also seen to occur. The latter is shown by the overall

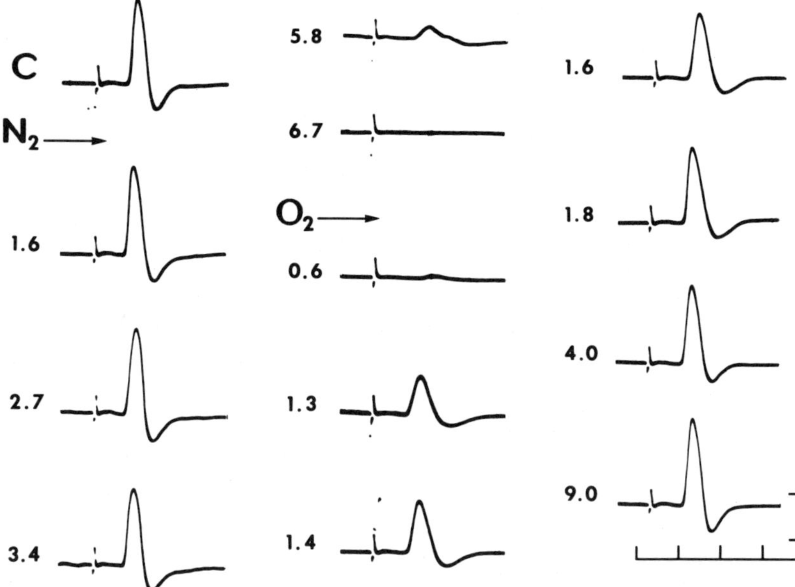

Fig. 7.17 Effect of nitrogen anoxia on action potentials. Control responses (C) represent maximal α action potentials. The distal recording electrode was near the injured cut distal end of the nerve and responses were close to being monophasic. At N₂ and the arrow at the left, the 95% O_2 + 5% CO_2 gas mixture was replaced with N_2. At the time (in minutes) shown at the left of the subsequent traces the responses began to diminish in amplitude. After 6 min they were completely blocked. The chamber was then reflushed with 95% O_2 + 5% CO_2 and action potential responses recovered back to control levels. From Ochs (1974c).

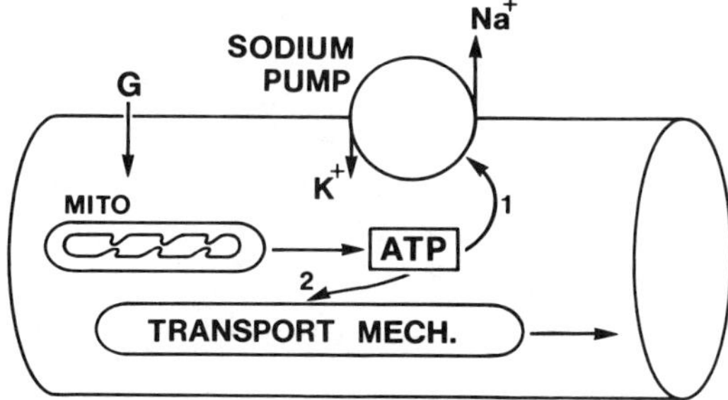

Fig. 7.18 Relation of ATP to the sodium pump and to the transport mechanism. The schematization shows glucose (G) entering the nerve and after subsequent metabolic steps (cf. Fig. 7.9) the mitochondria (MITO) produces ATP. This is used by the sodium pump (1) and the transport mechanism (2), the latter mechanism to be further described in Chapter 10 in conjunction with the transport filament model (cf. Fig. 10.2).

advance of the front of radioactivity which is accounted for by the time the nerve has been in the animal and oxygenated in the chamber before and after the period of anoxia (Fig. 7.19).

The effect of anoxia may be calculated from the position to which the front of radioactivity has been carried. As shown in this figure, arrow 1 designates the position estimated for downflow in the nerve after 2 hr *in vivo* plus 1 hr *in vitro*, using for the estimated downflow in the animal the usual rate of 17 mm/hr (Chapter 2). Arrow 2 shows the position to which the front has moved after 1 hr of anoxia with, in this example, a total time of 3.75 hr, during which 95% O_2 and 5% CO_2 was present *in vitro*. By comparing the fronts of arrow 2 and 3, the calculated position expected of the front of transport for all the time the nerve had been oxygenated, little difference between them was seen. This indicates that there was no lag in the resumption of transport when the nerve was reoxygenated. The rapid recovery of excitability on reoxygenation may be seen in the insert of this figure where the amplitudes of the maximal action potential responses are shown before, during, and after the period of anoxia.

Rapid recoveries of both transport and action potential responses were found after periods of anoxia lasting up to 1.5 hr. With longer periods of anoxia, however, an apparent failure of recovery of axoplasmic transport was seen (Fig. 7.20).

The total downflow of the front of radioactivity after a 2 hr period of anoxia indicated by arrow 2, is seen to remain close to that of arrow 1 which represents the downflow in nerves calculated to occur in the animal before the initiation of anoxia. This position falls far short of that of arrow 3, the downflow calculated on the basis of a full recovery of transport after reoxygenation of the nerves. On the other hand, action potential responses, as shown in the insert, promptly recovered their control amplitudes on reoxidation. This indicates that the mitochondria have not been damaged and that a resupply of ATP was available to adequately support the sodium pump for the return of action potentials. The $\sim$P measured after a period of anoxia when the nerve was reoxygenated showed a return to levels able to sustain both electrical responses and transport, even in nerves which had been made anoxic for times much longer than 1.5 hr.

The return of action potentials with an apparent failure of a recovery of transport has been termed "dissociation." The phenomenon may have its explanation in a failure of ATP utilization by the transport mechanism secondary to an effect on Ca^{2+} regulatory mechanisms (Chapters 8, 10, and 12).

The resumption of ATP production occurs in a surprisingly rapid time. Small action potential responses were seen within a minute after reoxygenation (cf. Fig. 7.17), indicating that the requisite ion asymmetry had been established in at least some of the nerve fibers within that time. The fibers contributing to the earliest return of potential responses are likely to be those at the surface of the nerve trunk where higher O_2 levels are more quickly established than in the depth of the nerve trunk. The growth of response

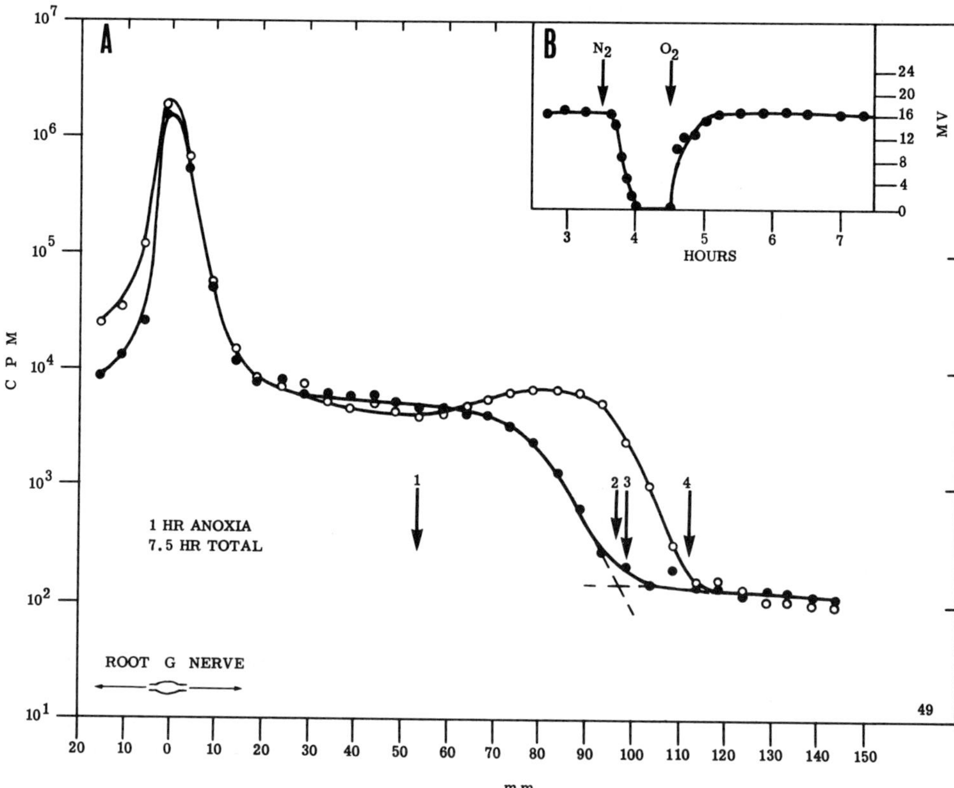

Fig. 7.19 Transport and electrical responses before and after 1 hr of N_2 anoxia. (A) The ganglia were injected with ^{3}H-leucine and after 3 hr of transport in the animal the nerves were removed and placed in chambers for a further time of *in vitro* downflow with 95% O_2 + 5% CO_2 present at 38°C. The control nerve (○) shows the usual pattern of outflow for the total downflow time of 7.5 hr, with arrow 4 at the front of the outflow of activity at the distance expected of the usual fast rate of transport. Arrow 1 shows the position of the front expected of outflow in the animal after 3 hr. Arrow 2 shows the downflow of the front of the other nerve (●) after 1 hr of N_2 anoxia *in vitro* and then switching back to 95% O_2 + 5% CO_2 for a recovery period. Arrow 3 shows the position calculated for the front if it had moved at the usual rate of axoplasmic transport minus the 1 hr when transport was blocked by anoxia. The good recovery of axoplasmic transport after anoxia is indicated by the closeness of arrows 2 and 3. (B) The insert shows amplitudes of maximal action potentials before and after initiation of N_2 anoxia (N_2) and then recovery after reintroduction of 95% O_2 + 5% CO_2 (O_2). Maximal α action potentials are shown before, during, and after the period of anoxia with amplitudes in mV in the ordinate, time in hours on the abscissa. From Leone and Ochs (1978).

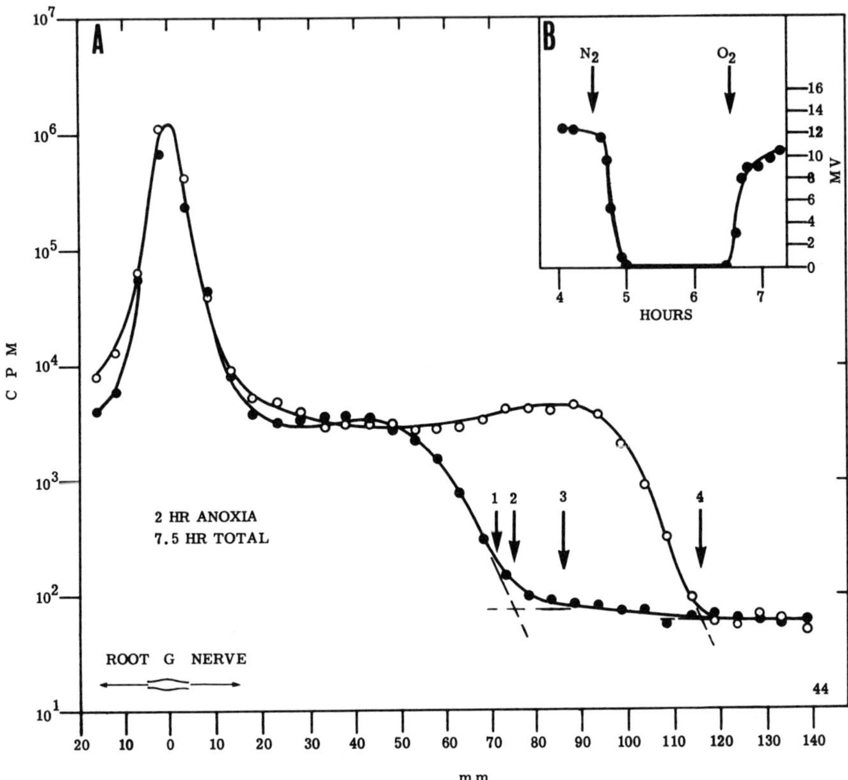

Fig. 7.20 Transport and response recovery after 2 hr of anoxia. The procedure was the same as that described in Fig. 7.19 except for a longer period of anoxia of 2 hr. (A) The downflow after N₂ anoxia (●) shown by arrow 2 is close to that of arrow 1, the position calculated on the expectation of a complete failure of recovery from axonia. Arrow 3 shows the expected position if recovery had occurred after anoxia, and arrow 4 shows the advance in the control nerve (○) not subjected to anoxia. Insert (B) shows that there was an early and a good recovery of action potential responses after anoxia. From Leone and Ochs (1978).

height over the next 6–10 min or so during recovery could represent the addition of responses to the action potential from fibers deeper in the nerve trunk, with a time-course depending on the diffusion of oxygen through the tissue. Such rapid recoveries of action potential responses were seen to occur even with longer periods of anoxia lasting 4–5 hr.

E. RECOVERY AFTER LONGER PERIODS OF ANOXIA PRODUCED BY PRESSURE

To assess the reversibility from longer periods of anoxia, longer recovery times are needed than are possible when using an *in vitro* system. For this, pressure cuffs were used to produce a period of anoxia of the sciatic nerve *in vivo*. After a designated period of anoxia and a determined period of recovery, transport was examined by injecting the L7 dorsal root ganglia with ³H-leucine and allowing for a transport time sufficient to carry the front of labeled proteins through the region of nerve made anoxic (Ochs, 1974a; Leone and Ochs, 1978). Cat limbs were compressed with cuffs at pressures of 300 mm while maintaining a temperature of 38°C throughout the time of compression. Following the compression, various recovery times were allowed before injecting ³H-leucine to assess axoplasmic transport in the nerves made anoxic. After a 2-hr period of pressure cuff ischemic and allowing one day for recovery, axoplasmic transport was seen to have fully recovered (Fig. 7.21).

This result is to be compared with the *in vitro* experiment shown in Fig.

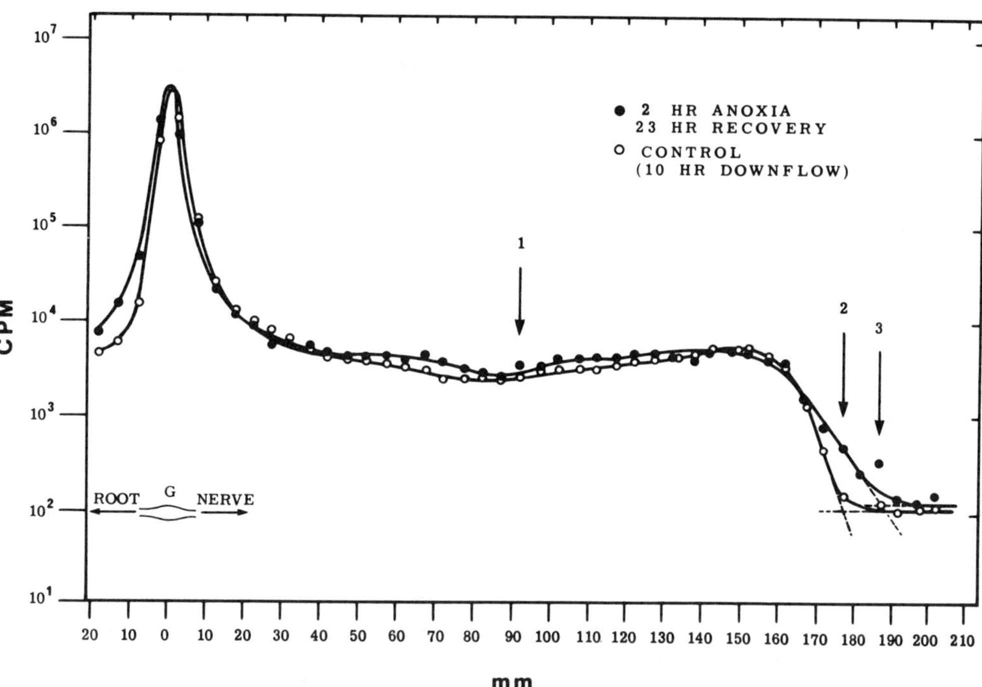

Fig. 7.21 Two-hour ischemic anoxia *in vivo* with 23 hr of recovery. A blood pressure cuff was placed high on the hind limb of the anesthetized cat and inflated to a pressure of 310 mm Hg to produce an ischemic anoxia lasting 2 hr while the limb temperature was kept as 38°C. The animal was allowed to recover for 23 hr and then reanesthetized. The L7 dorsal root ganglia were injected with ³H-leucine and a 10-hr downflow was allowed to carry labeled radioactive proteins through the region previously made anoxic. The upper part of the region cuffed is indicated by arrow 1. The outflow curve for the nerve made anoxic (●) is similar to the control (○), with the expected rate of fast transport indicated by the fronts and arrows 2 and 3. From Leone and Ochs (1978).

7.20 where a block of transport was seen with the shorter time allowed for recovery. A recovery of axoplasmic transport after a 5-hr period of pressure cuff ischemia was also seen when 2 days were allowed for recovery (Fig. 7.22).

While transport had for the most part recovered in this case, as is shown by the movement of the front through the region which had been made ischemia, the small amount of damming seen just above the region of cuffing indicates that some of the nerve fibers had been damaged by the cuffing or had still not recovered.

When longer times of anoxia were produced, 2 days were not sufficient for a recovery of axoplasmic transport. However, even after anoxias maintained for as long as 7 hr, transport was resumed if enough time, 6 days or more, was allowed for recovery.

Ischemic anoxias lasting more than 7 hr produced an irreversible failure of nerve function. The animals revealed a lack of reflex fanning, absence of

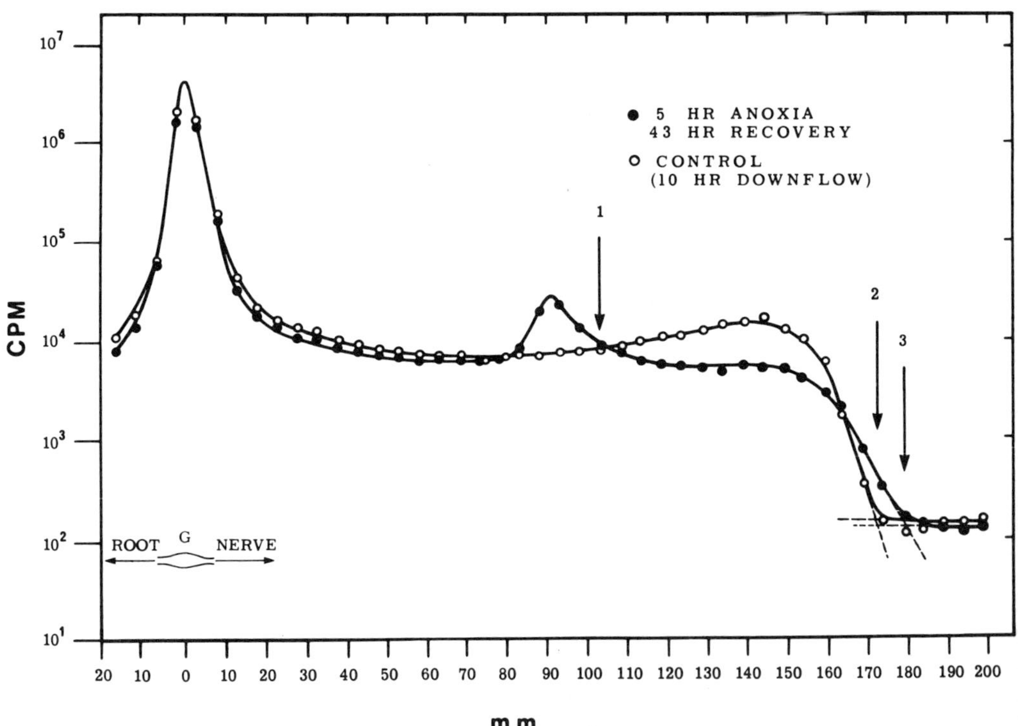

Fig. 7.22 Five hours of ischemic anoxia *in vivo* with 43 hr of recovery. Transport through a region made anoxic (cf. Fig. 7.21) for 5 hr shows a small amount of damming above the cuff site (arrow 1). The front of the transport (arrow 3) in the nerve made anoxic (●) is similar to that of the control nerve (○) (arrow 2), indicating that it has the normal rate after this relatively long time of recovery. From Leone and Ochs (1978).

contact placing, and a dragging of the dorsum of the paw of the limb on walking, behavior usually observed on complete transection of the sciatic nerve. A similar limitation of 7 hr of anoxia for recovery to occur was found by Lundborg (1970) using electrical responses as the measure of nerve function. After 7 hr of anoxia produced by cuff compression, Lundborg found those nerves to have undergone Wallerian degeneration (Chapter 6).

There is a relation between the deleterious effects of anoxia and body temperature. Recovery of transport after an ischemic anoxia lasting even longer than 7 hr was seen when the nerves were compressed at lower limb temperatures. A systematic study of the relationship between temperature and maximal compression time, however, has not been carried out.

While the pressure cuff compressions are considered to affect transport by producing an ischemia, it is also possible that some direct mechanical effects on the nerve fibers may be produced. These are particularly apt to occur at the edges of the cuffs where the pressure applied per unit surface area is greater. Such mechanical effects are most pronounced when a narrow cuff or a tourniquet is used and still higher pressures are applied (Gilliatt *et al.*, 1978). In the hind leg of baboons pressures of 800–1200 mm Hg sustained for 2 hr cause an acute demyelination and block of nerve conduction. The characteristic lesion produced is a demyelination and a nodal displacement outside the demyelinated zone (Gilliatt, 1980). Recovery of conduction may occur after several months when the fibers are remyelinated. Such high pressures do not lead to Wallerian degeneration, an indication that axoplasmic transport is maintained in the axons. The prolonged time needed for the recovery of electrical responses seen after such compressions is to be distinguished from the phenomenon of dissociation described above where the delay in the recovery of axoplasmic transport is a matter of several days or so after a rapid return of electrical responses.

A special chamber made with rubber diaphragms connected to a pressure system allowed graded pressures to be directly applied to isolated tibial and vagus nerves *in situ* (Sjöstrand *et al.*, 1978). With this chamber, a study could also be made of changes in the microvasculature resulting from direct nerve compression, as well as the capability of the nerve to support axoplasmic transport. Pressures as low as 50 mm Hg for 2 hr blocked transport, with a return of normal transport when 1 day of recovery was allowed. After a pressure of 200 mm Hg applied for 2 hr, transport was seen to be acutely blocked. Recovery occurred 3 days after such compression. After a period of compression of 400 mm Hg for 2 hr, 7 days were required for recovery. These experiments further show that the time needed for recovery depends on the magnitude of the pressure applied and the duration of compression. Sjöstrand considers these experiments to show that ischemic anoxia is the factor bringing about the block of transport following compression with this apparatus.

Hahnenberger (1978) used a fluid flow method to apply pressure to nerves *in vitro* while axoplasmic transport was going on. A special superfusion chamber allowed small sections of rabbit vagus nerves to be compressed to different degrees by varying the rate of flow of the superperfusion fluid. A pressure of 90 mm Hg produced a block of transport which apparently was not complete, as shown by the degree of accumulation of radioactive proteins at distal and proximal ligations, and in the nerve segments on either side of the superfused region. An accumulation of radioactivity below the superfusion gap also indicated that labeled material which had been transported down to a lower ligation had turned around and then been carried back to the gap by retrograde transport. Reversibility of the effect of compression was shown by the decline of accumulated radioactivity in the nerve segments on either side of the gap after nerves were exposed to a pressure of 60 mm Hg for 4 hr and a time of 12 hr was allowed for recovery.

It is difficult to make a direct comparison of compression studies made *in vivo* in the animal's limb to those made using direct cuff compressions of nerves *in situ*. There may be a greater sensitivity of nerves to direct pressure *in situ* or in an *in vitro* chamber. An interference of oxidative metabolism may be produced when using some of these local methods to produce a block, a possibility which could be determined by assessing the propagation of action potentials through the region subjected to such local pressures.

Rydevik *et al.* (1980) pointed out that in their system transport was occurring in the smaller unmyelinated fibers, these much less likely to be subject to mechanical deformation than the larger fibers in the studies of Leone and Ochs (1978). Hahnenberger (1978) speculated that the effect of compression might be to compress the endoplasmic reticulum of the fibers and so block transport. However, as he points out, such a view requires a separate transport of soluble materials inside the endoplasmic reticulum and transport of particles and mitochondria outside those channels. These, and other considerations of the underlying mechanism of transport are discussed in Chapter 10.

F.   BLOCK OF TRANSPORT BY INCREASED PARTIAL PRESSURE OF O_2 (pO_2).

Using the chamber shown in Fig. 7.16, studies of action potentials and axoplasmic transport *in vitro* were carried out at various levels of increased pressures up to 10 Atm using 95% O_2 + 5% CO_2, or O_2 with different amounts of an admixture of N_2 so as to vary the pO_2 (Ochs and Heckaman, 1972). A block of action potentials and axoplasmic transport was seen at 2–3 Atm, the block occurring at earlier times at the higher pressures. Both action potentials and axoplasmic transport were blocked within 1.5–2 hr at 5–7 Atm. Measurement of ATP and CP levels in nerves after a period of block at high pressures showed ~P levels to have fallen to about half of control levels at the time when action potentials had failed and fast transport was blocked. The correlation of the fall of ~P to the block of responses and transport with increased pO_2 corresponded to the results seen when using metabolic blocking agents (cf. Part B). This gives further evidence that a failure in the supply of ATP to the transport mechanism is the critical factor. The diminution of ATP production in nerve subjected to high partial pressure of oxygen is considered to occur as a result of oxygen poisoning (Haugaard, 1968), although the exact point in metabolism at which such poisoning occurs still remains uncertain.

The block of transport and electrical responses is not brought about by an increased partial pressure of CO_2. When 100% O_2 was used at high pressures instead of a 95% O_2 + 5% CO_2 mixture, the same block of transport and action potential responses was seen. Similarly, when a mixture of 20% O_2 + 80% N_2 at a pressure of 5 Atm was used, one where the pO_2 was calculated to be equivalent to that of 100% oxygen at normal atmospheric pressures, electrical responses and fast axoplasmic transport were well maintained. Thus, an increased gas pressure *per se* is not responsible for the block of transport; the increased pO_2 produces the effects seen.

8

Calcium and Other Ions Related to Transport

In discussing the action of agents used to block metabolism in nerves *in vitro* in the preceeding chapter, the problem of the low permeability of the perineurial sheath of nerve to different substances was noted. While it turns out that the permeability to agents used to block metabolism is adequate to carry out such studies (Chapter 7), the permeability of the sheath is low for ions, sugars and a number of other agents of interest for their action on the transport mechanism. To investigate the effect of agents on transport where the perineurium presents too great a barrier, a desheathed preparation is necessary. We will first describe the morphology, and some aspects of the perineurial sheath in relation to its permeability, and then the production of a desheathed nerve preparation. With its use, the essential role of Ca^{2+} in axoplasmic transport was demonstrated, as will be described. The related subject of Ca^{2+} regulatory mechanisms within the axon and the role of calcium binding proteins, particularly of calmodulin, is discussed. The action of other ions in relation to transport determined by the use of the desheathed nerve preparation will also be taken up in this chapter. Other studies of neurotoxins and specific agents affecting transport made using the desheathed nerve preparation will be discussed in Chapter 12.

A. NERVE SHEATHS

1. Nerve Trunk Sheaths

Three sheaths are distinguishable in a peripheral nerve bundle or nerve trunk. From the outside inward, these are the epineurium, perineurium, and the endoneurium. The outer sheath, the epineurium, is a loosely structured fibrous tissue investing the nerve trunk which contains collagen and elastin fibers running for the most part longitudinally along the surface, and having within it a variable amount of adipose tissue. The epineurium has been characterized by Sunderland (1978) as a matrix binding the nerve funiculi or nerve bundles and acting as a cushion to protect its contained fibers against deforming mechanical forces.

The epineurium embraces one or more funiculi or fascicules (nerve bundles) containing the nerve fibers, each funiculus enclosed by a perineurial sheath. The perineurial sheath is a well defined entity composed of concentric layers of flattened cells with tight junctions at their boundaries. These constitute the permeability barrier between the extracellular space and the endoneurial space within the funiculus containing the nerve fibers. The perineurial sheath is comprised of a few to as many as 15 such cell layers, the number of layers increasing with the size of the nerve bundle, each layer characterized by a basement membrane. The relation of the perineurium to the nerve fibers contained within it is shown for a small fiber in Fig. 8.1.

The endoneurial space within the perineurium contains, in addition to the nerve fibers, the endoneurium, a supporting connective tissue containing collagenous and reticular fibers which are mostly disposed longitudinally between the nerve fibers and the intrafunicular blood capillaries. Following Sunderland (1978), we distinguish inner and outer layers of the endoneu-

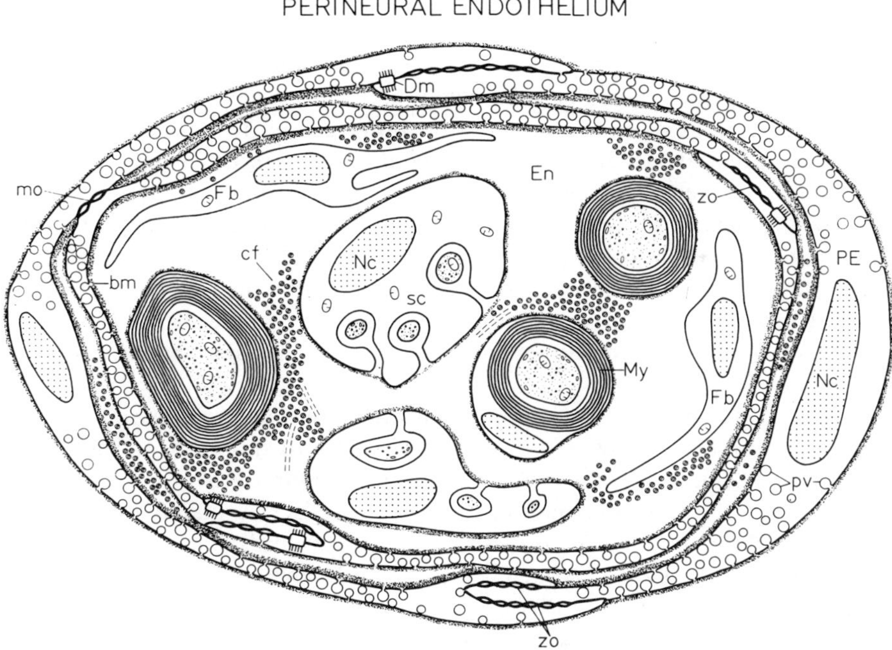

Fig. 8.1 Diagrammatic representation of the perineural sheath. A small nerve funiculus is shown with the perineural epithelium (PE) surrounding several myelinated and unmyelinated fibers. The basement membrane (bm), collagen fibers (cf), desmosome (Dm), fibroblast (Fb), endoneurium (En), macula occludens (mo), myelin (My), nucleus (Nc), Schwann cell (sc), and zonula occludens (zo) are shown. Note the large number of endocytotic vesicles (pv) in the cytoplasm of the perineurial cells. From Akert *et al.* (1976).

rium; the inner layer investing each of the myelinated fibers or a group of unmyelinated fibers to form the endoneurial sheaths or nerve tubes, the outer layer of the endoneurium filling up the remaining space within the funiculus.

2.   Low Permeability of the Perineurium to Ions

As noted above, the tight junctions between the cells of the perineurium and the low permeability of the perineurial cells constitute a permeability barrier to a variety of substances. With excised nerves *in vitro,* another possible path of permeation must be considered, namely, the severed blood vessels passing through the perineurial sheath. Permeation through this path *in vitro* is low, however, because, as described in Chapter 7, only relatively small precapillaries and capillaries pass through the perineurium and they do so at an angle. These vessels collapse when the nerve is removed from the animal in preparation for *in vitro* study and, as a result, the passage of materials via this route is minimal. In contrast, in the *in vivo* situation where circulation is present, the permeability of a given molecular species depends on the state of the capillary endothelium within the endoneural space. Large molecules such as albumin or horseradish peroxidase (HRP) which normally have a low permeability because of their size may pass out into the endoneurial space when the capillaries have been damaged, as for example by anoxia.

That the perineurium acts as a permeability barrier for a number of ionic and small molecular species in nerves studied *in vitro* has been recognized for some time (Feng and Gerard, 1930; Crescitelli, 1951; Krnjević, 1955). The low permeability of the perineurium *in vitro* accounts for the finding that mammalian nerves exposed to an oxygenated isotonic sucrose media for hours may still give rise to action potentials. This is due to the retention of Na^+ and other ions within the endoneurial space. When the sheath is removed and nerve fibers are directly exposed to the isotonic sucrose medium, the responses fail rapidly.

Earlier experiments which appeared to indicate that axoplasmic transport did not depend on Ca^{2+} were performed using sheathed nerves. In such nerve preparations fast axoplasmic transport was maintained *in vitro* with Ca^{2+} deleted from the external medium and/or with EGTA added to it (Ochs, 1972b; Hammerschlag, Dravid, and Chiu, 1975; Edström, 1974; Banks, Mayor and Mraz, 1973a; Abe, Haga and Kurokawa, 1973). Some experimental evidence suggesting that Ca^{2+} might be required was the finding that the addition of high concentrations of oxalate (30–50 mM) to the incubation medium blocked axoplasmic transport, presumably by its entry into the fibers to chelate Ca^{2+} (Ochs, 1972a). And, Kirkpatrick and Palmer (1972) in their microscopy studies saw a block in the movement of particles in single chicken nerve fibers after exposure of single nerve fibers to 10 mM EGTA. The participation of Ca^{2+} in transport was clearly shown by the use of the desheathed nerve preparation (Ochs *et al.,* 1977; Ochs, 1980b; Chan *et al.,* 1980a). We shall first describe the desheathed nerve preparation in the following section and then its use in *in vitro* studies where Ca^{2+} was deleted or its concentration in the medium altered to examine the effect on axoplasmic transport (Part B).

3.   The Desheathed Peroneal Nerve Preparation

An important requirement of a desheathed preparation adequate to carry out prolonged studies of the effect of changing the level of Ca^{2+} and other ions on transport *in vitro* is that a sufficiently long and uniform length of nerve fibers, 100–120 mm or longer, be desheathed. At the usual fast rate of transport of 410 mm/day or 17 mm/hr, such long lengths are required for longer-lasting *in vitro* studies where the effect on transport of exposure to an altered medium is assessed. For example, to reveal effects taking place over a time of 5 hr, we would require a length of at least 5×17 mm $= 85$ mm to be desheathed, where the figure 17 mm represents the hourly rate of fast transport (Chapter 2). As will be seen, approximately 3 hr of downflow are required to observe the full effects of a deletion of Ca^{2+} from the incubation medium and several more hours are required to assess the reversibility from such an exposure. Fortunately, the peroneal branch of the sciatic nerve in the adult cat has a sufficiently long length for this purpose. And, most importantly, the peroneal nerve nearly always is present as a single funiculus, making it easy to desheath in a uniform manner. The tibial nerve, on the contrary, has a number of branches with its complex pattern of funiculi making it difficult to desheath. It is usually left sheathed to compare the effects of a given medium on sheathed and desheathed nerves. Another advantage of using the cat sciatic nerve is that the peripherally directed sensory fibers of the L7 dorsal root ganglion supplying its peroneal and tibial branches cross over relatively close to the ganglion, with nearly all fibers doing so within 30 mm of the ganglion. Its two nerve branches can thus be readily separated by cutting the epineurial sheath binding the two branches up to that point without interrupting their individual nerve fibers. The peroneal branch of the sciatic nerve is then desheathed in the manner depicted in Fig. 8.2.

The sheath of the peroneal nerve is slit at a point 35 mm from the ganglion, cut down its length and stripped from the peroneal nerve to a point distally where the nerve branches, generally some 120-140 mm distal to the ganglion. The relatively long length of desheathed peroneal nerve and sheathed tibial branch constituting the preparation is shown in Fig. 8.3.

B.   CALCIUM REQUIREMENT FOR TRANSPORT

1.   Transport in Desheathed Nerves With and Without Ca^{2+} in the Medium

To assess the effect on transport *in vitro* in the desheathed nerve preparation of a given alteration in the medium, the L7 dorsal root ganglion is injected as usual with ^{3}H-leucine and after 2 hr of downflow of labeled

      Calcium and Other Ions

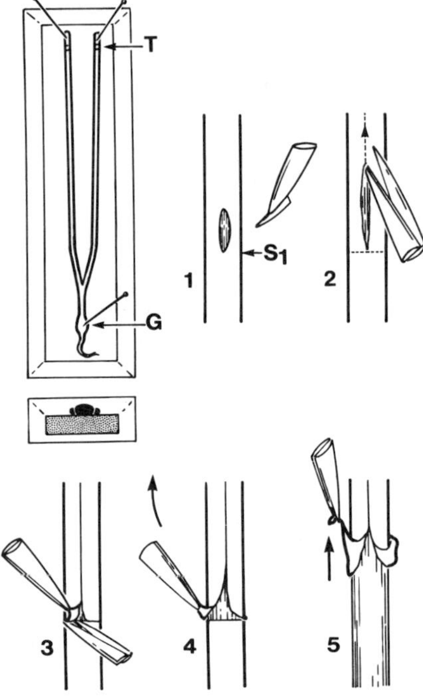

Fig. 8.2 Desheathing of the peroneal branch of the cat sciatic nerve. The nerve and its branches are pinned down on the chamber at the ganglion (G) and their ends. A tie (T) was placed on the ends of each branch to serve as a marker when stretched to a point where the bands of Fontana just disappear. Step 1: a slit is made in the perineurium of the peroneal branch approximately 35 mm distal to the ganglion by means of a knife made from a sliver of razor blade. Step 2: the blade of a pair of fine iris scissors is inserted and the perineurial sheath is cut distally. Step 3: the sheath is cut circumferentially at the proximal end of the slit. Step 4: the perineurium is grasped. Step 5: the perineurium is peeled back down the nerve to the point where branching occurs, usually 135–150 mm distal to the ganglion. From Chan, Ochs, and Worth (1980).

proteins in the animal, the sciatic nerve is removed and the peroneal branch desheathed as shown in Fig. 8.3. After 2 hr of transport at the usual rate of 410 mm/day, or 17 mm/hr, the front of radioactivity will have been carried down to a point close to 35 mm below the ganglion, the upper level at which the peroneal nerve branch is desheathed. The sciatic nerve with its sheathed and desheathed branches is then placed in flasks oxygenated with 95% O_2 + 5% CO_2, and kept at a temperature of 38°C for a specified period of *in vitro* transport.

With a mammalian Ringer solution as the incubation medium, the desheathed and sheathed branches both show the characteristic outflow pattern

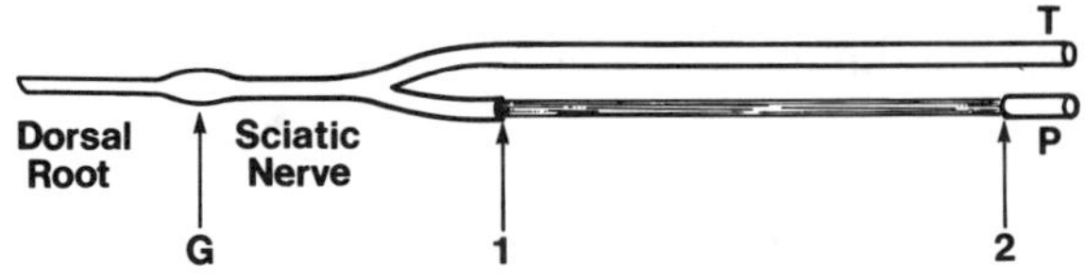

Fig. 8.3 Representation of the desheathed and sheathed branches of the sciatic nerve. The sciatic nerve with dorsal root ganglion (G) and tibial (T) and peroneal (P) branches. The latter is desheathed from approximately 35 mm distal to ganglion (1) to approximately 135–150 mm from the ganglion (2). From Chan, Ochs, and Worth (1980).

of fast transport, with the fronts of activity moved down to the distance expected at the usual fast rate of 410 mm/day (Fig. 8.4).

If the preparation had been placed in an incubation medium consisting only of an isotonic saline solution, with or without buffer present, transport in the desheathed peroneal branch was seen to slow fairly soon, within 20–30 min, the slope declining until a complete block occurs at 2.6 hr (Fig. 8.5).

Transport, on the other hand, was well maintained in the sheathed tibial nerve branch with its usual pattern and rate of outflow. The lack of Ca^{2+} was responsible for the block of transport in the desheathed branch shown in Fig. 8.5. With the addition of 5 mM Ca^{2+} to an isotonic saline medium a similar experiment showed a normal appearing pattern and rate of axoplasmic transport present in the desheathed nerve (Fig. 8.6).

The addition of various buffers to vary the pH over a range of 6–8, or even the use of an unbuffered medium, had little effect on transport. Trans-

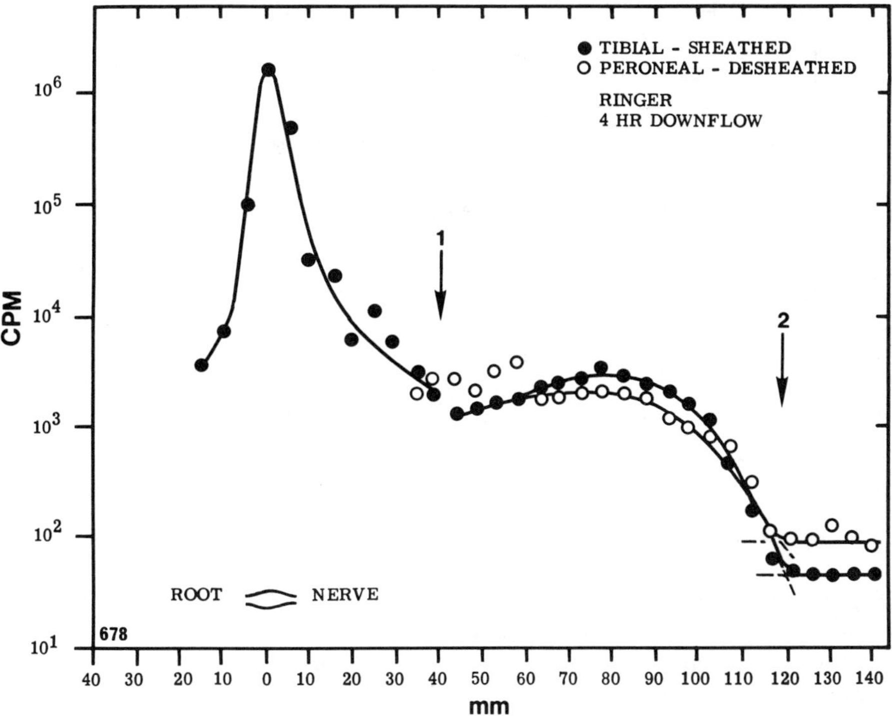

Fig. 8.4 Transport in the desheathed preparation in Ringer medium *in vitro*. After 2 hr of downflow of labeled protein in the animal, the nerve was placed in the *in vitro* Ringer medium after desheathing the peroneal branch (arrow 1). The expected transport (calculated at 17 mm/hr) *in vivo* plus transport for 4 hr *in vitro* in a Ringer medium was the same in the sheathed (●) and desheathed (O) branches (arrow 2). From Chan, Ochs, and Worth (1980).

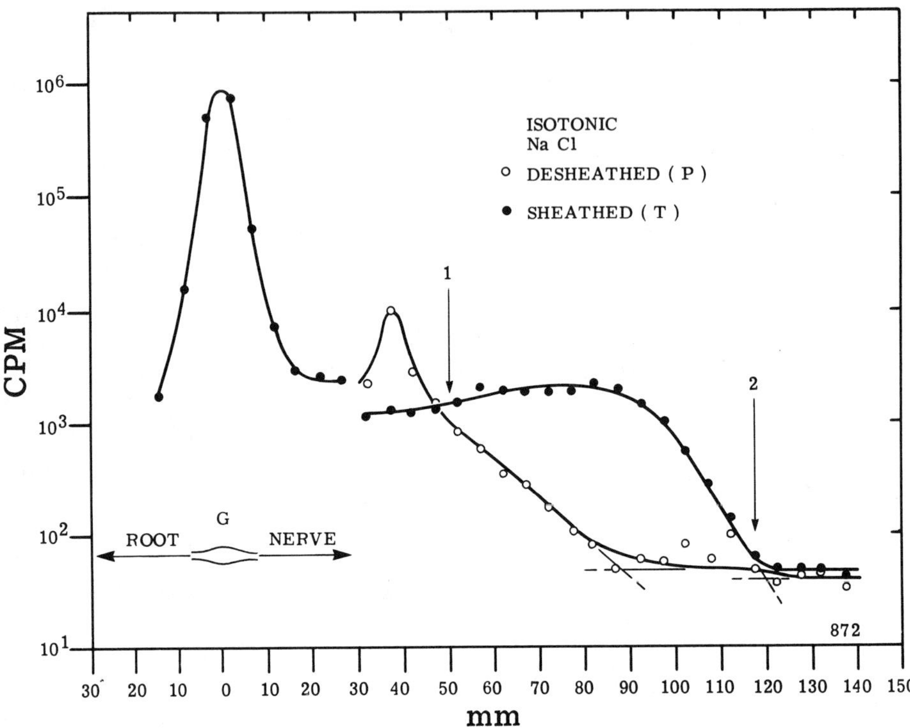

Fig. 8.5 Transport in the desheathed preparation in a Ca-free medium. After injection of the L7 ganglion with ^{3}H-leucine and downflow for 2 hr in the animal, the sciatic nerve was removed, the peroneal branch was desheathed, and the preparation was placed in an *in vitro* medium containing isotonic NaCl. Above the site where the peroneal nerve was desheathed (arrow 1), a peak of dammed activity is seen followed by a steep descent to the baseline level at a distance of about 85 mm. The sheathed tibial nerve shows a typical axoplasmic transport outflow to the expected distance (arrow 2). From Ochs, Worth, and Chan (1977).

port did not require Na^+; the same block of transport was seen in an isotonic sucrose medium. And, with the addition of 5 mM Ca^{2+} to an isotonic sucrose medium, the same normal appearing outflow pattern and rate of transport shown in Fig. 8.6 was found (Chan *et al.*, 1980).

This concentration of 5 mM Ca^{2+}, while it maintains transport, is higher than the free Ca^{2+} normally present in extracellular fluids or in mammalian Ringer solutions. When a concentration of 1.5 mM Ca^{2+}, one close to the normal level present in the extracellular fluid (Nicholson, 1980), was added to the isotonic saline medium, a partial defect of the outflow pattern was seen. The slope of the front of advancing activity appeared somewhat shallower, although the rate was not affected. The partial defect was overcome when 4 mM K^+, the concentration of K^+ present in a Ringer solution and

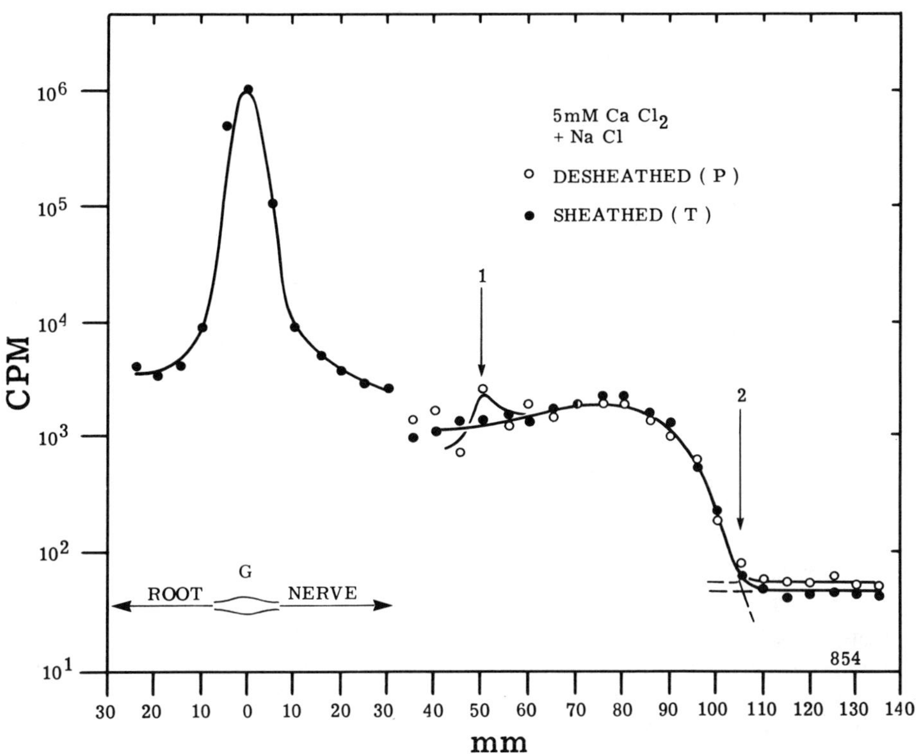

Fig. 8.6 Transport in a saline medium containing Ca^{2+}. Under conditions similar to that shown in Fig. 8.5, but with an addition of 5 mM $CaCl_2$ to the saline, the outflow in the desheathed nerve seen was similar to that in the sheathed branch and to the same distance (arrow 2). A small degree of damming of activity was seen at the upper site where desheathing was initiated (arrow 1). From Ochs, Worth, and Chan (1977).

close to that in extracellular fluids (Nicholson *et al.*, 1978; Somjen, 1979), was added to the incubation medium containing 1.5 mM Ca^{2+} (Fig. 8.7).

The effect of K^+ is to facilitate the action of Ca^{2+} at concentrations of 1.5 mM, and with K^+ present transport is maintained with concentrations of Ca^{2+} down to a lower limit of 0.5 mM (Chan *et al.*, 1980a). A concentration of 4 mM K^+ when added to isotonic NaCl without Ca^{2+} present had no effect in prolonging the time of transport beyond that seen with an isotonic solution of NaCl.

As a further indication that Ca^{2+} is essential for transport (Ochs, *et al.*, 1977), the addition of 4 mM of the Ca-chelator EGTA added to a Ringer solution caused the same block of transport in the desheathed nerve branch as occurs in a Ca^{2+}-free media (Chan *et al.*, 1980). We would expect that all constituents other than the Ca^{2+} chelated by the EGTA would remain present in their usual concentrations in the medium. Support for the finding

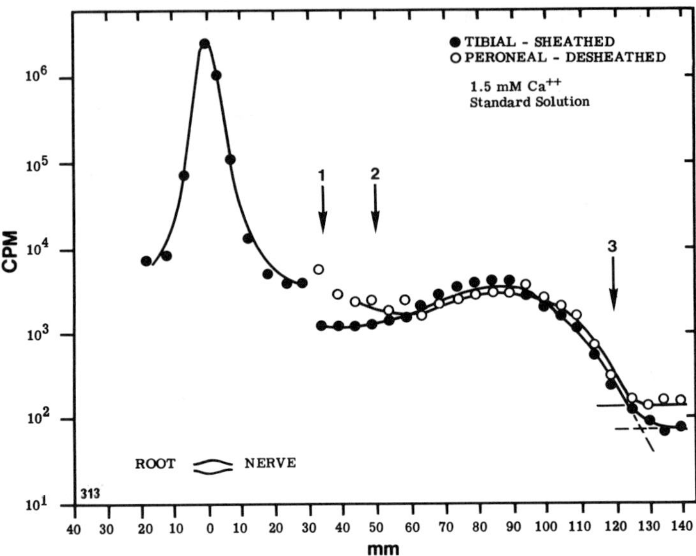

Fig. 8.7 Transport in a normal Ca^{2+} medium with addition of K^+. As in Fig. 8.6, arrow 1 indicates the point where desheathing was started; arrow 2 indicates the time when the desheathed preparation was placed in the medium and arrow 3 indicates the expected distance of normal transport. Both sheathed and desheathed branches show the same transport when 1.5 mM Ca^{2+} and 4 mM K^+ were present in the medium. From Chan, Ochs, and Worth (1980).

that Ca^{2+} is required to maintain axoplasmic transport came from the studies of Lavoie *et al.* (1979). Their studies were carried out in frog nerves where 4 mm length of perineurial sheath had been removed. Deleting Ca^{2+} from the medium caused a block of transport, and the addition of Ca^{2+} caused transport to continue. Additional support was given by the studies of Kanje and Edström (1981), and Kanje, Edström and Ekström (1982) who developed a novel method to increase perineurial permeability. Frog nerves treated for a short time with Triton X-100 showed an increased permeability to ions. Using that preparation, the removal of Ca^{2+} from the medium caused a block of transport.

The block of transport in the desheathed nerves placed in a Ca-free medium suggests that a depletion of Ca^{2+} from the fibers is the cause of the block of transport. This is presumably not compensated for quickly enough by the Ca-regulatory mechanisms present in the axons, those which act to maintain the normally low concentration of free Ca^{2+} in the axoplasm (Part C).

The depletion of Ca^{2+} is a reversible process. When nerves were exposed to a Ca-free medium for a time adequate to block transport and then subsequently transferred to a medium containing Ca^{2+}, a restoration of axoplasmic transport was seen (Fig. 8.8).

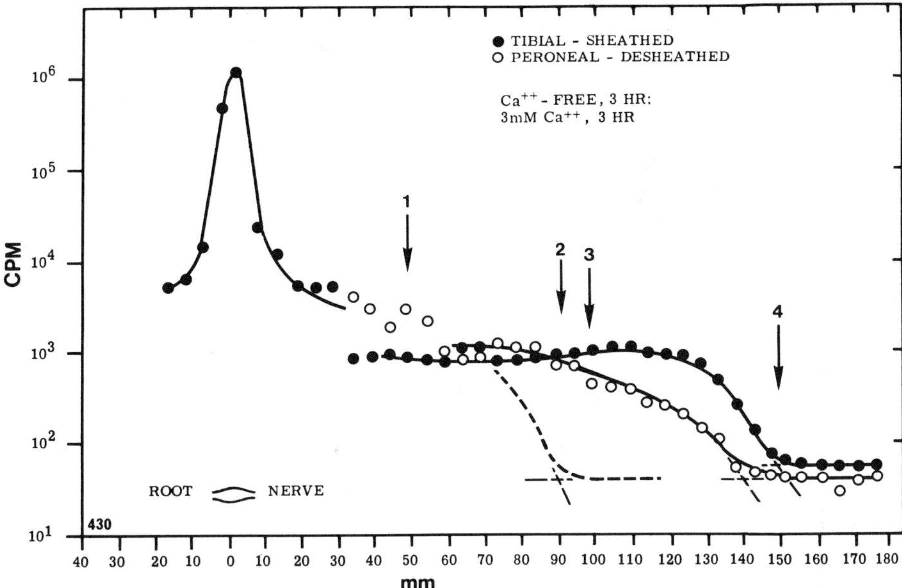

Fig. 8.8 Reversibility of block after exposure to a Ca^{2+}-free medium. Arrow 1 indicates the time the desheathed preparation was placed in a Ca^{2+}-free isotonic NaCl medium, and arrow 2 indicates the expected point to which transport in the Ca^{2+}-free medium would have been carried (dashed line). A further transport (○) occurred when the nerve was put back into a saline medium (arrow 3) containing 3 mM Ca^{2+} + 4 mM KCl. Arrow 4 shows transport in the sheathed branch (●). From Chan, Ochs, and Worth (1980).

In this example the nerve had been exposed to a Ca-free medium for 3 hr, a time sufficient to cause a block of transport at the point indicated by arrow 2 before it was placed in a Ca-containing medium for an additional 3 hr at the point indicated by arrow 3. The front of radioactivity was seen to have moved a further distance in the nerve. Less than an hour of transport time was lost as a result of the exposure to the Ca-free medium. Recovery of axoplasmic transport on transfer of the nerve to the Ca-containing medium was indicated by the front moving beyond the point represented by the dashed curve and arrow 2, the position at which it would have remained if the effect of Ca-depletion were not reversible.

2. The Effect of High Levels of Ca^{2+}

When desheathed nerves are exposed *in vitro* to higher than normal Ca^{2+} concentrations, of 20 mM or more, axoplasmic transport may appear to be unaffected until times longer than 4 hr. Exposure to still higher concentrations of Ca^{2+} causes a quicker block of axoplasmic transport (Fig. 8.9).

It is noteworthy that even with isotonic Ca^{2+} present in the medium, the

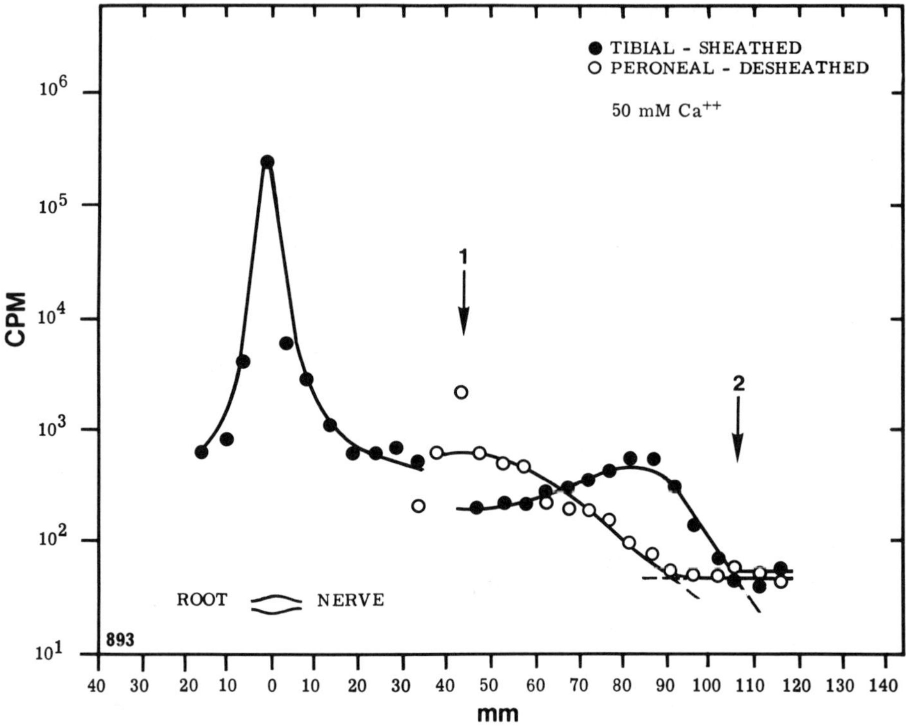

Fig. 8.9 Effect of a high-Ca^{2+} medium to block transport. At arrow 1 the preparation was placed into a 50 mM Ca^{2+} + NaCl medium and block occurs in the desheathed peroneal nerve (O). Arrow 2 shows transport to the normal distance in the sheathed tibial branch (●). From Chan, Ochs, and Worth (1980).

low permeability of the perineurial sheath allows transport to remain normal in the sheathed peroneal branch and, even with high concentrations of 50-60 mM Ca^{2+} present, the desheathed nerves might not be irreversibly blocked if the exposure time is not too long (Chan *et al.*, 1980a). An example of the reversibility of transport after an exposure of nerves to as much as 60 mM Ca^{2+} for several hours is shown in Fig. 8.10.

After transport in the animal brought the front to the point indicated by arrow 1, the nerve was removed and placed in an *in vitro* medium containing 60 mM Ca^{2+}. Arrow 2 and the dashed curve indicates the point at which a block of axoplasmic transport was expected with this concentration of Ca^{2+}. At the point indicated by arrow 3, after a time of 3 hr, the nerve was transferred to a 1.5 mM Ca^{2+} medium and transport recovered as shown by the further movement of the front to the distance shown by arrow 4.

The reversibility of transport in the desheathed nerves after exposure to high levels of Ca^{2+} such as those shown in Fig. 8.10, is explained by the relatively low permeability of the axolemma to Ca^{2+} and to mechanisms of

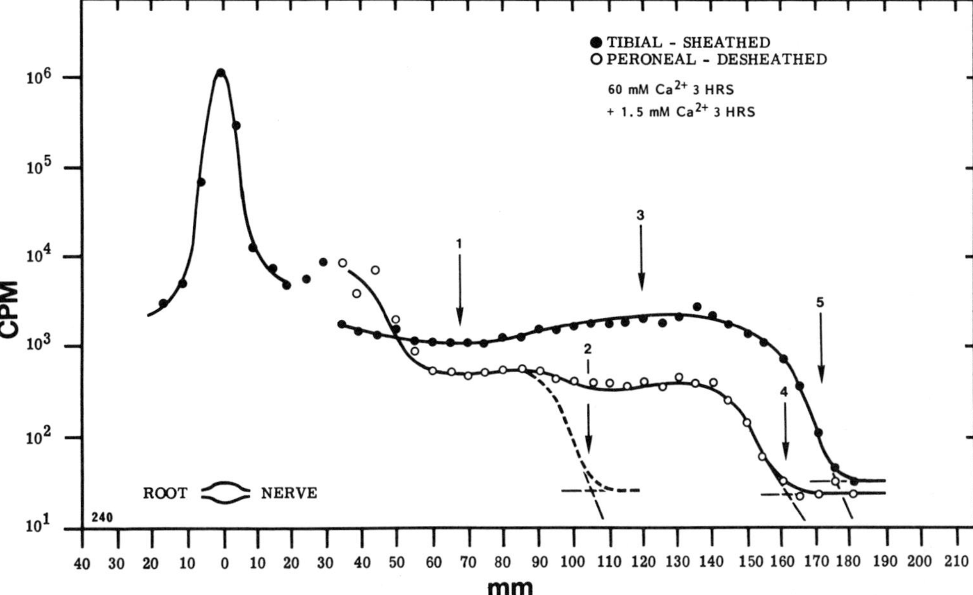

Fig. 8.10 Reversibility after exposure to a high-Ca^{2+} medium. At arrow 1 the nerve was put into a 60 mM Ca^{2+} + NaCl medium. Arrow 2 represents the expected transport block (dashed line) in the desheathed nerve. At arrow 3 the preparation was placed back into a 1.5 mM Ca^{2+} NaCl medium, and arrow 4 shows a further transport in the desheathed nerve. Arrow 5 shows normal transport to the expected distance in the sheathed branch. From Chan, Ochs, and Worth (1980).

Ca^{2+} regulation present in the fibers. Some of the latter mechanisms act to bind or sequester the excess Ca^{2+} which enters the fibers so as to maintain free Ca^{2+} at an optimal level (Part C). These Ca-regulatory mechanisms may, however, be overcome if nerves are exposed to high concentrations of Ca^{2+} for too long a period of time, the failure of Ca-regulation followed by irreversible functional and morphological changes. The latter appear as a loss of microtubules and neurofilaments with swollen mitochondria and distorted ER in the EM, changes similar to those reported by Schlaepfer (1971). In Schlaepfer's studies carried out using short, 0.3–0.5 mm lengths of cut rat sciatic nerve, those changes appeared in the fibers with concentrations of Ca^{2+} of 1 mM in the medium because of the ready access of Ca^{2+} to the axoplasm of the nerve fibers via the cut ends. In the desheathed nerve preparation, the relatively low permeability of the nerve axolemma to Ca^{2+} results in a very much lower amount of Ca^{2+} entering the axons. This may be altered by the Ca ionophore A23187 to increase Ca^{2+} permeability, resulting in a block of transport as the Ca^{2+} in the medium enters the fibers (Schlaepfer, 1977b). Such a block did not occur when Ca^{2+} was deleted from the medium (cf. Esquerro *et al.*, 1980). Similarly, A23187 and another Ca ionophore, X-537A, were seen to block the transport of ^{3}H-labeled proteins in frog nerves, with X-537A being more effective than A23187 in this regard (Kanje, Edström, and Hanson, 1981).

C. CALCIUM REGULATION

1. Mechanisms of Ca^{2+} Regulation

The need for a normal amount of Ca^{2+} in the *in vitro* medium to maintain transport was shown by the block of transport in nerves exposed to high levels of Ca^{2+} which was reversed when returned to a low Ca medium (cf. Fig. 8.10). The reversal of transport block in nerves exposed to a Ca-free media when the nerves are returned to a Ca-containing medium (Fig. 8.8) also indicates that some optimal level of Ca^{2+} is required to maintain normal transport in the axon. The level of free Ca^{2+} present in mammalian axons has yet to be determined by direct techniques, but most likely it is similar to the concentration of 10^{-7} M found present in the axoplasm of the giant axon of the squid (Baker, 1972; Blaustein, 1974). Actually, the total amount of Ca^{2+} in the giant axon is high, some 0.4 mM, with most of it sequestered in the mitochondria, in the ER, and in a bound form which is presumed to be associated with calcium binding proteins. The mitochondrion appears to be most important in regulating the level of free Ca^{2+} in the axoplasm (Borle, 1973), this organelle sequestering Ca^{2+} even in preference to its production of ATP (Lehninger, Carafoli, and Rossi, 1967; Carafoli and Crompton, 1978). The calcium regulatory mechanisms in the axon are schematized in Fig. 8.11.

The sequestration of Ca^{2+} indicated in the mitochondria and ER has been

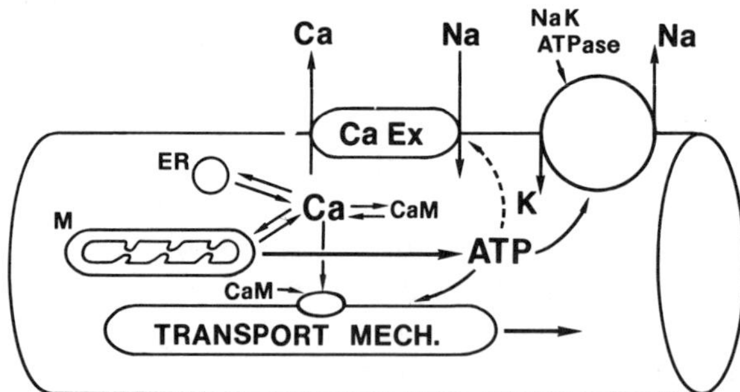

Fig. 8.11 Schematization of mechanisms of Ca^{2+} regulation in nerve. Calcium is sequestered in mitochondria (M) and endoplasmic reticulum (ER) or bound to calcium-binding protein (CaBP) or to calmodulin as indicated in this figure by CaM. Calcium is removed from the axon by a calcium exchange mechanism (CaEx) or an ATP-activated Ca pump (dashed line) in the membrane. Other symbols are the same as in Fig. 7.18 and Fig. 10.2A.

shown using pyroantimonate or oxalate. These agents, added in the course of preparing the tissue for EM, enter the fibers to bind to Ca^{2+} and form electron-dense particles (Theron *et al.*, 1975; Hindelang-Gertner *et al.*, 1976; Duce and Keen, 1978; and Henkart *et al.*, 1978). With pyroantimonate present, an increased number of electron-dense particles was seen in mitochondria and ER as well as in the axoplasm of desheathed nerves exposed to high concentrations of Ca^{2+} in the incubating medium (Chan *et al.*, 1980a). The electron-dense particles distributed in the axoplasm could represent Ca^{2+} bound to CaBP (Section C2). The increased number of electron-dense particles seen in the mitochondria, and to a lesser extent in the ER of the axons, is a picture expected of a sequestration of excess Ca^{2+} in these organelles on the basis of their Ca-regulatory role as indicated in Fig. 8.11. This possibility was further suggested by the reduction in the number of dense particles, particularly in the mitochondria, when nerves which had been exposed to high Ca^{2+} were subsequently transferred to a low-Ca^{2+} medium. The mitochondria were most labile with respect to particle changes, suggesting that this organelle takes a principal part in regulating the level of free Ca^{2+} in nerve fibers (Borle, 1973). Recently, it was discovered that the electron-dense particles binding Ca^{2+} were of two types. In one there was a relatively high Ca^{2+} level with respect to K^+ and in the other a higher K^+ with respect to Ca^{2+}. In both there was a large amount of antimony present from the pyroantimonate (Jersild and Ochs, 1982). These different combinations most likely reflect variations in the nucleation process leading to visible electron-dense particles. Sequestration of Ca^{2+} in the mitochondria and ER, and its binding to the CaBP of the axoplasm, does not exhaust all the mechanisms involved in Ca^{2+} regulation. Some process in the axonal membrane is required to eject the excess Ca^{2+} from the fiber so as to maintain a relatively low level of free Ca^{2+} in the axoplasm. Although the permeability of the axonal membrane to Ca^{2+} is low, the electrochemical gradient favors the continual entry of Ca^{2+} into the axon along with an additional amount entering as a result of activity (Baker, 1972; Blaustein, 1974; Brinley, 1976). An exchange mechanism in the membrane has been proposed for the removal of Ca^{2+}, with one mole of Ca^{2+} ejected for every 2–3 moles of Na^+ which enters (Baker, 1972, 1976; Blaustein, 1974; Brinley *et al.*, 1975). The sodium pump then pumps out the Na^+ which has entered so as to keep the level of Na^+ in the axoplasm low, a requirement for the maintenance of the normal resting membrane potential (cf. Fig. 8.11). Alternatively, evidence for an active Ca^{2+} pump in the giant axon membrane has been obtained, one requiring a direct supply of energy in the form of ATP as indicated by the dashed line in Fig. 8.11 (Baker and Glitsch, 1973; DiPolo, 1978). Such a Ca-pump was found present in the red blood cell membrane (Schatzmann, 1975).

An observation which may have some bearing on the means by which Ca^{2+} is extruded from mammalian nerve axons is the similarity of the block time of axoplasmic transport in desheathed nerves *in vitro* in Ca-free isotonic sucrose and Ca-free NaCl media. The apparent lack of sensitivity to Na^+ in the medium is not a behavior expected of a Na–Ca exchange mechanism and it suggests that the efflux of Ca^{2+} is mostly accomplished by a Ca-pump rather than a Na–Ca exchange mechanism. However, such evidence is as yet too meager to allow a definitive conclusion to be drawn on this point.

The kinetics of the sequestration or the binding of Ca^{2+} in the nerve fibers when an increased Ca^{2+} was present in the medium, or release from sequestration sites when nerves still transferred to a low- or zero-Ca media, is unknown. Changes of Ca^{2+} flux in relation to increased loading or to a depletion of Ca^{2+} from the axons are presently being studied using ^{45}Ca as a marker of Ca^{2+} flux (Chan and Ochs, unpublished experiments). In this technique portions of nerve are placed for a time in a medium containing ^{45}Ca and the ^{45}Ca-loaded nerves then transferred successively into a series of tubes containing unlabeled media (Kalix, 1971). After ^{45}Ca loading, desheathed pieces of peroneal nerves successively transferred to tubes containing normal amounts of Ca^{2+} or into a Ca-free media suggested that the ^{45}Ca efflux continues in the low-Ca medium.

A decline of axoplasmic transport starts within 20–30 min after placing desheathed nerves in a Ca^{2+}-free medium (cf. Fig. 8.5), and the question arises as to why the Ca^{2+} sequestered in the mitochondria and ER or bound to CaBP cannot supply sufficient free Ca^{2+} to the axon to maintain transport. These observations imply that these organelles act as long-term storage mechanisms and cannot act in the short-term case where an abrupt and relatively large decrease in the level of Ca^{2+} occurs.

Some additional insight into the mitochondrial regulation of Ca^{2+} may be had from a study of the isolated mitochondrial preparation. Mitochondria take up Ca^{2+} through the activity of the respiratory chain producing ATP, the Ca^{2+} uptake actually taking precedence over its production of ATP (Lehninger, 1975; Carafoli and Crompton, 1976). Changes in Ca^{2+} levels could therefore affect the production of ATP and the levels of ~P (ATP and CP) in the fibers. The level of ~P in desheathed peroneal nerves placed in a Ca^{2+}-free medium for several hours does in fact show some decrease, not however, to the degree that it in itself would account for the block of transport (Chapter 7). Sufficient ~P remained and action potentials could still be elicited after axoplasmic transport had been blocked by exposure to a Ca-free media (Chan et al., 1980a). A trace of Ca^{2+} was present in those experiments, an amount sufficient to allow action potentials to be elicited (Frankenhaeuser, 1957), but much below the concentration of 0.5 mM Ca^{2+} required to maintain axoplasmic transport (Section B1).

2. Calcium-Binding Proteins Including Calmodulin

Calcium-binding proteins have been found present in mammalian nerve (Iqbal and Ochs, 1978; 1979). These proteins have in recent years become a subject of intense interest because of the relation of calmodulin to Ca^{2+} as a second messenger in cells (Wasserman *et al.*, 1977; Cheung, 1980; Means and Dedman, 1980; Means, Tash and Chafouleas, 1982). The relation of calmodulin to transport will be described in Chapter 10. In this section the evidence for its presence along with other calcium binding poteins in nerve fibers is given.

A fast transport of Ca^{2+} in the chick retinal ganglion system (Knull and Wells, 1975), and in peripheral nerve was shown by Hammerschlag *et al.* (1975) after exposing the dorsal root ganglia of frog sciatic nerves *in vitro* to ^{45}Ca. The pattern of outflow of radioactivity in the nerves was one typical of fast axoplasmic transport. Calculation of its rate of transport when scaled to a temperature of 37°C using the Q_{10} known for frog nerve showed ^{45}Ca to be transported at a rate close to 410 mm/day, one typical of fast transport in mammalian nerves. Similar results were seen in the cat after injecting ^{45}Ca into L7 dorsal root ganglia (Iqbal and Ochs, 1975; 1978). The crest of fast outflow of ^{45}Ca radioactivity compared to proteins labeled with ^{3}H-leucine was much smaller, but it had the same fast rate close to 410 mm/day (Fig. 8.12).

The relatively small crest of outflow seen when using ^{45}Ca is due in part to the lower uptake of Ca^{2+} by the nerve cell bodies. As a result, more ^{45}Ca leaks back into the circulation to be taken up locally by peripheral tissues resulting in a higher background level present in the nerve tissue. This is shown by the relatively high level of base-line radioactivity in nerve segments taken forward of the front and in the nerve taken from the side not injected with $^{45}Ca^{2+}$ as shown in Fig. 8.12.

Portions of the nerve containing the crest and plateau of ^{45}Ca outflow indicated in Fig. 8.12 showed the ^{45}Ca to be associated with a 15,000-dalton protein. The crest and plateau portions of nerve were each homogenized, centrifuged, and the high-speed supernatant fraction passed through Sephadex G-100 columns. The ^{45}Ca radioactivity present in the nerve segments was found present in the Peak I eluate. To further determine the protein component to which the ^{45}Ca was bound, Peak I material was passed through Biogel A 5m columns and resolved as 3 subpeaks designated Peaks Ia, Ib, Ic (Fig. 8.13).

The first emerging peak, Ia, contains the highest-MW proteins, the second emerging peak, Ib, has a relatively large amount of tubulin (Chapter 9), with the third emerging peak, Peak Ic, containing the smallest amount of protein but with most of the ^{45}Ca bound to it. On comparing the amount of radioactivity in Peak Ic isolated from the crest portion of the nerve to that in the plateau, the crest was seen to contain relatively more ^{45}Ca than the plateau. The ratio of the specific activity of Ca^{2+} (cpm/mg protein) in the crest as compared to the plateau was 1.52, evidence that the Ca^{2+} binding protein (CaBP) is fast-transported. A similar higher ratio was also found for the ^{3}H-leucine-labeled 15,000 dalton-protein in Peak Ic obtained from the crest and plateau regions after injecting ^{3}H-leucine into the L7 dorsal root ganglia (Fig. 8.14).

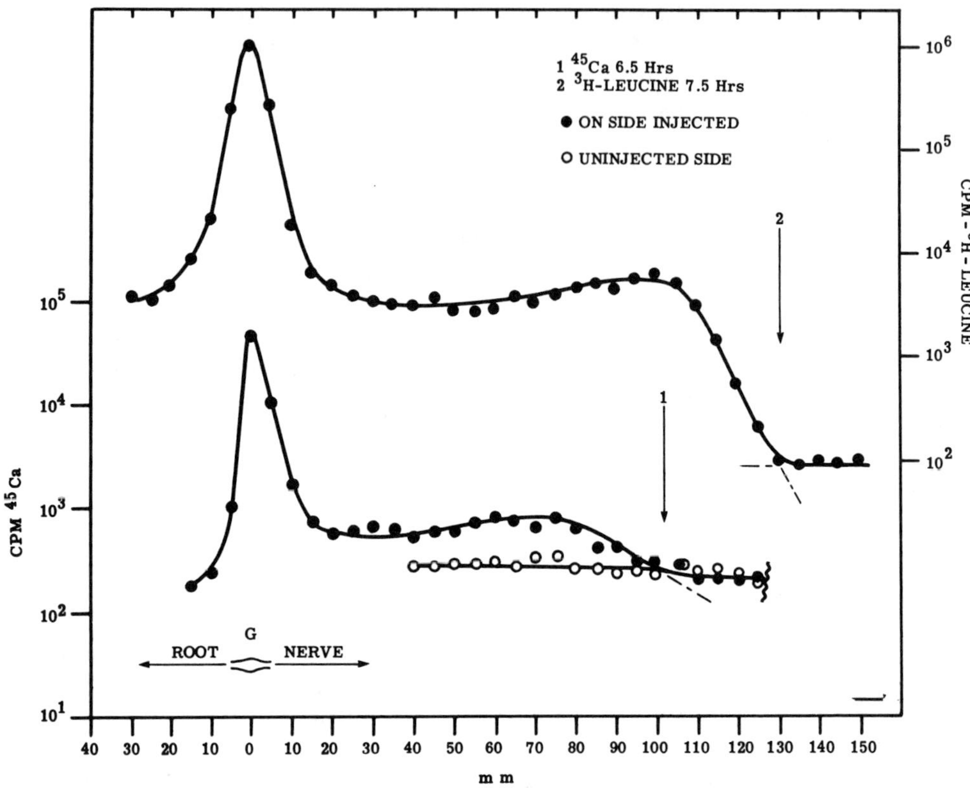

Fig. 8.12 Axoplasmic transport of Ca. Injection of ^{3}H-leucine (upper curve and arrow 2) and ^{45}Ca^{2+} (lower curves and arrow 1) was made into the dorsal root ganglia (G) of two animals to compare their outflows at 7.5 and 6.5 hr. Using ^{45}Ca^{2+} below, the unfilled circles (O) represent the level of activity in the nerve on one side not injected with ^{45}Ca^{2+} showing a relatively high background level with respect to the crest and plateau on the ^{45}Ca^{2+}-injected side (●). The outflow of Ca^{2+} is also seen to be low in comparison to nerve when ^{3}H-leucine was injected. From Iqbal and Ochs (1978).

The calcium-binding capability of this protein was assessed by equilibrium dialysis (Sandberg *et al.*, 1966). A K_d of 6.66 × 10^{-5} M was obtained from a Scatchardt plot, showing it to have a Ca^{2+} binding characteristic of CaBPs in general (Kretsinger, 1980). The CaBP was found to be closely similar to calcium dependent regulatory protein calmodulin (Iqbal and Ochs, 1980a). This was determined by comparing the CaBP isolated from nerve using sucrose homogenization and columns with calmodulin isolated as a heat-stable component and identified by its activation of calmodulin-dependent cyclic AMP phosphodiesterase. On isoelectric focusing the calmodulin was found to have a pI of 4.1–4.2, one similar to the CaBP. A further indication of its presence was obtained using trifluoperazine (TFP). This agent binds to calmodulin to block its enzyme activation (Weiss and Levin, 1978). It blocked

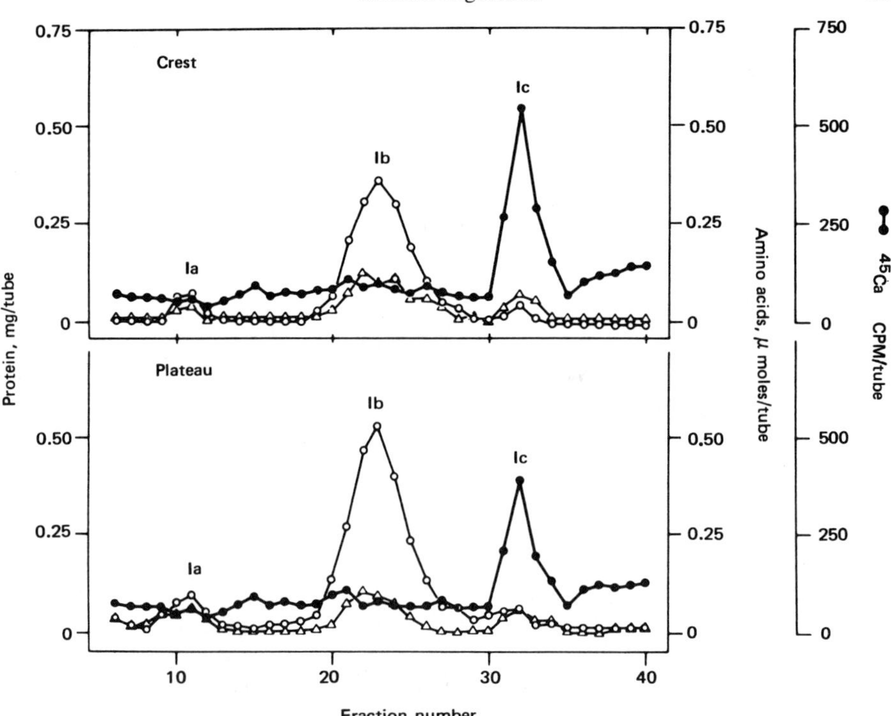

Fig. 8.13 Comparison of Ca^{2+}-associated components in the crest and plateau. Elution profiles of Sephadex G-100 Peak I proteins containing $^{45}Ca^{2+}$-bound materials taken from crest and plateau regions passed through Biogel A 5-m columns. The concentration of protein (O), amino acids (△), and radioactivity (●) in Peaks Ia, Ib, and Ic are shown. The crest contains comparatively more Peak Ic radioactivity than the plateau with a higher specific activity. From Iqbal and Ochs (1978).

the activation of phosphodiesterase when it acted on the material prepared from nerve.

Such evidence, however, does not discriminate between a true calmodulin which is fast-transported from the wider range of calcium-binding proteins. It was imperative to use other separation methods. For this purpose SDS–PAGE (Chapter 5) was employed. Cat L7 dorsal root ganglia were injected with high levels of ^{3}H-leucine or labeled aspartic and glutamic acid to label calmodulin. After several hours of outflow the nerves were ligated below the ganglia and a further outflow of several hours allowed to clearly separate the peak and plateau regions. These when sectioned into 5-mm pieces and prepared by SDS–PAGE were seen to comigrate with samples of calmodulin placed on tracks to the side. The calmodulin bands were seen to have a sharp peak moving at the usual fast transport rate close to 410 mm/day (Iqbal and Ochs, 1980b, and in preparation).

However, those studies showed that only a small proportion of labeled activity associated with the band comigrating with calmodulin. This was

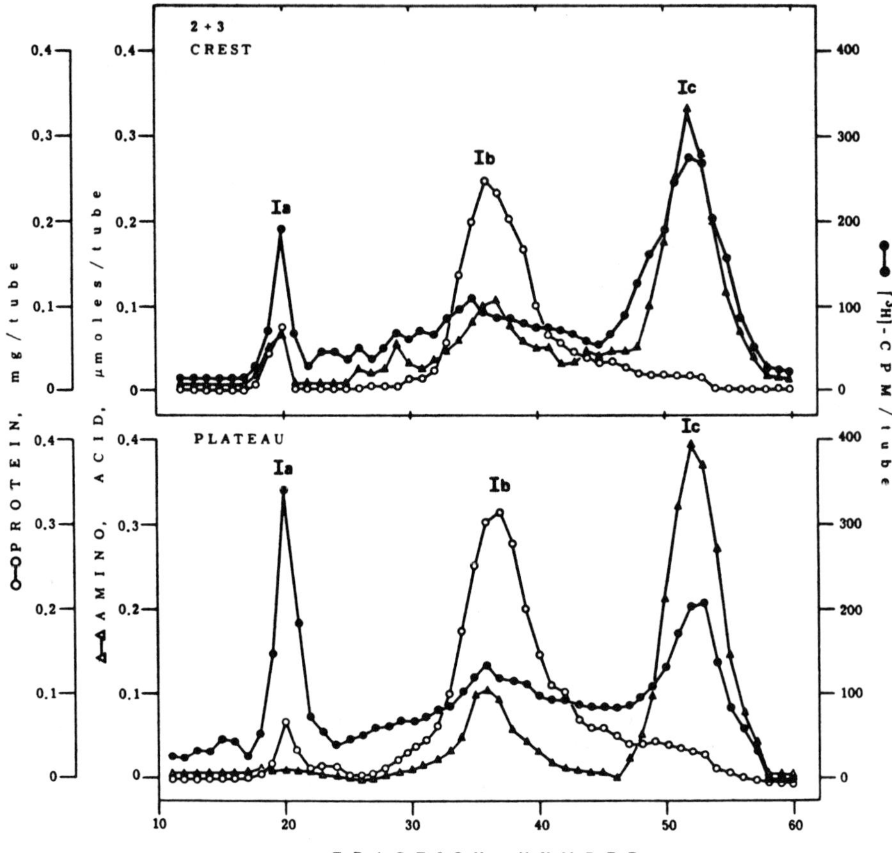

Fig. 8.14 Comparison of labeled proteins in the crest and plateau. Elution profiles of dialyzed supernatants prepared from crest and plateau regions of nerve taken 5 hr after injection of ³H-leucine into the ganglion. Two hours after injection, the nerve was ligated just distal to the ganglion and an additional 3 hr was allowed for transport. The 105,000-g supernatants of the crest and plateau regions (cf. Fig. 8.12) were filtered through Amicon Diaflo membrane DM-5 and the retentate was dialyzed and passed through Biogel A 5-m columns. Protein (○), amino acids (△), and ³H-counts (●) were determined in aliquots taken from each fraction. A higher specific activity was found in proteins of Peak Ic in the crest as compared to the plateau. From Iqbal and Ochs (1978).

further seen by first using isoelectric focusing to isolate calmodulin (Chapter 5) from the fast-transport crest and then separating the calmodulin by SDS–PAGE (Iqbal and Ochs, unpublished experiments). Erickson *et al.* (1980) recently studied the transport of calmodulin using 2-step electrophoresis first with and then without SDS, and found that the bulk was slow-transported, with 10% or less possibly fast-transported. The significance of transported calmodulin in relation to transport will be further discussed in Chapter 10.

The multiple roles which calmodulin can play in nerve requires further attention. Calmodulin acts to regulate a large number of enzymes which are important for cell function, including, in addition to the activation of cyclic AMP-dependent phosphodiesterase, an activation of protein kinases, the assembly and disassembly of microtubules (Chapter 9), the activation of a Ca–Mg ATPase of membranes, and a number of other enzymes (Wasserman *et al.*, 1977; Cheung, 1980; Means and Dedman, 1980). Its activation of Ca-ATPase, and in particular the Ca–Mg ATPase present in the microtubule side-arms (Ochs and Iqbal, 1980), is of key importance with respect to transport (Chapter 10). Another role to consider for calmodulin is its presence in the nerve terminals (Iqbal and Ochs, 1978; Blaustein *et al.*, 1978a–c). Ca^{2+} regulation is important at this site for maintenance of neurotransmission. Presumably it and/or other calcium regulatory proteins present in the terminal help regulate the processes needed to keep Ca^{2+} levels low so as to allow release mechanisms to function (Chapter 12).

D.   EFFECT OF IONS OTHER THAN Ca^{2+} ON TRANSPORT

1.   Comparison of Mg^{2+} with Ca^{2+} in Maintaining Transport

When 5 mM Mg^{2+} was added to a saline medium, axoplasmic transport in the desheathed nerve preparation was not as long maintained as with the addition of Ca^{2+} (Section B1). However, Mg^{2+} acted in the same direction as Ca^{2+} as shown by the somewhat longer downflow before a complete block appeared as compared to an isotonic NaCl medium. The time to block transport in an isotonic saline solution was 2.6 hr and with 5 mM Mg^{2+} added it was 3.3 hr (Chan *et al,* 1980). The addition of 4 mM KCl had little effect on the action of Mg^{2+} to prolong transport.

The difference between Mg^{2+} and Ca^{2+} in maintaining axoplasmic transport was more apparent when lower concentrations of the two cations were compared. A concentration of 0.75 mM Ca^{2+} with 4 mM KCl present was capable of maintaining a normal pattern of transport, while this was not the case with 0.75 Mg^{2+} present. It may be that Mg^{2+} acts to retain Ca^{2+} in the axons. Such a facilitating role is indicated by the ability of Mg^{2+} added to a medium containing a Ca^{2+} of 0.25 mM to maintain transport, a level of Ca^{2+} too low by itself to maintain transport (Chan *et al.*, 1980).

La voie *et al.* (1979) confirmed that Ca^{2+} is necessary to maintain transport in frog nerve fibers using a desheathed nerve preparation. The removal of Ca^{2+} from the incubation medium caused a block of transport and when Ca^{2+} was present there was a maintained transport. It appeared in their studies, however, that Mg^{2+} could also maintain transport in the frog nerve. Those results were obtained with a short 4-mm length of sciatic nerve desheathed. The partial maintenance of transport by Mg^{2+} may allow labeled materials to be transported through that relatively short length of desheathed nerve into the sheathed region of the nerve where normal transport can continue.

The block of transport following removal of Ca^{2+} from the medium seen in frog nerves in which the permeability of the perineurium was increased by exposure to Triton X-100, (Kanje, Edström, and Edström, 1982) could be relieved by addition of Ca^{2+} but not by Mg^{2+} (Kanje and Edström, 1981). This suggests that there is little effect of Mg^{2+} at all in this preparation.

2.   Effect of K^+ on Transport

As described in Section B1 above, the addition of 4 mM K^+ to media containing a concentration of Ca^{2+} of 1.5 mM will produce a normal appearing outflow. This level of K^+ appears to be facilitatory as described in Section B1. Another phenomenon appears when much higher levels of K^+ are present. With substitution of K^+ for Na^+, a normal appearing transport is maintained when a concentration of K^+ as high as 50–100 mM is present (Chan *et al.*, 1980a), as shown in the example of Fig. 8.15.

The rounding of the crest signifies that transport falls a little short of full

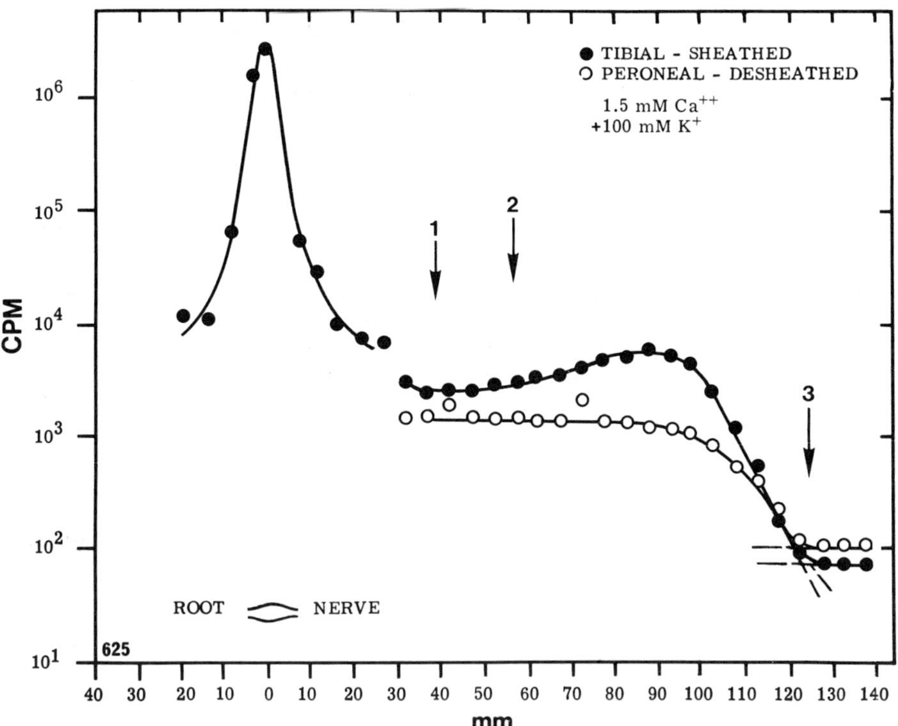

Fig. 8.15 High level of K^+ and maintained transport. Nerves removed after 2 hr of downflow in the animal were placed in a medium containing 1.5 mM Ca^{2+} + 100 mM K^+ + NaCl for an additional 4 hr of *in vitro* downflow. Arrow 1 indicates where desheathing started; arrow 2 indicates the point to which downflow moved when the nerve was put into the medium. Arrow 3 shows the transport distance normally expected in the sheathed (●) and desheathed (○) nerves. From Chan *et al.* (1980).

normalcy. The presence of transport in the face of the depolarization of the fibers brought about with such high levels of K$^+$ in the medium is in accord with other studies showing that axoplasmic transport does not depend on a functional membrane (Chapter 12). Another phenomenon revealed when high concentrations of 50–100 mM K$^+$ are present in the *in vitro* medium is that Ca^{2+} need not be present. The most likely explanation for this phenomenon is that these high levels of K$^+$ act to retain Ca^{2+} within the axons (Part B above). With a still further increase in the concentration of K$^+$ in the medium, up to 130 mM, transport was impaired (Chan *et al.*, 1980a). At present little is known as to how this effect of K$^+$ on transport comes about.

3.   Na$^+$ and Transport

Either an isotonic NaCl or isotonic sucrose medium to which Ca^{2+} is added are both capable of maintaining transport, indicating that the presence of Na$^+$ is not essential for transport. An increased Na$^+$ within the axon is, however, another matter. In that case transport is blocked. An increase in intra-axonal Na$^+$ is effected by the use of agents which increase the permeability of the membrane to Na$^+$ as will be described in Chapter 12. Another means of increasing Na$^+$ in the axon is through the use of ouabain to block the sodium pump. Ouabain causes a block of transport (Ochs, 1972a). Studies made using this agent and others which point to an increased axonal concentration of Na$^+$ as the cause of a block of transport will be described in Chapter 12.

4.   Sr^{2+} Substitution for Ca^{2+}

Deletion of Ca^{2+} from the medium results in a block of transport (Section B1), most likely when Ca^{2+} falls below the level required by calmodulin to activate the Ca–Mg ATPase providing the $\sim$ P required to support transport (cf. Part B above and Chapter 10, Section A2). A similar block of transport occurs when Ca^{2+} blockers cobalt, nickel, lanthanum, or verapamil are present to decrease the influx of Ca^{2+} (Ochs, 1982).

Strontium (Sr^{2+}) can substitute for Ca^{2+} and maintain transport in Ca^{2+}-free media when Sr^{2+} is present in concentrations of 1.5–50 mM (Ochs, 1982). After exposure of nerves *in vitro* to high concentrations of Sr^{2+}, the addition of pyroantimonate to the tissue during preparation for EM reveals, as in the case of exposure to high Ca^{2+} (Section C1), increased numbers of electron-dense particles distributed in the axoplasm along the axolemma and in the mitochondria (Ochs and Jersild, 1982, unpublished experiments). Electron probe X-ray microanalysis of the dense particles in these sites showed Sr^{2+} present in them. Those findings demonstrate that Sr^{2+} can enter the axon and have a similar localization as Ca^{2+}. Sr^{2+} could serve the same function in maintaining transport when substituting for Ca^{2+} by binding to calmodulin and allowing it to activate enzymes (cf. Cox *et al.*, 1981), including presumably Ca–Mg ATPase.

9

Microtubules, Neurofilaments, and Microfilaments

The fibrillar structures seen with light microscopy in the nerve fibers stained with silver, methylene blue, and other agents have been resolved in EM preparations as well-defined linearly organized organelles; neurofilaments, microtubules, and the more irregularly oriented microfilaments. In cross-sections of glutaraldehyde-fixed axons, the microtubules appear in EM as hollow-walled tubules, approximately 250 Å in diameter, and the neurofilaments as rods 80–100 Å in diameter (Fig. 9.1).

Also present in axons is the irregularly shaped endoplasmic reticulum (ER), more or less axially oriented, with usually somewhat wider and thinner-walls than the microtubules. The mitochondria are identified by their relatively large diameter of 0.3–0.5 μm, and, in longitudinal sections, their relatively great lengths of some 10–20 μm in the axial direction, and the organization of their cristae.

The proportion of microtubules to neurofilaments generally depends on the axonal diameter, with a greater density of microtubules present in the smaller-diameter myelinated fibers than in the larger ones. In the unmyelinated fibers, the density of microtubules is still higher and few neurofilaments are present.

Particular importance has been placed on the role of microtubules in axoplasmic transport. For example, in the transport filament model, the microtubules are considered to act as the stationary element along which the transport filaments and their bound transported components are moved (Chapter 10). This proposed role for microtubules derives in large measure from the action of tubulin-binding agents such as colchicine and the vinca alkaloids to bring about the dissassembly of microtubules and to block axoplasmic transport. The role of microtubules in transport will be reserved for Chapter 10 and the effects of various agents on microtubules in Chapter 12. In this chapter we shall describe the assembly of the microtubules from their tubulin subunits, the process of disassembly and assembly of microtubules in purified *in vitro* preparations and in the axons.

The neurofilaments and their protein subunits, and the microfilaments and the microtrabecular network will also be discussed in this chapter.

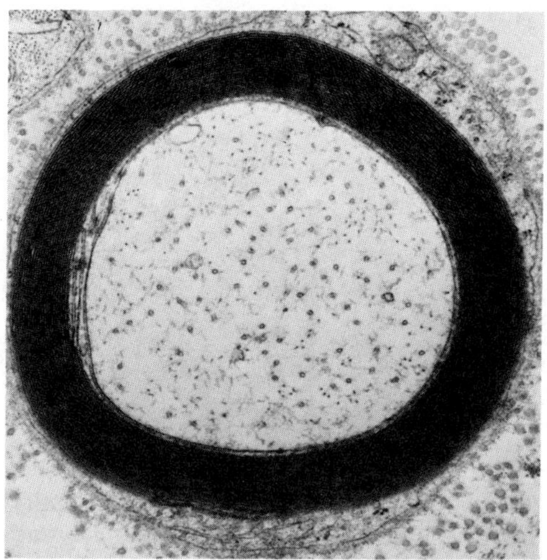

Fig. 9.1 Cross-section of myelinated nerve fiber. Microtubules are distributed throughout the axoplasm as hollow-walled profiles. Interspersed are smaller neurofilaments and the often irregularly shaped endoplasmic reticulum. Smooth compact myelin sheath is enclosed by the Schwann cell.

A. MICROTUBULES

1. Form and Composition

Much of our information on the molecular structure of microtubules has come originally from studies of ciliated organisms and from cells in the process of mitosis. The microtubules present in a variety of cell types studied, including those of nerve fibers, are generally similar in structure, appearing as unbranched cylinders. Using special high-resolution techniques, the wall of the microtubules is seen to be composed of globular proteins, roughly 50×50 Å, as indicated in Fig. 9.2.

Two types of tubulin monomers have been recognized by their slightly different amino acid compositions and charge, the α and β tubulins. These tubulin monomers are paired to form dimers aligned axially in the wall of the microtubules (Fig. 9.2).

The dimeric tubulin subunits join end-to-end to form the protofilaments oriented in the axial direction, with usually 13 such protofilaments linked by side-bonds to make up the wall of the microtubule. As can be seen in Fig. 9.2, the dimers are aligned laterally with their neighboring congeners, but are staggered with an α tubulin adjoining a β tubulin, giving rise to a helical pitch as corresponding monomers are traced laterally around the wall.

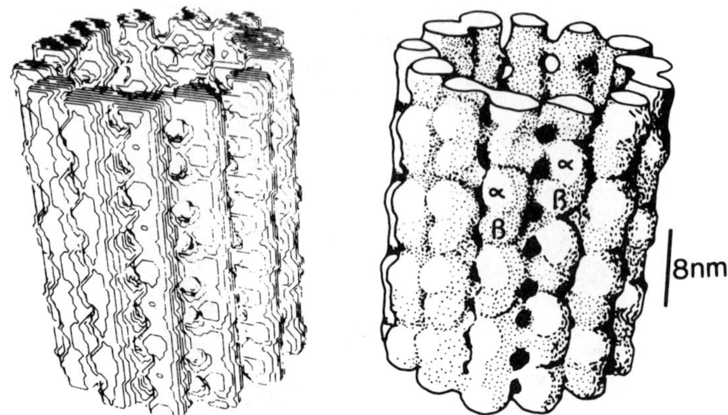

Fig. 9.2 Reconstruction of microtubules. At the left, a Fourier transform of a microtubule is computer-drawn on the basis of protein density distributions present in EM sections. The basal end of the microtubule is at the bottom. At the right, a schematic drawing based on the contoured surfaces is shown. From Amos, Linck, and Klug (1976).

2. Microtubule Assembly and Disassembly *In Vitro*

Tubulin can be isolated from nerve and brain in relatively pure form *in vitro* by taking advantage of the property of microtubules to disassemble into their tubulin subunits at low temperatures and to reassemble and reform microtubules when the temperature is raised and the proper assembly conditions are present (Weisenberg, 1972; Shelanski *et al.*, 1973; Borisy *et al.*, 1975). A microtubule-rich tissue such as the brain is homogenized at low temperatures near 0°C to depolymerize the microtubules. This material is centrifuged and the pellet containing various non-tubulin components discarded. The supernatant containing the soluble tubulin is warmed to 37°C in the presence of guanosine triphosphate (GTP), Mg^{2+}, and EGTA to chelate the Ca^{2+} in the medium. Under these conditions, tubulin rapidly assembles to form microtubules. The solution is then centrifuged to bring down the microtubules as a pellet and the supernatant containing non-tubulin soluble proteins discarded. After several such cycles of cooling and heating, a purified solution of tubulin is obtained.

The process of purification by cycles of cooling and heating may be followed by passing the resultant materials through DEAE-Sephadex gel columns. The efflux from the columns shows a prominent peak at 110,000–120,000 daltons, this representing tubulin in its dimeric form.

The tubulin dimers are separated into their α and β monomers by the use of SDS–PAGE as described in Chapter 5 where the tubulins have a MW close to 55,000 (Feit *et al.*, 1971; Bryan, 1974). The monomers appear as separate peaks when chromatographed on Biogel A columns (Fig. 9.3).

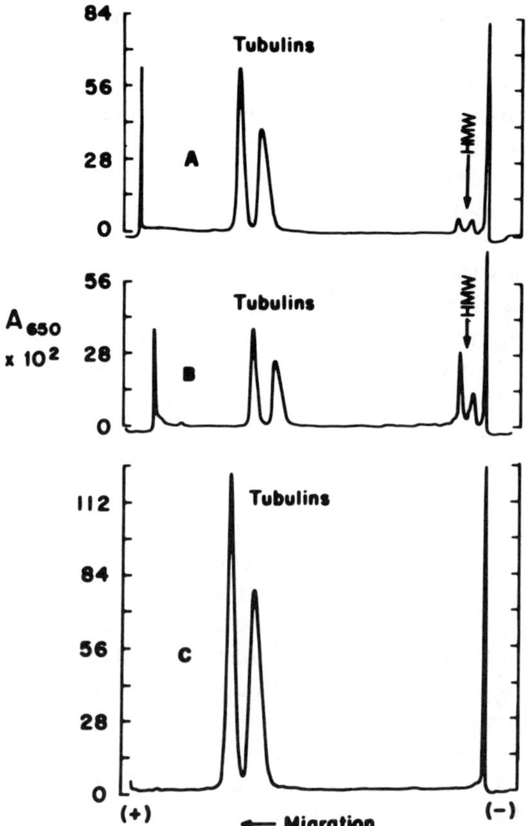

Fig. 9.3 Tubulins and microtubulin-association proteins (MAPs). These MAPs, here designated as high-molecular-weight proteins (HMWs), are shown in gel scans (fast green-stained Laemmli SDS acrylamide). Column fractions obtained from 2× polymerized tubulin chromatographed on 1.5 M Biogel A: (A) tubulins and HMW after chromatography; (B) the first peak eluted from the column rechromatographed showing enrichment of the HMW proteins and the presence of some tubulin; and (C) the second peak eluted from column showing only tubulin subunits without HMW proteins. From Rosenbaum *et al.* (1975).

When separated by SDS electrophoresis, the β tubulin is the more rapidly migrating of the two monomers (Bryan and Wilson, 1971; Eipper, 1972). The small charge difference between the α and β tubulin subunits (Luduena and Woodward, 1975) gives rise to the different mobilities with positions corresponding to 57,000 and 53,000 daltons seen on SDS–PAGE (Hoffman and Lasek, 1975; Forgue and Dahl, 1978).

Higher molecular weight (HMW) components copurify with the tubulins. These, as will be seen in a later section, are microtubular-associated proteins (MAPs) bound to tubulin and associated with microtubules as side-arms.

The MAPs of 350,000 and 300,000 daltons, termed MAP 1 and MAP 2, constitute respectively some 6% and 20% of the total microtubular protein associated with α tubulin (B of Fig. 9.3). Gel electrophoresis of β tubulin gives rise to tubulin peaks with little, if any, comigrating MAP components present (C of Fig. 9.3). Several functions are ascribed to the MAPs, e.g., the assembly of microtubules and a role in transport. The relation of MAPs to assembly will be discussed in this section and that related to axoplasmic transport in Chapter 10.

Calcium inhibits the assembly of microtubules, thus EGTA is added to the assembly medium to chelate Ca^{2+} and reduce it to low levels. On the other hand Mg^{2+} is required for assembly. Magnesium binds to tubulin in a ratio of 1 mole of Mg^{2+} to 1 mole of tubulin (Olmsted and Borisy, 1975), at some site other than that to which Ca^{2+} binds. Apparently, when Mg^{2+} binds to tubulin it changes the conformation of the molecule, making it more likely to assemble. Another factor important in assembly is GTP. The nucleotide undergoes a dephosphorylation with a bound molecule of GDP remaining on each tubulin monomer as it assembles into the microtubule. The nucleotide ATP can also promote assembly (Weisenberg, 1972) after its transphosphorylation to GTP. Assembly may occur without nucleotide if a sufficiently high concentration of glycerol is present (Shelanski *et al.*, 1973). Glycerol apparently causes a conformational change of tubulin enabling it to assemble.

Glycerol protects the sulfhydril groups of tubulin (Mellon and Rhebun, 1976). Approximately 7 free sulfhydril groups are present per 55,000-MW tubulin subunit when tubulin is prepared in a glycerol-containing medium, and only 4 free sulfhydrils when prepared without glycerol. Assembly requires the presence of free sulfhydril groups which are close to the tubulin-binding sites. An inactivation of the sulfhydrils interferes with the assembly process. Diamide, an agent effective in oxidizing sulfhydril groups (Kosower and Kosower, 1969; Kosower *et al.*, 1972), causes a marked inhibition of microtubular assembly (Mellon and Rhebun, 1976).

The assembly of microtubules from a solution of tubulin is promoted by a 30S substance (Weisenberg, 1974; Olmsted *et al.*, 1974; Kirschner and Williams, 1974) which was discovered to consist of rings or discs of tubulin (Fig. 9.4).

This figure shows several models of microtubular assembly. In one such model the rings open up lengthwise to form the protofilaments which then line up side by side. These can spontaneously curve inward until the 13 protofilaments join to complete the wall. A wall with 13 protofilaments apparently represents the lowest energy level for the lateral binding of the protofilaments into the completed wall of the microtubule. Additionally, assembly occurs by the addition of dimers to the ends of the protofilaments to thus extend the length of the microtubule.

Two other factors promoting assembly are those of the MAPs (Sloboda *et al.*, 1976) and the *tau* proteins, a group of 4 or 5 polypeptides of 50,000–70,000 daltons (Penningroth, Cleveland, and Kirschner, 1976; Weingarten

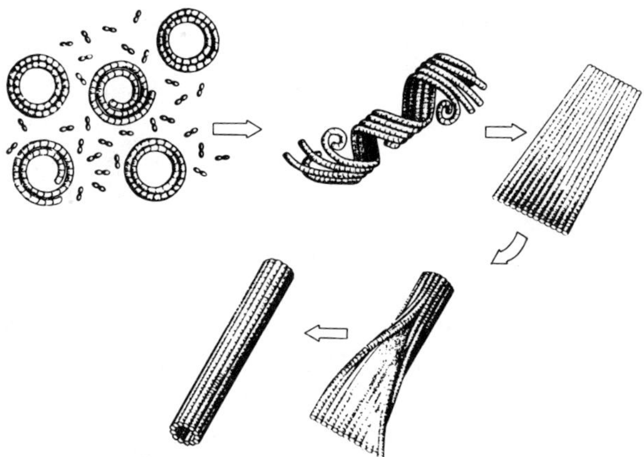

Fig. 9.4 Schematization of microtubule assembly. Rings open to form sheets of protofilaments which curl laterally to form the microtubule wall. From Penningroth, Cleveland, and Kirschner (1976).

et al., 1975). It appears from the work of Herzog and Weber (1978) that while both the MAPs and the *tau* proteins are capable of stimulating microtubule assembly, the MAPs are more active. On the other hand, Lockwood (1978) has isolated from the *tau* factor a tubulin assembly protein (TAP) of 67,000 daltons which he considers to be specifically required for microtubular assembly. Kim *et al.* (1979) found a MAP$_2$, one which has the property of being heat-stable, that is able to stimulate microtubule assembly from concentrations of tubulin ordinarily too low for self-assembly. Those microtubules were seen to contain 14 protofilaments in their wall instead of the more usual 13, a deviation which requires further investigation.

3. Polypeptide Factors Preventing Assembly *In Vitro*

On the basis of their relation to the colchicine-binding site of tubulin, several protein species have been found which promote the disassembly of microtubules. A description of how colchicine is considered to act in relation to its binding to tubulin will be reserved for discussion in Chapter 12. We here assume colchicine-binding as a property of tubulin protein.

The question may be raised as to why colchicine, a plant substance, binds specifically to an animal protein. A similar question was raised with respect to the morphine receptors of brain where by its use substances acting like morphine, the endorphins and enkephalins, were isolated from the brain (Snyder and Childers, 1979). Following this lead, endogenous proteins were isolated from brain which were able to bind to the colchicine-binding sites of tubulin (Lockwood, 1979; Sherline *et al.*, 1979). In the studies carried out by Lockwood, tubulin was coupled to a gel for affinity column extraction of endogenous substances. Materials from the brain were passed through the columns and those components specifically binding to the affinity column were afterwards eluted with KCl. Several proteins were found by this technique, including those which bind specifically to tubulin and inhibit its binding of colchicine. One of the inhibitor proteins found by this method had a MW in excess of 15,000, another a MW of less than 5,000. The trypsin sensitivity of the larger species indicated it to be a protein. The smaller-MW species proved to be heat stable and only partially trypsin sensitive, as might be expected of a low-MW polypeptide.

These endogenous components inhibit microtubule assembly. In contrast to colchicine, which is slow acting, temperature-dependent, and essentially binds by an irreversible process (Wilson and Bryan, 1974), these endogenous inhibitory components appear to bind to a hydrophobic pocket in tubulin. They are eluted from the binding site by increasing ionic strength, a procedure which does not affect colchicine binding, indicating an ionic binding of the endogenous substances to tubulin. Apparently, when the endogenous inhibitor occupies a hydrophobic site on tubulin, it overlaps onto the hydrophilic binding site responsible for microtubule assembly.

In the studies of Sherline and his colleagues, a somewhat different method of isolating the endogenous inhibitor substance was used. They found it to be associated with the microsomal pellet prepared by differential centrifugation. The inhibitor was partially solubilized and found to have an apparent MW of 250,000. As in the case of Lockwood's inhibitor substance, it also has the property of inhibiting the binding of colchicine to tubulin and the assembly of tubulin into microtubules. The inhibitor was inactivated by heat and by trypsin, indicating its protein nature. The relation of these different inhibitory proteins to one another remains to be resolved, as well as the control over the process of microtubule assembly these substances may have in the axon (cf. Section A5 below).

4. The Directionality of Microtubule Assembly *In Vitro*

The assembly of microtubules occurs by an addition of the tubulin subunits to the ends of the growing microtubules as pictured in Fig. 9.5.

The rate of assembly is controlled by the kinetics of association and dissociation of the tubulin at the microtubule ends. Assembly at the two ends of the microtubules have different rates (Bergen and Borisy, 1980), thus giving rise to the possibility of a directionality in the assembly of the microtubules. Such a directionality of assembly was shown by Rosenbaum *et al.* (1975). He used short pieces of labeled microtubules as initiating sites for the assembly of microtubules *in vitro* from a solution of unlabeled tubulin (Fig. 9.6).

As indicated in this schematization, if tubulin adds at a similar rate to either end of the short, initiating piece of microtubule, a bidirectional growth would occur with the labeled segment remaining in the center. If tubulin adds to one end of the growing microtubule at a greater rate, an essentially

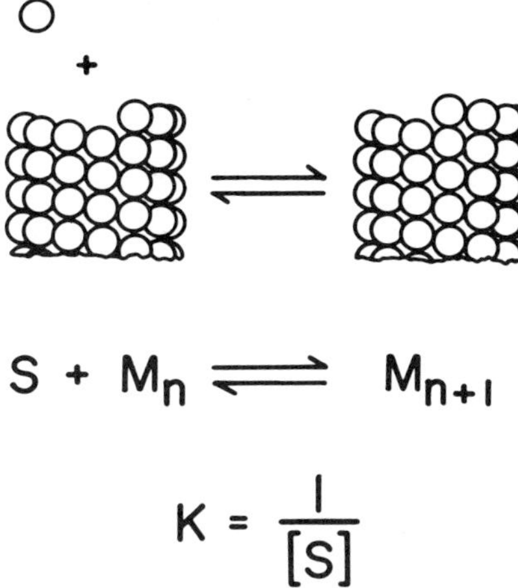

Fig. 9.5 Assembly of tubulin dimers at the ends of microtubules. The association of 6S tubulin dimer subunits at the end of the microtubule is shown by the equilibrium relation where S, the subunit, joins to the microtubule M_n to give M_{n+1} an increased length of microtubule. Rate constants for association and dissociation result in an overall equilibrium constant K which is inversely related to the subunit concentration $[S]$. From Borisy *et al.* (1975).

unidirectional assembly will take place. The results of this experiment indicated that tubulin adds to one end, namely, that growth is unidirectional (Fig. 9.6).

Wilson and his associates used ³H-GTP as a marker to assess the assembly and disassembly of microtubules *in vitro*. Tubulin has an exchangeable GTP site which, as it becomes assembled into the microtubule, contains non-exchangeable GDP (Margolis and Wilson, 1978). Using the labeled GDP bound to tubulin as a marker, tubulin was seen to add to the growing microtubules at one end, the marker gradually being moved down the microtubule to the other end where disassembly takes place, a process called "treadmilling." The advance of the labeled GDP as a marker indicated a rate of microtubule assembly of 0.69 μm/hr, or approximately 16.6 μm/day. The difference between the sites of assembly and disassembly at the two ends was also shown by the tubulin-binding agent podophyllotoxin (cf. Chapter 12). This agent appears to preferentially bind to the assembly end of the microtubules and cause a block there while disassembly at the other end was unaffected by the agent, this resulting in the disassembly of the microtubules.

The *in vitro* undirectional assembly of microtubules shown by Margolis and Wilson might have some relation to slow transport. However, the rate of *in vitro* assembly indicated by the GDP marker movement is almost several orders smaller than the rate of approximately 1 mm/day usually considered typical of slow transport, as will be discussed in Chapter 10.

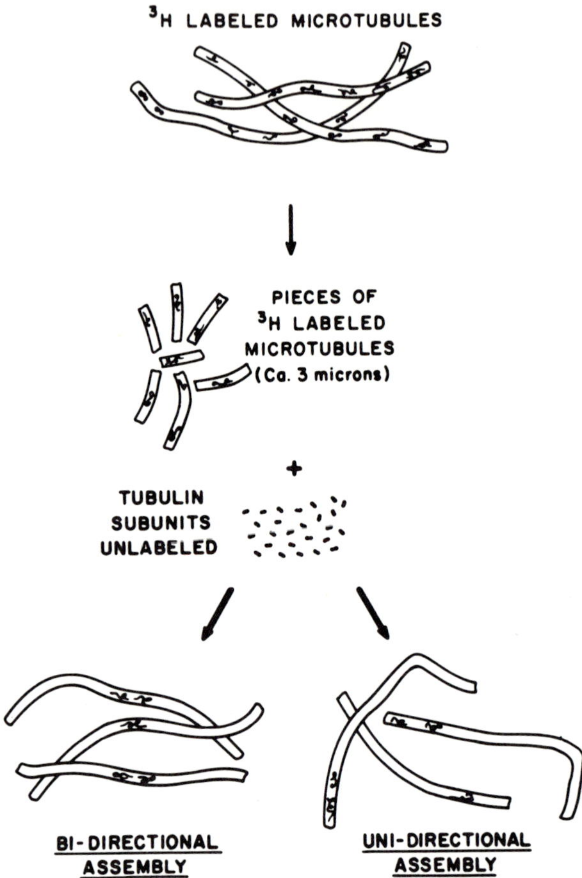

Fig. 9.6 Directionality of microtubule growth. Directionality experiments were designed to show either a unidirectional growth with unlabeled tubulin adding to one end or, if bidirectional growth is occurring, to both ends. Experimental evidence for unidirectional assembly was found. From Rosenbaum *et al.* (1975).

Heidemann and McIntosh (1980) have shown that extra tubulin can be added to the wall of the microtubule. Isolated *Tetrahymena* ciliary axonemes were lysed and exposed to high concentrations of tubulin and the direction of the hook-like additions of tubulin was found to depend on the polarity of the microtubules. Those studies and similar ones made in cut renal nerves (Heidemann, 1980) and more recently by Burton and Paige (1981), further indicate that the majority, if not all microtubules, are similarly oriented in the nerve fiber. The implications of this finding for transport will be discussed in Chapter 10.

5. Assembly and Disassembly of Microtubules *In Vivo*

While additions of tubulin are made at one end of the microtubule with disassembly at the other *in vitro*, it is possible that *in vivo* other processes

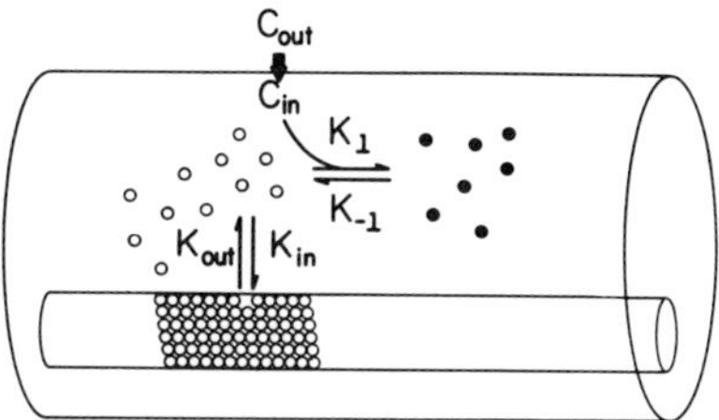

Fig. 9.7 Microtubule turnover and colchicine. The dissociation of microtubules in nerve fibers in the presence of colchicine is shown schematically. The large arrow represents diffusion of colchicine from the extracellular pool (C_{out}) into the axon and into an intracellular pool (C_{in}). The forward rate constant for the formation of the colchicine–tubulin complex in the axon and the rate constant for the dissociation of the bound complex are K_1 and K_{-1}, respectively. Free tubulin is indicated by the small unfilled circles (○) and the tubulin–colchicine complex by the solid circles (●). The equilibrium between the free tubulin subunits and the polymerized microtubule is indicated by the association rate constant (K_{in}) and dissociation rate constant (K_{out}). A shift to bound tubulin decreases free tubulin and in turn depletes tubulin from the microtubule. From Paulson and McClure (1975).

are at work whereby tubulin undergoes a local turnover all along the length of the microtubules (cf. Chapter 10). Microtubules in the axons are disassembled by low temperature and can rapidly reassemble on rewarming (Chapter 12) and assembly occurs in the regenerating fiber (Chapter 15). We would expect that there must be some mechanisms present in the axon whereby tubulin can turn over in the microtubules or the microtubules can quickly become reformed from the pool of tubulin present within the axons after their disassembly (Chapter 5). A local turnover of tubulin in the microtubules of dividing cells had been proposed by Inoué and Sato (1967) and by Dietz (1972) to explain the action of tubulin-binding agents to cause a disassembly of microtubules, a concept diagrammed for the axon by Paulson and McClure (1975) as shown in Fig. 9.7.

The action of colchicine to bind to tubulin, as noted in this model, will be taken up in more detail in Chapter 12. One phenomenon pertinent to our present considerations is that colchicine does not bind to any great extent to intact microtubules *in vitro*, presumably because the colchicine binding sites are hidden or inaccessible (Wilson and Meza, 1973). Yet colchicine and other tubulin-binding agents do cause a disassembly of microtubules in nerve *in vivo*. Such a behavior suggests that the agents are acting at sites on microtubules normally present *in vivo*, presumably where there are mechanisms for the turnover of tubulin in the wall of the microtubule. Perhaps the process can occur at the ends of the staggered short lengths of microtubule measuring 1.2–10.7 μm found in the serial EM reconstructions of the nematode *Caenorhabditis elegans* axons by Chalfie and Thomson (1979), the longer average lengths of 108 μm (assuming a similarity of their lengths) present in dorsal root neuron fibers in the tissue cultures of Bray and Bunge (1981), and the 370–706 μm short lengths measured in mouse saphenous nerve by Tsukita and Ishikawa, (1981). The implications of this organization for transport will be pointed out in Chapters 10 and 11.

The MAP and *tau* factors described as promoting assembly of microtubules *in vitro* (Section A2) could very well be involved in the assembly–disassembly of microtubules normally taking place continually *in vivo* all along the length of the nerve fiber at the ends of the short lengths of microtubules. The distribution of MAPs along the microtubules was shown by the use of monospecific antibodies to the MAPs. In neuroblastoma cells exposed to the antibodies to MAP an extensive array of coated filaments (microtubules) passing from the cell nucleus into the neurites was seen. Colchicine, which disassembles microtubules, caused a marked reduction in that reaction (Sherline and Schiavone, 1977). Similarly, the *tau* factor was shown by immunological means to be associated with the filaments (microtubules). Antibodies to the TAP component isolated by Lockwood (1978) were also found distributed along the length of the microtubules. Pretreatment of the cells with colcemid or cold to cause microtubular disassembly destroyed that distribution pattern. Immunological studies also revealed that the purified TAP factor is not the same as, or a part of, the higher molecular weight MAPs.

We would suppose that MAPs including *tau* factor or TAP are present in a pool in the axon along with a pool of tubulins, and that as the tubulins are assembled into the microtubules, the MAPs are added on as well. Takenaka and Inomata (1981) found a ^{32}P labeled 310,000 dalton protein with the electrophoretic mobility of MAP2 transported in rat sciatic nerve with an apparent rate of 6.6-10.6 mm/day. The MAP2 protein could be moved down to enter a pool in the axon from which it turns over along with the tubulin turning over in the microtubules.

This view differs from the one where the microtubules are considered not to undergo a turnover of their tubulin in the axon, but that the microtubules are assembled in the cell bodies and the organelle moved down within the axon as such, a concept to be dealt with in Chapter 11. Here we note that this latter view would imply that the MAPs would be moving down within the nerve fibers at the same rate. Tytell, Brady, and Lasek (1980) found that while *tau* protein was present along with the tubulins in the SCa component of slow transport, very little of the MAPs were seen to be transported. This the authors consider might be explained by the heterogeneity of microtubules in the preparations examined. Additionally, those findings are also in accord with different downflows into pools of the various proteins which participate in the turnover of tubulins and MAPs in the microtubules within the axon.

The assembly–disassembly process is activated by calmodulin (Chapter 8). This was indicated in studies of the disassembly of microtubules in the spindles of dividing cells (Marcum *et al.*, 1978; Means and Dedman, 1980) where indirect immunofluorescence showed calmodulin to be specifically associated with the chromosome-to-pole region of the mitotic apparatus in dividing cells during the metaphase–anaphase stage, i.e., in the microtubule-rich area. Calmodulin, in the presence of 10^{-5} M Ca^{2+}, was able to inhibit and reverse microtubular assembly *in vitro*, while in a concentration of 10^{-6} M, it was ineffective. A variation in Ca^{2+} concentration thus can act as the messenger allowing calmodulin to exert its action on the microtubules when the level of free Ca^{2+} in the cell rises. Evidence that tubulin is prevented from assembling by means of a calmodulin-activated process was given by Kumagai and Nishida (1979).

Possibly the calmodulin present in nerve (Iqbal and Ochs, 1980a) has, as one of its functions, a modulation of the turnover of tubulin in the microtubules, at disassembly and/or assembly sites all along the length of the microtubules, presumably at the assembly ends of the short lengths of microtubules described above. Under conditions wherein the Ca^{2+} regulatory mechanisms are inadequate to keep Ca^{2+} levels low (Chapter 9), and free Ca^{2+} rises too high, calmodulin may activate enzymes which bring about the disassembly of microtubules. Thus, there are several means by which the microtubules undergo assembly–dissassembly in the axons *in vivo*. How endogenous control proteins and/or a calmodulin-activated process are related to such assembly–disassembly processes remains at present unknown.

Taxol, an agent derived from the plant *Taxus brevifolia*, interacts with tubulin to enhance its polymerization *in vitro* (Schiff, Fant and Horowitz, 1979) and *in vivo* (Schiff and Horwitz, 1981). The agent will induce polymerization without MAPS when GTP is present. With MAPs and taxol present *in vitro*, polymerization was dramatically augmented, occurring as a wave at 0° with GTP present. A second wave of polymerization was seen on warming to 38° (Hamel *et al.*, 1981). Further study of the effects of taxol and other agents augmenting and inhibiting assembly promises to better reveal the underlying mechanisms.

6. Location of Microtubular Associated Proteins (MAPs) and ATPase Activity

The MAPs are present in the side-arms positioned at intervals along the microtubules in EM preparations. This identification was made by comparing the microtubules assembled from tubulin freed of MAPs with the microtubules assembled from tubulin in the presence of MAPs (Fig. 9.8).

As can be seen in this figure, side-arms are seen when MAPs are present. Kim, Binder, and Rosenbaum (1979), using the superimposition technique of McIntosh (1974) to enhance the appearance of the side-arms, also showed the side-arms to have a regular periodicity of 320 Å axially along the length of the microtubules (Fig. 9.9A).

The side-arms on microtubules have ATPase activity, as indicated by their incorporation of ^{32}P from ^{32}P-ATP (Sloboda *et al.*, 1976). Bovine brain microtubules contain a Ca-ATPase (White, Coughlin, and Purich, 1980).

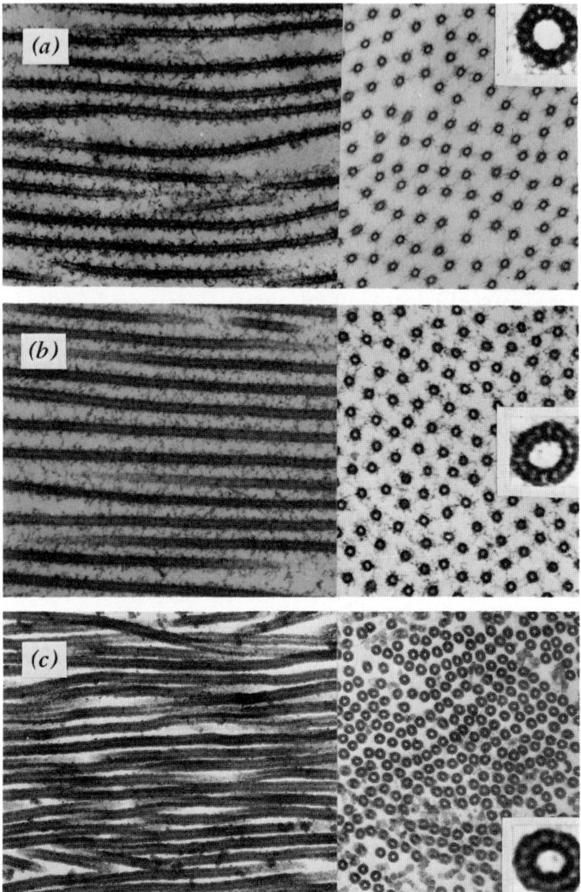

Fig. 9.8 Microtubules with and without side-arms. Microtubules in longitudinal section (left) and in cross-section (right) with a single microtubule in the insert showing 13 tubulin subunits in the wall. In (a) microtubules were assembled from tubulin with a purified MAP$_2$ fraction present. Side-arms are present. (b) Microtubules are assembled with saturated unfractionated MAPs present. Some side-arms are present. (c) Microtubules assembled from tubulin only show smooth-walled microtubules without side-arms. From Kim *et al.* (1979).

However, as the authors indicate, under their conditions of isolation the ATPase they isolated may not be the only one present *in vivo*. Hiebsch, Hales, and Murphy (1979) showed the presence of dynein-like ATPase activity associated with tubulin prepared by several cooling–heating cycles. Using this method, Ca–Mg ATPase activity was also found associated with tubulin (Ochs and Iqbal, 1980). A broad peak of Ca–Mg ATPase activity was seen at Ca^{2+} concentrations of 10^{-8}–10^{-5} M, a level at which free Ca^{2+} is likely to exist in the axon. The sensitivity of the enzyme to Ca^{2+} in this low range

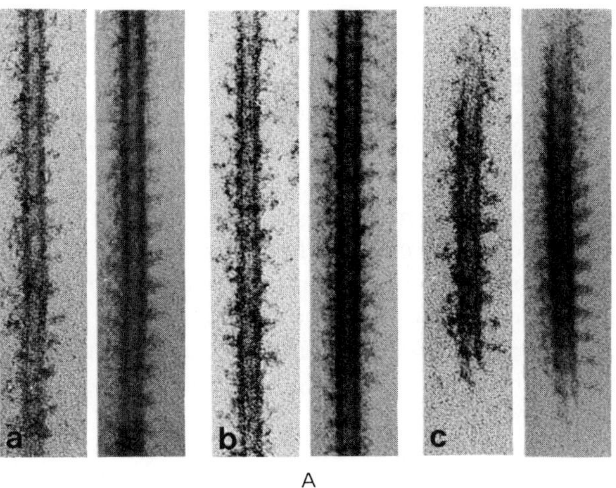

Fig. 9.9A Side-arm spacing on microtubules. Electron micrographs of thin sections of MAP$_2$-saturated microtubules before and after translational superposition are shown. In each pair of microtubule images, the one on the right was produced by translational superposition along the axis of the microtubule to a distance determined by measuring the spacing between projections on the original microtubule in the left image. Portion (a) was photographed three times with a translation increment equivalent to 29 nm; portion (b) was photographed three times with the translation increment equivalent to 28.2 nm; portion (c) was photographed four times with the translation increment equivalent to 27.5 nm. From Kim *et al.* (1979).

of Ca^{2+} concentration indicates a calmodulin activation of the ATPase, a possibility further suggested by the effect of trifluoperazine (TFP) to block the Ca–Mg ATPase activity. This agent acts to interfere with calmodulin's activity (Chapters 8 and 12). The Ca–Mg ATPase isolated with tubulin by cycles of cooling and warming may not necessarily be those seen as structural entities in EM (e.g., Figs. 9.8 and 9.9). They may be bound to or only associated with the side-arms. That the enzyme–calmodulin complex is only bound to the side-arms and is not an integral part of them is suggested by the irregularity of the Ca–Mg ATPase activity in such preparations and the loss of activity after a number of cycles of purification. The association of Ca–Mg ATPase with the side-arms and its role in relation to transport will be further discussed in Chapter 10 where transport mechanisms are dealt with. We should, however, note here the similarity between the microtubule side-arms and the dynein of cilia and flagella considered to be the cause of a sliding of the microtubules and in turn the bending movements of the cilia and flagellae. This is achieved by a cyclic attachment of the side-arms between the adjacent microtubules, the dynein of the side-arms containing an ATPase capable of utilizing ATP for the energy required for such sliding movements (Satir, 1968; McIntosh, Helper, and Van Wie, 1969; Summers and Gibbons, 1971).

Haimo, Telzer, and Rosenbaum (1979) prepared dynein from the microtubules of the flagella of *Chlamydomonas* and combined it with purified brain tubulin free of MAPs. The dynein became bound to the tubulin and, on assembly of the tubulin subunits into microtubules, the thick projections characteristic of dynein side-arms were seen present all along the length of the microtubules with a spacing related to the tubulin dimer and at intervals similar to those of the side-arms in Fig. 9.9B. The dynein side-arms retained the properties characteristic of their flagellar source. The addition of ATP caused an aggregation of the microtubules, brought about by the side-to-side cross-bridging of microtubules by dynein. The fact that dynein binds to sites on brain microtubules shows that it shares this binding property with the MAPs of nerve microtubules. The dynein on the nerve microtubules maintained an orientation at an angle of 55° to the axis of the microtubule, indicating a directionality of the binding sites which could be related to the directionality of the side-arms normally present on the microtubules, or it may represent a property of dynein.

A clear zone is often seen around the microtubules in EM cross-sections. In part this represents a coating on the microtubules. Such coatings were seen on microtubules isolated by differential centrifugation from brain tissue in a medium containing hexylene glycol at an acid pH (Kirkpatrick *et al.*, 1970), this procedure preventing the disassembly of the microtubules which normally occurs at low temperatures. Similarly, an amorphous coating was seen on microtubules isolated from brain homogenates in a medium containing glycerol and DMSO (Behnke, 1975). The coating appears to consist of a series of stacked disks with an overall thickness measuring 360–390 Å. Such coatings appear to consist in part of acid glycoproteins and mucopoly-

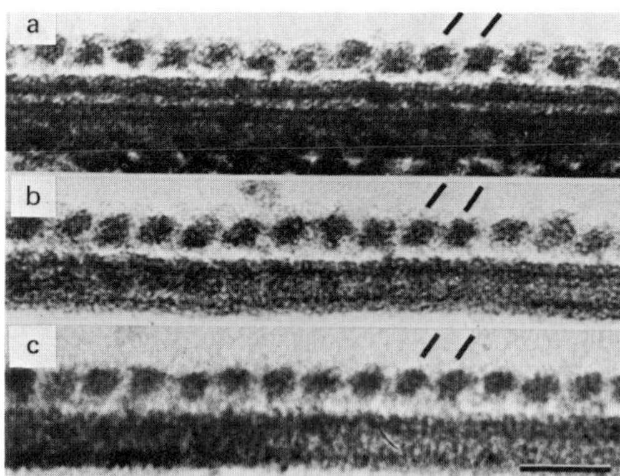

Fig. 9.9B Dynein addition to microtubules. (a) Longitudinal sections of outer doublet of an intact axoneme of *chlamydomonas* shows a row of dynein arms. In the lower sections (b,c), microtubules from brain polymerized with dynein present shows the same type of regular addition as side-arms from brain (cf. Fig. 9.9A). From Haimo *et al.* (1979).

saccharides (Section C below) which have been suggested to play a role in transport in one model of transport (Chapter 10).

B.   NEUROFILAMENTS (INTERMEDIATE FILAMENTS)

1.   Form of Neurofilaments in Axons

The other prominent linear organelle extending axially in the nerve fiber is the neurofilament (Wuerker and Kirkpatrick; 1972; Goldman *et al.*, 1976). It appears as a rod-like structure 80–100 Å thick and has been termed an "intermediate filament" because it falls in diameter between the microtubules and the thinner 30–50-Å microfilaments (Part C below). The neurofilaments are usually more numerous than the microtubules in the larger myelinated nerve fibers but are relatively sparse in the dendrites and in unmyelinated fibers (Potter, 1971). On the other hand, neurofilaments are the only filamentous elements present in the giant axons of *Myxicola* and squid. Thus, while emphasis has been placed on the role of microtubules with regard to transport (Chapter 10), the possibility that neurofilaments may also play such a role, while less likely, cannot be completely discounted (Ochs, 1971b).

The neurofilaments appear to be composed of globular subunits 35 Å in diameter with 3–6 such units forming the wall of the neurofilament. This gives rise to the beaded appearance of the neurofilament seen in high-resolution EM. The overall diameter of the neurofilaments of approximately 80–100 Å leaves room for only two such units across its width. Some possible spatial arrangements of the subunits to account for its morphological appearance have been given by Wuerker (1970), as shown in the following two models (Fig. 9.10).

Alternatively, neurofilaments may consist of two or three strands of helically twisted polypeptide chains. Neurofilaments like microtubules are also seen in EM to have projecting side-arms (Peters and Vaughn, 1967; Metuzals, 1966; Tennyson, 1970).

2.   Subunit Composition of Neurofilaments

Intermediate filaments are present in both glia and neurons. It is therefore necessary to select tissues enriched with neurofilaments in neurons to obtain identified material for biochemical analysis. One such source is the highly structured bundle of neurofilaments present in the giant axon of the marine worm *Myxicola* (Gilbert, 1972). Neurofilaments can be rapidly extracted as an entity from these axons, and when analyzed by SDS gel electrophoresis, were found to contain two dominant polypeptides of 160,000 and 152,000 daltons (Gilbert *et al.*, 1975). An important finding was that a Ca-dependent protease is also present in these axons and, if enough Ca^{2+} is present in the course of extraction, the enzyme is activated to cleave the polypeptide chains

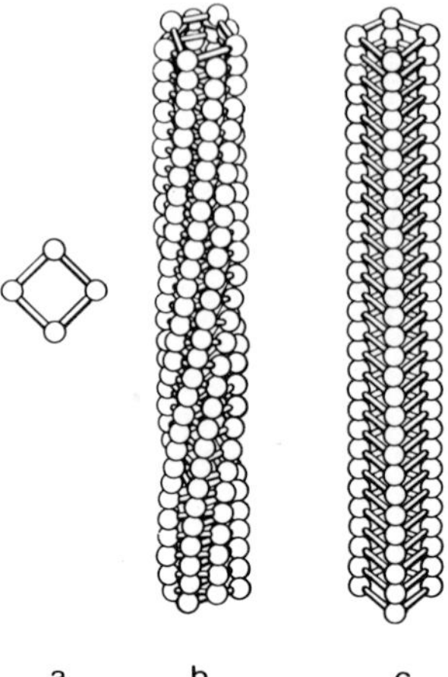

Fig. 9.10 Theoretical arrangements of subunits in neurofilaments. (a) The basic unit is seen from an end view. (b) Units shown in a longitudinal view exhibit spiral rotation. (c) Units are shown stacked directly on each other. From Wuerker (1970).

to lower-MW polypeptides of 70–90,000 and 58,000 daltons. The higher-MW polypeptide components of the neurofilaments in *Myxicola* do not appear to be uniformly present in all axons. Squid axons, for example, which also contain neurofilaments and not microtubules, were shown by Lasek *et al.* (1979) to contain mainly polypeptides of 60,000 and 200,000 daltons. More recently, using squid brain as a source, the polypeptide composition found for neurofilaments was a dominant 60,000-dalton component (constituting 70% of the total protein) with three minor polypeptides of 74,000, 100,000, and 220,000 daltons (Zackroff and Goldman, 1980). This result is of special importance in that the composition was determined in neurofilaments purified by an assembly–dissassembly procedure, a technique heretofore not available for the study of this material.

A triplet of labeled polypeptides of 200,000, 145,000 and 68,000 daltons was found in rat nerve using SDS–PAGE by Hoffman and Lasek (1975) and Lasek and Hoffman (1976) and identified as the neurofilament subunits (cf. Chapter 5). A similar identification was made by Schlaepfer and Freeman (1978). The latter took rat peripheral nerve and spinal cord and subjected it to osmotic shock in media containing a Ca-chelator to isolate neurofilaments, identified as such in EM preparations. The washed neurofilaments were subjected to SDS–PAGE and shown to contain labeled polypeptides of 200,000, 150,000 and 69,000 daltons. When exposed to Ca^{2+}, the neurofilaments were disrupted, the ion stimulating a Ca-activated proteolytic enzyme which is present in the axons. The proteolysis could be prevented by incubating the nerve tissue with p-chloromercuric-benzoate to inhibit the proteolytic enzyme (Schlaepfer and Hasler, 1979). The Ca-activated protease appears to account for earlier reports that a 50,000–54,000-dalton polypeptide is the dominant neurofilament subunit, this polypeptide appearing as a degradation product of the action of the protease (Shelanski and Liem, 1979). Further evidence that the triplet polypeptides represent the subunits of the neurofilaments was found using a special control over proteolytic degradation and the isolation of the triplet polypeptides 200,000, 150,000 and 68,000 daltons with a high degree of purity (Liem *et al.*, 1978).

Using an immune-affinity EM technique, Willard *et al.* (1979) found a protein termed "H" of 195,000 daltons, which appears to be the high MW component of the triplet proteins, along with the other two triplet polypeptides of 145,000 and 73,000 daltons, transported in nerves at a slow rate of about 0.7–1.1 mm/day. Antigen raised to the H protein obtained from rabbit spinal cord was plated on EM grids and these were found to specifically adsorb the 100-Å neurofilaments isolated from spinal cord extracts. This finding adds strong confirmation to the view that the neurofilaments do in fact contain this protein or at the least that "H" protein is associated with them.

Further study of vertebrate neurofilaments by the assembly and disassembly technique of Zackroff and Goldman (1980) should throw more light on the composition of the triplet components and their relative amounts within the neurofilaments. It is likely that the 60,000-dalton polypeptide found by Zackroff and Goldman as the dominant component in squid neurofilaments is analogous to the 68,000-dalton component in vertebrate neurons (cf. Schlaepfer, 1977a). The 68,000-dalton component could be the principal subunit forming the linear core of the neurofilaments, the other proteins of the triplet acting as associate components similar to the MAPs of the microtubules (cf. Dahl *et al.* 1980). Evidence that the 68,000-dalton component is essential to the assembly of the neurofilaments (intermediate filaments) has been presented by Liem (1982) and Liem *et al.* (1982).

The studies of Willard and Simon (1981) have given strong support for such a view. Neurofilaments were decorated with antibodies raised to the three polypeptides designated as H (MW = 195,000), 45 (MW = 145,000), and 46 (MW = 73,000). The latter species, equivalent to the afforementioned 68,000-MW components, appears to be the "central core," the linearly organized part of the neurofilaments. Helically wrapped around it is the H polypeptide decorated with anti-H protein. The Helix appears to be displaced from the surface by some 100 Å over most of its length, and it also appears in some cases to be cross-linked with nearby filaments. This could represent various functional configurations of side-arms or changes in its form during the preparation of the material. The structure could, if it is similar to the side-arms of microtubules, collapse onto the central core in the process of preparation of the material. As in the case of the microtubules, whether the neurofilaments are assembled in the cell body as a coherent organelle and moved down the axon as such, or whether the triplet polypeptides are moved down the axon as subunits which are then turned over in the neurofilaments by a process of assembly–disassembly all along their lengths, is a question to be discussed further in Chapter 11.

3. Regulation of Neurofilament Density in Axons

Early morphological studies suggested that the neurofilaments and microtubules are interconverted from a basically similar subunit (Vaughn and Peters, 1967). Evidence marshalled for that view was the observation that in the course of development microtubules decrease in density in optic nerve fibers with a concomitant increase in the density of neurofilaments. Also, nerves treated with colchicine often show an increased density of neurofilaments concomitant with a decreased density of microtubules (Chapter 12). In any case, as indicated above, the molecular weights of the subunits of microtubules and neurofilaments differ as do their amino acid compositions.

There remains to be explained the inverse relation often found for the density of microtubules and neurofilaments under various conditions. Friede (1970) showed that the combined densities of microtubules and neurofilaments usually remains fairly constant for a nerve fiber of a given diameter, a constancy holding in the course of maturation. He suggested that some mechanism in the cell bodies serves to control the proportion of microtubules and neurofilament subunits produced. In addition, some mechanism of interaction or regulation locally present in the axons is required to explain the rapid reciprocal changes in microtubules and neurofilaments seen after local exposure of nerves to tubulin-binding agents.

Just as the microtubules are considered to exist as relatively short segments, the neurofilaments do not extend all along the length of the axon. The discontinuity of neurofilaments had been noted by Weiss and Mayr (1971), who stressed the lability of this organelle. Kreutzberg and Gross (1977) reported that 15% of the cross sections of pike olfactory nerves did not reveal their presence, the variability in the number of neurofilaments per cross section indicating that they are not continuous. More recently, Tsukita and Ishikawa (1981) found in myelinated fibers of the mouse saphenous nerve that the neurofilaments did not extend through the nodes.

These observations are compatible with a disassembly and assembly of neurofilament subunits at the ends of the neurofilament segments, allowing for a turnover of their component proteins (cf. Chapters 10, 11).

C. MICROFILAMENTS AND THE MICROTRABECULAR NETWORK

The third linearly organized organelle present in nerve, the microfilaments, 50–60 Å in diameter, are found located mainly near the axonal membrane and most prominently in the growth cones of regenerating fibers (Chapter 15). Actin is associated with or in part constitutes the microfilaments. The interaction of actin with myosin is responsible for muscle contraction and motility in lower cells, suggesting that these components might also account for transport in axons (Chapter 10).

Actin, extracted from brain by acetone, has the property of polymerizing from its globular (G) to the linear or filamentous (F) form. One mole of ATP is hydrolyzed and ADP bound per actin monomer in the course of the polymerization of the actin chain. A MW of 45,000 daltons has been assigned to the globular actin subunit (Pollard and Weihing, 1974). Actin, as in the case of the thin filaments of muscle, binds myosin or their meromyosin heads at intervals along its length, giving rise to a typical "herringbone" decoration of the filament (Ishikawa *et al.*, 1969; Nachmias and Huxley, 1970). The herringbone pattern has been used to identify the presence of actin filaments within neurons (Ishakiwa, Bischoff, and Holtzer, 1969; Burton and Kirkland, 1972; Lebeux and Willemot, 1975a,b; Alonso *et al.*, 1981). In those experiments, the glycerination of tissues and cells allows a permeation of the meromyosin. However, it also causes a disruption of nerve fiber structure which makes for some uncertainty as to the exact localization of actin in the cells, a problem which should yield to further technical advance.

While a number of studies have shown that actin is present in microfilaments mainly located near the axonal membrane, in some cases the band-like arrays of parallel fibers appearing after using special methods of fixation have indicated that actin also extends throughout the cell as a "cytoskeleton" (Webster *et al.*, 1978). The flocculant, amorphous structure seen in axons ordinarily considered to be the "ground" substance, has been related to the "microtrabecular network" visualized with stereo high-voltage EM after special preparation of the tissue (Wolosewick and Porter, 1976); Ellisman and Porter, 1980). Nerve tissue is fixed with gluteraldehyde and post-fixed with osmium after undergoing critical-point drying at liquid CO_2 temperatures to reveal the network in EM without the usual embedding matrix. Mitochondria, microtubules, microfilaments, endoplasmic reticulum, and ribosomes appear to be connected together by thin strands forming a meshwork, the infra-structure referred to as the microtrabecular network or the "cytoskeleton" (Fig. 9.11).

The question at issue is whether this lattice-like structure is brought about by the process of critical-point drying (Gray, 1976), or if such a network is normally present and is destroyed by the usual methods of histological preparation (Porter, Byers and Ellisman, 1979). That the critical-point drying technique can cause some change is suggested by the reduction in the diameter of the microtubules from 250 Å to about 130 Å. The loss of ribosomes in cells and changes in the structure of the ER seen when using this technique also points to this possibility. Some of the changed appearance may be related to the tendency of actin to gel, a process which requires ATP and Mg^{2+} (Pollard, 1976). As a general proposition it would appear likely that the removal of water in the process of critical-point drying would cause an

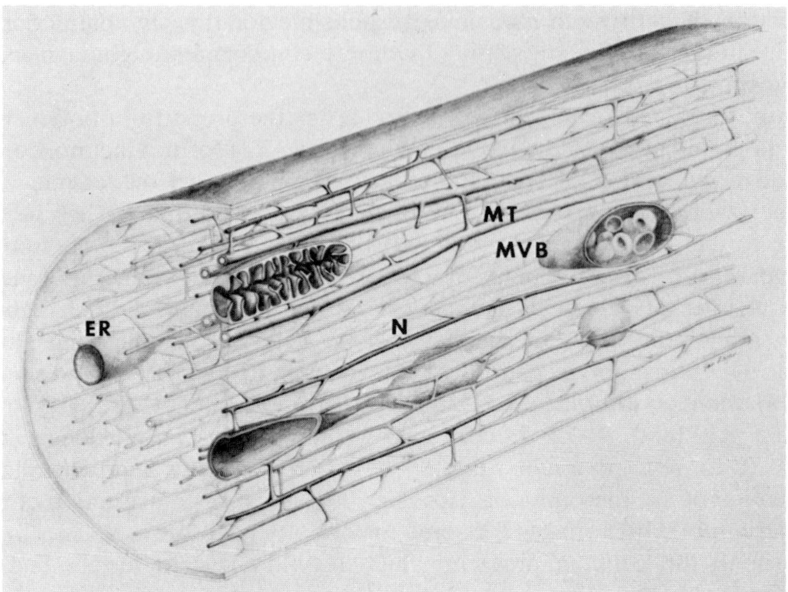

Fig. 9.11 Diagrammatic representation of microtrabecular network in a fiber. Filamentous connections between organelles are shown: a mitochondrion, a multivesicular body (MVB), and endoplasmic reticulum (ER) are connected to the microtubules (MT) and neurofilaments (N) forming the microtrabecular network. From Ellisman and Porter (1980).

aggregation of solutes in the axoplasm around structures which are normally present, for example, the side-arms of microtubules and neurofilaments. The side-arms are seen to be more deeply stained by the use of phosphotungstic acid (LeBeux and Willemot, 1975a,b) or lanthanum (Burton and Fernandez, 1973), substances which have been used to better reveal the lattice-like appearance. Ruthenium red which stains the lattice and the substance coating the microtubules (Tani and Ametani, 1970) precipitates acid polysaccharides, indicating that the microtubules are coated with this substance (Samson, 1971). Burton and Fernandez (1973) found the microtubules of axons in the crayfish ventral cord coated with lanthanum. This agent stains acid mucoproteins, indicating the presence of this material on the microtubules. A variety of stains was used to reveal the nature of the coating seen around the microtubules by Stebbings and Bennett (1975). They concluded that the coating was a mucopolysaccharide which also extended on the side-arms and appeared to form the bridges between the microtubules contributing to the appearance of the microtrabecular network when using lanthanum.

Metuzals (1969) has shown the appearance of a microtrabecular network in the giant nerve fibers of squid using various preparation methods. He believes the network to be composed of 30-Å-thick unit filaments intercoiled in strands 70–250 Å thick oriented longitudinally in the axon. These filaments follow an often sinuous course with numerous cross-connections. Using an indirect immunoferritin procedure, Webster *et al.* (1978) found the cytoskeleton to be decorated by an antibody raised to actin. In addition to the F-actin in the microfilament bundles, the cytoskeleton appears to be composed of a finer actin network that interconnects all the various axoplasmic structures.

The work of Maupin-Szamier and Pollard (1978) throws further light on the matter. They showed that actin filaments are fragmented by osmium tetroxide. Thus, this filament can be converted from a straight unbranched filament into what looks like a network. Cross-linking with glutaraldehyde, the method commonly used for fixation, does not prevent such destruction by osmium. The best method used to preserve the form of the axon and its organelles, the quick-freezing method introduced by Van Harreveld, has shown that much of the empty space between microtubules and neurofilaments seen with usual fixation methods is occupied by linearly oriented filaments with the dimension of actin microfilaments (Kadota; Tsukita and Ishikawa; personal communications).

D. PHYSICAL PROPERTIES OF AXOPLASM AND ITS CONTAINED STRUCTURES

Insight into the interrelation of the organelles within the axoplasm, the nature of the microtrabecular network or cytoskeleton matrix, may be gained by a study of the rheological properties of axoplasm. The axoplasm of squid giant axons may be readily expressed and placed in glass tubes. In response to an imposed pressure, the axoplasm was seen to act as a Bingham body, namely, a substance having a high-yield stress and plastic viscosity (Rubinson and Baker, 1979; and cf. Reiner, 1949). This behavior would fit with a network having relatively strong internal bonds, as in the case of a microtrabecular network. The axoplasm within the interstices has a relatively low viscosity. This was shown using the technique of electron spin resonance (Haak, Kleinhans, and Ochs, 1976; Rubinson and Baker, 1979). In this method, a spin label molecule is used which has an unpaired electron able to resonate in a high-frequency radio wave field while under the influence of a magnetic field. The net result is that the spin label lines up to show differences in their energy levels at resonance. In the case of the spin label, tempone, two triplet signals appear, one representing the presence of the spin label in the water phase, the other the label in the lipid phase. By suitable subtraction techniques the energy levels of the water signal of tempone in the nerve can be used to give a measure of its microviscosity. In mammalian nerves slit to bypass the perineurial sheath, a value of 5 centipoise (cP) was found, i.e., one 5 times that of water (Haak, Kleinhans, and Ochs, 1976). In similar studies, Rubinson and Baker (1979) determined a value of 2 cP for the axoplasm of the giant fiber of the squid. These figures are reasonably close to the low viscosities of 4–5 cP found for frog sartorius muscle (Belágyi, 1975), and 5–10 cP for the myoplasm of barnacle muscle fibers (Sachs and Latorre, 1974). The viscosities measured over a range of temperatures in mammalian nerve vary smoothly with the temperature reflecting changes of water viscosity as shown in Fig. 9.12.

The curve shows no unusually large increase in viscosity at low temperatures, particularly at 11°C where cold block occurs, one of the possibilities which could account for the phenomenon of cold-block (Chapter 2).

The rheological properties and the low values found for microviscosity contrast with the properties of axoplasm described by Biondi *et al.* (1972). The latter studied axoplasm expressed from myelinated nerve and arrived at a value of 10^5 cP for the viscosity of the axoplasm. This proceedure, how-

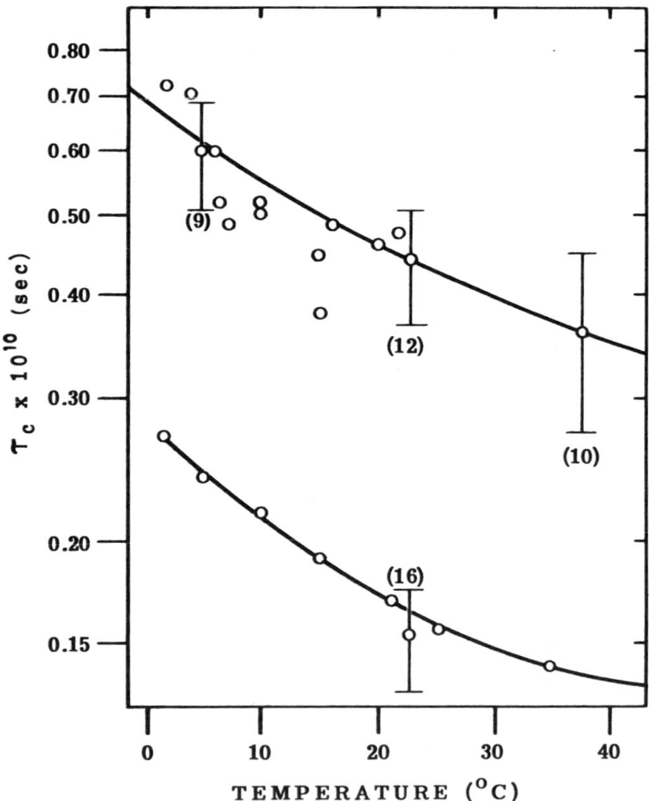

Fig. 9.12 Variation of viscosity in nerve with temperature. On the ordinate, τ_c represents a measure of viscosity of the axoplasm of cat sciatic nerve (upper curve) and for water (lower curve) as a function of temperature. Viscosity shows a similar gradual increase for both as temperature is lowered. Error bars represent one standard deviation above and below the mean for the number of independent experiments enclosed in parentheses. From Haak *et al.* (1976).

ever, is less certain of giving results uncontaminated with myelin as compared to the expressed axoplasm of the giant axon.

The question remains regarding the degree of rigidity of the connections forming the matrix of the microtrabecular network and how it fits with transport in the nerve fiber as discussed in Chapter 10. This question has to do with the movements of such relatively large bodies as the dense core vesicles (cf. Chapter 3) or the mitochondria axially in the nerve fibers and the connections of the microtrabecular network to them. The giant nerve fiber axoplasm shows a high viscosity near the surface (Rubinson and Baker, 1979), indicating that the strength of the cross-bonds of the matrix to the axolemma would be very great in the relatively smaller diameter myelinated nerve fibers. The phenomenon of beading (Ochs, 1965), on the other hand, suggests that the cross-connections of the matrix are likely to be weak and labile. Beading was observed by stretching nerves and while under stretch, quickly freezing them at low temperatures close to $-160°C$ to prepare the nerves for histological examination by the method of freeze-substitution (Ochs, 1965). The changes in form resulting from weak force of stretch are labile and are only held in place during the process of histological preparation with this method. Whereas the fibers of unstretched nerves have a cylindrical appearance, stretched nerves show the series of constrictions and swellings along their length termed beading (Fig. 9.13).

Beading results from constrictions produced at intervals along the length of the nerve fibers. These constrictions are not related to any particular

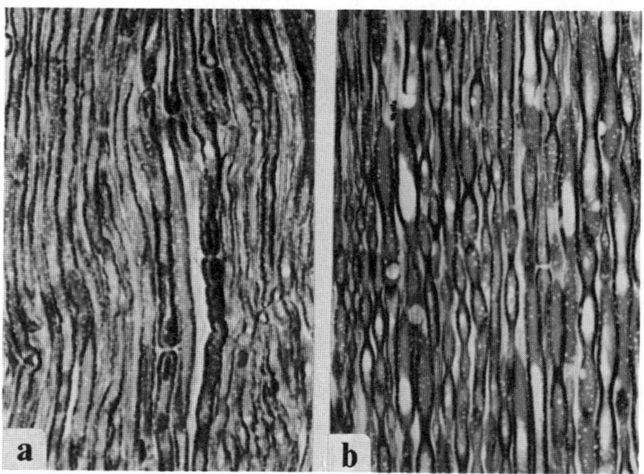

Fig. 9.13 Beading of stretched fibers. (a) Unstretched nerve shows its contained fibers as parallel-walled cylinders. (b) These become transformed to beaded fibers with a number of constrictions along their lengths on stretching as shown. The tissue was prepared by freeze-substitution. Fine light dots are seen scattered throughout. These represent ice crystals which arise during freezing but do not affect the morphology of the fiber. From Ochs (1965).

feature of nerve, such as the Schmitt–Lanterman incisures. Also, the axolemma appears to separate from the myelin sheath, indicating that the constrictions are brought about by a contraction of the axoplasm rather than by an external compression of the Schwann cell and myelin sheath.

Recovery from beading rapidly occurs on release of the imposed stretch. The apparent readiness with which the constrictions occur and recover indicates that the cross-links of the microtrabecular matrix are relatively labile.

Another phenomenon which points to a lability of the interconnections of the matrix is the flow of axoplasm found the cut end of nerve fibers (cf. Chapter 1). This suggests some radial contraction in the axons causing the outflow, one similar to the underlying mechanism responsible for the constrictions causing beading in stretched nerves. Such a mechanism may be excited in the early stages of Wallerian degeneration (Chapter 6), to give rise to the ovoids in the fibers. What that constriction mechanism might be remains at present unknown. It is, however, an attractive hypothesis that the actin which is dispersed throughout the axoplasm and seen in greater amounts near the membrane could cause a radial contraction, one giving rise to the constrictions of beading nerve fibers. Perhaps the actin acts in conjunction with the myosin also known to be present in the fibers.

10

Mechanisms of Transport

In an early attempt to account for transport, the view was advanced that the "fibrillary components" of axons could serve as stationary elements along which transported materials are moved in "bucket brigade" fashion (Ochs *et al.*, 1960). When the microtubules were discovered to be a ubiquitous organelle in a wide variety of cells, including nerve fibers (Chapter 9), Porter (1966) related the function of transport to the microtubules. In this chapter we will discuss theories of axoplasmic transport based on the microtubules, as well as other theories involving the other organelles present in the fibers (Chapter 9).

Theories advanced to account for both "fast" and "slow" transport will also be discussed in this chapter. The question as to whether slow transport is mediated by a separate mechanism other than the one responsible for fast transport, or whether both fast and slow transport are due to the operation of the same transport mechanism, the concept termed the "unitary hypothesis," will be taken up in Chapter 11. In addition, the phenomenon of "routing" and the implications drawn from that phenomenon for some of the theories of transport will be described in Chapter 11.

A. FAST TRANSPORT MODELS

1. "Rolling-Vesicle" Model

Schmitt (1968) envisioned all fast-transported materials to be carried down the nerve fibers in the form of vesicles. The vesicles were viewed as having specific projections on their surfaces which temporarily bind to complementary sites on the microtubules; the making and breaking of those bonds in some fashion causing the vesicles to roll down along the microtubules (Fig. 10.1).

The binding sites on the microtubule would, by combining with an ATPase or a GTPase present on the vesicle binding site, utilize the ~P of either ATP or GTP to provide the required energy for movement. The Ca–Mg ATPase found present in nerve (Khan and Ochs, 1974) is not, however, fast-transported; the enzyme accumulates above nerve ligations with a time lag even

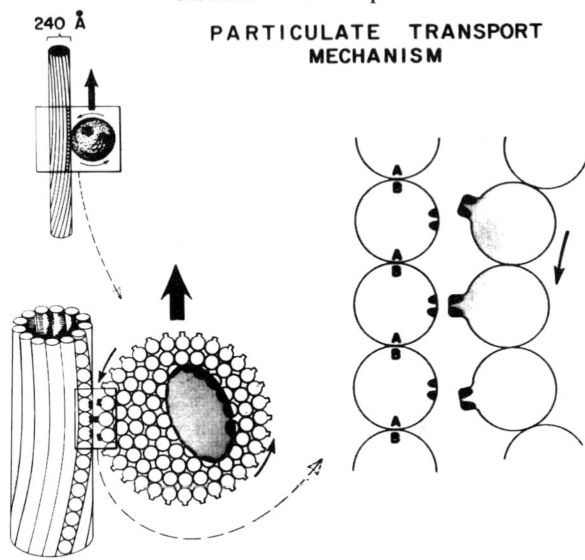

Fig. 10.1 Vesicle transport model. The transported vesicles are considered to be composed of units with specific attaching sites to the microtubules; the making and breaking of those bonds cause the vesicles to roll down along the microtubules. At the right, details of the postulated attaching sites of tubulin to the microtubule sununits are shown. A and B represent bonds of tubulin forming the microtubules. From Schmitt (1968).

greater than that usually seen for most slow-transported components. This would suggest, rather, that the Ca–Mg ATPase is associated with the microtubules (Chapter 9).

A matter to consider in Schmitt's model is that the vesicles would be temporarily attached at only a single point as it rolls forward to the next point of attachment. This would require some mechanical force to maintain the rotation of the vesicle against the viscosity of the axoplasm. While the viscosity is relatively low, 5 cP, as measured in mammalian nerve by Haak *et al.* (1976), it would be enough to hinder the movement postulated for a rolling vesicle, particularly if a microtrabecular network (Chapter 9) is present in the axoplasm. An even greater impediment is posed by the abnormally increased density of neurofilaments produced by exposure of nerves to aluminum and certain neurotoxins, a situation in which fast transport may have its usual rate (Chapter 12). The vesicle model does not take into account the evidence that a wide range of soluble proteins of different molecular weights, polypeptides, free amino acids, as well as large organelles are being moved down the fiber at the same fast rate. The latter include the fast-transported dense-cored vesicles (DCVs) in adrenergic nerve fibers, the serotonin-containing vesicles (Chapter 4), the hormone containing DCVs moved down within the fibers of the hypothalamo-neurohypophysial system (Chapter 13), and the mitochondria, which are known to be temporarily fast-transported (Chapter 3).

2. The Transport Filament Model

The transport filament model originally advanced to account for fast transport (Ochs, 1971b, 1972c) was based on analogy to the sliding filament theory of skeletal muscle contraction (Huxley, 1969). The model took into account the fundamental findings that axoplasmic transport is closely dependent upon oxidative metabolism supplying ATP, and that a wide range of components and organelles are transported at the same fast rate, implying a common carrier. These include large and small MW proteins, polypeptides, amino acids, membrane components, and short lengths of ER, the relatively large vesicles, and, temporarily, the mitochondria (Chapters 3 and 4). In the model, the common carrier, the transport filament, binds the various materials to be transported. The filaments are moved along the microtubules, the stationary element, by means of the side-arms projecting from the microtubules (Fig. 10.2A).

The side-arms contain, or have associated with them, a Ca–Mg ATPase (Chapter 9) which utilizes the ATP supplying the energy necessary for the movement of the transport filaments. The directionality of movement of the transport filaments in the axon is determined by the orientation of the side-arms, one set on microtubules subserving anterograde transport, another set on microtubules with the opposite orientation for retrograde transport. Or, side-arms with different orientations may be present on different aspects of the same microtubules. The need for a participation of Ca^{2+} in transport (Chapter 8) and the properties of the Ca–Mg ATPase in nerve led to the modification of the model (Ochs, 1980b; 1981a,b) as shown in Fig. 10.2A.

A large number of lower organisms exhibiting motility contain a Ca^{2+}-Mg^{2+} stimulated ATPase (Ca–Mg ATPase), as has been described by Jahn and Bovee (1969). The ATPase found in the giant axon by Libet (1948) and that in crayfish nerve by Bowler and Duncan (1966) attributed by them to membrane excitability, and the ATPase with actomyosin properties which Berl and Puszkin (1970) considered to play a role in the release of neurotransmitter from nerve terminals (Chapter 13), may be related to transport. The relatively high level of Ca–Mg ATPase activity found present in mammalian sciatic nerve (Khan and Ochs, 1974) has a specificity for ATP which is not absolute; other nucleotides are hydrolyzed. A similar Ca–Mg ATPase was seen in squid nerve by Shecket and Lasek (1977) and in frog nerve by Edström *et al.* (1980) and Hammerschlag and Bobinski (1981). The enzyme is maximally activated by Ca^{2+} at a concentration of 1–3 mM, a level much higher than the levels of free Ca^{2+} present in the axoplasm (cf. Chapter 8). A localization of Ca–Mg ATPase to the side-arms of the microtubules is required in the transport filament model and evidence for such a localization was recently given by Sharp, Fitzsimmons, and Kerkut (1978), and by R. L.

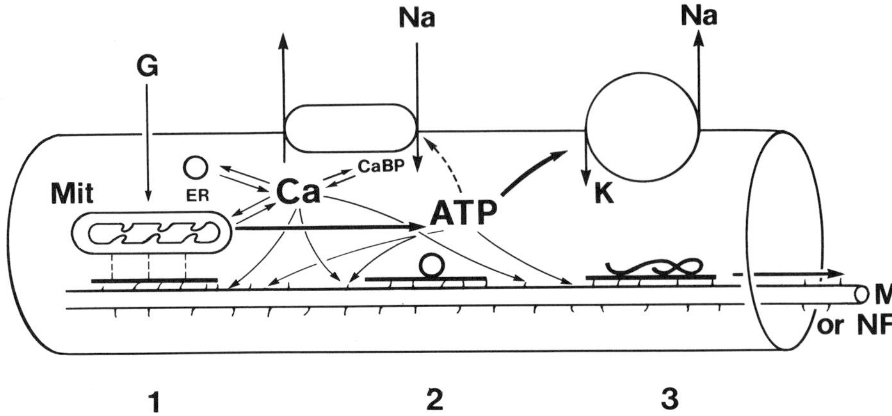

Fig. 10.2A Transport filament model. Glucose (G) enters the fiber and after glycolysis, oxidative phosphorylation in the mitochondrion (Mit) gives rise to ATP. The ~P of ATP supplies energy to the sodium pump, controlling the levels of Na⁺ and K⁺ in the fiber and energy to move the side-arms on the microtubules and in turn the movement of transport filaments. The transport filaments are shown as black bars to which the various components transported are bound and thus carried down the fiber by side-arm activity. The components transported include: the mitochondria (Mit) attaching temporarily as indicated by dashed lines to the transport filament (1); vesicles (2); and soluble proteins (3), the latter shown as a folded globular protein. Small polypeptides are also transported. The wide range of all these components transported at the same fast rate indicates a common carrier, namely, that represented by the transport filaments. Calcium (Ca) is shown, its level regulated in the axoplasm to an optimal level. Calcium-binding protein (CaBP), the endoplasmic reticulum (ER), and a Ca–Na exchange (or Ca pump requiring ATP) also participate in such regulations (cf. Fig. 8.11). Calcium is needed at the right level by calmodulin to activate the Ca–Mg ATPase associated with the side-arms. The enzyme can then utilize ATP as the source of energy to move the transport filaments. A separate orientation of side-arms subserves retrograde transport.

Ochs and Burton (1980) using an EM cytochemical technique. A purified preparation of tubulin obtained by several cycles of assembly and disassembly (Chapter 9) showed the presence of dynein-like ATPase (Hiebsch, Hales, and Murphy, 1979). Further purification of the ATPase revealed it to be a particle with a diameter of approximately 10 nm. The enzyme is not a contaminating membrane ATPase and it has a lower affinity of association for tubulin than for the MAPs. A similar study showed Ca–Mg ATPase activity associated with the tubulin, the enzyme activated by Ca^{2+} present in concentrations of 10^{-8}–10^{-5} M (Ochs and Iqbal, 1980), a range expected for free Ca^{2+} in the axons. The sensitivity of the Ca–Mg ATPase to these low levels of Ca^{2+} is accomplished through the mediation of calmodulin. This was indicated by the finding that trifluoperizine (TFP), which binds to calmodulin (Chapter 8) and thus prevents its action, causes a block of Ca–Mg

ATPase activity. It is significant that TFP also blocks transport *in vitro* in intact nerve fibers (Chapter 12).

The exact relation of calmodulin to Ca–Mg ATPase is unknown. As one possibility we can consider that calmodulin is bound to, or is an integral part of, the transport filament, and that as the transport filaments are moved forward, the calmodulin on them locally activates the Ca–Mg ATPase of the side-arms. In this way ATP is utilized just at its point of action to cause a micromechanical conformation of the side-arms needed to pull the transport filaments forward (cf. Fig. 10.2B). Another possibility shown in this figure is that calmodulin is present on the microtubular side-arms bound to the Ca–Mg ATPase. The transport filaments could contain a regulatory component which, as it meets the side-arms, allows a calmodulin activation of the Ca–Mg ATPase to occur when the transport filament comes into contact with the side-arms. With regard to that possibility, it has been observed that calmodulin is not readily removed from the tubulin associated Ca–Mg ATPase (Ochs and Iqbal, 1980). Hiebsch *et al.*, (1979) also found the Ca–Mg ATPase to have a lower affinity than the MAPs to tubulin; the ATPase activity was much diminished or lost after 2 cycles of purification.

The use of antibodies raised to calmodulin in an EM technique (Lin *et al.*, 1980) could show the localization of calmodulin (cf. Chapter 8) and help resolve the question as to which of these possible mechanisms is present. It should be noted, however, that in addition to its participation in transport, calmodulin participates in a number of enzyme reactions in the axon, which requires that it be continuously transported down into the axons from the cell bodies and dropped off in the axon.

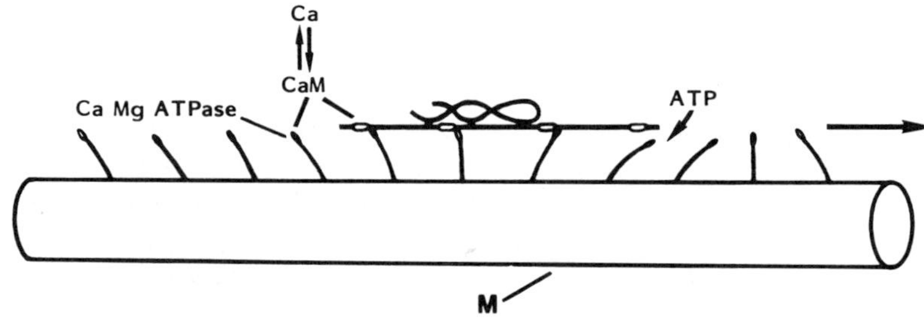

Fig. 10.2B Hypothetical mechanisms for transport-filament propulsion. The manner in which the side-arms on the microtubules engage the transport filaments is detailed. In this model a globular protein bound to the transport filament is shown. The Ca–Mg ATP as associated with the side-arms are pictured engaging calmodulin (CaM) on the filaments, this activating the Ca–Mg ATPase which then utilizes ATP allowing the side-arms to drive the filaments forward. Alternatively, calmodulin may be associated with Ca–Mg ATPase on the side-arms, and some regulator protein on the transport filaments meeting the side-arms allows the calmodulin-activated ATPase to utilize ATP when the activator meets the side-arms. Ca^{2+} is shown activating calmodulin at the low levels present in the axoplasm. From Ochs (1981a); cf. earlier versions of the model (Ochs, 1971b, 1972c).

Some theoretical considerations point to a filament as a carrier. A transport filament oriented in the axial direction would offer the least frictional resistance to movement. A filament would also allow the temporary attachment of a number of microtubular side-arms to it and thus allow more force to be exerted on the transport filament to move it forward, as in the case for the sliding filaments of muscle. A relatively large force appears required to move a vesicle against the resistance of a microtrabecular network (Chapter 9) or a proliferation of neurofilaments as occurs in certain neurotoxicities (Chapter 12). The transport filaments bridged by a number of side-arms would also be directed in their movement axially in the fiber. If, for example, a vesicle or a small protein were to bind to one side-arm without the intermediation of a transport filament, it is difficult to see how it could be directed in its movement to the next side-arm, particularly if its size relative to the side-arm spacing of 320 Å (Chapter 9) is small. A transport filament also seems needed to bridge over the ends of microtubule segments estimated to be 1.2–10.7 μm in length in nematode neurons (Chalfie and Thomson, 1979) and averaging 108 μm (if of equal lengths) in dorsal root ganglion neurons (Bray and Bunge, 1981) or longer (Tsukita and Ishikawa, 1981) (cf. Chapter 9).

The main problem with respect to the transport filament model is the isolation and characterization of the transport filament. One possible candidate is actin in its F (filamentous) form, on analogy to the actin–myosin interaction of muscle. The recent evidence that DNAase I, which binds to F actin to depolymerize it, causes a block of axoplasmic transport when injected into nerve cell bodies (Isenberg, Schubert, and Kreutzberg, 1980; Goldberg *et al.*, 1980), is most suggestive. On the basis of the transport filament model, the binding of DNAase I to the actin to depolymerize it would prevent it from acting as a transport filament and block transport. However, the actin present in the cell may well have other actions which are indirectly related to a role as the transport filament. For example, it may act to initiate transport from the cell to the fiber. A better kind of evidence would be to determine if it blocks transport when injected into nerve fibers. Recently, just such evidence was gained. Goldberg (1981) injected DNAase I into the giant axon of *aplysia* to cause a block of axoplasmic transport. It remains to be determined if this is due to a direct action on transport filaments or a general disruption of nerve structure.

While the transport filament hypothesis has some analogy to the sliding-filament theory of muscle, there are significant differences. Skeletal muscle contains several complex regulatory proteins, the troponins and tropomyosin, which mediate the interaction of the thick and thin filaments. One of the troponins, troponin C, is acted on by the Ca^{2+} released from the sarcoplasmic reticulum to cause a conformational change allowing the union of the actin and myosin filaments and the expression of Ca-ATPase activity. Such triggered complex sequences are unlikely to occur in the nerve. There are, nevertheless, some interesting similarities between axoplasmic transport, cell motility, and skeletal muscle contraction (Huxley, 1973). The rate at which the thick and thin filaments slide past one another during a muscle twitch ranges from 4–10 μm/sec (Close, 1965). Fast axoplasmic transport with its rate of about 4.5 μm/sec falls within the lower end of this range. Axoplasmic transport may thus be viewed as a "continuous contraction" in contrast to the triggered brief action of a skeletal muscle contraction, one which requires subsidiary mechanisms for Ca^{2+} release and resequestration in order to start and stop the contractile process.

In a further analogy to muscle, axoplasmic transport and muscle contraction appear to have a similar Q_{10}. The temperature coefficient of 2–3.5 found for axoplasmic transport (Chapter 2) is similar to the Q_{10} values found for the contraction of the frog sartorius muscle by Hartree and Hill (1921a,b). A Q_{10} of 2.5 was found for the rate of increase of tension of muscle, a characteristic with dimensions related to velocity, and a Q_{10} of 2.8 was found for heat output, a function also related to velocity. A more recent analysis of the effect of temperature on the velocity of sarcomere contraction of rat extensor digitorus longus muscle gave a Q_{10} of just under 2 over the temperature range of 25–35°C, and a Q_{10} of 2.6 for rat cardiac muscle (Close, 1965).

Vesicles appear to be closely associated with the microtubules in EM (Chapters 4, 13 and note especially Fig. 13.2) . A close association of mitochondria to the microtubules was reported by Raine *et al.* (1971), and a positive association shown statistically by Smith (1975). In an interesting example given by Smith *et al.* (1975), the microtubules appear to have their side-arms in close apposition to a nearby mitochondrion (Fig. 10.3).

The close proximity of mitochondria to microtubules was also strongly suggested by the *in vivo* dark field microscopic studies of Cooper and Smith (1974). They observed mitochondria bending in the course of their movements along what appeared to be the curvature of the microtubules in the axon (cf. Fig. 3.18). Cooper and Smith suggested that different affinities of the microtubules for transported components could account for their anterograde and retrograde transport, a model which differs from that of the transport filament hypothesis where differing orientations of the side-arms are considered to account for anterograde and retrograde transport.

3.   The Microstream Hypothesis

Gross (1975) postulated that a fast transport of materials takes place within low-viscosity channels surrounding the microtubules. The energy needed to drive the fluid stream and, in turn the components transported, is considered to be supplied by ATP (Fig. 10.4).

In this model the channels are postulated to have a radial gradient of viscosity which increases with distance from the microtubules. The channel thus formed could then act like a chromatography column and thus cause a differential distribution of components within the axon in the course of their axial movement. The smaller components would, in this model, become locally deposited in the axon early in its axial progress to a greater extent

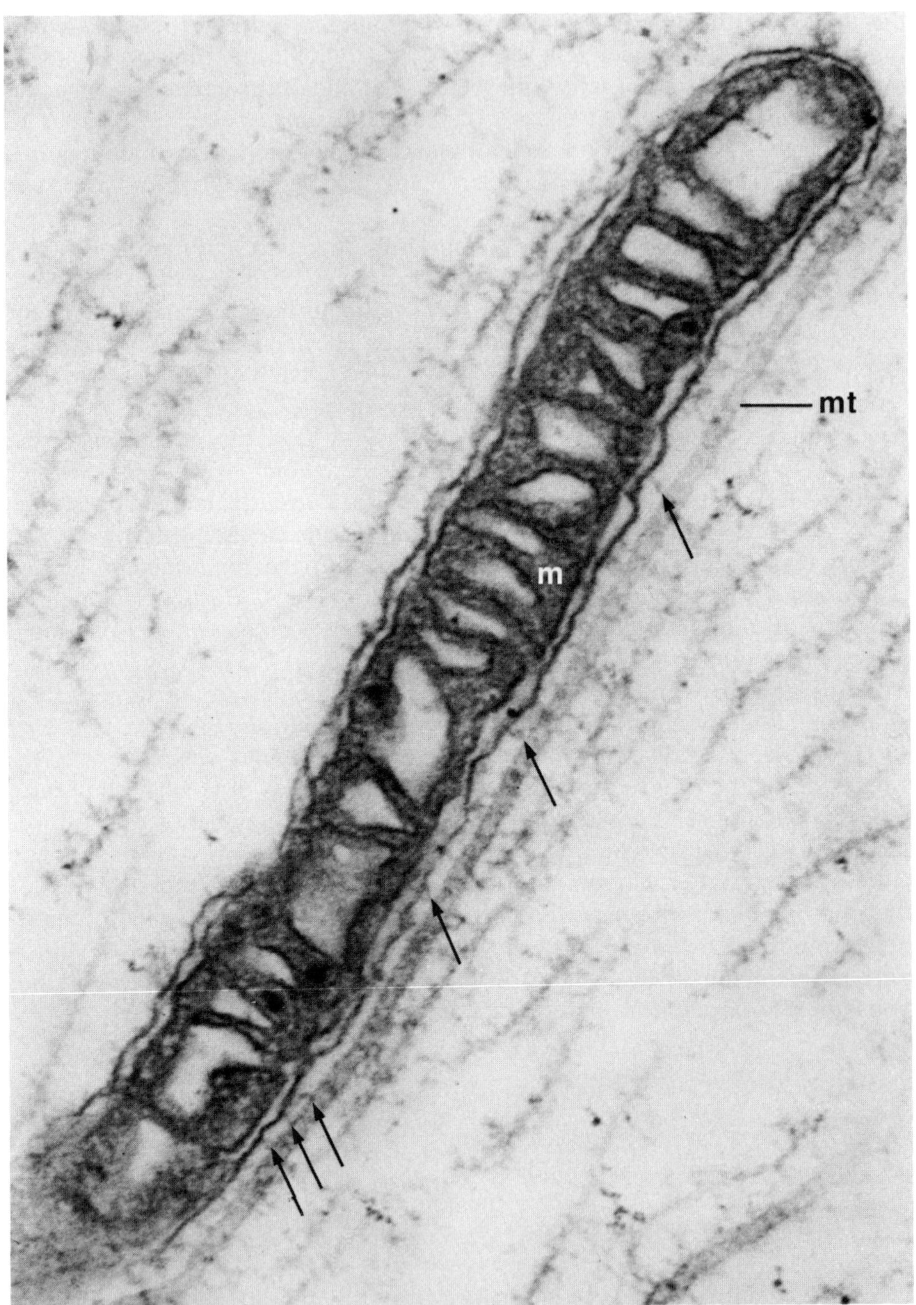

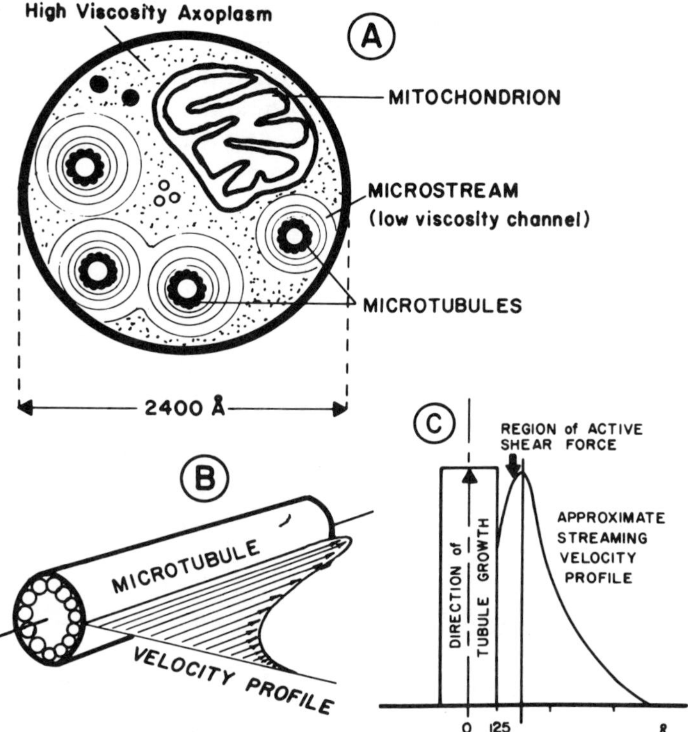

Fig. 10.4 Microstream hypothesis. Channels around the microtubules are considered to act like chromatography columns and cause a length-dependent distribution of components. (A) Schematic representation of postulated microstreams within a C fiber of garfish olfactory nerve is shown in a cross-section containing one mitochondrion and four microtubules with low-viscosity channels around them. The movement of molecules by fast axoplasmic transport is thought to occur within the low-viscosity channels surrounding the microtubules by a shear force generated by ATPase reactions at the surface of the microtubules. (B) A hypothetical streaming velocity profile laterally is depicted and (C) shows the postulated streaming velocity decreasing laterally from the microtubule plotted vertically. Average transport velocities would be determined by the amount of time a particle or molecule spends in different regions of this profile. All anterograde transport, including that of the mitochondria, is assumed to depend on microstreams. From Gross (1975).

than larger ones such as vesicles. The latter would therefore be carried down to the ends of the fibers. However, a wide range of soluble proteins and polypeptides ranging from high to low MWs, as well as particulates and temporarily the mitochondria, are seen to move at fast-transport rates in the crest. This was shown by comparing the composition of fast-transported labeled species in the crest to those in the plateau in cat sciatic nerve (Sabri and Ochs, 1972). Both portions contained a similar composition of high-

and low-MW substances. A similar finding was reported by Cancalon, Elam, and Beidler (1976) in the garfish olfactory nerve.

The essential finding which the microstream hypothesis was devised to explain, the drop-off of labeled materials behind the advancing crest, is also contained in the "unitary hypothesis" to be discussed in Chapter 11.

4. Polyelectrolyte Contraction

The material coating the microtubules contains mucopolysaccharide (Chapter 9). This is an acidic polyelectrolyte which was considered by Samson (1971) to have contractile properties resulting from its interactions with small ions or as a consequence of pH changes. In his model of transport based on that phenomenon, these proteins, by their expansion and contraction, are considered to bring about transport. How this would be related to the required supply of ATP to maintain axoplasmic transport (Chapter 8) was not specified. The hypothesis does not account for the uniformity in the rate of transport of the wide range of different substances and the transport of relatively large organelles (cf. Section A2 above).

5. Transport Within Microtubules

Weiss (1970) proposed a mechanism of fast transport based on a convection of materials within the microtubules. Energy supplied at one end of the microtubule was postulated to produce a change in its molecular conformation, a closer molecular packing of the globular tubulin comprising the microtubule wall. This would trigger a similar change at the adjoining region and so spread down the length of the microtubule. The wave of constriction so produced would be equivalent to a peristaltic movement of the microtubules. No evidence for such contractility of tubulin exists. The model also cannot account for the transport of DCVs or the temporary fast movement of the mitochondria which would not fit inside the microtubules. The most difficult aspect of the theory is that the hydraulic pressure required to move materials down the inside of such narrow channels of such lengths would be so large as to make this an energetically unlikely mechanism.

6. Transport in the Axon Membrane

Integral proteins are mobile within the bilipid layers of the membrane model advanced by Singer and Nicholson (1972). This has suggested the possibility that the axolemma may be a route by which proteins are transported down the fibers (Marchisio *et al.*, 1975). In support of that view, Byers (1974) found, after a period of transport of labeled proteins, autoradiographic evidence that grains of radioactivity are distributed to a greater degree near to or at the axolemma as compared to the axoplasm (Fig. 10.5).

Insofar as those studies were done in unmyelinated fibers exposed to

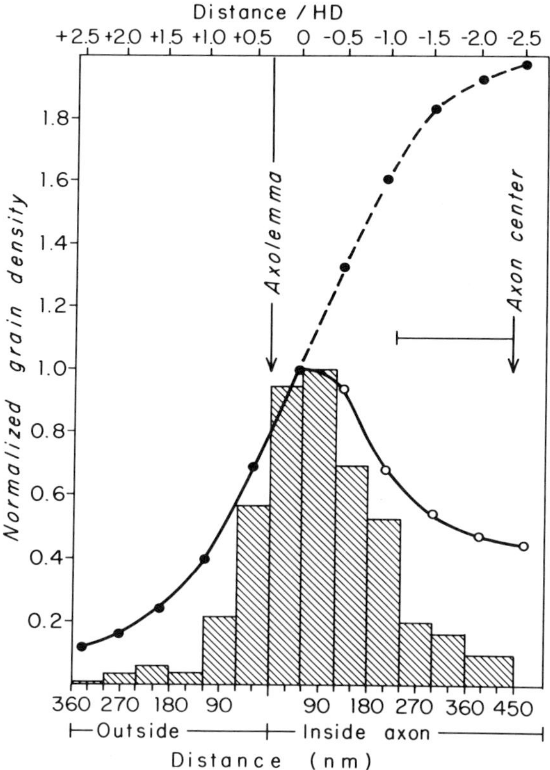

Fig. 10.5 Transport in the axolemma. This hypothesis was based on a distribution of grains with apparently a high density near the axolemma. The histogram shows the grain density over 497 labeled unmyelinated axons from 13 nerve samples (1 control, 3 proximal, 4 middle, and 5 distal colchicine-treated samples). The data were normalized to unity at a zone 60–120 nm inside the axon for each sample. The mean silver grain position was 53 ± 19 nm inside the axon; the mean axon center was 450 ± 230 nm. Superimposed on the histogram is the theoretical curve for grain density around a radioactive unfilled circle (O) or a solid circle (●) of radius = 2.5 HD (1 HD = 165 nm). The observed density corresponds to that of an unfilled circle whose circumference is located approximately 60 nm inside the axon. From Byers (1974).

colchicine, the possibility exists that the labeled proteins may have moved radially from their position within the axon toward the membrane as a result of exposure of the nerves to colchicine. A later shift may also occur without exposure to the agent after the front of fast-transported radioactivity has moved down the nerve. When nerve segments were taken from the fronts of fast-transported radioactivity and prepared for autoradiography, grains were found distributed over the entire cross-sectional area of the myelinated axons (Ochs, 1972a), as shown in Fig. 2.17. A similar localization was also seen in unmyelinated nerve fibers using EM autoradiography (Ochs and Jersild,

1974) as indicated in Fig. 2.18. Studying transport in regenerating nerve fibers and using the technique of Byers (1974), Griffin *et al.* (1981) found a lateral shift of labeled proteins in the axon, to a location closer to the axolemma. Such evidence is consistent with a transport of labeled components within the axon with later a translocation of labeled components radially into the axolemma (cf. Chapter 11).

A more lateral location of radioactive grains was observed in autoradiographs of myelinated nerve fibers of the newt (Lentz, 1972). However, the majority of grains were located not at the axolemma but displaced a short distance from it medially within the axon. This particular distribution of grains of radioactivity is of special interest in that it corresponds closely to the more lateral location of microtubules in the nerve fibers of the newt in contrast with the fairly uniform distribution of the microtubules seen in mammalian nerve fibers. The granule movements in amphibian nerve fibers observed using dark field microscopy adds further support to a generalized transport throughout the axon (Cooper and Smith, 1974). Granules were seen to move in every plane throughout the whole cross-sectional area of the nerve fiber (Smith, 1979). In crayfish nerve fibers the microtubules are found in greater density near the axolemma, and in them most of the granules were seen moving in that location (Smith, 1979).

In tissue culture, Koda and Partlow (1976) reported that the lectin concanavalin A binds externally to the axon membrane and this external marker was seen to move in the retrograde direction at a rate of 1.2 mm/day, a phenomenon most likely related to the growth of the fiber. This observation would argue against an anterograde movement of materials going on within the membrane in the opposite direction at the same time. The mobility of AChE in the membrane appears to be low (Brimijoin *et al.*, 1978), a finding incompatible with a fluid membrane. In any event, the concept of a transport in the axolemma cannot account for the transport of vesicles or the temporary fast movement of the mitochondria.

7.   Transport Within the Endoplasmic Reticulum

Using high-voltage electron microscopy, Droz *et al.* (1975) described the ER as a continuous network which passes from the cell bodies down to the nerve fiber terminals and considered the ER to be the route by which fast-transported materials are carried in the axons (Fig. 10.6).

This hypothesis was based on the finding that labeled grains of transported radioactive materials were seen in EM autoradiographs in, and closely associated with, the ER. However, such accumulations of grains in the ER were found in nerve segments taken just above a compression, i.e., at the site of damming. This localization in the ER may be the result of a translocation of labeled components into the ER at the site of damming, as was suggested by Byers (1974). In the unconstricted nerve fibers, as may be seen in the

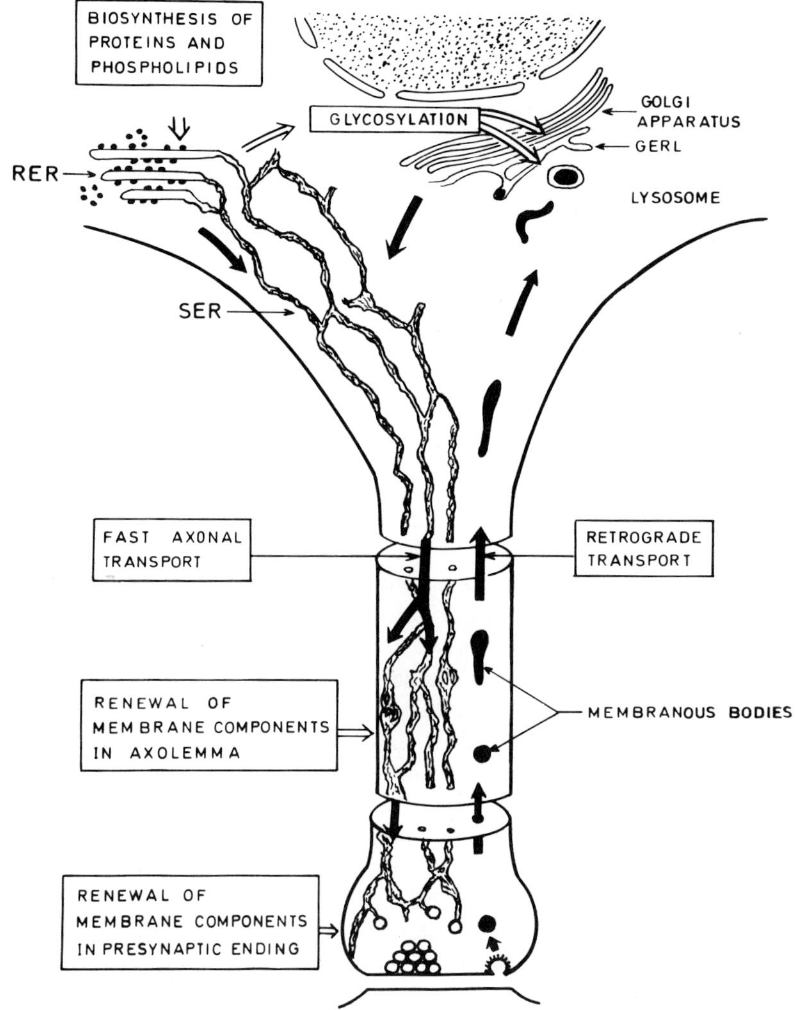

Fig. 10.6 Transport within the endoplasmic reticulum (ER). The ER is shown diagrammatically as a continuously connected network within which materials are considered to move down the fibers at a fast rate of transport. The ER is postulated to exist as a continuous system extending from the perikaryon to the axon terminal. The convoluted tubules of the smooth ER runs from the initial segment of the axon parallel to the long axis of the axon. Loosely anastomosed to the ER are "subaxolemmal plates" made up of smaller tubules. It is postulated that an exchange of fast-transported macromolecules takes place at the contact between the subaxolemmal plates and tubules. At the terminals, the synaptic vesicles appended like blebs at the tip of thin tubular branches are shown to result by fission from the smooth ER. Microtubules are considered to serve for slow transport. From Droz, Rambourg, and Koenig (1975).

example given by Droz *et al.* (1975), only a few scattered grains are seen in, or in close proximity to, the ER.

A finding not consistent with the ER as the route for fast-transported components comes from studies of pike olfactory fibers carried out by Kreutzberg and Gross (1977). They found the ER to be discontinuous axially in sections of the nerve fibers. At least 14% of the cross-sections made along the length of the pike olfactory nerves showed no ER profiles present in the axons. Kreutzberg and Gross concluded that such discontinuities are incompatible with a transport in the ER. It should be noted that the usual pattern of outflow of labeled proteins typical of fast transport is present in the pike olfactory nerve (Gross and Kreutzberg, 1977). An additional consideration which makes it unlikely that the ER serves as the route for fast-transported components is that the DCVs and the mitochondria, which temporarily have a fast movement, are too large to move down within such a channel. Even if they did, it is difficult to see how those organelles inside the ER could carry out their functions: the supply of ATP and regulation of Ca^{2+} by the mitochondrion (Chapters 7 and 8) and the turnover of amines in the DCVs of the adrenergic fibers (Chapter 4). On the other hand, Špaček and Lieberman (1980) have observed in serial sections of nerve an apparent continuity of the outer membrane of the mitochondria with the ER. They consider as a hypothesis that the outer membrane represents a part of a continuous ER channel and that the mitochondria can move within the ER. The authors note, however, that their evidence for such a continuity of the outer membrane and the ER is not definitive and that others have not found such continuity (e.g., Droz, Rambourg, and Koenig, 1975; Tsukita and Ishikawa, 1976). In any case, the means by which such movements of mitochondria within the ER channel could occur remains unknown.

In a variant of this hypothesis, Hammerschlag (1980) proposed that proteins, phospholipids, and cholesterol synthesized into the membrane of the ER in cell moves down the axon, the ER budding off so that these components supply the axolemma along the axon and eventually the nerve terminals (cf. Fig. 5.12). Here too, the fast transport of non-membrane components and the larger organelles, the vesicles, and mitochondria, were not considered.

Short segments of the ER may be moved retrogradely in the nerve by the fast transport mechanism, as indicated in the studies of LaVail and LaVail (1974). The HRP transported in the retrograde direction (Chapter 3) was found in EM sections of the fibers within vesicles and short lengths of ER. However, the smooth ER is not a continuous channel for retrograde transport of HRP (LaVail *et al.*, 1980) . The HRP taken up into the nerve terminals by endocytosis (Chapter 11) and retrogradely transported (Chapter 3), initially within vesicles, may coalesce to form what appears to be short segments of ER. Some of the HRP found was non-membrane bound, this component presumably bound to transport filaments for retrograde transport. Colchicine was effective in blocking the retrograde transport of HRP, as would be expected of the similarity of the anterograde and retrograde transport mechanisms including the action of tubulin-binding agents to block (Chapter 12). An interesting observation made by LaVail (1975) relative to the transport filament model was that in some EM sections cross-bridges from the microtubules appeared to contact the vesicles containing HRP.

8. The Microtrabecular Network

The microtrabecular network (Chapter 9) has been suggested as playing a role in axoplasmic transport (Porter, 1976). We can conceive of the network opening before and constricting behind transported components. However, a specific model whereby the microtrabecular network could give rise to a movement at the same fast rate of small peptides, soluble proteins, vesicles, and temporarily mitochondria is not in hand. In particular, the manner in which the trabeculae make and break their connections between the microtubules, neurofilaments, the ER, mitochondria, and membrane remains unspecified, as well as how the requirement for ATP to maintain transport fits with an action of the microtrabecular network. If the microtrabecular bonds have any strength, enough to act as a force-generating mechanism, this would appear to militate against the movement of the larger organelles through their narrow interstices. The cross-linkages forming the microtrabecular network would have to be labile enough to allow the larger structures such as DCVs and mitochondria to move axially in the nerve fibers and yet be strong enough to exert force to bring about such movement. The constriction of fibers in the process of beading (Section D, Chapter 9), suggests that the radially directed contractile force is able to readily break the side-to-side connections of the network, and thus that such binding is weak.

9. A "Worm Screw" Mechanism

Considering the helical appearance of the H component of the neurofilament triplet proteins wrapped around the central filamentous core (cf. Section B2, Chapter 9), Willard and Simon (1981) proposed that the H protein could, by its constant turning around the central core filament, propel the various components in the nerve down the neurofilament in the axial direction. Those components would be bound to some degree to the neurofilament in the groove formed by the helix. This mechanism would require that the components not be bound to the H components but that they be pushed by the helix. Otherwise, materials would move helically around the filament, a type of movement not seen in microscopic studies of particle movement. There must also be some association or temporary binding of transported components with the filament core, a binding sufficient to hold them in place, yet, not so tightly bound that the components cannot be pushed. This would also imply some kind of "railing" to keep the components centered axially in the groove in the course of their movement. In addition to these special requirements, the similarity of the rates of transport of molecular species extending from molecules to such large structures as vesicles and mitochondria, which would not fit in the grooves, needs an explanation.

B.   SLOW TRANSPORT MODELS

1.   Bulk Flow of Axonal Contents

Ramon y Cajal (1928) in his studies of nerve degeneration and regeneration observed the effect of a partial ligation of nerve. In rabbit sciatic nerve so prepared and examined several weeks afterwards, the nerve fibers above the partially constricted region show a series of beaded enlargements with thin fibers present in and beyond the constrictions (Fig. 10.7).

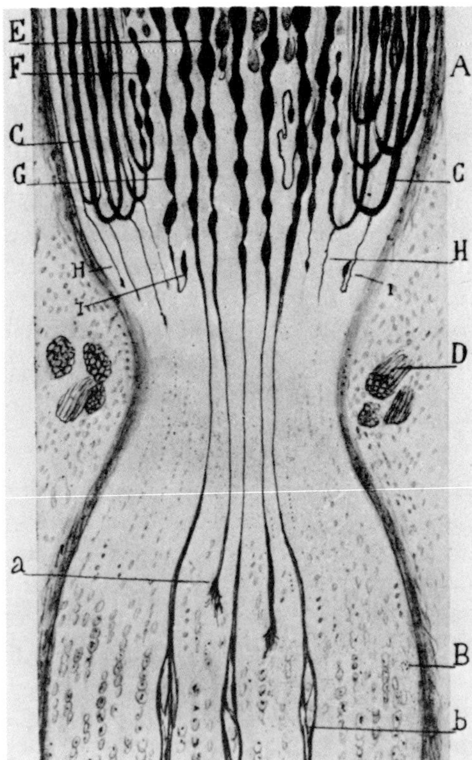

Fig. 10.7 Starting and stopping of regenerative growth. A nerve ligated to partially construct the sciatic nerve of a rabbit is shown 13 days afterwards: A, central stump; D, central region in which the ligature was placed; B, peripheral stump; C, retrogradely growing fibers; F, arrested retrogradely growing fibers; E, anterogradely growing fibers appearing as a string of beads with finer growing branches penetrating the compressed region; H, fine exploratory fibers; a, terminal cone; b, "nervous reunion" in the peripheral stump. From Ramon y Cajal (1928).

The thin fibers were taken by Cajal to represent regenerating fibers because of the growth cones at their distal ends (*a* of Fig. 10.1). The beads in the fibers in Cajal's interpretation are brought about by alternating periods of regenerative growth and rest. In similar studies of partial constriction of nerve, Weiss and Hiscoe (1948) cut rat sciatic nerves distally and slipped short segments of artery proximally up over the nerve. The segments of artery were of a size which, in response to circulating or applied catecholamines, caused a partial constriction of the nerve trunk. After several weeks or more of such partial constriction, the nerves taken for histological examinations showed the fibers just proximal to the constrictions to exhibit the beading and other form irregularities shown in Fig. 10.8. These form changes were explained as due to a continually "flow" or "growth" of the axonal contents down within the nerve fibers (Weiss, 1961, 1972). In Weiss' view, "the 'axis cylinder' of the mature neuron keeps growing forth throughout life from its base in the cell body, its macromolecular substance being . . . conveyed as a cohesive semisolid mass toward the distal ending . . . at a standard rate of the order of 1 mm/day, driven by a microperistaltic wave generated at the axonal surface" (Biondi, Levy, and Weiss, 1972; Weiss and Mayr, 1971b).

The beading, swellings, tortuosities, and other deformations of the fibers were considered to be caused when the moving "semisolid column" of axoplasm meets with the resistance imposed by the partial constriction of the fibers as diagrammed in Fig. 10.9.

Spencer (1972) looked for such form changes in sections of dammed regions of nerve fibers in EM preparations. Large axonal swellings were not seen in the region of damming above partial ligations. Instead, the nerve

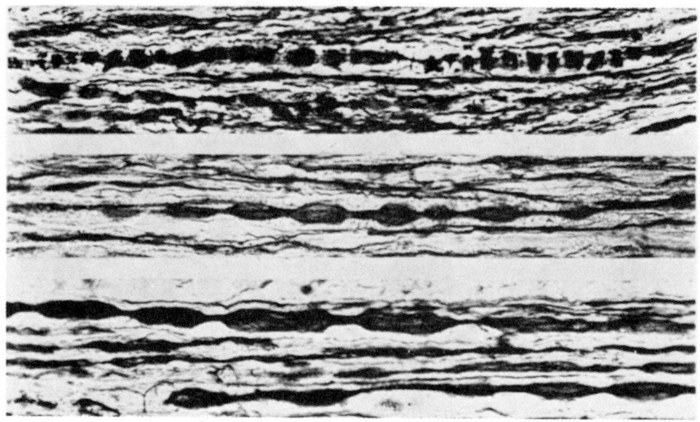

Fig. 10.8 Damming of fibers above a region of partial constriction. Convolutions and beading of fibers above a site of partial constriction are shown in three panels. The upper panel represents a portion of nerve taken closest to a partial constriction of a nerve trunk, the other panels are from nerve somewhat further away. From Weiss and Hiscoe (1948).

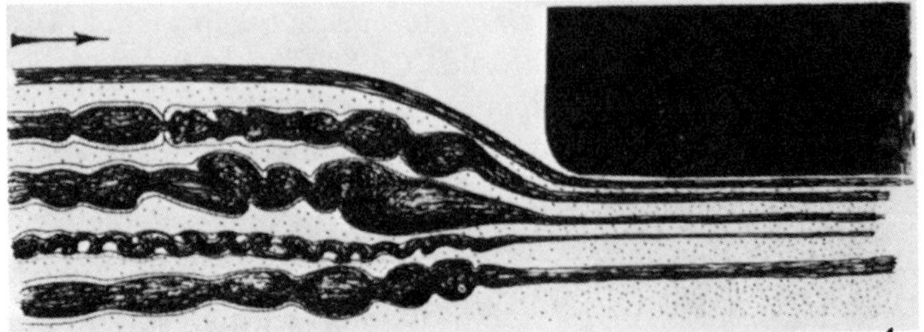

Fig. 10.9 Damming and continuous flow hypothesis. The black bar represents a region of partial compression. A damming of flow in fibers at the left shown diagrammatically causes fibers to become convoluted and beaded. From Weiss (1969).

fibers showed sprouts of regenerating fibers (Chapter 15) mixed with normal appearing nerve fibers (Fig. 10.10).

Regenerating fibers arise in large numbers from the region of damming and they could appear as tortuousities, swelling, and convolutions in silver-stained preparations. Similar observations showing a lack of fiber swellings or tortuosities in the myelinated fibers in the region above ligations or cut

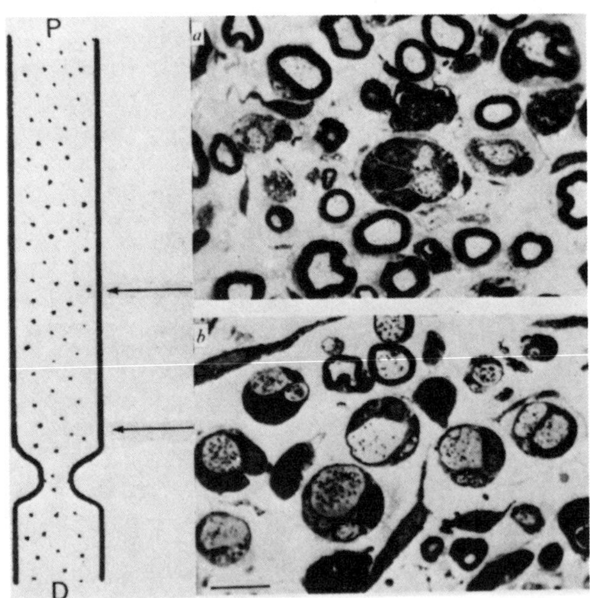

Fig. 10.10 Fibers in a dammed region examined electron microscopically. The diagram at the left illustrates a nerve partially constricted for 8 days (P and D indicate the proximal and distal ends). At the right (a) a cross-section in the EM taken from the upper portion of nerve (upper arrow) shows clusters of regenerating axon sprouts amidst a population of relatively normal uninterrupted fibers; (b) from a site closer to the constriction (lower arrow), a larger number of regenerating clusters are seen, many of which possess markedly swollen axons. A few myelinated fibers in the population are not swollen. Bar equals 10 μm. From Spencer (1972).

ends of fibers were also reported by Berthold and Skoglund (1967). Using freeze-substitution, the fibers above a region of constriction showed little if any signs of swellings and tortuosities (Ochs, unpublished observations).

Further evidence bearing on this point comes from the work of Friede and Bischausen (1980), who reconstructed the form changes of 5 fibers taken from the region just above a transection made 4 days previously from EM serial sections (Fig. 10.11).

In two fibers (*D* and *E*), a gradual increase of axonal volume was seen toward their cut end, but the increase did not have the form expected of Weiss' theory. The other 3 fibers did not increase in diameter at all; instead, a number of regenerating axons were seen sprouting from them. The total axonal volume of the original fiber and the sprouted regenerating axons showed an overall increase, but this was mostly due to the regenerated fibers. These can be profuse, as shown by Perroncito (1905) and by Ramon y Cajal (1928); also, cf. Chapter 15.

Weiss (1961) interpreted the thinning of nerve fibers within and distal to the partially compressed region of nerve as due to a reduction in the supply of axoplasm to the fibers distal to a partial obstruction. Another explanation is likely. The compression causes a degeneration of fibers, particularly of the larger myelinated fibers which have a greater susceptibility to mechanical compression than do the smaller fibers (Gasser and Erlanger, 1929; Aguayo *et al.*, 1971). This leaves the hardier, smaller fibers remaining within, and distal to, the region of compression along with regenerating nerve fibers (Chapter 15).

If nerves are gradually compressed, degeneration may not occur at all. This was first shown by Duncan (1948), who placed tantalum ligations snugly around sciatic nerves of rats 10 days of age. As the animals grew, the nerves became partially compressed. Nerves taken after 164–196 days were seen to be reduced to 1/4 or 1/5 of control diameters without, however, the swellings, tortuosities, beading, and other form changes in the fibers above the compression characterized as signs of damming. Much of the swelling in the nerve trunk proximal to ligations is due to an increase in the endoneural space (Weiss and Davis, 1943). Aguayo *et al.* (1971), using Duncan's technique placed ligations on the medial popliteal nerve of rabbits aged 3–5 weeks. They found a shift to smaller fiber diameters within the region of compression after a period of months which was mostly the result of a demyelination of the thicker fibers. Fibers distal to the compression might regain their normal diameters.

When the compression was removed in Weiss' experiments, a decrease in the size of the ballooned fibers above the site of compression was seen, along with an apparent enlargement of the thinner fibers below that region. This was interpreted as a released flow of the dammed up contents, the convoluted and ballooned axons becoming "straightened and drained of their axonal material while the myelin, as a more rigid container, retaining for a time its contorted configuration" (Weiss, 1961). As against this interpretation, the removal of the constriction would allow regenerating fibers to grow into that

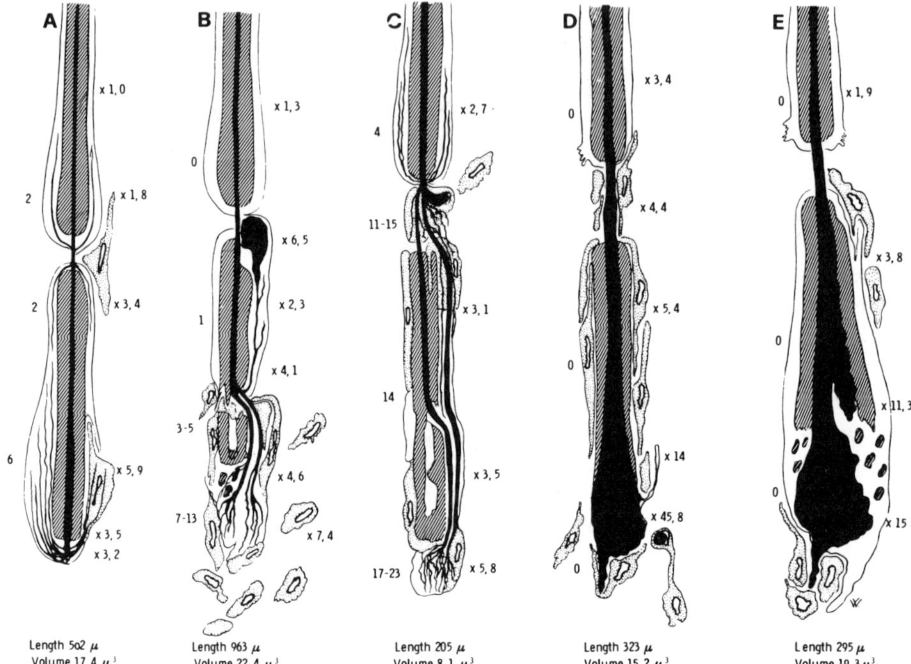

Fig. 10.11 Reconstruction of fibers from serial section taken above a nerve transection. After making a nerve transection 4 days previously, the axonal organization of the stumps of fibers A–E is shown reconstructed from serial electron microscope (EM) sections made above the transection. Numbers on the right side of each fiber indicate the extent of the increase in overall axoplasmic area of the fiber and its axon sprouts at that level, as compared to the mean value of the area taken from that same fiber at a more proximal position. The numbers at the left of each fiber indicate the number of axon sprouts at that level (in fiber C some were not drawn). The numbers at the bottom indicate the length of the first internode; some were shortened in the drawing. "Volume" indicates the total amount of axoplasm (sprouts and swelling) within the first internode in μm^3. The thickness and flaring of the myelin sheath in fibers D and E illustrate one type of change where sprouts were not seen. In these two fibers, the behavior of retracted nodal Schwann cell pseudopodia is shown at the first internode. The positions of migratory cells (dotted cytoplasm) are shown with some simplification. From Friede and Bischausen (1980).

region, reducing the tangle of regenerating fibers in that region (cf. Chapter 15).

2. Microperistalsis

The flow of fluid from the freshly cut end of the giant axon or frog nerve fiber (Chapter 1), would suggest that the axoplasm is normally under some degree of lateral tension (cf. Chapter 9). Young (1944a,b) proposed in an early model of axonal flow that the continual production of axoplasm by the cell body creates a pressure moving the axoplasm down within the fiber, a pressure on the fiber wall considered to be opposed by the surface tension of the myelin in nerve fibers. In support of this concept, Young pointed to the beaded appearance of mammalian nerve fibers seen in the early stages of Wallerian degeneration (Chapter 6), a phenomena analogized to the unduloid Plateau figures produced by surface tension (Thompson, 1942). However, the lamellated structure of the myelin sheath (Geren, 1954) militates against its acting as a fluid, one with such surface tension properties ascribed to it. Weiss (1961) also pointed out as an objection to Young's hypothesis that a considerable force would have to be exerted from the cell body to move axoplasm down the longer lengths of nerve fibers.

A local microperistaltic mechanism of propulsion all along the length of the fiber was proposed by Weiss (1972) to account for the slow transport in his flow model. In time-lapse cinematography of nerve fibers taken from sensory nerve explants, he saw indentations of the surface of the fibers which he took to be peristaltic-like waves successively moving down along the surface of the fibers at an average rate of 2.14/hr.

Weiss (1972) found no evidence for an internal contractile change in the axon which might be related to the indentations and he concluded that the indentations were externally produced by the Schwann cells. However, as Szentágothai (1963) pointed out, the indentations pictured by Weiss are similar to those seen in the early stages of Wallerian degeneration and Weiss (1972) later referred to the indentations as "incipient Wallerian degeneration." Heslop (1975) made the cogent observation that the indentations indicated by Weiss were asymmetrical. If they did represent a true microperistalsis, they should appear as annular constrictions of the nerve fibers. Samson (1976) also pointed out that such a model of microperistalsis would not account for transport in the unmyelinated fibers.

The question as to whether mature nerve fibers in their normal environment *in vivo* do in fact show such a microperistalsis was answered directly by the microscopic transillumination technique developed by Williams and Hall (1970). Observations were made of nerve fibers within their perineurial sheaths *in vivo* over long periods of time and with remarkable clarity. Fibers of the common peroneal nerve of the mouse observed for hours each day over.a period of days gave no evidence of any alterations in the form of nerve fibers which could be described as a microperistalsis (Williams and Hall,

1971). They found "no change in either external or internal myelin sheath contour: the majority of internodes appeared to be smooth-walled and cylindrical."

Other kinds of movements are seen in cultured nerve fibers *in vitro* and in fibers which have become recently myelinated (Murray and Herrmann, 1968), but with none of the indentations indicative of a microperistalsis. Studies of single fibers isolated from frog nerve and transilluminated *in vitro* (Singer and Bryant, 1969) showed openings and closings of the Schmitt–Lantermann clefts and some undulatory movements of the myelin sheath, but no microperistalsis.

3. Beading and Flow

An attempt was made to relate the phenomenon of beading (Chapter 9) to axonal flow (Ochs, 1965). The exponentially declining outflow characteristic of slow transport (Chapter 2) was considered to be brought about by a random series of fiber constrictions which, with the continued production of proteins and other materials by the cell bodies, was supposed to cause a proximodistal outflow of axoplasm. However, as discussed in the above section, there is little evidence that constrictions appear normally in nerve fibers *in vivo*. It should be noted that some radially directed contractile mechanism is present in the fibers, one perhaps due to the actin and myosin present in the fibers (cf. Section D, Chapter 9). Such constrictions may come into play in the normal fiber under unusual circumstances, such as stretching of the nerve, or in the early phase of Wallerian degeneration.

4. Outgrowth of Microtubules and Neurofilaments

As part of the bulk flow of all the axoplasm moving down within the nerve fibers, Weiss and Mayr (1971b) visualized an assembly of tubulin in the cell body with the outgrowth of the organelle in the axon. An outgrowth of the microtubules and the neurofilaments assembled in the cell bodies was also proposed by Hoffman and Lasek (1975). On injecting the L5 motoneuron region of rat cord with ^{3}H-leucine, they found a wave of labeled proteins (SCa) which, from the proximo-distal translation of its peak with time (Fig. 11.10), they estimated to be moving down the sciatic nerves at a rate of 1–1.2 mm/day. The problem of peak-like waves related to later outflow will be discussed further in Chapter 11. In any case, using ^{35}S-methionine as a precursor, and allowing sufficient time for slow downflow, Hoffman and Lasek took small 2-mm segments all along the length of nerve and analyzed their content of labeled polypeptides by SDS-polyacrylamide gel electrophoresis (SDS–PAGE). As described in Chapter 5, two groups of polypeptide bands were found which accounted for the major portion (70–85%) of the radioactivity moving down as the slow-wave SCa: the 57,000- and 53,000-dalton bands corresponding to the α and β tubulins, the subunits of the micotubules; and a triplet of proteins of 200,000, 145,000, and 68,000 daltons which represents the subunits of the neurofilaments. These can be seen in the fluorogram of Fig. 5.17.

Lasek and Hoffman consider the tubulins and triplet polypeptides to be moved down within the nerve fibers in a coherent fashion, namely, as completely assembled microtubules and neurofilaments. These organelles were visualized as being continually assembled from their respective subunits in the cell body with the formed microtubules and neurofilaments moved down within the axon by means of an actin–myosin interaction between these organelles and other components in the fiber (Fig. 10.12).

The force generated by the actin interaction with the membrane would create the waves of surface indentations reported by Weiss. However, as described in the preceding section, such indentations are not seen in normal nerve fibers. A necessary feature of the theory is that the microtubules and neurofilaments are disassembled in the nerve terminals at the same rate that they are formed in the cell body and moved down the fiber (Fig. 10.13).

As indicated in this schematization, a Ca^{2+} activated protease is shown in the terminals disassembling the organelles. Recent evidence indicates that Ca^{2+} is present in nerve endings in micromolar quantities (Baker, Knight, and Whitaker, 1980). The levels calculated were similar to those likely present in the axon, 0.1 μM, a concentration of Ca^{2+} too low to selectively activate proteases disassembling microtubules and neurofilaments in the terminals. Microtubules in fact may be present in the nerve terminals as indicated when a special fixation technique was used by Gray (1976, 1978). The microtubules are either fixed to dense structures within the inner membrane of the terminals, or form loops in the terminals (Bird, 1976; Chan and Bunt, 1978) as described in Chapter 13.

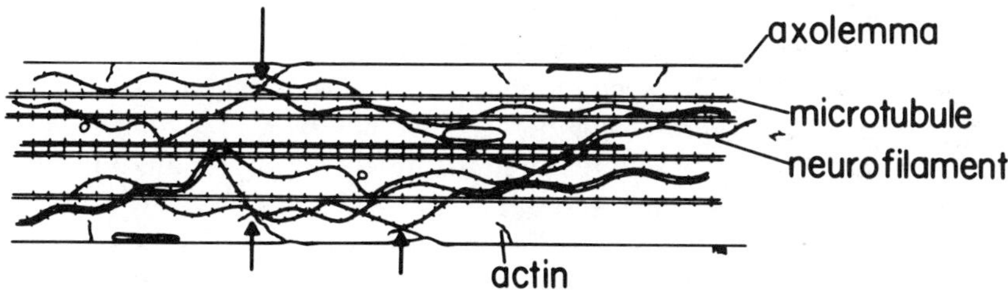

Fig. 10.12 Hypothetical movement of microtubules and neurofilaments in the axons. In this model the microtubules and neurofilaments are proposed to move down the axon as preformed organelles. The myosin attached to the neurofilaments interacts with actin attached to the axolemma. The neurofilaments, which course through the axon in a wavy or helical fashion, contact the actin associated with the plasma membrane at some points to form a network with the microtubules. Movement of the network is considered to occur by unidirectional contractions of the actin–myosin complexes, resulting in its unidirectional displacement. From Lasek and Hoffman (1976).

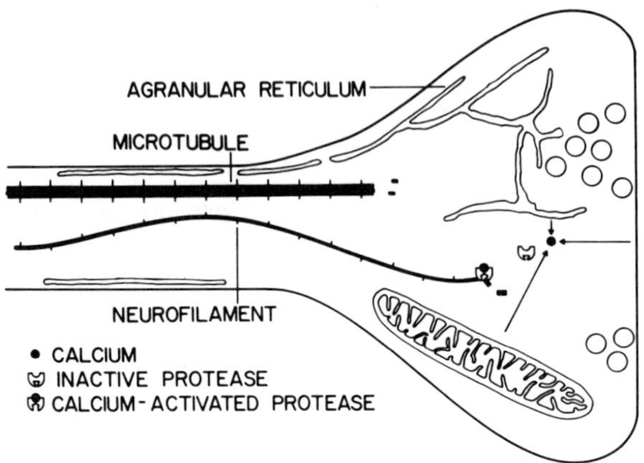

Fig. 10.13 The disassembly of microtubules and neurofilaments in the terminals. In the model of Fig. 10.12, where the microtubules and neurofilaments are continuously being moved down as formed structures, the organelles require to be disassembled in the terminals. The intracellular Ca^{2+} concentrations in the axon terminal are considered to be greater than in the axon. As the microtubules and neurofilaments advance into the axon terminal, the increased Ca^{2+} concentration there activates a protease which cleaves the neurofilaments. Microtubules are also depolymerized in this model under conditions of elevated Ca^{2+} concentration. From Lasek and Hoffman (1976).

If microtubules move down the axons coherently in the fashion suggested, one might expect to find an increased number of microtubules curled up above nerve constrictions. In the reconstruction studies of nerve fibers just above transections, Friede and Bischausen (1980) pointed out that the microtubules showed their usual straight axial position in the axons without a twisting or curling of microtubules as might be expected of accumulation. The observations of Friede and Bischausen support a process of assembly of microtubules in the regenerating fibers just above the transection, a process which would be due to the accumulation of the tubulin subunits at that site. The increased number of microtubules, as much as $10\times$ in regenerated fibers above a transected fiber, would require a sufficient supply of tubulin subunits for the assembly of those microtubules at that site.

The neurofilaments also show discontinuities (Weiss and Mayr, 1971a; Kreutzberg and Gross, 1977), with their absence noted in the nodes by Tsukita and Ishikawa, (1981). Such findings are difficult to reconcile with a continual outgrowth of neurofilaments from the soma.

A constant movement of formed neurofilaments and microtubules within the fibers at a rate of 1 mm/day (in SCa), also does not agree with the measured regeneration rate of nerve fibers of 3.5–4.5 mm/day (Gutmann *et al.*, 1942; Bisby, 1978; Griffin *et al.*, 1976). As described in Chapter 15, microtubules and neurofilaments are present in the fine regenerating fibers extending to, and into, the base of the growth cones. This difference in the two rates would result in the growing regenerating fibers outdistancing the microtubules if they were supplied by coherent outgrowth in the regenerating nerve fibers.

This criticism could be answered by an increased rate of transport during regeneration (Hoffman and Lasek, 1980). In a recent study along this line, Hoffman, Griffin, and Price (1981) found an increase in the rate of actin and tubulin, to 7 and 6.1 mm/day, respectively, in developing and regenerating fibers as compared to rates of 3.9 and 2.8 mm/day in mature control animals. An increase in neurofilament triplet proteins might be expected, but it fell 50%.

On the other hand, McQuarrie, King, and Lasek (1981) found no change in the SCb of rat sciatic motor fibers from a normal rate of 3.4 mm/day to 3.3 mm/day in regenerating fibers. The SCa component moves at a rate of 1.2 mm/day. Increases in the amount of tubulin and actin in SCa and as well in SCb were found (McQuarrie *et al.*, 1980). The neurofilament protein did not increase in its normal rate of 1.2 mm/day. The picture of rate changes during regeneration is complicated by variations in amounts of the various components transported (Chapter 9) and the different rates seen in various types of nerve and in nerves taken from different species (cf. McQuarrie, Brady, and Lasek, 1980).

An alternative explanation is that the subunits of the microtubules and neurofilaments are carried down as soluble proteins, the subunits adding to the growth of these organelles at their distal ends to supply the regenerating fibers. The presence of subunits all along the fibers would also allow for their constant turnover in the organelles all along the length of the fibers, presumably at the assembly ends of the short microtubule segments (Chapter 9). The short half-life of α and β tubulins, of 6.2 days, found by Forgue and Dahl (1978) is consistent with such a turnover. On the basis of an outgrowth of the microtubules as formed organelles, their proteins in long lengths of nerve would require half-lives of months or years.

Grafstein, McEwan, and Shelanski (1970) found that ^{3}H-colchicine injected into goldfish eyes is transported to the tectum bound to tubulin. Such binding would likely interfere with the assembly of the labeled tubulin into microtubules (Wilson *et al.*, 1974; Chapter 9) and suggests that the outflow measured is one of tubulin subunits. The MAPs, including tau factors and TAP, would also be present in a pool in the axon for turnover in the microtubules (Chapter 9).

It is possible to "partition" the axonal contents by stretching nerves to cause them to bead (Ochs, 1963). In nerves stretched and frozen quickly at low temperatures for freeze-substitution and EM preparation, the filamentous components (microtubules and neurofilaments) were seen concentrated in the constricted regions of the beaded fibers (Ochs, 1965). Nerves were taken after injecting ^{3}H-leucine into the spinal cord and sufficient time for the labeled proteins to move down into the motor fibers by slow transport allowed before being subjected to stretch to cause the fibers to bead. After preparation by freeze-substitution and making sections for EM autoradiography, those fibers showed no greatly increased concentration of grains of radioactivity in the constricted regions as compared to the expanded parts of the beaded nerve fibers. Those studies indicated that the filamentous organelles are not more heavily labeled as might be expected of their slow outflow. Labeled components are thus not fixed or part of the filamentous elements.

5.   Actin–Myosin Interaction and Slow Movement of the Microtrabecular Network

Durham (1974) proposed that actin attached to the membranes interacts with myosin under the influence of cyclic AMP under the control of Ca^{2+}. The Ca^{2+} was considered to enter the fibers as a wave along with the action potential. The actin chains are pictured as attaching to vesicles which are then pulled along the membrane in cyclic fashion as actin and myosin chains interact. This theory does not explain the movement of the larger mitochondrion axially throughout the whole cross-section of the axon, nor does it account for retrograde transport. The observation most difficult to explain on this theory is that transport is independent of the excitability or the state of polarization of the axonal membrane (Chapter 8,12).

The actin present in the microtrabecular network is considered to move down within the fibers at a rate of 3–4 or 4.3 mm/day (Willard *et al.*, 1979). This rate is similar to that of the slow wave designated SCb by Lasek and Hoffman (1976) and the slow movement of actin as determined by SDS–PAGE after labeling with ^{35}S-methionine (Lasek, 1980). However, the apparently close binding of the microtrabecular network to microtubules and neurofilaments, mitochondria, and to the cell membrane (cf. Chapter 7) makes it difficult to envision the microtrabecular network as a defined structure moving in the fibers, one having a rate which differs from that of the microtubules and neurofilaments moving at close to 1 mm/day (cf. Section 4 above). How microtrabecular connections are made and broken to account for a slow anterograde transport of this network requires an explanation.

6.   A "Mitochondrial Shuttle" Hypothesis

An attempt was made to account for slow flow on the basis of a to-and-fro movement of mitochondria (Ochs, 1974c). The idea proposed was that the movement of the mitochondria would pull or drag soluble components with it. However, the mitochondria are seen to move at relatively infrequent intervals and it is not likely to be the motive force for the various sizes of components transported. An hypothesis which accounts in a more satisfactory manner for the wide variety of material transported is described in the following chapter, the "unitary hypothesis" based on the transport filament model.

11

The Unitary
Hypothesis and Routing

When dealing with slow transport at various points throughout this book, particularly in Chapter 10, note was taken of the two fundamentally different interpretations advanced to account for it. In one view, separate mechanisms are considered responsible for fast and slow transport. Alternatively, slow transport has been accounted for as an aspect of the same mechanism subserving fast transport (Ochs, 1975c; Gross and Beidler, 1975). The "unitary hypothesis" (Ochs, 1975c,e, 1976) holds that slow transport is due to the drop-off of certain classes of materials from the transport filaments locally in the axon, the materials dropped off undergoing a turnover in the various organelles of the fiber. We shall describe the evidence for the unitary hypothesis in the first part of this chapter and in the second part the phenomenon of "routing," the asymmetrical transport of materials in different branches of the same neuron. How routing can be accounted for by the transport-filament model and the unitary hypothesis in comparison to some of the other models of transport will also be discussed in this chapter.

A. THE UNITARY HYPOTHESIS

In the unitary hypothesis three main processes are considered responsible for the outflow pattern of slow transport. These are; (a) a later efflux of incorporated materials from a compartment in the cell body into the axon following pulse labeling (cf. Chapter 5), (b) the early drop-off of materials from the transport mechanism in the axons, and (c) the redistribution of materials within the fibers. In the model of Gross and Biedler (1975), components drop out of the microstream channel (Fig. 10.4) on the basis of their molecular weights, whereas in the transport filament model, the drop-off of components depends on the nature of the binding of those components to the transport filaments and the turnover of the various components in organelles of the axon as pictured in Fig. 11.1.

Those components which drop off the transport filaments enter a pool in the axoplasm which supplies the membrane, the microtubules, and other

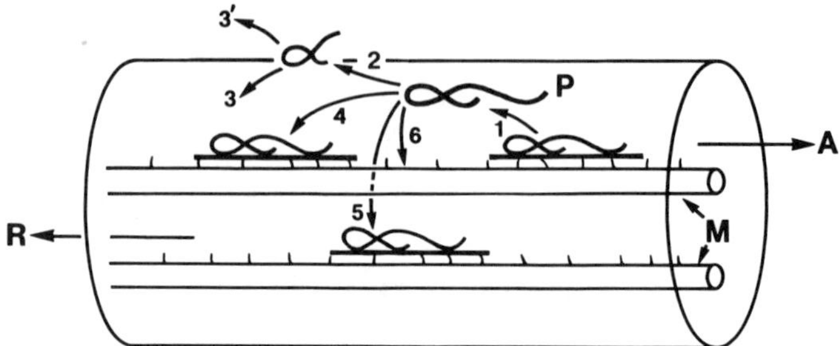

Fig. 11.1 The unitary hypothesis based on the transport filament hypothesis. Components which drop off from the transport mechanism enter into an axonal pool (P). Their subsequent modes of redistribution contributing to slow transport are shown: (1) drop-off; (2) insertion in membrane; (3) and (3′) return or loss from membrane; (4) return to transport mechanisms for anterograde transport (A) or (5) retrograde transport (R), with (6) representing components turning over in microtubules (M) (or in neurofilaments not shown).

organelles of the axon. Components may reattach to transport filaments for further anterograde transport, or retrograde transport back to the cell body where presumably those materials may be reutilized. In addition, components turning over in the membrane may leave the axon, possibly to supply the Schwann cell, or re-enter the axon to undergo redistribution.

1.   Generality of Downflow on the Basis of the Unitary Hypothesis

In the transport filament model, materials and vesicles bound to the transport filaments move down to the nerve terminals as fast-transported components (Chapter 13). Those which contribute to slow transport are considered to have a weaker binding to the transport filaments so that they more readily dissociate and drop off to contribute to a local turnover in the axon. Thus, according to the unitary hypothesis, all components would at some point in the course of transport be carried by the transport filaments. A particular substance may not appear in the crest of fast transport if the drop-off occurs too soon and the amount carried over a long distance is too small to be detected. Thus, an inadequate technique may fail to detect fast transport. A number of cases are known of transported components which earlier had been considered to be uniquely slow-transported and then subsequently found to be fast-transported as well. An example is the enzyme tyrosine hydroxylase (TH), which Brimijoin and Wiermaa (1977a) demonstrated to be carried both as fast and slow waves. Various other examples in adrenergic and cholinergic neurons are known (Chapter 4).

The bulk of tubulin and triplet subunits (Chaper 9) are slow-transported (Chapter 10). However, a small amount appears to be fast-transported as well (Stromska, Iqbal, and Ochs, 1979). Cancalon (1979b) also found a 58,000-dalton polypeptide which comigrated with tubulin in the fast-transported crest in garfish olfactory nerves. On the other hand, however, two-dimensional SDS electrophoresis studies show only a small amount of tubulin and triplet proteins to be fast-transported and this point requires further evaluation.

A class of compounds commonly considered to be carried down exclusively by fast transport are the glycoproteins. Their fast outflow was shown using the sugars ^{3}H-fucose and ^{3}H-glucosamine as precursors (Forman *et al.*, 1971, 1972), these glycosylating the proteins (Chapter 5). However, glycoproteins are not only fast-transported, they also show a drop-off in the axon, presumably for their turn-over in the membranes (Elam, 1979). Using ^{3}H-fucose as a precursor, Griffin *et al.* (1981) have shown in EM autoradiographs the downflow and insertion of glycoproteins in the axolemma (Chapter 10). Some differences in the degree of drop-off may be related to the type of nerve fiber. Using either ^{3}H-fucose or ^{3}H-glucosamine as markers of glycoprotein in comparison with ^{3}H-leucine for labeling of proteins, a difference between the outflow of glycoproteins in motor and sensory fibers was found, with less labeled glycoproteins seen remaining in the plateau in the sensory fibers as compared to the motor fibers (Stromska and Ochs, 1978).

Another example of an organelle showing both fast and slow transport is the mitochondrion (Chapter 3). Dark field or Nomarski optics microscopy of nerve fibers shows these rod-shaped particles to be at rest for the most part, with only at infrequent times their movement in the anterograde or retrograde directions at rates approaching that of fast transport (Chapter 3). They appear to move in response to some signal, such as the local need for a supply of ATP (Chapter 7), or the level of free Ca^{2+} (Chapter 8). On receipt of the signal the mitochondria would attach to the transport filament for rapid movement. Because the mitochondrion spends only a relatively short time attached to the transport filaments moving in either the anterograde and retrograde directions (Chapters 2 and 3), the result is the slow net movement in the anterograde direction.

2. Drop-off of Materials in the Axon

The drop-off of components from the fast-transported crest is shown by the decrease in crest amplitude with distance in the garfish olfactory nerve (Fig. 2.30), and in the mammalian nerve (Ochs, 1972c) as shown in Fig. 11.2.

The use of a logarithmic ordinate scale in the mammalian example does not show the decrease in the crest amplitude as dramatically as when a linear scale is used as shown in the garfish olfactory nerve or frog nerve given in Fig. 2.27. The decrease in the amplitude of the crest is associated with a trail of labeled activity behind the peak of fast-transported activity (Fig. 11.3).

In the studies carried out by Muñoz-Martínez, Nuñez, and Sanderson (1981), an outflow of labeled protein of 3 hr was allowed before making ligations just below the ganglion and allowing a further time of downflow

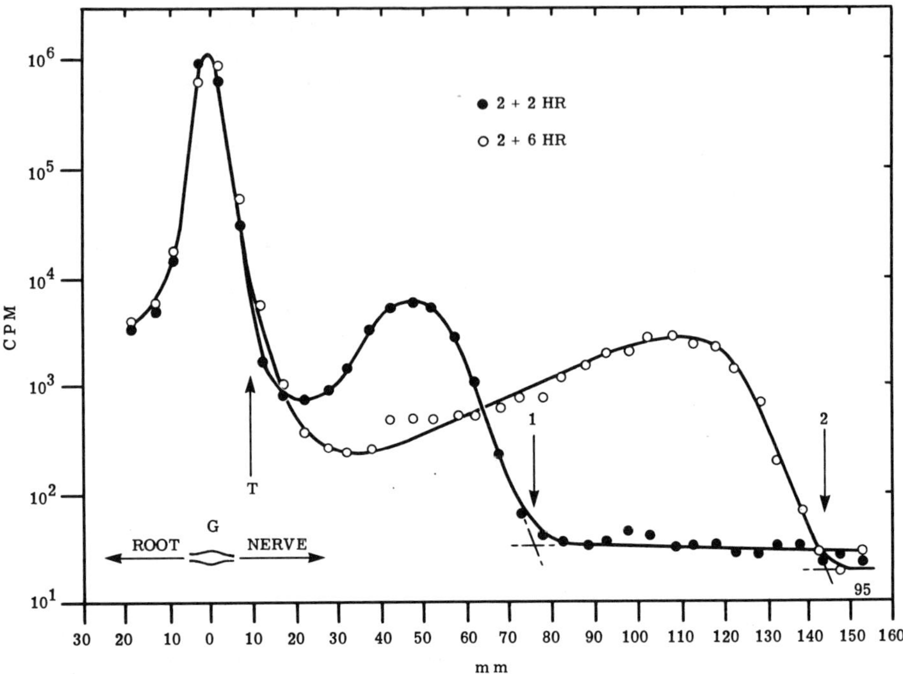

Fig. 11.2 Decrease in crest height with transport distance. Two hours after equal injections of ³H-leucine into the L7 dorsal root ganglia of a cat, ligations were placed below each ganglion. An additional downflow of 2 hr is allowed in one nerve before it was removed (●) with its front transported to the expected position shown by arrow 1. The nerve on the other side was removed after an additional 6 hr of downflow (○) with transport to the position indicated by arrow 2. The latter peak shows a smaller height and more trailing behind the crest than in the nerve in which the shorter time of downflow had been allowed. From Ochs (1975c).

from 0.5 to 5 hr. The slope of the trail behind the peak gradually became shallower with distance, a pattern expected of materials dropping off from the fast-transported peak. The material dropped off gives rise to a concomitant increase in plateau height as shown in Fig. 11.4.

The level of radioactivity in the plateau is higher after 2 hr of downflow as compared to 1 hr may be as seen in this figure, and is to be compared to the further increase in the plateau height after 7 hr (Fig. 2.19). Muñoz-Martínez *et al.* (1981) further analyzed the level of the plateau with respect to the decrease in amplitude of the crest as it passes down the length of the nerve fibers (Fig. 11.5).

The drop-off in the plateau and its decrease in the plateau with distance can be correlated directly with the decrease in the crest amplitude. The line through the points in Fig. 11.5 represents the normalized distribution of labeled materials in segments of the plateau taken after ligatures made 18 hr

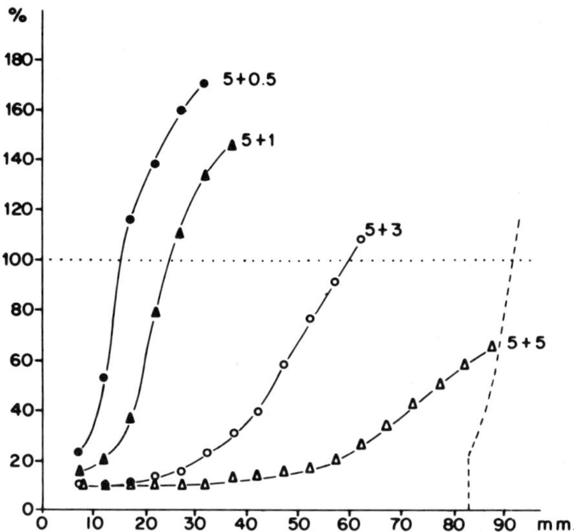

Fig. 11.3 Trailing behind the crest. After 5 hr of downflow, ligations were made behind the ganglia and a further downflow of from 0.5 to 5 hr was allowed. An increasing shallow slope of trailing behind the crest is seen with the longer subsequent downflow times (cf. Fig. 11.2). From Muñoz-Martínez *et al.* (1981).

previously. The following equation giving the theoretical distribution indicated by the solid line is calculated by a constant amount of crest material dropped off per unit length of nerve:

$$N_r = N_o \, (1 - \alpha \Delta x)^r$$

where r = the number of segments, N_o = radioactivity in the initial segments, N_r = radioactivity in segment r, Δx = unit length, α = the fraction of radioactivity dropped off.

The labeled proteins dropped off in the axons to constitute the plateau are similar in their composition to those which remain on the transport mechanism contributing to the crest. This was determined by assessing the composition of labeled components in the crest and plateau regions separately at various times after injection and ligation (Sabri and Ochs, 1973; Cancalon and Beidler, 1975). At later times of a day and longer, changes appear (Chapter 5) as a result of differential drop-off, turnover and the redistribution of various components in the fibers, as well as differences in the labeled components contributed from the cell bodies to the fibers at later times (cf. Section 4 below).

3. Redistribution

The shift to a shallow slope of the exponential outflow characteristics of slow transport (Chapter 2) is due in large measure to the redistribution of the

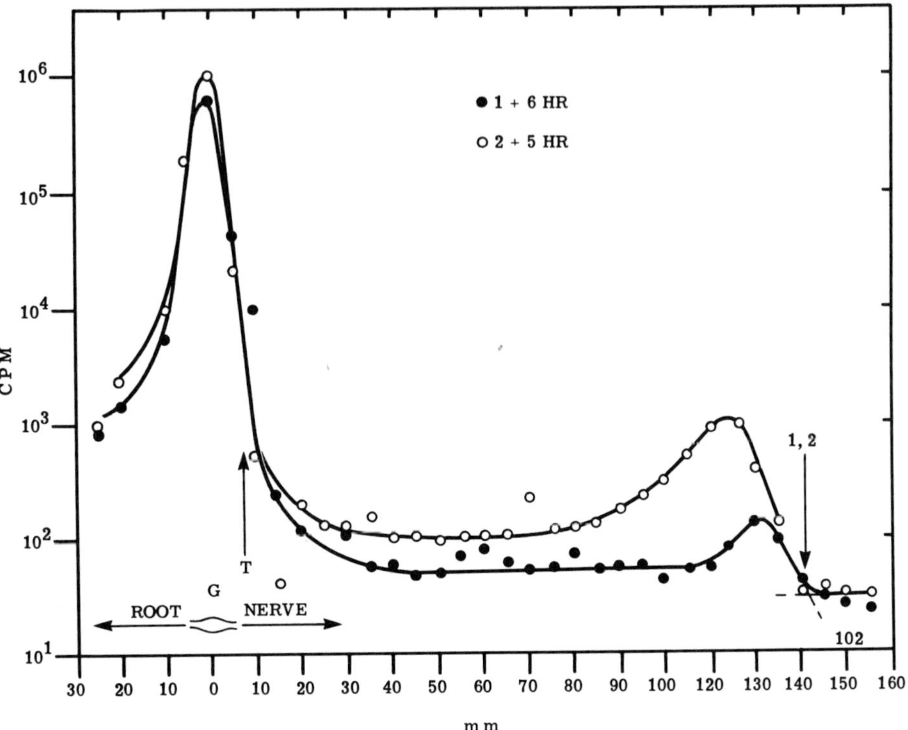

Fig. 11.4 Change in plateau height with different times of outflow. Ligations were made 1 and 2 hr after injection with a further period of downflow so that total downflow times were similar. One nerve (●) was ligated just below the ganglion 1 hr after ^{3}H-leucine injection into the L7 ganglion and a further 6 hr of downflow was allowed. The other nerve (○) was ligated 2 hr after injection and a further 5 hr of downflow was allowed. Arrows 1 and 2 show a similar advance of the fronts of the crests. An increase in the plateau level is seen when comparing 2 hr of outflow with that of 1 hr. From Ochs (1975c).

components dropped off in the fibers. Redistribution was shown by making ligations in the nerves at different places and times after injecting ^{3}H-leucine into the ganglia. The radioactivity found in the nerve below ligations made just below the ganglion 2 hr after injecting ^{3}H-leucine shows incorporated radioactivity remaining in the plateau after a downflow of 20 hr (Fig. 11.6).

The relatively small amount of activity left behind in the plateau after the crest of fast-transported activity had swept down the fibers is to be contrasted with the much higher level of radioactivity present in the plateaus in nerves which had not been ligated and where there is a further addition of locally deposited labeled materials (cf. Fig. 11.7).

A small declining slope of the radioactivity is seen in the ligated nerve which represents labeled materials dropped off as the crest moves down the

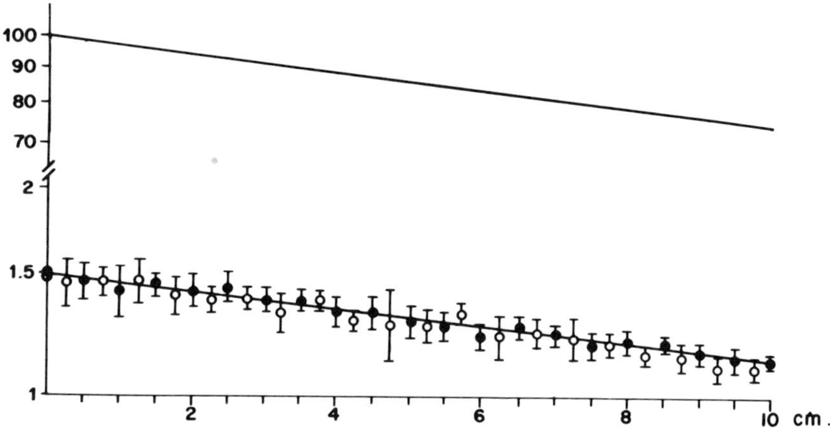

Fig. 11.5 Analysis of plateau height as a function of outflow. The upper curve represents the theoretical declining amount of labeled proteins along 10 cm of nerve following a peak of outflow (cf. text). The lower curve shows the distribution of radioactivity in nerves found 18 hr after making ligations and after either ganglion injection (●) or spinal cord injection (○). The line represents a theoretical retention of 1.5% of the protein in the fiber dropped off from the peak as it moves down the nerve, the slope of the line matching the distributions of radioactivity found. From Muñoz-Martínez *et al.* (1981).

length of the nerve fibers (cf. Fig. 11.6). The small proximo-distal slope is converted to a line parallel to the abscissa when the nerve was additionally ligated distally in the nerve.

In this example, a ligation was made in one nerve below the ganglion 2 hr after the injection of ^{3}H-leucine and a declining slope due to drop-off was seen 22 hr later. In the other nerve, a second ligation was made distally in the nerve 7 hr later in addition. The labeled components trapped within the double-ligated segment of nerve became level with the abscissa as a result of redistribution. As part of this process radioactive labeled proteins which had accumulated above the distal ligation were carried in the retrograde direction to produce the leveling of the plateau. Such a retrograde movement of labeled components shown by the accumulation of labeled materials below an upper ligation was reported by Lasek (1968), Bray *et al.* (1971), Edström and Hanson (1973a), and Bisby and Bulger (1977).

When ligations are made to study redistribution, Wallerian degeneration in the amputated part of the nerve limits the time of observation. Within this experimental limitation, a shift in the slope of slow transport indicative of redistribution may be observed. A day after ^{3}H-leucine is injected, a ligation was made to block further entry of labeled components and another day of downflow allowed. The change in slope shows the "clearance" of labeled components below the ligation due to redistribution (Fig. 11.7).

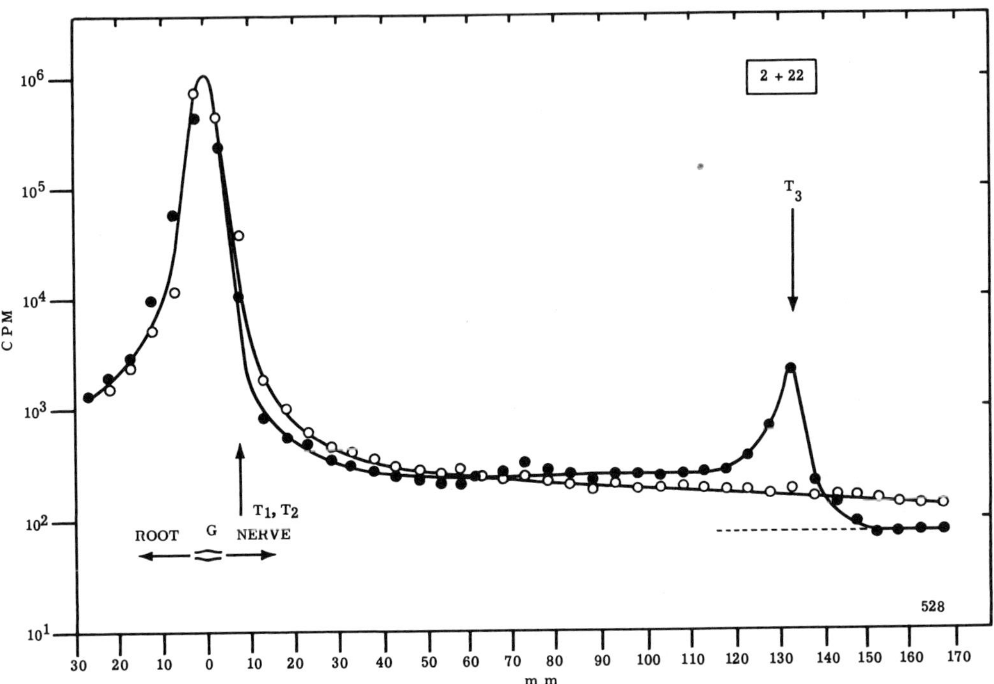

Fig. 11.6 Redistribution within a double-ligated nerve. Outflow of radioactivity over a 22-hr period after ligation 2 hr following injection of the ganglion with ^{3}H-leucine shows a decreasing slope distally in the nerve due to drop-off (O) after it was ligated below the ganglion 2 hr after injection at T_1. The other nerve was similarly ligated at T_2 and distally as well at T_3 7 hr after injection, when the front had passed that point. A further outflow of 13 hr was allowed (●). The dashed line extending from the baseline distal to T3 shows the baseline activity. The level of radioactivity is seen to become horizontal between the two ligations as a result of redistribution. From Ochs (1975c).

The capacity for such redistribution, however, diminishes rapidly and it is small even a day later (Fig. 11.8).

Frizell, McLean, and Sjöstrand (1975) carried out similar experiments which essentially showed the same kind of results. Two days after labeling the motor nuclei of the vagi with ^{3}H-leucine, further transport was prevented by making a ligation or applying colchicine and another day allowed for the downflow of labeled components which had entered the fibers. These were seen to move down the nerve to a distal ligation made in the nerve. The capacity for such clearance was soon lost. Redistribution decreases with time between the labeling of the cells and the interruption of downflow (Fig. 11.9).

The radioactivity accumulated at the distal ligation of the doubly-ligated nerves decreases with time, indicating that the labeled components which drop off locally in the fibers gradually become bound in various organelles of the axon losing the capacity for further redistribution.

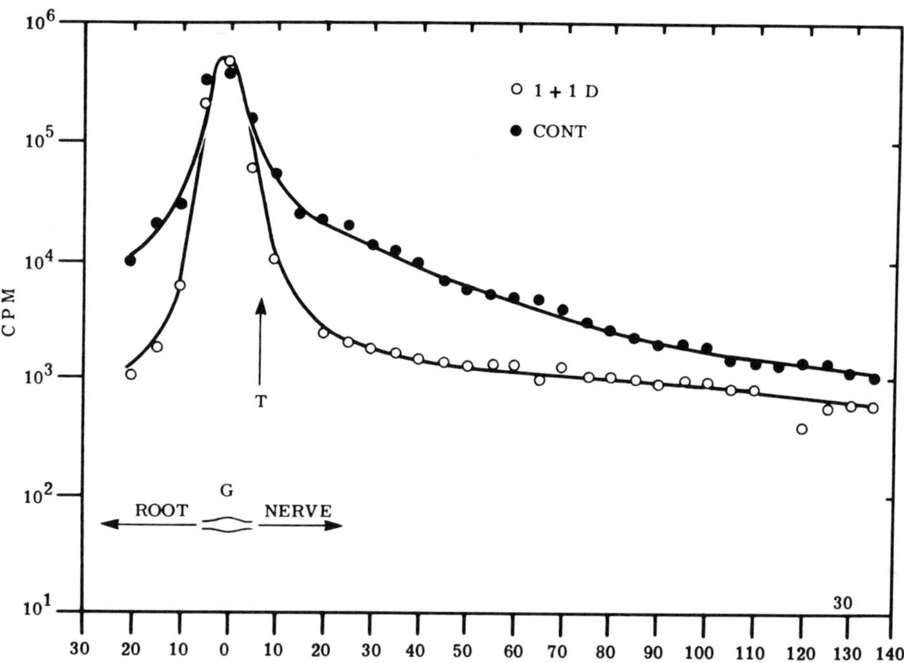

Fig. 11.7 Redistribution of later outflow. A control nerve (●) shows the outflow 2 days after injection of the precursor. The other nerve (○) was ligated just below the ganglion 1 day after injection and a further outflow of 1 day allowed. A depletion of radioactivity was seen in comparison to the control. From Ochs (1975c).

4. Compartmentation and Later Outflow

Several factors account for the pattern associated with slow transport. One already noted was the compartmentalization of proteins synthesized in the cell bodies (Chapter 5). Proteins incorporating labeled leucine may remain in the nerve cell bodies for days before eventually entering the axon for axoplasmic transport. The proteins released later differ in composition from those labeled materials transported at early times, with a relatively larger amount of high-MW proteins than low-MW proteins exiting at the later times.

5. "Waves" of Slow Outflow

While the outflow pattern of labeled material at the later times identified as slow transport is characteristically seen as a declining exponential curve (cf. Figs. 2.3, 2.4, and 2.6), later wave-like variations appear in addition. These variations were taken as indicative of a separate slow transport by Hoffman and Lasek (1975). One such wave termed SCa was considered to move at a

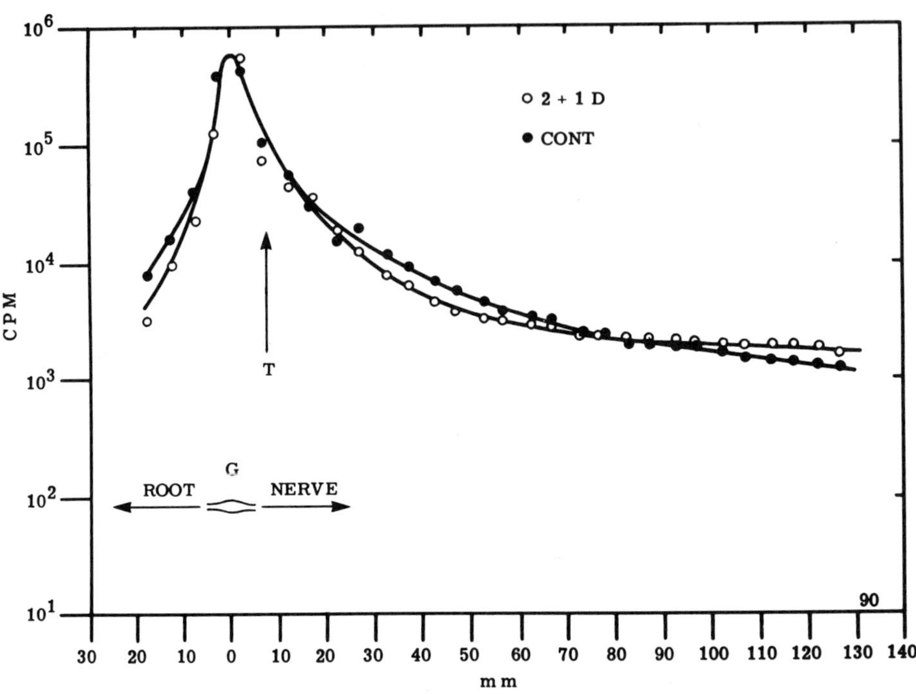

Fig. 11.8 Decrease of redistribution at later times. This experiment is similar to that shown in Fig. 11.7 with 1 more day of outflow added before making a ligation. In the control nerve (●), an outflow 3 days after injection of the precursor is shown. In the other nerve (○), a ligation was made just below the ganglion 2 days after injection and a further outflow of 1 day was allowed. The similarity of the two curves (cf. with Fig. 11.7) indicates a greater retention of components dropped off in the fibers from the transport mechanism with a reduction in the capacity of their redistribution. From Ochs (1975c).

rate of 1.0–1.3 mm/day (Fig. 11.10). Another slow wave termed SCb was considered to move at 3–4 mm/day (Lasek and Hoffman, 1976; Black and Lasek, 1979; and cf. Chapter 9). A number of other such waves have been reported, some moving at faster or intermediate rates (Chapters 2 and 9).

Slow waves show wide variations in the regularity of their pattern or may not even appear to have a regular wave movement at all. This was seen to be the case in the sensory fibers of the peroneal nerve in the cat in a study of slow outflow carried out over long periods of time (Stromska and Ochs, 1981). While waves were suggested in nerves taken 50 and 67 days after ^{3}H-leucine injection, these did not move down with a clearly defined rate. Instead, broad and irregular patterns of radioactivity were found at the later times (Fig. 11.11).

These irregularities were found to be similar in the nerves taken from the two sides in any given animal, indicating that they reflect features of trans-

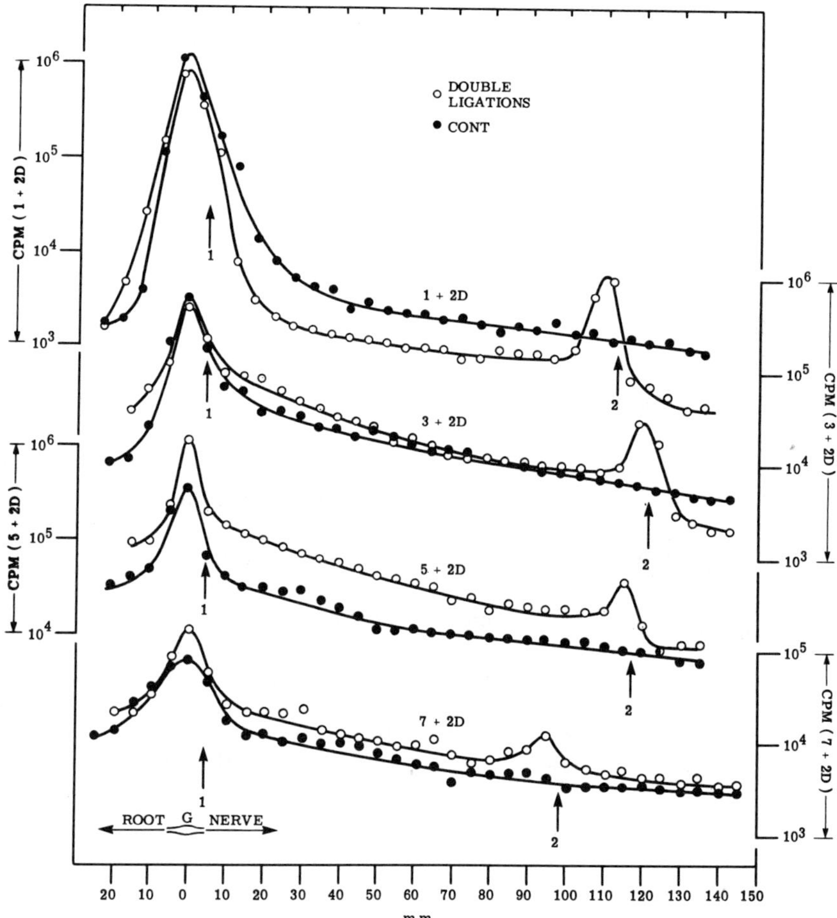

Fig. 11.9 Decreased redistribution at later times shown by accumulations in double-ligated nerves. Four pairs of nerves are shown. In the uppermost pair, one nerve (O) was ligated just below the ganglion (arrow 1) 1 day after injection and also distally in the nerve (arrow 2). A further time of 2 days outflow was allowed and accumulation at the distal ligation is seen. The other nerve (●) was not ligated distally. In the lower pairs of nerves the same procedure and the same symbols apply, with delays of 3, 5, and 7 days after injection before making ligations below the ganglion. In each case a further downflow of 2 more days was allowed. The accumulation at the distal ligations of the double-ligated nerves was seen to decrease with additional time, indicating a lesser capacity for redistribution. Partial ordinate scales for each of the 5 pairs of nerve are identified by the days given after ligation plus 2 more days allowed for outflow. From Ochs (1975c).

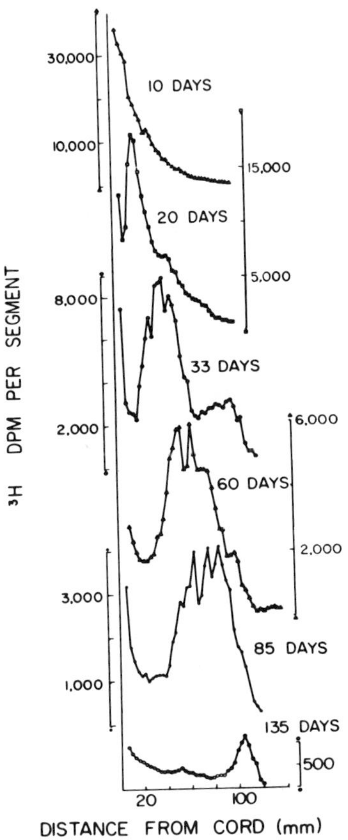

Fig. 11.10 Waves of slow outflow. An injection of ^{3}H-lysine into the ventral horn of the rat spinal cord shows a peak which appears to have a somatofugal movement in the rat sciatic nerve at a rate of 1.0–1.2 mm per day. The radioactive content of consecutive 3-mm segments of the L5 and L6 roots and their extensions into the sciatic, tibial, and peroneal nerves were plotted against the distance of each segment from the spinal cord at the time (in days) between the injection of ^{3}H-labeled amino acids and the removal of the nerves for sampling. Each data point represents the mean of five values. From Hoffman and Lasek (1975).

port rather than sampling errors. Similar results were found for the sensory fiber of the rat sciatic nerves (Stromska and Ochs, 1981). On the other hand, remarkably regular slow waves were recorded in the garfish olfactory nerve (Cancalon, 1979a,b), as shown in Fig. 11.12.

The regular slow waves found by Cancalon most likely reflect the homogeneity of the unmyelinated fiber population of the garfish olfactory nerve as compared to vertebrate nerves with its mixed population of fiber sizes. As can be seen in Fig. 11.12, the slow waves in the garfish nerves fall off in amplitude and broaden with time, both in the anterograde and retrograde directions. These changes can be interpreted on the basis of the unitary hypothesis, namely, a redistribution of components in the fibers.

The apparent rate of slow wave movement in the garfish olfactory nerves also shows a strong temperature dependence with a Q_{10} of 3.3–3.7 (Cancalon, 1979a,b). At a higher temperature the peak movement and fall-off of amplitude develops more quickly. A high Q_{10} may be explained on the unitary hypothesis by the kinetics of drop-off, the time course of turnover of those components locally in the fiber, the rate with which those components can

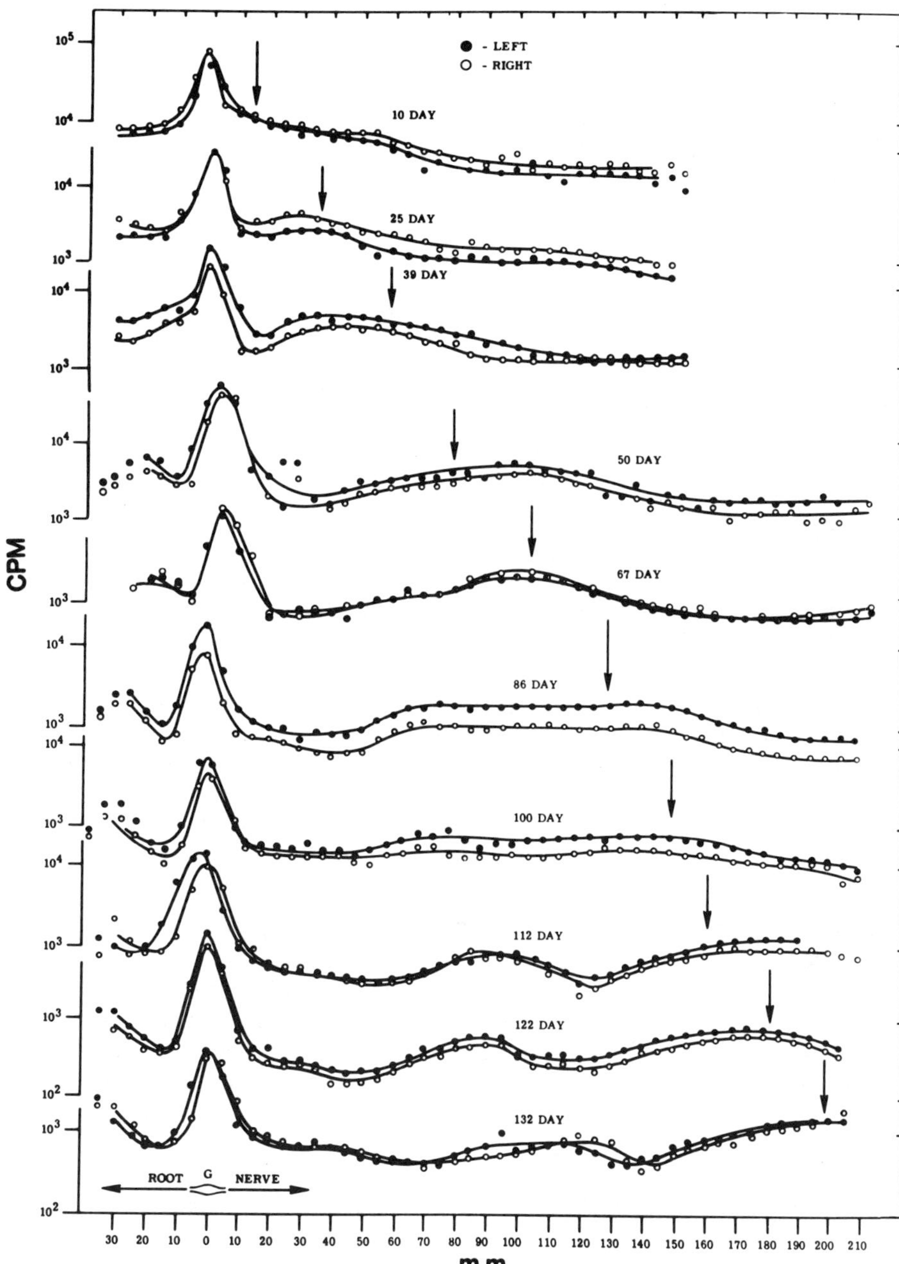

Fig. 11.11 Distribution patterns of slow outflow in cat sensory fibers. At the times indicated (in days) after L7 dorsal ganglia were injected, nerves from each side were taken and their outflow patterns were assessed. The arrows indicate where the peak would be expected to appear at a hypothetical slow rate of 1.5 mm/day. Note the similarity of radioactivity in the nerves each taken from the individual animals. An absence of a clearly defined slow transport wave is evident (cf. Fig. 11.10). From Stromska and Ochs (1981).

263

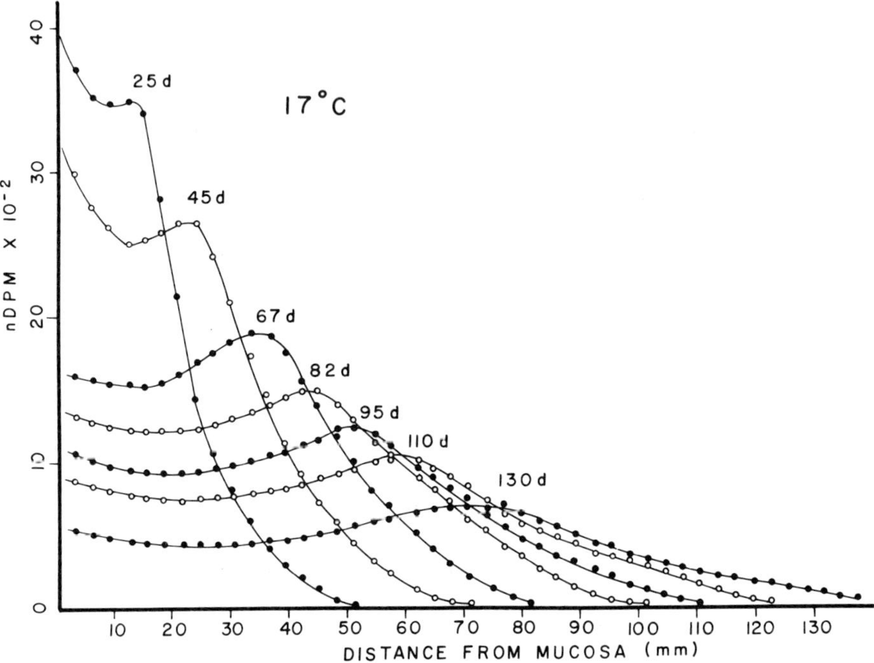

Fig. 11.12 Slow transport waves in garfish olfactory nerves. The normalized curves of TCA-insoluble radioactivity representing protein were measured along the nerve after downflows at 17°C between 25 and 130 days following ^{3}H-leucine application to the olfactory nerve cell bodies. The radioactivity (nDPM = normalized disintegration per min) shows a decrease in amplitude and a spread of slow waves consistent with a redistribution of proteins. From Cancalon (1979a).

reattach for redistribution, and the redistribution transport in the anterograde and retrograde directions.

A slow wave is also observed after a local type of "pulse labeling," the injection of nerve subperineurally with ^{3}H-N-succinimidyl proprionate (^{3}H-N-SP). As shown by Fink and Gainer (1980a,b), the ^{3}H-N-SP is taken up by the nerve fibers to alkylate polypeptides within the nerve fibers which then shows a declining outflow characteristic of slow transport in both the anterograde and retrograde directions (Fig. 3.15).

B.   ROUTING IN NERVE FIBER BRANCHES

1.   The Phenomenon of Routing

The neurons of the mammalian dorsal root ganglion are T-shaped with the labeled proteins incorporated in the cells transported into their dorsal root and sensory nerve fiber branches. Taking advantage of the relatively long length of dorsal root present in the rhesus monkey, the rate of fast transport in the dorsal root was compared to that in the sciatic nerve by the positions of their fronts of radioactivity after injecting the L7 dorsal root ganglia with ³H-leucine (Ochs, 1972a). The outflow in each branch had the usual pattern of fast outflow (Fig. 11.13).

The distances to which the fronts of radioactivity had moved into the dorsal roots and sciatic nerves showed fast transport in each to be close to 410 mm/day. However, as can be seen in this example, approximately 3 to 5x more radioactive material was carried in the crest of the sensory fibers as compared to the dorsal root fibers. Similar differences in the amounts of labeled activity fast transported into the sensory nerve and dorsal root branches of the dorsal ganglion neurons have been reported for the cat (Anderson and McClure, 1973) and the rat (Komiya and Kurokawa, 1978), and for the superior and distal portions of the vagus nerve after injecting the nodose ganglion with labeled precursor (Watson *et al.*, 1975).

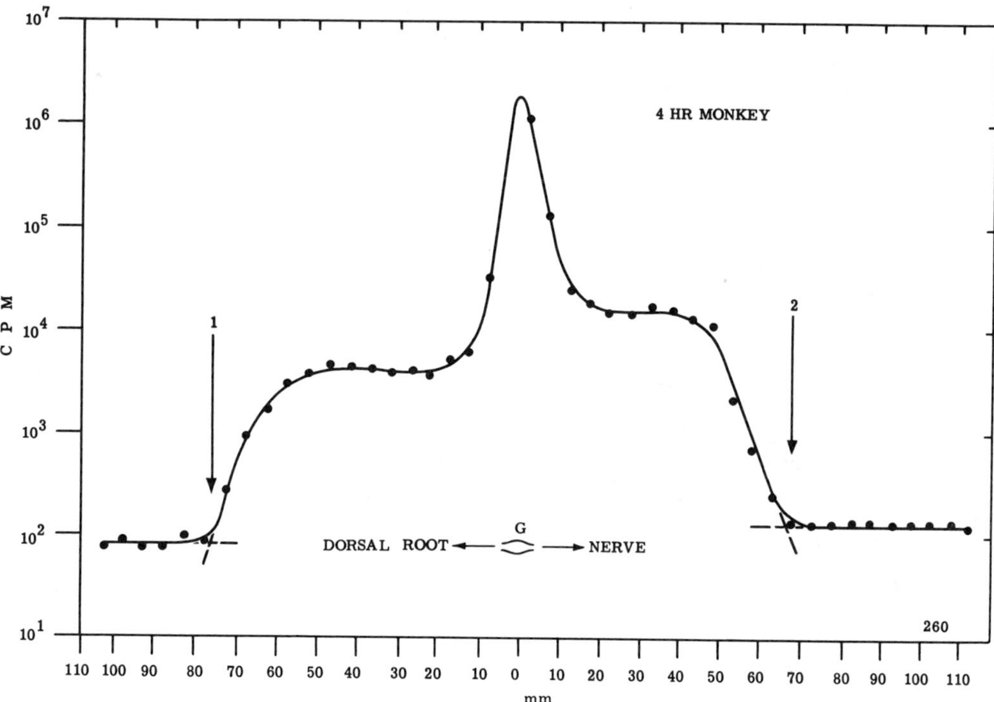

Fig. 11.13 Asymmetry of transport in dorsal root and sciatic nerve sensory branches of the dorsal root ganglion. The L7 ganglion of a monkey injected with ³H-leucine shows 4 hr later that the rate of transport in the long dorsal root is comparable to that in the sciatic nerve. Arrow 1 at the foot of radioactivity in the dorsal root is displaced to the same extent as arrow 2 at the foot in the sciatic nerve. The amplitude of the crests, however, are 3–5× greater in the sensory nerve than in the dorsal root. (Note logarithmic scale for cpm on ordinate.) From Ochs (1972a).

How can such large differences in the amounts of labeled materials transported into the two branches of the same neuron be accounted for? It has been established that the numbers of myelinated fiber branches from the dorsal root ganglion supplying the sensory nerve are equal to those of the dorsal root (Dale, 1900; Ramon y Cajal, 1909; Barnes and Davenport, 1937; Duncan and Keyser, 1936; Lieberman, 1976). While exceptions have been noted for unmyelinated fibers (Coggeshall, 1980), nearly all the large dorsal root ganglion neurons contributing myelinated fibers divide in T-shaped fashion with only one daughter branch supplying the sensory nerve, the other branch the dorsal root.

A possible explanation for the discrepancy in the amounts of labeled components transported might be that the myelinated fibers supplying the sensory nerves have a greater diameter than those of the dorsal roots (cf. Crosby, Humphrey, and Lauer, 1962). Licberman (1976), on the other hand, concluded from his review that they were the same. This point was examined by constructing histograms of the myelinated nerve fibers on both sides of the L7 dorsal root ganglion from histological sections prepared by freeze-substitution (Ochs et al., 1978), of the dorsal root and sensory nerves at the sites indicated in Fig. 11.14.

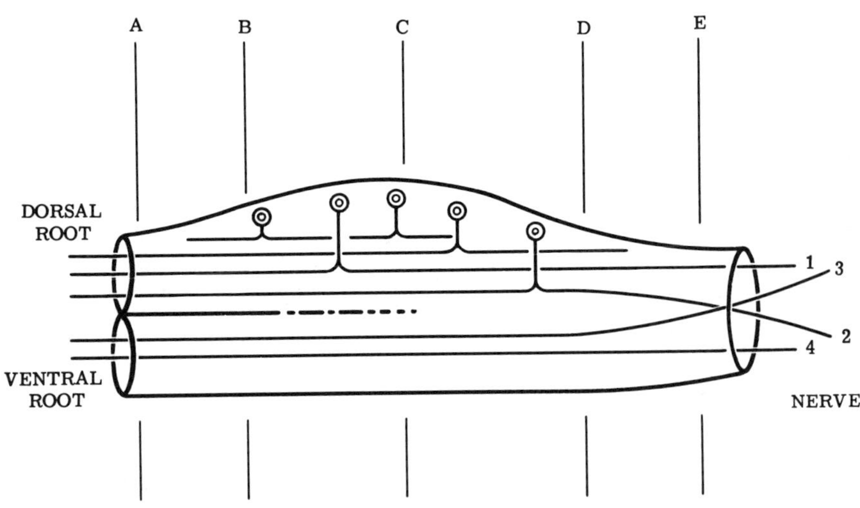

Fig. 11.14 Dorsal root ganglion showing the T-shaped dorsal root neurons and sampling sites. Branches of the ganglion neurons are seen ascending in the dorsal root and descending in the sciatic nerve, either directly (1) or curving down (2) to mix with motor fibers contributed from the ventral root (4). A portion of the ventral root fibers curves up to mix with sensory fibers (3). A to E show sites at which sections were taken to construct histograms of myelinated fiber branches on the two sides (cf. Fig. 11.15). From Ochs et al. (1978).

To eliminate the admixture of motor fibers on the distal side which could distort the sensory fiber population, the ventral roots were cut some 30–40 days beforehand to cause their degeneration (Eccles and Sherrington, 1930). Histograms of the fiber populations in the sensory nerves remaining and in the dorsal roots, showed their fiber populations to be closely similar (Fig. 11.15).

Another possible explanation for the asymmetry in the amounts of down-flow on the two sides might be that there is a greater density of microtubules

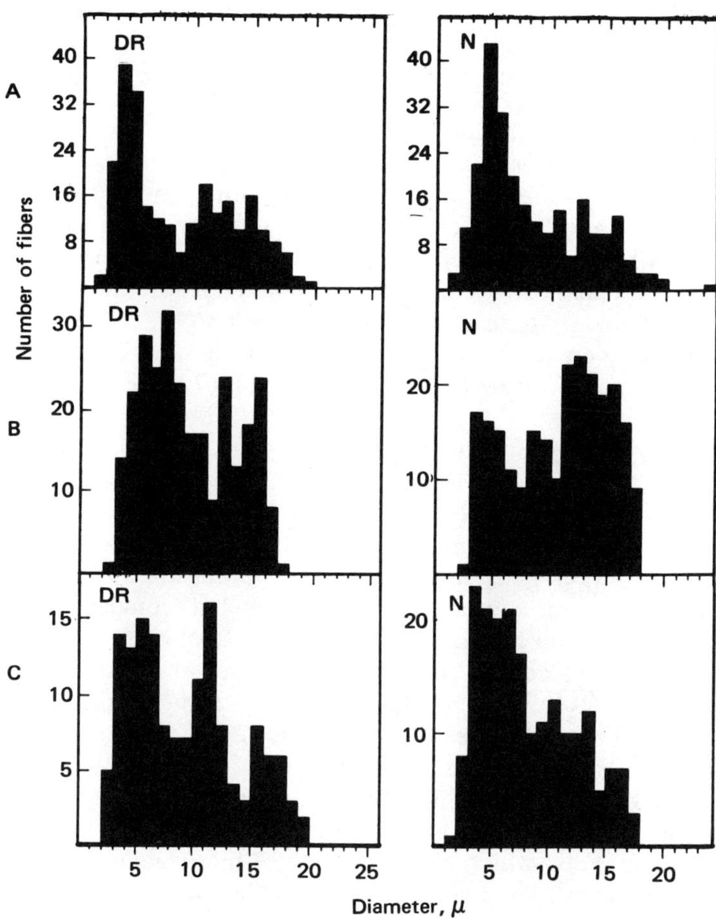

Fig. 11.15 Histograms of fiber branches of the ganglion neurons. The fibers were examined in sections taken from either side of the ganglia at the positions indicated in Fig. 11.14 in row A, dorsal root (DR) and sensory nerve (N) histograms of cat; B, similar histograms of a monkey 45 days after cut of its ventral root; row C, dorsal root (DR) and nerve (N) of a cat with a ventral root section made at an earlier time. In these histograms, the ordinates show the numbers of fibers at the nerve fiber diameters given in μm on the abscissa. From Ochs *et al.* (1978).

in the fibers of the sensory nerve branches than in the dorsal root fibers. In locally perfused glutaraldehyde-fixed preparations, comparably sized fibers from the dorsal root and sensory nerve taken at the sites indicated in Fig. 11.14 were counted to determine their microtubular densities and the two branches found to be similar in this respect (Ochs *et al.*, 1978) as shown in Figs. 11.16A and 11.16B.

Smith (1973) reported a lower microtubular density in the dorsal root as compared to the sensory nerve fiber branches of *Xenopus laevis* with a suggestion that dorsal root fibers would have a much lower density than the sensory fibers. Zenker, Mayr, and Gruber (1973) found a 20% higher density of microtubules in the ventral root fibers as compared to the dorsal root fibers in the rat, again suggesting some difference with respect to sensory fibers. Variations in the densities of microtubules in the nerves of different species and variations in fibers which depend on the technique of fixation are known. The lability of microtubules with respect to fixation is also a recognized property of these organelles. In any case, the similarity of the microtubular densities shown in the branches of the dorsal root ganglia neurons of the cat and monkey when fixed by local perfusion in the living animal indicates that differences in the numbers of microtubules could not account for the different amounts of labeled proteins transported into the two branches. Additionally, it was seen that the neurofilaments have similar densities in the fibers of the sensory nerve and dorsal root branches (Figs. 11.17A and 11.17B).

2.　A Functional Model for Routing

The lack of an obvious morphological basis for the asymmetry of outflow in the two fiber branches of dorsal root ganglion neurons led to a functional explanation based on the transport filament hypothesis. More transport filaments moved down the sensory nerve fiber branch as compared to the dorsal root fiber branch could give rise to the asymmetry of outflow seen. Alternatively, more materials could be loaded on one set of transport filaments as compared to the other. This hypothesis requires some sorting or "routing" mechanism in the cell bodies whereby one group of transport filaments and the various kinds of materials bound to them are moved down one set of microtubules in the sensory fiber branch, while other materials are transported in the dorsal root fiber branch (Fig. 11.18).

One mechanism by which such a routing of materials may be carried out is indicated in the left-hand panel (*A*). Two different sites of synthesis L_1, L_2 are shown in the cell body, each supplying a separate set of components and the transport filaments carrying them down separate sets of microtubules leading into each of the daughter nerve fiber branches of the neuron. Alternatively, as shown in the right-hand panel (*B*), gates G_1, G_2, select the components and the transport filaments carrying them along either of the two sets of microtubules passing into the two daughter fiber branches.

The term "routing" is used to refer not only to the asymmetry in the amount of material moving down the branches of a neuron (both axons and

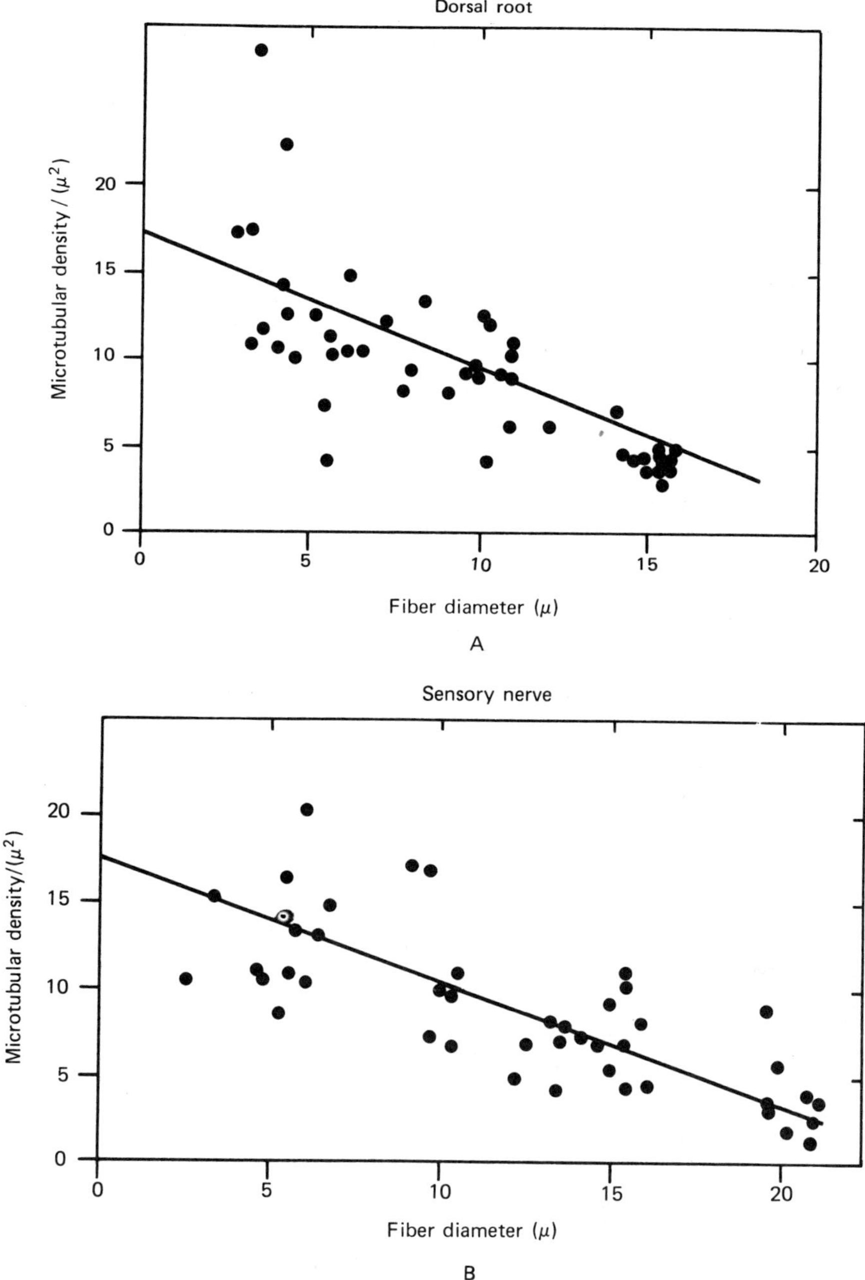

Fig. 11.16 Microtubular densities and fiber diameters. (A) Microtubular densities in fibers of the dorsal root are plotted on the ordinate as a function of fiber diameters on the abscissa. The regression line shows a decrease in microtubular density (number/μm^2) with an increase in fiber diameter. The fibers were examined in sections taken at A, Fig. 11.14. (B) Microtubular densities in sensory nerve fibers are shown as a function of fiber diameter from the same preparation (taken at D, Fig. 11.14). From Ochs *et al.* (1978).

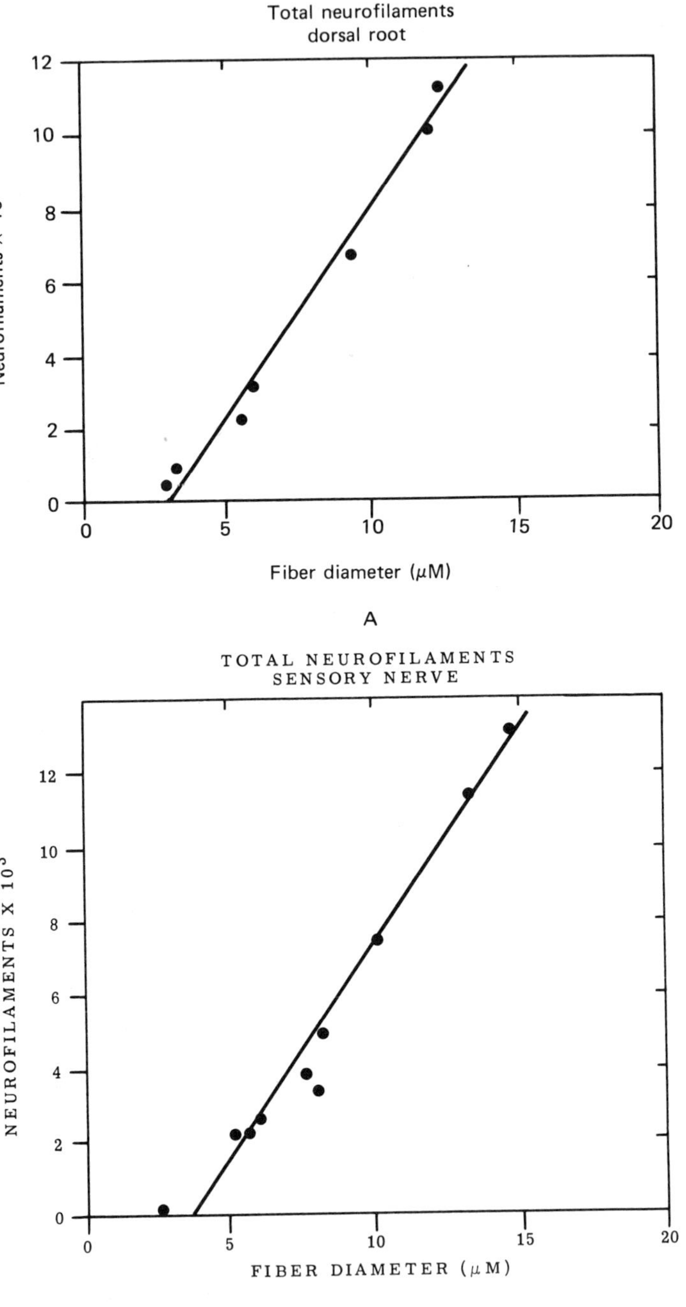

Fig. 11.17 Total numbers of neurofilaments. (A) The total number of neurofilaments in myelinated fibers of the dorsal root (taken at A, Fig. 11.14) are plotted on the ordinate as a function of fiber diameter. The regression line shows an increase in neurofilaments with diameter. (B) The number of neurofilaments in sensory nerve fibers taken from the same animal (taken at D, Fig. 11.14) shows the same increase in the slope of their number with diameter. From Ochs *et al.* (1978).

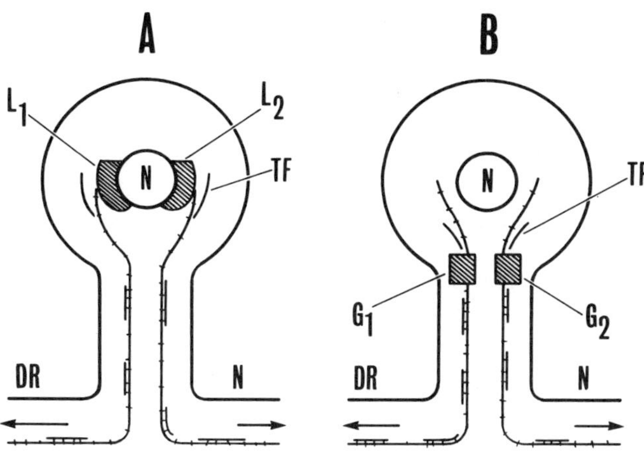

Fig. 11.18 Routing mechanisms. To account for the finding that more material is transported in one fiber branch than another of the same neuron, the transport filament model is invoked with two possible mechanisms: (A) Two synthetic sites L1 and L2 near the nucleus (N) from which microtubules take their origin to pass down the stem portion of the neuron and then out into the dorsal root (DR) and nerve (N) branches of the neuron. The transport filaments (TF) to which the various materials are bound for transport are led onto each of the sets of microtubules and thus sent on their individual ways. More such transport filaments are considered to be moved down the sensory nerve microtubules per unit time than the dorsal root fiber branches (or more components bound to each filament). (B) An alternative possibility is that a selection is exercised by two different gates (G_1 and G_2) over the different kinds and amounts of material permitted to exit from the cell and to be carried over one or the other set of microtubules leading into the two fiber branches. From Ochs *et al.* (1978).

dendrites as well), but also to differences in the kinds of materials passing into the branches. Thus, there is a routing in the dorsal root ganglion neuron of neurotransmitter and neurotransmitter-related components which are supplied to the nerve fiber terminals in the spinal cord entering via the dorsal root, while components subserving sensory transduction are transported into the sensory nerve fibers branch to supply its terminals. Routing could account for a possible supply of different neurotransmitters to different nerve fiber terminals of the same neuron (Burnstock, 1978; Dismukes, 1979; Potter *et al.*, 1981). And, routing could account for the selective growth of some of the multiple branches of regenerating nerve fibers, and not of others (Chapter 15).

Except for minor differences, SDS–PAGE analysis (Chapter 5) has shown that the bulk of the monomeric polypeptides supplied to the two branches by the dorsal root ganglion neurons are similar. More sensitive techniques are required to reveal the differences in enzymes, neurotransmitters, polypeptides, small peptides, and amino acids we expect are differentially routed.

3.   Origin and Independence of the Microtubules Subserving Routing

By following the early stages of neurogenesis of the avian and mammalian dorsal root ganglion (Ramon y Cajal, 1911), we can see how different sets of microtubules and neurofilaments come to supply their daughter branches. In the early maturation of the avian dorsal root ganglion, Ramon y Cajal (1911) showed the originally bipolar neurons becoming transformed into their adult T-shaped form (Fig. 11.19).

The neurofibrils arising from the two ends of the bipolar cell become drawn together as the two daughter fibers merge to form the stem portion of the T-shaped neuron in the course of maturation. In the adult the stem portion is seen as a long, often convoluted fiber with a branch point at the T where the daughter fibers pass into the dorsal root and sensory nerves (Zenker and Högl, 1976). In EM preparations of the T junction, the microtubules and neurofilaments are seen as two separate groups passing from the stem fiber

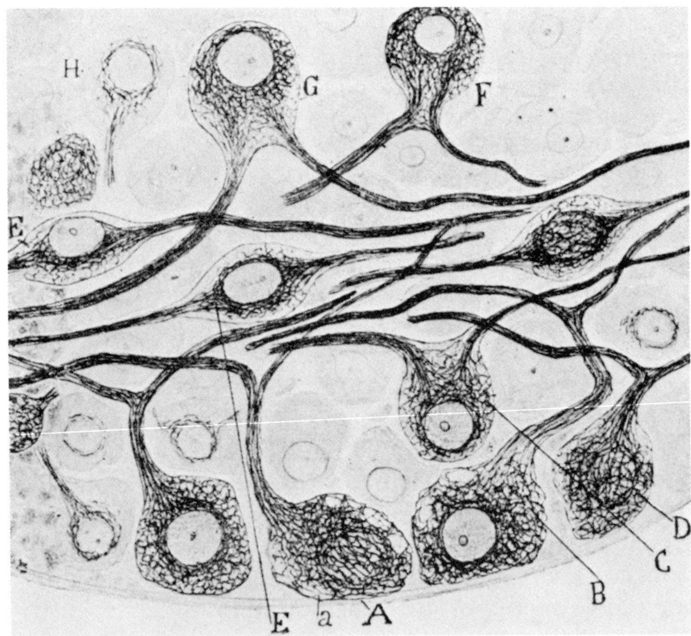

Fig. 11.19 Ontological development of dorsal root ganglion neurons. The dorsal root ganglion cells are originally bipolar (E). The neurofibrillary fibers (microtubules + neurofilaments) are seen emerging early from the two ends. Then, in the course of the first weeks of development as shown in the avian preparation, the cell body remains peripherally situated in the growing ganglion (a) and the fiber branches merge to form the stem of the T-shaped neuron (A). This process is indicated by the various cells shown at different stages (A–H). A similar process occurs in the mammal. From Ramon y Cajal (1911).

into the two daughter branches without any cross-over of their elements (Ha, 1970).

The separation of the two sets of microtubules reflect their original bipolar origin. This is suggested by the finding that HRP taken up by the sensory nerve and retrogradely transported to the cell bodies accumulates in the ganglion cells without crossing over to ascend into the CNS via the dorsal root fibers (Neuhuber, Niederle, and Zenker, 1977). However, such apparent complete independence is relative and depends on the concentration of HRP retrogradely transported to the cell bodies as well as the sensitivity of the method used to detect HRP. When a more sensitive method for detection of HRP was used (Mesulam and Rosene, 1979), some of the marker protein was seen to cross over to be carried via the dorsal root fibers to their terminal ends in the spinal cord and medulla (Mesulam and Brushart, 1979). A cross-over of HRP in the dorsal root ganglion cells, both in the disto-proximal and proximo-distal directions, was confirmed by Zenker, Mysicka, and Neuhuber (1980). Additionally, they showed that it was necessary for the HRP to enter the soma for such cross-over transfers to occur.

The question as to whether the microtubules are continuous throughout the whole length of the axon was addressed by Weiss and Mayr (1971a,b). They counted microtubules in nerve fibers above and below a point of branching and found the sum of the microtubules in the two daughter branches to equal that of the parent fiber above the branch point. This would suggest that they are continuous, all things being equal. A similar result was found by Nadelhaft (1974) in crayfish abdominal nerve cord fibers. However, Nadelhaft also found some exceptions to an equality in the counts of parent and daughter fiber microtubules. Another indication that this rule does not hold true everywhere in the neuron was found in fibers close to their terminations. The overall calculated cross-sectional axonal area might be as much as 10x greater than that measured at a more central point but with similar microtubular densities in both parts of the fibers (Zenker and Hohberg, 1973). This indicates a greater number of microtubules in the distal fibers. Friede and Bischausen (1980) also found an increase in the total number of microtubules in regenerating nerve fibers just above a transection as compared to a more central site (Fig. 10.11).

The microtubules do not appear to be continuous all along the length of the axon. This was indicated in studies of nerve fibers in the nematode *Caenorhabditis elegans* (Chalfie and Thomson, 1979), where the microtubules were found to consist of staggered, short lengths varying from 1.2–10.7 μm, in cultured rat sensory axons by Bray and Bunge (1981) where they averaged 108 μm on the tentative assumption of their similarity in length and 370–760 μm in mouse saphenous nerve fibers on the same assumption (Tsukita and Ishikawa, 1981) as described in Chapter 9. If, as suggested by such studies, this is how microtubules are organized in the fibers, we can account for local increases in microtubule numbers as well as functional changes in the hypothalamo-neurohypophysial neurons with differences in microtubular densities of their nerve fibers (Chapter 12). Such variations fit with the evidence discussed in Chapter 9 that a constant turnover of tubulin occurs in the microtubules at the ends of their short segments. With respect to the transport filament model, the microtubules would have some specificity so that two sets of microtubules present in one fiber selectively route transport filaments into each daughter branch. The amount of component routed along one set of microtubules may be changed by forcing more material into a fiber branch. The amount of ^{3}H-5HT carried in the lip nerve of *Aplysia* was increased when the cerebro-buccal connective was cut close to the cell body (Goldberg, Goldman, and Schwartz, 1976), as shown in Fig. 11.20.

The experiment suggests that the serotonin vesicles backed up along the microtubules of the cut nerve branch can, as a result of their accumulation, be loaded onto the transport filaments of the lip nerve in greater number and subsequently transported in that branch.

In those experiments a change in rate of transport was suggested. However, as a general rule the rate of fast transport does not change with the amount of component transported (Chapter 4), and the control rates given in those experiments by Goldberg *et al.* (1976) were lower than those subsequently reported (Goldberg *et al.*, (1978). In another experimental approach where a temperature differential was used to create an overload, moderate increases of 2–3x in the DCVs transported did not appear to cause an increase in rate (Brimijoin, 1979). However, more recently, Litchy and Brimijoin (personal communication) did find an increase in rate. It is likely that such rate increases are short-lived and that the normal rate is soon resumed after the overload has been relieved. Overloading of transport filaments on the microtubules, or "queuing," would require the specificity of the transport filaments to the microtubules supplying a given nerve branch to be maintained. In the case of the *Aplysia* neuron, the two nerve branches normally carry the serotonin-containing vesicles and a similarity of transport in the two branches of the neuron would be expected. The cross-over of transport filaments from one set of microtubules to another, as in the case of dorsal root neurons, seems to be minimal. Cross-over does occur after accumulation at nerve fiber interruptions or at the nerve terminals where turn-around occurs with a subsequent retrograde transport.

4. Asymmetry and Routing of Slow Transport

An asymmetry of the amounts of labeled materials slow-transported into the dorsal root and sensory nerve fiber branches of cat dorsal root ganglia neurons would be expected on the basis of the unitary hypothesis. The amounts locally dropped off in slow transport should reflect the difference in the amounts fast-transported. The slopes of outflow in the sensory nerves and

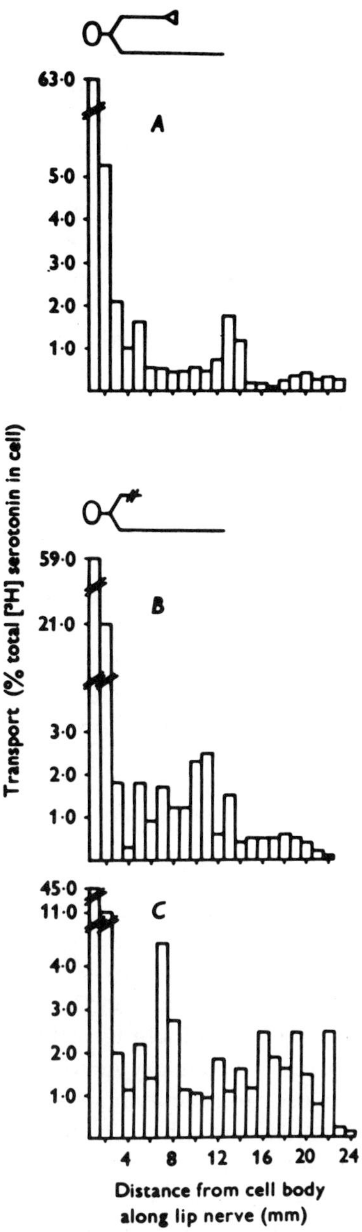

Fig. 11.20 Redistribution of transport in the branches of an *Aplysia* neuron. (A) Distribution of ³H-serotonin-labeled radioactivity along the lip nerve 6 hr after intrasomatic injection. (B) The cerebro-buccal connective was cut 0.5 hr before ³H-serotonin injection with an increased amount seen transported in the lip nerve. (C) When the connective was cut 3 hr before injecting the precursor, an increased accumulation was seen, indicating more of a shift of labeled components from the cut cerebro-buccal nerve to the lip nerve. From Goldberg *et al.* (1976).

dorsal roots seen at various times after injection of the dorsal root ganglia do in fact show differences in their slopes in accord with that expectation (cf. Fig. 2.6). In experiments designed to assess the amounts of labeled radioactivity in the dorsal root and sensory nerve, the integration of radioactivity in equal lengths of nerve showed that approximately 2x more labeled material was carried into the sensory nerve as compared to the dorsal root (Ochs and Stromska, unpublished observations). The differences found by Komiya and Kurokawa (1978) in the amounts of labeled material transported into the daughter branches of rat dorsal root ganglion neurons were considered due to a lower rate of slow transport in the dorsal root as compared to peripheral nerve fibers. This interpretation, however, was based on the presence of apparent waves of slow transport, waves which are not consistently seen in all preparations (cf. Section A5 above). The difficulties of assigning different rates of slow transport in the two nerve branches are exemplified by the findings of Mori *et al.* (1979). They reported that the triplet proteins, tubulin, and actin move in the peripheral sensory nerve branch at rates of 2–3 mm/day, 9–13 mm/day, and 19 mm/day, respectively, while in the dorsal root branch the corresponding rates were 1–2 mm/day, 3–4 mm/day, and 4 mm/day. Rather than invoking 6 different rates and presumably 6 different mechanisms of slow transport, we can account for such apparent variations in rate on the basis of the unitary hypothesis, namely, to differences in the kinetics of drop-off and redistribution of the various components carried by the transport filaments in the two branches.

5. The Implication of Routing for Theories of Transport

The phenomenon of routing is difficult to harmonize with some of the theories advanced to account for axoplasmic transport (Chapter 10). For example, if transport were due to downflow within the axolemma (Chapter 10, Section A6), there would have to be some means by which the various components are sorted out in the membrane of the stem portion of the dorsal root ganglion neuron when the site of the fiber branching at the "T" is reached so that more materials are carried into the sensory nerve branch than the dorsal root branch. A similar need for a sorting mechanism at the "T" would also apply to the selective routing of one or other special component into the two branches at the dividing point.

On the bulk flow hypothesis of slow transport (Chapter 10, Section B1), we would have to envision the axoplasm flowing down the stem portion of the nerve fiber as a semisolid column splitting into two portions at the point where the stem portion divides into its two daughter branches with a greater amount of axoplasm entering the sensory nerve fiber branch. The mechanical forces required to split the semisolid structure envisioned in this theory makes such a division of axoplasm difficult to conceive.

A difficulty also pertains to the hypothesis that the microtubules and neurofilaments are slowly growing down within the fibers as coherent entities

(Chapter 10, Section B4). We would expect on that basis a greater number of microtubules and neurofilaments to be present in the sensory nerve fiber branch than the dorsal root branch. As noted in Section B1 above, these organelles are seen to have similar densities in the two branches. The microtrabecular network concept (Chapter 10) also entails conceptual difficulties when confronted with the routing phenomenon. At the junction between the stem portion of the dorsal root ganglion neuron and the two daughter branches, some mechanism would be required to divide the microtrabecular network so that more of it passes into the sensory branch as compared to the dorsal root branch. There is little evidence for an asymmetry in the microtrabecular network components present in the two branches.

Before leaving this subject, it should be noted that the concept of routing has the virtue of offering an explanation for a puzzling phenomenon; namely, the growth of neurites toward a target with the resorption of aberrant branches in the course of maturation or regeneration (Ochs, 1981c and Chapter 15, Part A).

12

Agents and Pathological States
Affecting Transport

Agents affecting axoplasmic transport form a heterogenous group. They may be divided into those which; (a) block metabolism and the supply of ATP needed to maintain axoplasmic transport, (b) act on microtubules or other elements of the transport mechanism, or (c) act on the membrane to change permeability to ions and cause an altered ionic composition in the fibers, in turn affecting the transport mechanism (Ochs, 1975a).

The list of agents to be described in this chapter is not exhaustive; some have been dealt with in other chapters. For example, metabolic blocking agents have been dealt with in Chapter 7, and the effect of an alteration of Ca^{2+} and other ions *in vitro* in Chapter 8. In this chapter we will discuss the action of tubulin-binding agents for their effect to block transport, a subject briefly touched on in Chapter 9 with respect to the relation of tubulin to the microtubules. We have noted the effects on transport of low temperatures in Chapter 2 and the phenomenon of cold-block in Chapters 3 and 9. In this chapter we will discuss the effect of low temperature further, especially with respect to the recovery from cold-block.

Finally, neurological diseases and several neurotoxic agents (Spencer and Schaumberg, 1980) which affect axoplasmic transport will be briefly noted, though the explanation as to how they affect the transport mechanism is as yet largely unknown.

A. TUBULIN-BINDING (ANTIMITOTIC) AGENTS BLOCKING TRANSPORT

1. Tubulin-Binding Agents *In Vivo*

While much recent information on the behavior of tubulin-binding agents has been obtained using purified tubulin preparations (Chapter 9), the effects of these agents on microtubular function within the nerve fiber is still imperfectly known. The first information concerning this class of agents came from studies on the effect of colchicine on cells undergoing mitosis where it

278

causes a disassembly of the microtubules forming the mitotic spindles (Dustin, 1978). In the model originally proposed by Inoue and Sato (1967), colchicine binds to tubulin which is in equilibrium with the microtubules. The binding causes a shift of tubulin from the microtubules and in turn, their disassembly. A more potent class of tubulin-binding agents causing microtubular disassembly are the vinca alkaloids: vincristine (VCR), vinblastine (VLB), and a recently synthesized amide derivative, vindesine (VDS) (Gerzon, 1980). The vinca alkaloids are used clinically in the treatment of neoplastic diseases with, as one of their untoward side effects, the production of neuropathies (Weiss, Walker and Wiernik, 1974). This side effect was related to the effect of the vinca alkaloids on tubulin in the microtubules of the nerve fibers with a resultant block of axoplasmic transport. Such an action of the tubulin-binding agent colchicine on axoplasmic transport was first found by Kreutzberg and by Dahlström. Using the fluorescent stain for catecholamines (Chapter 4), Dahlström (1968) found an accumulation of noradrenaline (NA) in the cell bodies and fibers in ganglia exposed to colchicine with as a result, a block of transport and an accumulation of NA above the block site in the fibers. Relatively high concentrations of colchicine were needed to block transport, the highest concentrations producing a degeneration of the nerve fibers. However, a block of axoplasmic transport was seen with concentrations of colchicine which did not give rise to such signs of nerve degeneration. Kreutzberg (1968, 1969) used the accumulation of the enzyme acetylcholinesterase (AChE) above nerve crushes as the measure of axoplasmic transport and found transport of the enzyme to be blocked when colchicine was injected into the nerve above that site. Both NA and AChE are transported at the fast rate of transport, NA in the dense core vesicles (DCVs) of adrenergic nerve fibers and AChE in a particulate fraction of the colchinergic nerve fibers (Chapter 4). Those early studies showed that colchicine causes a block of fast axoplasmic transport. Subsequently, a block of slow transport by colchicine and the other tubulin-binding agents was reported, a result we would expect on the basis of the unitary hypothesis (Chapter 11). In crayfish fibers, fast transport was found to be more sensitive to block by colchicine than slow transport (Fernandez *et al.*, 1971). This has also been reported for the rabbit optic system (Karlsson and Sjöstrand, 1969; Karlsson *et al.*, 1971; Karlsson and Sjöstrand, 1971c), hypoglossal and vagus nerves of the rabbit (Sjöstrand *et al.*, 1970), chicken optic system (Marchisio and Gremo, 1971), and the pigeon optic system (Boesch *et al.*, 1972). Some quantitative differences might be expected, considering the relatively larger amounts of materials slow-transported as compared to those fast-transported and the time needed for the effects of colchicine to develop.

More difficult to explain is the apparent lack of effect of colchicine on slow transport altogether, where there is a block of fast transport. This was indicated by Kreutzberg (1969) on the basis of a continued accumulation of diphosphopyridine nucleotide diaphorase at a crush below the site of injection. This enzyme is a marker for mitochondria and its continued accumulation below the site blocked by colchicine injections was taken to show a remaining slow transport due to the column of axoplasm "growing" or "flowing" as a whole within the fibers (Chapter 10). A constant "growth" or "flow" of axoplasm as a coherent whole is unlikely (Chapter 10), and we can account for Kreutzberg's finding by the known ability of mitochondria to continue to move for a time both in the anterograde and retrograde directions in amputated portions of nerve fibers (Chapter 3). The fact that the mitochondria move within isolated nerve segments, and in both directions before Wallerian degeneration occurs, was shown by their accumulation at the distal and upper transection sites of nerve (Friede and Ho, 1977). Additionally, the movement of mitochondria was seen microscopically in isolated portions of nerve fiber and its continued movement also inferred from accumulations of mitochondrial markers at ligations *in vitro* studies (Chapter 3).

James *et al.* (1970) reported colchicine to have a greater effect on slow than on fast transport. In their studies, ^{3}H-leucine was injected into chicken spinal cords 72 hr after injecting colchicine, and nerves were taken after a 5-hr downflow to assess fast, and after 22 hr to assess slow transport. The apparently greater effect of colchicine on slow transport could be due to the increasing effect of colchicine at the longer time intervals, perhaps as a result of its diffusion from a site of injection somewhat removed from the major distribution of motoneuron cell bodies. Griffiths and McLean (1980) recently reinvestigated this point using a cuff technique and found colchicine to be equally effective in blocking fast and slow transport in rabbit vagus nerves.

2.   Effect of Tubulin-Binding Agents on Transport *In Vitro*

Using the *in vitro* vagus nerve preparation, Hinkley and Green (1971) reported a reduction in the density of microtubules in colchicine-treated nerves. Colchicine did not affect the energy metabolism in the nerve or the properties of the membrane and the block of axoplasmic transport was not accompanied by a failure of electrical excitability. Those results were thus in accord with the concept that the block of transport is due to a disruption of microtubules as expected of their relation to the transport mechanism (Chapter 10).

A comparative study of several different vinca alkaloids was initiated to help assess the relationship of microtubules to transport (Ochs and Worth, 1975). The precursor ^{3}H-leucine was injected into the L7 dorsal root ganglia and after a period of downflow of incorporated proteins in the animal for 2 hr, the nerves were removed and transferred to flasks or chambers containing tubulin-binding agents for a further period of downflow. The time at which transport was blocked in these *in vitro* studies was used as the measure of the comparative potency of several vinca alkaloids with the order of effectiveness found to be; VCR>VLB>VDS. The more rapid block of transport produced by VCR is in accord with its greater neurotoxicity. A different order of effectiveness was found by Donoso *et al.* (1977) using the vagus nerve exposed *in vitro* to these vinca alkaloids on the basis of the morphological changes produced by these agents. The order of effectiveness they found was VLB>>VCR>VDS. However, in both these experiments the perineurial sheath was intact, adding a complication because of its relatively low permeability to these agents (Chapter 8). This could account for variations of outflow in nerves blocked by the vinca alkaloids (Ochs and Worth, 1975), the peaks in the outflow seen in the course of transport block by Byers (1974) using colchicine, by Donoso *et al.* (1977) using vinca alkaloids, and by Hanson and Edström (1977) with vinca alkaloids. The latter considered that the peaks may be explained by a differential block in fibers of different diameters by the vinca alkaloids. This may possibly be the case in the vagus nerve where a sizeable proportion of the labeled activity is transported in the smaller diameter fibers. In the cat sciatic nerve, however, the bulk of the radioactivity appears to be transported by the medium and larger-diametered myelinated nerve fibers, making this an unlikely explanation. Irregularities in the course of block are more likely due to variations in the permeability of the perineurial sheath to the tubulin-binding agent at various points along the length of the nerve trunk. This was inferred from studies using the desheathed nerve preparation (Chapter 8) where no such irregularities in the shape of the outflow of labeled proteins were seen in the course of the block of transport produced by the vinca alkaloids *in vitro* over a wide range of concentrations (Chan *et al.*, 1980).

Desheathing the nerve markedly reduced the concentration of tubulin-binding agent required to block transport. This would be expected from the low permeability of the sheath to these agents, a finding reported by Byers (1974) using colchicine on the desheathed vagus nerve. She found concentrations as low as 7 mM effective in blocking transport, a level much lower than that required to block the sheathed nerve. Similarly, when using the desheathed cat sciatic nerve preparation, a concentration of 10 mM colchicine was effective in blocking axoplasmic transport *in vitro* in contrast to the weak effect seen with amounts of as much as 50 mM or more required in sheathed nerves (Ochs, unpublished observations). A marked reduction in the minimal effective concentration of vinca alkaloid required to block transport in the desheathed nerve was seen by comparing the minimal concentration of 1–3 mM required to block transport *in vitro* in the sheathed nerves with the minimal concentration of 3 μM required in the desheathed nerve preparation (Chan *et al.*, 1980b), a difference of more than 1,000x. The rapidity with which transport was blocked by VLB in a concentration of 25 μM is shown in Fig. 12.1.

Comparing the time at which the fronts of outflow had been blocked in the desheathed preparation, the vinca alkaloids were shown to have the following order of potency: VCR>VLB>VDS (Table 12.1).

The best discrimination between the three vinca alkaloids with respect to their difference in blocking transport was seen at a concentration of 25 μM. At higher concentrations of 100 uM and greater, all three vinca alkaloids

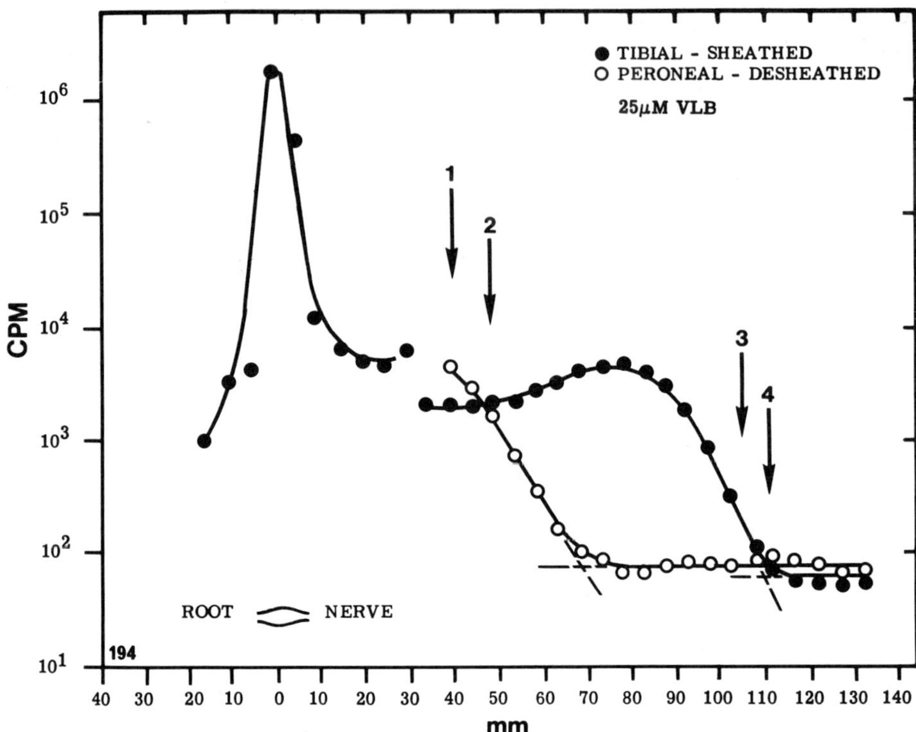

Fig. 12.1 Block of transport by vinblastine. After a period of downflow of labeled proteins in the animal following ganglion injection of ^{3}H-leucine, the nerve was desheathed and the preparation was placed in a flask to which a concentration of 25 μM of vinblastine (VLB) was added (arrow 1). Transport in the desheathed portion *in vitro* is seen to be blocked (arrow 2). At this relatively low concentration of vinca alkaloid, transport in the sheathed branch was not much affected (arrow 3); the front had moved to the position normally expected after this time of downflow (arrow 4). From Chan *et al.* (1980b).

TABLE 12.1   Vinca Alkaloids and Block Time[a]

Agent	n	Block time $\pm$ SD, hr
VDS	6	2.94 $\pm$ 0.32
VLB	9	2.41 $\pm$ 0.43
VCR	9	1.96 $\pm$ 0.47

[a]Block time of transport given by front of outflow at time in hours (hr). Means $\pm$ standard deviation (SD).
n = number of experiments, concentration in all cases 25 μM.

blocked transport so rapidly as to make a differentiation between them on the basis of their block times statistically difficult to ascertain. At lower concentrations the times of block were too long for good *in vitro* measurements and statistical evaluation.

The greater effectiveness of VCR in blocking transport in comparison to VLB and VDS is paralleled by the greater relative uptake of VCR by the nerve fibers. This was shown by adding ^{3}H-labeled vinca alkaloids to an incubation media in which desheathed nerves were placed (Iqbal and Ochs, 1980c), as shown in Fig. 12.2.

Two main factors determining the different uptakes of vinca alkaloids by nerves are the different membrane permeabilities of the fibers to the various agents and the affinities with which the alkaloids bind to components within the axon to effect a block, principally to tubulin (cf. Chapter 9). Attempts to increase the pool of free tubulin within the axons by lowering the temperature in order to dissociate the microtubules resulted in only small increases of uptake, suggesting that the major factor accounting for differences in their uptake is likely to be the differing permeabilities of the axonal membrane to

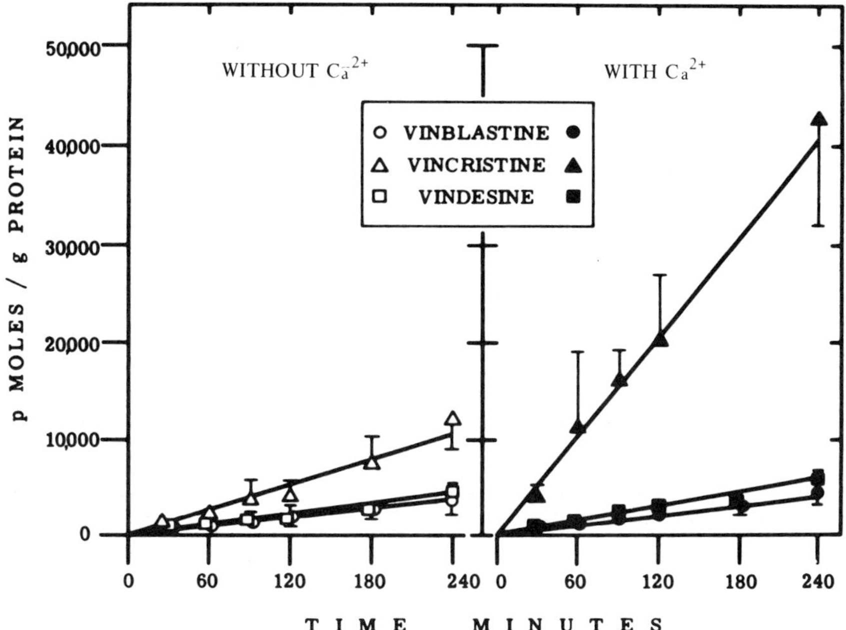

Fig. 12.2 Uptake of labeled vinca alkaloids by desheathed nerve. A greater uptake of ^{3}H-vincristine by the nerve than of the other labeled vinca alkaloids, vinblastine and vindesine, is seen. Incubation of desheathed nerves was carried out in a sucrose medium with 3.0 mM Ca^{2+} (●) or without Ca^{2+} (O). Uptake is given on the ordinate as pM of the alkaloid per gram of protein in the soluble fraction. The points represent means ± S.D. obtained from four to six experiments. From Iqbal and Ochs (1980c).

those alkaloids. A comparison of the partition coefficients to the binding affinities of the three vinca alkaloids (Gerzon, Ochs, and Todd, 1979), showed an inverse relation to lipid solubility. Those results suggest that some other factor is responsible for the variations seen in their uptake by the nerve fibers.

3.   Block of Transport by Tubulin-Binding Agents and Microtubules

The block of axoplasmic transport by tubulin-binding agents is generally related to a decrease in microtubular density in the nerve fibers (Banks and Till, 1975; Hanson and Edström, 1978). This may not invariably be the case. There are instances where a block of axoplasmic transport was reported with apparently no decrease in the number of microtubules (Dustin, 1978). Conversely, Byers (1974) saw very few microtubules present in nerve fibers exposed to colchicine while there was an apparent continuation of transport. The interpretation made was based on the irregular patterns of outflow seen at the time of block which suggested some downflow still occurred. As noted above, however, irregularities of outflow in the course of block by such agents may be ascribed to variations in the permeability of the sheath to the blocking agent. It is not known in those experiments when block of transport did in fact occur. Some fibers are more greatly effected by the tubulin-blocking agent than are others, particularly unmyelinated fibers (see below). It is necessary to take those factors into account in carrying out quantitative studies relating microtubular density to transport block. One such study was carried out using the antimitotic blocking agent maytansine (Ghetti and Ochs, 1978). Maytansine, an ansa macrolide, causes a disassembly of microtubules which most likely occurs by an action similar to that of other tubulin-binding agents (Remillard, *et al.*, 1975). In cells undergoing mitosis, maytansine appears to be more effective than the vinca alkaloids while in nerve it blocks axoplasmic transport in the desheathed nerve preparation *in vitro* in concentrations similar to those of the vinca alkaloids (Ghetti and Ochs, 1978). The minimal effective concentration of maytansine found effective in blocking transport was 3 μM. As with the vinca alkaloids, the time at which transport was blocked was related to the concentration of maytansine present in the *in vitro* medium. The block of transport by maytansine in desheathed nerves occurred at earlier times when higher concentrations of the agent were used (Fig. 12.3).

The microtubular density in the larger myelinated fibers was found to decrease in relation to the concentration of maytansine present in the medium. The medium and larger-diameter myelinated fibers carry the bulk of radioactive proteins in the sciatic nerve and these were taken to determine microtubule densities after exposure to maytansine. At a concentration of maytansine where transport was blocked, the microtubular density was found reduced to approximately half of control values. A much greater reduction of microtubule density was seen in the unmyelinated nerve fibers at that time.

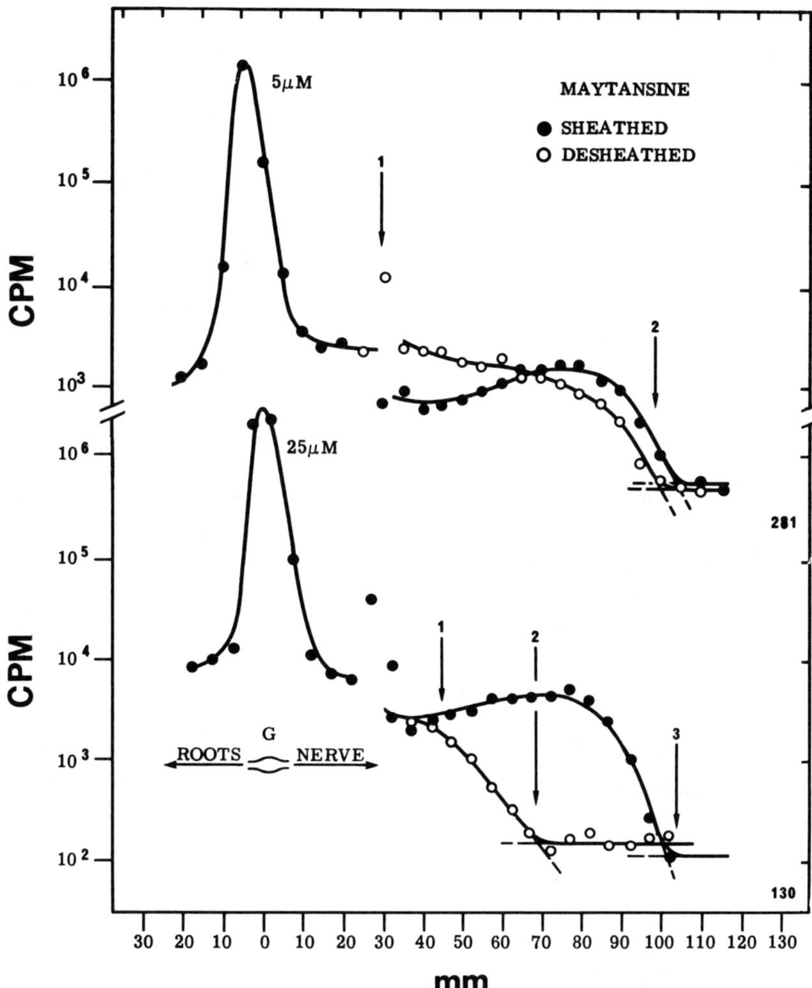

Fig. 12.3 Transport block by maytansine. Two hours after downflow of labeled protein after ^{3}H-leucine injection, the nerve was removed, peroneal nerve was desheathed, and the preparation was placed in a flask (arrow 1) with a medium containing 5 or 25 μM maytansine for 4 hr. Arrow 3 indicates the position to which the fronts of the crests are calculated and would be transported at the usual rate of 410 mm per day. With 5 μM MYT present, the front of the crest of the desheathed branch lags only a little behind the front of the sheathed branch (arrow 2). With 25 μM present, the front of the desheathed branch shows an earlier block of transport (arrow 2). From Ghetti and Ochs (1978).

285

The much greater sensitivity of the unmyelinated fibers to the agent underlines the necessity of correlating the block of axoplasmic transport with the microtubular density in those fibers carrying the bulk of the components transported.

The finding that the microtubular density was reduced to half when axoplasmic transport was blocked raises the question as to why those microtubules still remaining were not able to maintain a sizeable amount of transport. Several possible explanations may be advanced. One is that the microtubular population is heterogenous, the portion more readily dissassembled being the one carrying the bulk of transported material. Another explanation is that the tubulin-binding agent causes random defects along the length of all the microtubules (Paulson and McClure, 1975; Ochs *et al.*, 1978), making them inoperative. If these agents produce breaks in the microtubules at infrequent intervals along their length, cross-sections taken through the nerve fibers to determine the microtubular density would not be likely to pass through the affected regions and the apparent decrease in microtubular density would appear to be minimal. Only when such microtubular breaks become long enough would the probability of a given cross-section passing through the defective portion result in a measured reduction in microtubular density. Such breaks may perhaps start at a site along the microtubules where processes controlling the turnover of tubulin in the microtubules take place (Chapter 9). A model which is conceptually related to that possibility takes into account the likelihood that the microtubules consist of relatively short lengths as described above and in Chapter 9. When assembly is blocked the microtubule segments dissemble and shorten to a critical length until a block of transport occurs when the transport filaments can no longer bridge the gaps between the segments.

Some observations made by Hammond and Smith (1977) using the dark field microscope on the effects of tubulin-binding agents on particle movement are of interest in this regard. They found a reduction in the range of particle movement appearing 1–4 hr after an exposure to a tubulin-binding agent. This was then followed by a complete and irreversible cessation of movement. When the particles moved, they did so at their normal rates with no gradual slowing as the agent took effect. The particles were seen to remain stationary more frequently and for progressively longer periods of time. Often they were seen to oscillate over short distances of 1 μm for a time without a sustained longitudinal progression before they became immobile. This would be the case if the short microtubule segments had shortened too far. Hammond and Smith also found that as the nerve fibers were exposed to increasing concentrations of the tubulin-binding agent, the density of microtubules in the fibers fell from a control value of $10-15/\mu m^2$ to $3/\mu m^2$.

Recent studies by Horie, Takenaka, and Inomata (1981) shed light on how antimitotic agents act. Cultured dorsal root ganglia neurons exposed to colchicine or vinblastine show a block of particle movement when observed with Normarski optics microscopy. Regions of swelling occurred along the neurites with a block of transport into those regions and a reduction of microtubules in the swellings. Thus, the spotty interruption of microtubule continuity produced by antimitotic agents can block axoplasmic transport with apparently little change in microtubule density seen by examining cross-sections of nerve which would, on a statistical basis, be less likely to pass through regions which are not swollen.

A fairly large number of tubulin-binding agents effective in blocking axoplasmic transport in nerve are known (Paulson and McClure, 1975; Banks and Till, 1975; Dustin, 1978; Hanson and Edström, 1978; Hanson, 1981). Using inhibition of NA vesicle movement as the measure of fast transport, Banks and Till (1975) found the following order of potency: podophyllotoxin > colchicine > griseofulvin. Podophyllotoxin caused a more rapid loss of microtubules than colchicine. The authors consider this as evidence for an integral role of microtubules in transport. It would be of interest to carry out further *in vitro* quantitative studies of a number of these agents to determine if they have a common mode of action, namely, to cause a similar decrease in microtubular density when transport is blocked.

B. COLD-BLOCK AND RECOVERY FROM COLD-BLOCK

As the temperature is reduced, the rate of axoplasmic transport *in vitro* slows with the appearance of a decrease in the slope of the advancing front of radioactivity at temperatures between 28 and 13°C (Fig. 2.23). At temperatures close to 11°C, the cold-block temperature (Chapter 2), transport is completely blocked (Fig. 2.24). One possible explanation for cold-block might be that at low temperatures axoplasmic viscosity is greatly increased, as has been reported for various other cells (Heilbrunn, 1952). However, a large increase of viscosity correlated with the cold-block temperature was not found when the spin label tempone and electron spin resonance was used to assess viscosity (Fig. 9.12). Another explanation for cold-block is a disassembly of microtubules at low temperatures, a process perhaps similar to that observed for purified microtubules *in vitro* (Chapter 9). A disassembly of microtubules at low temperatures was not found in crayfish nerves (Fernandez, Huneeus, and Davison, 1970). This invertebrate, however, does not show the same fast transport rates as that present in the nerves of mammals or non-mammalian species such as the frog, and there are known variations with respect to the sensitivity of nerve microtubules of different species to cold-disassembly (see below). A decrease in the density of microtubules at low temperature was demonstrated in *Heliozoa* by Tilney and Porter (1967), in frog nerves by Rodríquez-Echandia and Piezzi (1968), in leg nerve axons of the crab *Carcinus meanus* by Rome-Talbot *et al.* (1978), and in rabbit and bullfrog nerve by Brimijoin, Olsen, and Rosenson (1979). The latter investigator found a 30% reduction of the microtubules in the myelinated fibers at temperatures below 13°C, the cold-block temperature they determined in the rabbit. Smaller gradual decreases of microtubular density in the fibers were found as the temperature was lowered down toward the critical cold-block temperature (Fig. 12.4).

Bullfrog axons were seen by Brimijoin *et al.* (1979) to be more sensitive to cold, with some 35% of the microtubules lost at a temperature of 15°C and 65% at 10°C, the latter near the cold-block temperature for this nerve.

The same question arises as discussed above with respect to tubulin-binding agents, namely, why transport is completely blocked when a considerable number of microtubules is left remaining in the cold-blocked nerves. The same two possible explanations previously advanced may apply to the effects of lower temperature. Namely, a portion of the microtubules in the fibers is more sensitive to cold, those carrying the bulk of the transported materials, or, as the temperature is lowered random breaks appear at various points along the length of all the microtubules or there is a shortening of the small staggered lengths of microtubules (Chapter 9) to the point where trans-

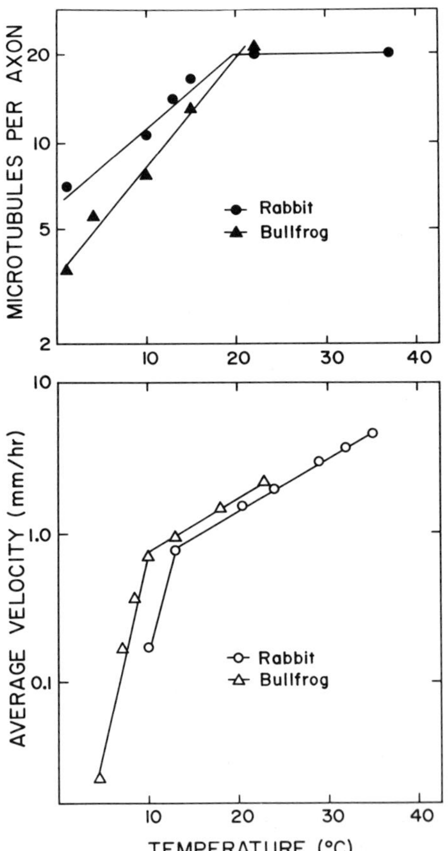

Fig. 12.4 Effect of lower temperatures on microtubule density and transport rate. The upper panel shows that as temperatures were reduced to 11°C, microtubule density fell to about half of the control density in the rabbit (●) with a greater fall in density seen in frog nerve (▲). In the lower panel, a gradual fall of transport velocity (DBH) was seen as temperatures were reduced until, at about 11–13°C, an abrupt fall occurred in the rabbit nerve (○). A lower cold-block occurred, the temperature at which was observed in the bullfrog (△). Regression lines were fitted to the data by the method of least squares, with separate calculations for the regions above and below the breaks in the curves. From Brimijoin *et al.* (1979).

port is impaired and then blocked. The number of microtubules measured is determined by the length of the breaks or short lengths remaining and the probability of a given section taken for EM passing through the missing portions of the microtubules. Random breaks or excessive shortening of microtubule segments can account for the increasingly shallow slope of the front at temperatures ranging between 28 and 11°C (Fig. 2.23) as well. In this range, as the temperature is lowered, the number and length of such local gaps would increase, until, at the critical cold-block temperature, they will have become large enough to completely block axoplasmic transport.

Differences in the mechanisms responsible for tubulin turnover in various fibers may account for the known variation in the susceptibility of microtubules to cold-disassembly. A lesser susceptibility to cold might well be expected of the microtubules in the nerve fibers of poikilothermic animals able to function at low temperatures. For example, axoplasmic transport is maintained in garfish olfactory nerves at 10°C (Gross, 1973) and at 9°C in the goldfish (Grafstein *et al.*, 1972). In a molluscan nerve axoplasmic transport was seen to be maintained at temperatures below 4°C (Heslop and Howes, 1972), well below the cold-block temperature of mammalian nerve.

Accepting a disassembly of microtubules as the basis for cold-block, the question arises as to the speed of reassembly and resumption of transport when mammalian nerves are restored to a temperature of 38°C. Nerves maintained *in vitro* at cold-block temperatures for periods of several hours and then rewarmed to a temperature of 38°C show a rapid resumption of axoplasmic transport. An example of a nerve kept at 0°C for 1.5 hr and then returned to a temperature 38°C for a further period of downflow of 1.5 hr is shown in Fig. 12.5.

If no recovery of transport had occurred on rewarming, the front would have remained at the position indicated by arrow 1. The further movement of the front, to the point shown by arrow 2 after the nerve was rewarmed to 38°C, indicates that transport occurred to the distance expected for all the time it had been kept at 38°C. The front fell short of the control at arrow 3 by only the time during which the nerve had been kept at the cold-block temperature. There was, therefore, no additional block seen as would be the case if there were a lag in the recovery of transport from the cold-block. We may infer from this that the reassembly of microtubules to bridge gaps or restore the short segments to the length needed to maintain transport could have taken no more than 10–15 min at most. A rapid recovery of axoplasmic transport from cold-block was also indicated in the studies of Edström and Hanson (1973b) and Brimijoin (1975) using cold-block to produce a stop-flow (Chapters 2 and 4). The advance of the waves of the dammed material when the temperature was raised to 37°C showed no lag. A rapid reassembly of microtubules after disassembly at low temperatures had also been seen in *Heliozoa* by Tilney and Porter (1967) and the reassembly of microtubules from purified tubulin *in vitro* occurs within a matter of minutes on raising the temperature (Weisenberg, 1972; Gaskin, Cantor, and Shelanski, 1975).

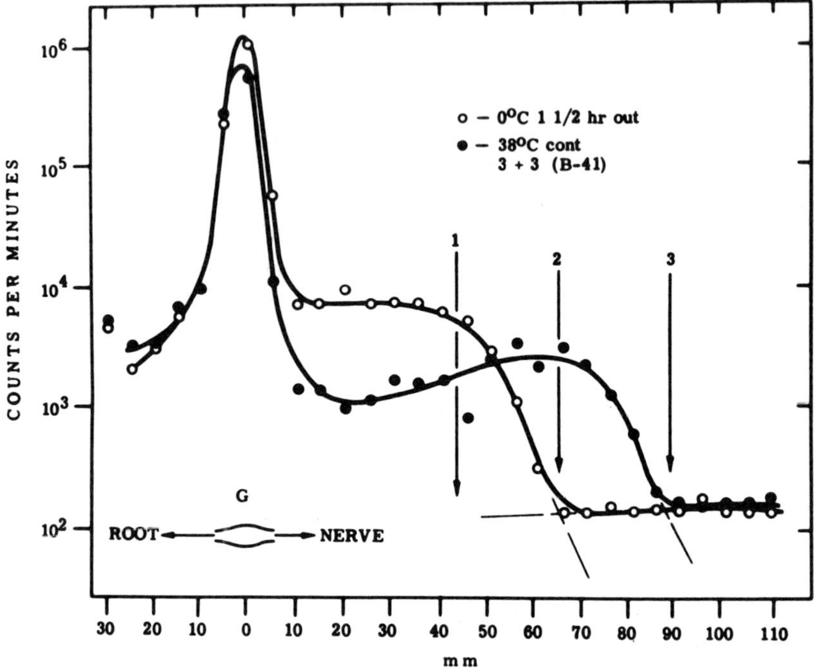

Fig. 12.5 Recovery of transport after cold-block. After 3 hr of downflow of labeled protein *in vivo*, the nerve was removed and placed *in vitro* for a 1.5-hr period of cold-block at 0°C, then transferred to a chamber at 38°C for a further period of 1.5-hr downflow (○). Arrow 1 indicates the distance the front would have been carried in the cold-blocked nerves if no further transport occurred beyond that taking place in the animal. Arrow 2 indicates transport resulting from the 1.5-hr period of transport at 38°C after recovery from cold-block. The control nerve (●) was kept at 38°C *in vitro* for the entire 3-hr period (arrow 3). From Ochs and Smith (1975).

A rapid recovery in transport was also seen when nerves were subjected to longer-lasting periods of cold-block of 14–20 hr. However, in those nerves the position of the front fell short of the distance calculated from the usual fast rates of transport (Fig. 12.6).

This discrepancy was not due to a lag in the resumption of axoplasmic transport on rewarming, but rather to a lower rate of axoplasmic transport in nerves which had been subjected to cold-block temperatures for such long periods of time. The rate change was shown by varying the times allowed for transport after the nerves were rewarmed to 38°C. On this basis an estimated rate of 14.4 mm/hr (318 mm/day) in the rewarmed nerves was determined (Ochs and Smith, 1975) as compared to the normal rate of 17 mm/hr or 410 mm/day.

Additional studies of nerves kept at cold-block temperature for long times and then rewarmed to 38°C for 2 and 5 hr for further transport gave similar results (Ochs, unpublished experiments). The change in the transport mech-

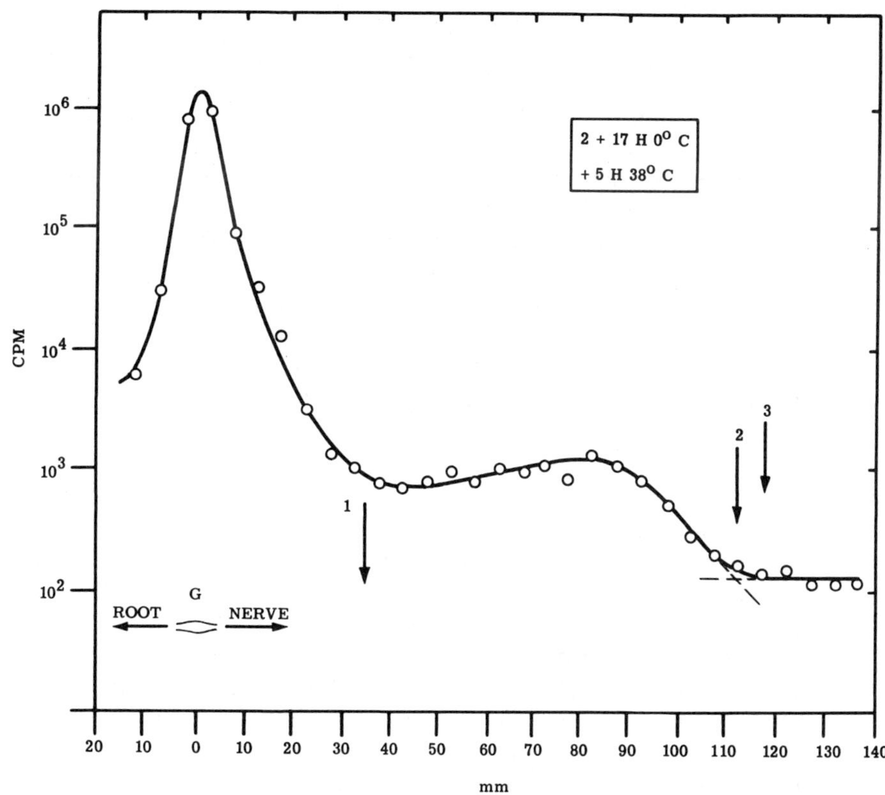

Fig. 12.6 Recovery of transport after a long period of cold-block. The nerve was taken after 2 hr of downflow of labeled proteins in the animal and kept at 0°C for 17 hr *in vitro*. Following this period of cold-block, it was transferred to a chamber warmed to 38°C for a further 5 hr of downflow. Arrow 1 shows the point at which *in vivo* transport would have remained under cold-block conditions, arrow 2 shows the additional downflow which took place after rewarming, and arrow 3 shows the distance expected of the usual fast rate of transport. From Ochs and Smith (1975).

anism signalled by the lower rate of transport is as yet unknown. It may be due to an altered ionic composition in the nerve fibers as the result of their prolonged stay in the *in vitro* medium. Possibly the interruption of the sodium pump by low temperature with a gain of Na^+, as is seen in frog muscle (Harris and Ochs, 1966), would, as a result of the augmented Na^+ in the axons (cf. Chapter 8 and Section F4 below), give rise to inhibition of axoplasmic transport via an indirect effect on Ca^{2+}

C. MICROTUBULE "STABILIZATION"

Heavy water (D_2O) is classically considered to stabilize microtubules. This was first shown in Arbacia eggs in mitosis when exposed to high hydrostatic pressure to cause a disassembly of the spindle microtubules and the disassembly was blocked when D_2O was substituted for H_2O in the medium (Marsland and Hiramoto, 1966). Anderson *et al.* (1972) found D_2O to cause a block of axoplasmic transport in nerve *in vitro*. In those studies, an accumulation of labeled activity at distal ligations was taken as the measure of transport. The block was partial with a 50% replacement of water by D_2O and it was almost complete with a 90% replacement with D_2O. Similar studies showed D_2O to block the front of transported labeled proteins (Ochs, unpublished experiments). Another possibility is that D_2O acts on metabolism or effect ionic exchange causing an altered intra-axonic composition in the fiber (cf. Chapter 8).

The agent dimethylsulfoxide (DMSO), a polar solvent which readily permeates tissues (Rammler and Zaffaroni, 1967; MacGregor, 1967), has also been reported to have a stabilizing action on microtubules. This was indicated by its ability to lower the temperature at which polymerization of tubulin *in vitro* occurs and to increase the resistance of microtubules to disassembly by high ionic strength media and by colchicine (Dulak and Crist, 1974). Transport was reported by Donoso, Illanes, and Samson, (1977) to be blocked in vagus nerves *in vitro* by a concentration of DMSO of 5%, a full block appearing only with concentrations of DMSO of 10%.

In those studies, nerve fibers exposed to 10% DMSO showed inconsistent swellings or shrinking with a preservation of their microtubules. Nerves exposed to this concentration of DMSO for as long as 2.5 hr could recover a substantial, though not all, of their normal axoplasmic transport. In similar such studies, the cat desheathed nerve preparation appears to be somewhat more sensitive to the action of DMSO (Garg and Ochs, unpublished experiments). The point is of practical importance in that DMSO is sometimes used as a solvent and its effects as a solvent vehicle for agents and must be kept in mind when it is used in transport studies.

Hexylene glycol at an acid pH stabilizes microtubules (Kirkpatrick *et al.*, 1970). In addition to its stimulation of microtubule assembly (Chapter 9), taxol is also considered to be a microtubule stabilizer (Schiff and Horwitz, 1980).

D. SULFHYDRIL OXIDIZING AGENTS AND TRANSPORT BLOCK

Oxidation of the 8 to 11 sulfhydril (half-cysteine) groups present on each tubulin monomer reduces their ability to polymerize and form microtubules (Mellon and Rhebin, 1976). Diamide, a potent sulfhydril oxidizing agent (Kosower *et al.*, 1972; Kosower and Kosower, 1969), is effective in blocking axoplasmic transport in nerve (Ochs, unpublished observations).

It is as yet unknown, however, whether a microtubule disassembly occurs in the nerve fibers blocked by diamide. The agent could perhaps act more readily on some essential sulfhydril-containing enzyme to reduce the supply of ATP needed to maintain transport. Sulfhydril blocking agents act to block

Ca–Mg ATPase, a sulfhydril containing enzyme, to different degrees (Khan and Ochs, 1974). Mersalyl and p-chloromercuribenzene sulfonic acid (PCMBS) are more effective in inactivating this enzyme than IAA. As would be expected from its action on sulfhydril groups, mersalyl is effective in blocking axoplasmic transport. While these agents affect sulfhydrils generally, IAA appears to have a more selective effect on the key glycolytic enzyme, GAPD (Chapter 7), than on other sulfhydril enzymes and the sulfhydril groups of tubulin in the nerve fibers. This was indicated by the almost complete reversal of the block of transport produced by IAA when pyruvate or lactate was also supplied in the incubating medium (Chapter 7). Sabri, Moore, and Spencer (1979) have recently pointed to the critical position of GAPD in connection a number of neurotoxins blocking transport and producing neuropathies.

Metallic ions appear to act on the sulfhydril groups of a number of enzymes involved in glycolysis and oxidative phosphorylation (Mildvan, 1970). Edström and Mattsson (1975) reported that Zn^{2+} blocks transport at a concentration of 10^{-4} M, Cd^{2+} at 5×10^{-5} M, Hg^{2+} at 5×10^{-6} M, and Cu^{2+} at 10^{-4} M. Interestingly, still lower concentrations of Cd^{2+} and Zn^{2+} were reported to increase the amount of material fast-transported without affecting the rate of axoplasmic transport (Edström and Mattsson, 1976). The explanation may be that at the low concentrations these metals act to prevent the drop-off of materials from the transport filaments that contribute to slow transport (Chapter 11), thus allowing more material to be carried down by fast transport. Stimulation of a greater degree of synthesis of fast-transported components in the cell bodies by these ions may also occur.

The organo-metal compound methyl-Hg, which acts on sulfhydrils, blocks transport (Abe *et al.*, 1975), as do the sulfhydril blocking compounds N-ethylmaleimide (NEM) and PCMBS (Edström and Mattsson, 1976). At a lower concentration of 10^{-6} M, NEM and PCMBS were reported to increase the amount of transported labeled proteins, probably by the same mechanisms described above in connection with Cd^{2+} and Zn^{2+}. The notion of an increased amount of protein synthesis in the cell bodies by these agents seems borne out by the finding that more proteins are transported when the ganglia were exposed to these agents. However, the effect was also reported present when the agents were allowed to act only on the nerve fibers. Thus, the explanation advanced above for the effect of low concentrations of Ca^{2+} and Zn^{2+} may apply, namely, that NEM and PCMBs may prevent some of the materials which normally drop off the transport mechanism in the axons and contribute to slow transport from doing so, thus allowing more of the downflow to appear as fast-transported material.

E. NEUROFILAMENT PROLIFERATION

Colchicine, the vinca alkaloids, and maytansine not only cause a disassembly of microtubules, but they generally produce an increased density of neurofilaments, a phenomenon first noted in the cell bodies by Wiśniewski and Terry (1967) when using colchicine. The early possibility advanced that the neurofilaments are formed from the same subunits constituting the microtubules was discounted when it was shown that the microtubular and neurofilament subunit proteins differ in their amino acid composition and molecular weights (Chapter 9). Another explanation for the proliferation of neurofilaments is a shift in the character of protein synthesis in the cell bodies (Ghetti, 1980). Neurofilament proliferation can occur *in vitro* within a portion of peripheral nerve isolated from the cell bodies. Thus, it appears likely that both central and peripheral factors controlling the proportion of microtubules and neurofilaments in the fibers must be considered.

In the course of maturation there is a shift from the higher proportion of neurofilaments to microtubules characteristic of the immature fiber to the higher proportion of microtubules to neurofilaments typical of the mature fiber (Chapter 6). When a ligature is loosely placed on the sciatic nerves of immature rats, the growth of the nerve brings about its gradual partial constriction (cf. Duncan, 1948; Chapter 10) with, as a result a greater proportion of neurofilaments to microtubules remaining in the fibers distal to such constrictions (Friede, 1971). This was interpreted to be due to the lack of some substance provided by the cell bodies required to increase the relative proportion of microtubules to neurofilaments in the nerve fibers, thus leaving them in an immature state. The increased density of neurofilaments often seen in chromatolytic cells (Chapter 6) might also imply a shift in the synthesis of materials in the cell bodies toward an increased production of neurofilaments, a regression to an immature state, a condition to which the chromatolytic cell has been equated.

Further insight into this process may be gained from studies of other agents known to produce a proliferation of neurofilaments, classically aluminum (Al). When Al-phosphate was injected into the brain or spinal cord (Klatzo *et al.*, 1965; Wiśniewski *et al.*, 1966), it was seen to give rise to a marked production of neurofibrillary tangles in the cell bodies and dendrites. These were later shown to be morphologically identical to the neurofilaments of normal neurons (Terry and Wiśniewski, 1970).

In spite of the massive overproduction of neurofilaments induced in the cell bodies by Al, surprisingly little effect on axoplasmic transport was seen. The outflow of labeled proteins in the sciatic nerves of Al-treated animals after injecting the spinal cord of rabbits with ^{3}H-leucine was seen to have its normal form with a rate of transport of 440 mm/day, one close to the usual 410 mm/day rate of controls (Liwnicz *et al.*, 1974). The amount of labeled proteins transported shown by their accumulation above ligations was also the same in Al-treated as in control nerves.

An agent which increases the amount of neurofilaments in nerve fibers, methyl-*n*-butyl ketone (MBK), was reported to affect the rate of axoplasmic transport (Mendell *et al.*, 1977). Chronic exposure of rats to MBK caused a distal giant axonal neuropathy characterized by an increased density of neurofilaments with focal areas of axonal swelling and a thinning of the myelin sheath (Spencer and Schaumburg, 1976). The neuropathological picture has some similarity to the "dying-back" seen in humans accidentally exposed to the agent. We shall defer a further discussion of dying-back to Part I. In the MKB-treated rats, a reduction in the downflow of labeled proteins in the motor fibers after lumbar cord injection with ^{3}H-leucine was correlated with the paralysis produced by the agent. The pattern of axoplasmic transport in sensory nerves seen after injecting the dorsal root ganglia with ^{3}H-leucine was comparable in controls and in the nerves of the MBK-treated animals. However, the rate of axoplasmic transport determined 3 and 4 hr after injection was 283 mm/day in the MBK-treated animals as compared to a rate of 400 mm/day in the controls.

Microtubular density was found not to be much changed in the nerve fibers of the MBK-treated animals in spite of the much increased density of neurofilaments. This suggested to Mendell and his colleagues that the increased density of neurofilaments produced by MBK might hinder transport. Adriamycin causes a similar slowing of rate in sensory fibers (Sahenk and Mendell, 1979), and it will be interesting to know if it produces similar effects.

Another agent which produces a picture similar to MKB is acrylamide. There are, however, conflicting reports on its effect on axoplasmic transport. Acrylamide has been reported to reduce the rate of transport by Pleasure *et al.* (1969), while Bradley and Williams (1973) reported no change in the rate of transport. When injected into the dorsal root ganglion, acrylamide caused a reduction in the amount of material transported, as shown by a decreased accumulation at a distal transection, with no effect on the rate of axoplasmic transport (Griffin *et al.*, 1976). Others, on the other hand, reported a deleterious effect on the rate of fast transport (Sumner *et al.*, 1976; Weir *et al.*, 1978). After 4–6 days of oral treatment with acrylamide, cats developed head tremors and then hind-leg weakness and the rate of fast transport found in the sciatic nerves of those animals dropped to 294 mm/day as compared to the control rate of 424 mm/day (Weir *et al.*, 1978). Morphological changes in the ER seen in the nerve fibers were considered responsible for the slowing of the rate of fast transport. Souryi, Chretien, and Droz (1981) showed a multifocal lesion caused by acrylamide with a retention of labeled proteins in the fibers, much of it in the ER. However, the ER changes found could be a concomitant effect of the agent and not the cause of the decreased rate of transport. Considerations which make it unlikely that the ER is the route of fast-transported components have been given in Chapter 10.

Another agent producing a picture characteristic of the dying-back phenomenon with an associated increased proliferation of neurofilaments in nerve fibers is tri-orthocresyl phosphate (TOCP). No effect on transport was found with this agent by Pleasure *et al.* (1969). The neurotoxin found in the fruit of Tullidora *(Karawinski humboldtiana)* causes a dying-back neuropathy, more so in the motor than in the sensory nerve fibers (Muñoz-Martínez and Chavez, 1979; Weller *et al.*, 1980). In addition to a loss of conduction of the nerve impulse and signs of demyelination, there is a more pronounced failure of transport into the distal portion of the motor nerve fibers compared to the sensory fibers (Muñoz-Martínez and Ochs, unpublished observations).

Yet another agent bringing about dying-back is *p*-bromophenylacetylurea (BPAU). Jakobsen and Brimijoin (1981), using ^{35}S-methionine injections and timed placement of ligatures on the sciatic nerve, found that BPAU causes a delay and/or a decrease in the turn-around of transported proteins in sensory nerve fibers. Such defects of turn-around were seen in the absence of changes in transport velocity and were apparently unrelated to the dramatic increase in tubulomembranous material (presumably in part ER) in distal and preterminal axons (Blakemore and Cavanagh, 1969; Ohnishi and Ikeda, 1980). Brimijoin (1982a) considers that BPAU (and possibly other agents causing dying-back) produces an accumulation of ER with in turn a resulting "plugging" of the axon preventing the transport of key components required to maintain function to the distal part of the axon.

Another type of neuropathological alteration produced by neurotoxic agents is a ballooning of the motor nerve fibers close to their cell bodies with associated weakness and other signs of motor disturbances. Iminodipro-prionitrile (IDPN) injected systemically into rats causes such a ballooning of axons with an increased content of neurofilaments in the balloons (Chou and Hartmann, 1964). The authors considered that an "axostasis" of axonal flow is brought about by the increased amount of neurofilaments. It was shown in later studies, however, that IDPN does not block fast axoplasmic transport. In animals injected some 21 days beforehand with IDPN, a normal appearing front of radioactivity moving down in the nerves at a rate of 396 mm/day was found, compared to a control rate of 392 mm/day (Griffin *et al.*, 1978). On the other hand, slow transport was clearly blocked by IDPN (Fig. 12.7).

The block is shown by the greater retention of the outflow of radioactivity in the nerve fibers close to the cord in the IDPN-treated animals (b) as compared to controls (a). This localized retention included the triplet polypeptides, the subunits of the neurofilaments (Chapter 10), these found in greater amount close to the cord in SDS–PAGE preparation of the nerves of the IDPN-treated animals as compared to controls (Fig. 12.8).

Only later, after 2 to 9 months, was the outflow of tubulin and actin also impeded.

The authors consider that IDPN causes a halting of the "slowly-transported column of axoplasm," the neurofilaments still being synthesized and moved out as coherent structures into the axons (cf. Chapters 10 and 11) with as a result of the block of slow transport, these organelles becoming dammed up to accumulate in the fibers close to the cord. In the discussion of models of slow transport (Chapter 10), reasons were given for an alternative view, the unitary hypothesis (Chapter 11), wherein the triplet protein subunits of neurofilaments and the tubulin subunits of microtubules are moved down as soluble proteins. The agent IDPN, by causing an earlier drop-off

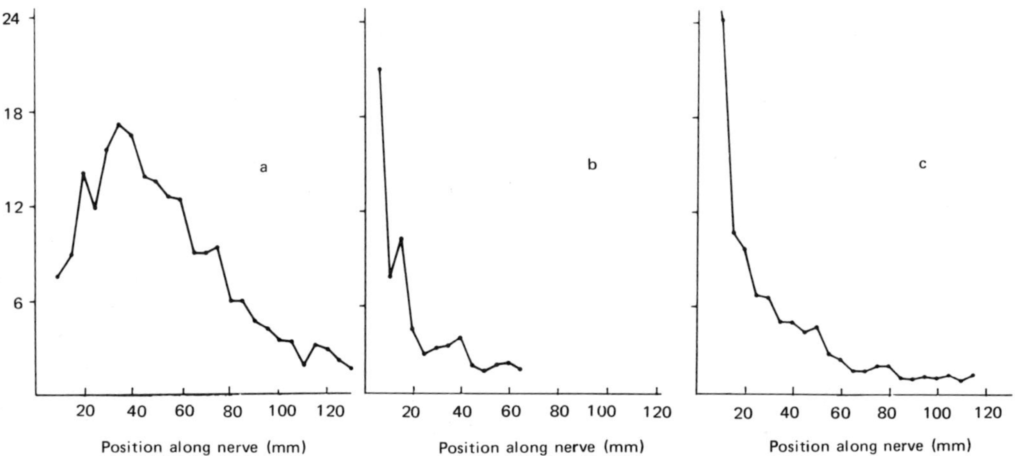

Fig. 12.7 Block of slow-transported proteins with IDPN. After intraspinal injection of ^{35}S-methionine in the rat, the outflow pattern in the sciatic nerve 21 days later is shown in (a). An animal that received IDPN (2 g/kg) intraperitoneally 1 day after labeling with ^{35}S-methionine gave rise to the distribution shown in (b). The nerve from an animal that received 0.05% IDPN in its drinking water for 9 months before being labeled with ^{3}H-leucine is shown in (c). In taking 5-mm nerve segments for counting, the origin of the L5 ventral root from the spinal cord was designated as O. The major slow peak (20–50 mm along the nerve) in (a) appears to be retained in nerve segments close to the cord in the IDPN-treated animals. From Griffin *et al.* (1978).

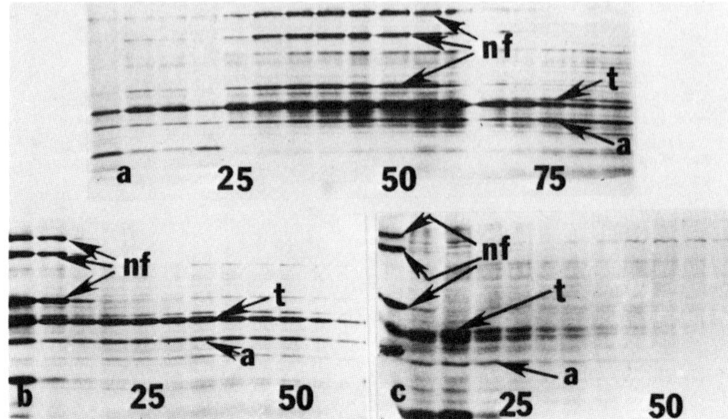

Fig. 12.8 Distribution of proteins after IDPN treatment. Fluorograms of nerve segment after IDPN treatment (cf. Fig. 12.7) and electrophoresis on polyacrylamide slab gels are shown. Each track represents proteins in a 5-mm nerve segment, with the most proximal segment at the far left. The numbers on the abscissa identify the distance of the segments along the nerve. (a) Control 21 days after labeling with ^{35}S-methionine shows the neurofilament triplet proteins (nf) (68,000, 145,000, and 200,000 daltons) present primarily in an outflow peak 25–50 mm from the cord (with a rate estimated at 1–2.5 mm/day) as indicated in Fig. 12.7. Tubulin (t) and actin (a) appear to be transported with a wide spectrum of velocities (0.5–5 mm/day). (b) Nerve taken 21 days after intraspinal injection of ^{35}S-methionine and 20 days after injection of IDPN shows neurofilament triplet proteins mainly in the initial two segments. Tubulin and actin are transported more distally in the nerve, but to an impaired extent when compared to the control nerve. (c) The animal given IDPN in its drinking water (as in c of Fig. 12.7) also shows a marked impairment in the slow transport of those components. From Griffin *et al.* (1978).

from the transport filaments of triplet protein subunits in the proximal regions of the axons, could, as a result of the increased amount of the subunits present, give rise to an augmented assembly of neurofilaments in the proximal portion of the nerve fiber.

Further study of the effects of IDPN has shown that the neurofilaments are segregated peripherally in the axon with the microtubules moved into the central portion of the axon where they are still able to support transport and maintain the viability of the fiber (Griffin *et al.*, 1981; Papasozomenos, Autilio-Gambetti, and Gambetti, 1981a,b). The latter investigators have shown that IDPN causes this reorganization all along the length of the axon. A similar reorganization is seen when IDPN was injected locally into the nerve in the endoneurial space (Griffin *et al.*, 1981). There was no overall change in the density of neurofilaments and microtubules even though the downflow of the 68,000 MW component of the neurofilament was shown to be decreased when using a specific immunochemical assay for this protein (Papasozomenos *et al.*, 1981). Thus, the effect is mainly a reorganization of organelles in the nerve fiber. Possibly IDPN might affect the side-arms connecting the neurofilaments to the microtubules (Shelanski, Leterrier, and Liem, 1981). A failure of turnover of triplet proteins in the neurofilaments, particularly those which may constitute the neurofilament side-arms (cf. Chapter 9), may be at the root of the reorganization of the organelles in the nerve produced by IDPN.

F. MEMBRANE-ACTIVE AGENTS AND TRANSPORT

1. Tetrodotoxin (TTX)

Although membrane excitability may be completely blocked by tetrodotoxin (TTX), it has no effect on axoplasmic transport *in vitro* (Ochs and Hollingsworth, 1971). A similar lack of effect of TTX on transport was shown by LaVoie, Collier, and Tenenhouse (1976) and by Pestronk, Drachman, and Griffin (1976) in chronic *in vivo* studies where the agent was applied either by means of a silastic cuff containing the agent placed around the nerve or by repeated sub-perineurial injections of the agent. More recently, a capillary implant technique was developed whereby a prolonged release of TTX sub-perineurally could be provided and again, no effect was seen on transport after a chronic block of excitability lasting weeks (Bray, Hubbard, and Mills, 1979). Retrograde transport was also found not to be blocked by TTX (Boegman and Riopelle, 1980).

2. Anesthetic Agents

The local anesthetic procaine can block membrane excitability without affecting transport *in vitro* (Ochs and Hollingsworth, 1971). However, this differential effect depends on the concentration of procaine; if too high, procaine can also block transport. Procaine, and other members of this class of local anesthetics, has a lipid solubility which depends on its pK and the pH. In an alkaline medium it is more lipid-soluble and can more readily pass through the axonal membrane. A block of transport by local anesthetics has been reported by Fink *et al.* (1972), Anderson and Edström (1973), and Byers *et al.* (1973). The permeability of the perineurial sheath to procaine is yet another important factor (cf. Chapter 8). Procaine blocks transport in the desheathed nerve preparation *in vitro* at a much lower concentration than it does in the sheathed nerve. (cf. effect of local anesthetics in silastic cuffed, nerves in Chapter 14).

Lidocaine, a member of this group of local anesthetics having a higher lipid solubility than procaine, shows the dependence of transport block on concentration (Byers *et al.*, 1973). At low concentrations it blocked membrane excitation without affecting axoplasmic transport. At somewhat higher concentrations it also blocked axoplasmic transport with, at still higher concentrations, a decrease of microtubular density in the fibers. The block of transport produced by lidocaine was reversible if the concentration used was not too high and the time of exposure short. For example, nerves exposed to a concentration of 0.5% lidocaine for 60 min showed a good recovery of transport, while nerves incubated for 90 min in 0.6% lidocaine solution did not. In the latter case an almost complete loss of microtubules appeared in both the myelinated and unmyelinated fibers (Byers *et al.*, 1973). If the time of nerve exposure was not too long, the disassembled microtubules could reassemble. However, transport remained blocked, indicating that other factors were involved. Haschke and Fink (1975) suggested that lidocaine might produce its blocking effect by an uncoupling of oxidative phosphorylation (cf. Chapter 7).

The volatile anesthetic halothane arrests cell division by causing a disassembly of the mitotic spindles, and on this basis its anesthetic action was linked to an action on the microtubules (Nunn *et al.*, 1969). However, as in the case of lidocaine, lower concentrations of halothane were found effective in blocking membrane excitability without causing a microtubular disassembly or having an effect on axoplasmic transport (Kennedy *et al.*, 1972). At moderately high concentrations, halothane blocked axoplasmic transport without disassembling microtubules, and with still higher concentrations a reduction of microtubular density occurred as well. The observation that axoplasmic transport is blocked without obvious signs of microtubular disassembly, led to the possibility that halothane might act by interfering with oxidative metabolism (Kennedy *et al.*, 1972).

An as yet unexplained finding made by Hinkley and Green (1971) was that exposure of rabbit vagus nerve to halothane in lower concentrations might cause an increased microtubular density; from a control level of 40/μm^2 to about 55/μm^2. This suggests that the agent may affect mechanisms of assembly/disassembly of the microtubules (cf. Fig. 9.7 and Section A5, Chapter 9).

3. Depolarization by High K$^+$ Concentrations

Further evidence that a normal membrane potential is not required for transport was seen when high levels of K$^+$ replaced Na+ in the *in vitro* incubation medium. High K$^+$ concentrations depolarize the fibers and it was reported that this had no affect on axoplasmic transport (Partlow *et al.*, 1972; Ochs, 1972a). However, those studies had been carried out using sheathed nerves *in vitro* and the low permeability of the perineurial sheath to ions vitiates the conclusion drawn on that basis. When the experiments were repeated using the desheathed nerve preparation (Chapter 8) with concentrations of K$^+$ up to 100 mM present in the media and where an almost complete depolarization would occur, axoplasmic transport was see to remain unaffected (Fig. 8.15). Those findings substantiated the proposition that the state of the membrane does not directly affect axoplasmic transport.

4. Increased Intra-axonal Na$^+$ Produced by Batrachotoxin (BTX)

The possibility was raised that a changed membrane permeablity leading to
an increased concentration of Na$^+$ in the axon could act to block transport.
This was first inferred from the block of transport seen when using ouabain
to block the sodium pump (Chapter 8). Further study of this phenomenon
using the desheathed nerve preparation showed that the block of axoplasmic
transport by ouabain was reduced when Na$^+$ was deleted from the *in vitro*
medium (Ochs, unpublished experiments). Further support for an increased
level of Na$^+$ within nerve fibers blocking transport came from the use of
batrachotoxin (BTX) to increase the level of Na$^+$ in the axons. This agent is
known to block excitability by opening and keeping open the Na$^+$ channels
in the membrane (Narahashi, Albuquerque, and Deguchi, 1971; Albuquerque and Daly, 1976). Batrachotoxin was found to be the most effective agent
so far known to block axoplasmic transport (Ochs and Worth, 1975). *In vitro*
it blocks transport in sheathed nerves in concentrations as low as 0.2 μM,
the time of block decreasing with higher concentrations of BTX. Boegman
and Albuquerque (1980) showed BTX to be very effective in blocking transport *in vivo* when it was used to perfuse the spinal cord. Retrograde transport
as shown using ^{125}I-NGF was also blocked by BTX at low concentrations in
in vivo experiments (Boegman and Riopelle, 1980).

In the earlier studies made with sheathed nerves, the BTX block appeared
to be independent of its action on Na$^+$ channels; an identical block of transport was seen when nerves were exposed *in vitro* to BTX in a NaCl medium
or a sucrose medium without Na$^+$ present (Ochs and Worth, 1975). However,
in sheathed nerves there is a retention of Na$^+$ in the endoneural space of the
nerves (Chapter 8) which could enter the fibers treated with BTX. A re-
examination of this point was made using the desheathed nerve preparation
(Chan *et al.*, 1980; Worth and Ochs, in preparation). In desheathed nerves
BTX blocked transport at even lower concentrations, 17–190 nM. Over this
range of BTX concentrations, the block produced by BTX depended on the
presence of Na$^+$ in the medium. Transport was not blocked when the nerves
were placed in a low Na$^+$ or a Na$^+$-free sucrose medium and exposed to
BTX over that same range of concentrations (Fig. 12.9).

Those results show that BTX blocks transport by bringing about an increased intra-axonal concentration of Na$^+$. The block of particulate movement by BTX observed microscopically in neuroblastoma cells, was also
found to depend on Na$^+$ in the medium with a block due to an increased
intracellular Na$^+$ concentration (Forman and Shain, 1981).

When much higher concentrations of BTX were used, the block of transport occurs more quickly along with a small reduction of $\sim$P levels, a finding
suggesting that BTX might have in addition an internal action (Ochs and
Worth, 1975). The effect, on $\sim$P levels in itself was not large enough to
account for the block of transport, but indicated an internal action of BTX.

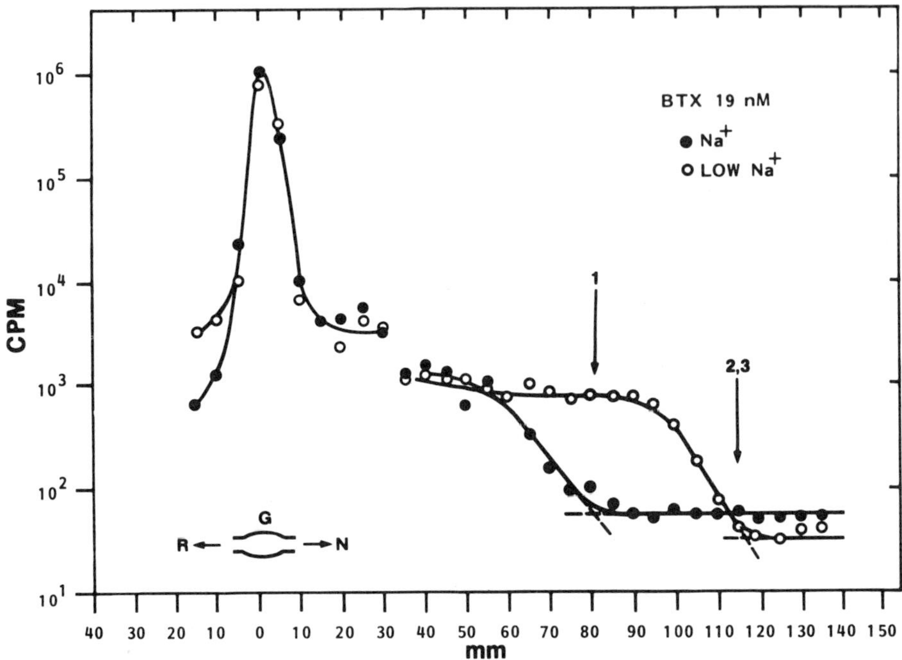

Fig. 12.9 Block of axoplasmic transport *in vitro* by BTX. After ^{3}H-leucine injection of the L7 dorsal root ganglion of the cat and 2 hr of *in vivo* downflow, the nerves were removed and desheathed. One nerve (●) was exposed to 19 nM of BTX *in vitro*, with the usual high level of Na$^+$ present in the medium, and the front of the crest (arrow 1) showed a block. The other desheathed nerve (○) was exposed to 19 nM of BTX with a low level of Na$^+$ present (10 nM). Its front moved to a position (arrow 2) close to that expected of normal downflow (arrow 3), indicating no block. From Worth and Ochs (unpublished experiments).

Evidence for a direct internal action of BTX on the transport mechanism was given by Kumara-Siri (1979), who injected BTX intraneurally and found a block of transport with little change of membrane potential.

Calcium protects against the blocking action of BTX on transport. In earlier studies made using sheathed nerves (Ochs and Worth, 1975), this had been observed only when high levels of Ca^{2+} were present in the *in vitro* medium because of the low permeability of the perineurial sheath to Ca^{2+}. In the desheathed nerve preparation much lower concentrations of Ca^{2+}, of as little as 5-10 mM, were seen able to protect against BTX (Worth and Ochs, in preparation). Possibly Ca^{2+} protects against BTX by preventing its binding to or near the Na$^+$ channel in the membrane. A relation of BTX to the sodium channel action was indicated by the finding that tetrodotoxin (TTX) interfered with the blocking effect of BTX on transport (Ochs and Worth, 1975).

Additional support for the concept of a deleterious effect of increased intra-axonal Na$^+$ was gained through experiments with the alkaloid veratridine which acts to increase the permeability of the membrane to Na$^+$. In low concentrations, of the order of 10 μM, veratridine blocks transport in the desheathed nerve preparations and this effect was diminished by removing Na$^+$ from the medium (Ochs and Gaziri, unpublished experiments).

The increased concentration of Na$^+$ in the axon may block through its action on the mitochondria causing higher levels of free Ca^{2+} to appear in the axon. Exposure of isolated mitochondria to levels of Na$^+$ as low as 10 mM may give rise to a release of Ca^{2+} from the organelle (Carafoli and Crompton, 1976; 1978). In turn, the Ca^{2+} may effect a block of transport (cf. Section H2 below).

G.   INCREASED MEMBRANE ACTIVITY AND TRANSPORT

An early study showed that a high rate of stimulation of nerves *in vitro* reduced the rate of fast axoplasmic transport (Ochs and Smith, 1971a). On the other hand, Jankowska *et al.* (1969) reported little apparent effect of a high rate stimulation of nerves *in vitro* on AChE translocation taken as the measure of fast transport. Subsequently, it was shown that a maximal stimulation of nerves at the high rate of 350 pps for 4 hr reduces the rate of axoplasmic transport by some 17% (Worth and Ochs, 1976). An example of the effect of such stimulation is indicated in Fig. 12.10.

With increased duration of stimulation, a corresponding decrease in transport was found. One possible explanation for this phenomenon is that an augmented amount of Na$^+$ in the fibers resulting from activity would cause an increased activity of the sodium pump and in turn lead to a decrease in the amount of ATP available to the transport mechanism. Assay of $\sim$P in nerves (Chapter 7) stimulated at these high rates, however, showed little depletion of $\sim$P. Similarly, little change of $\sim$P in frog nerves subjected to long-maintained repetitive stimulation was reported by Okada and McDougal (1971). Thus, we may consider that the effect of a high rate of increased activity to decrease transport may be due to the effect of increased intra-axonal Na$^+$, the effect of which may be to increase Ca^{2+} as has been discussed in the preceeding section.

H.   NEUROTOXIC EFFECTS RELATED TO ALTERED Ca^{2+} REGULATION

1.   Trifluoperizine (TFP)

The requirement of Ca^{2+} for transport, and the Ca^{2+}-regulatory mechanisms present in the axons to maintain a normal level of Ca^{2+} (Chapter 8), have been discussed in relation to the transport filament model in Chapter 10. As

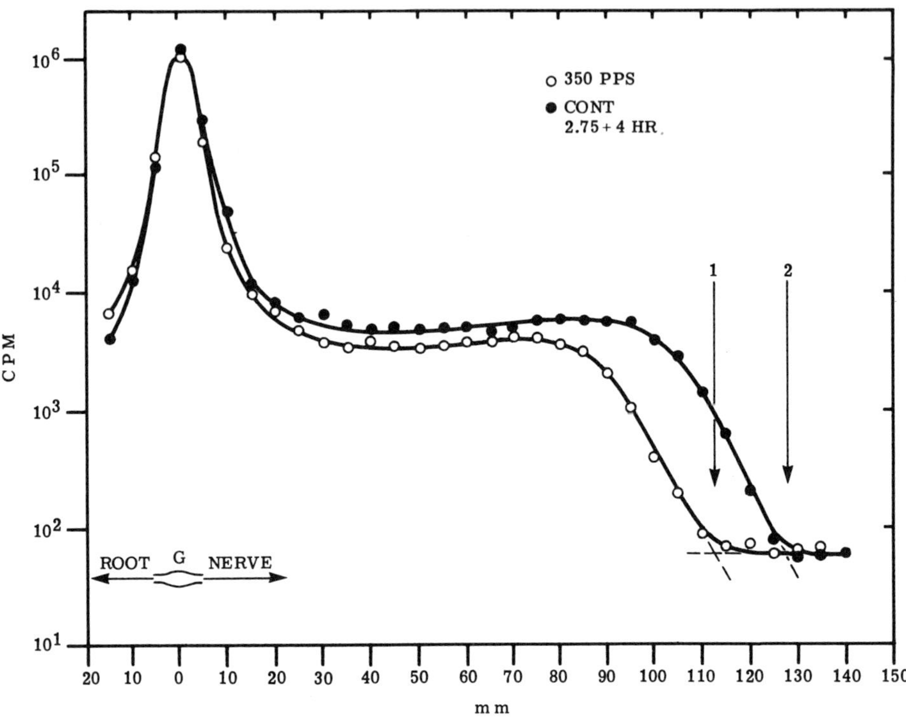

Fig. 12.10 Effect of high rates of stimulation on transport. Nerve removed after a 2-hr downflow of labeled protein in the animal was maximally stimulated in a chamber at 350 pps for 4 hr *in vitro* (cf. Fig. 7.16). The front of transport in the stimulated nerve (arrow 1) falls short of that of the control nerve (arrow 2). From Worth and Ochs (unpublished experiments).

was noted, trifluoperizine (TFP) binds to calmodulin preventing it from activating enzymes, in particular the Ca–Mg ATPase located on the side-arms of the microtubules hydrolyzing ATP to provide the energy needed to move the transport filaments. Calmodulin could be present on or be part of the transport filaments, or be closely associated with or be part of the Ca–Mg ATPase on the side-arms (Ochs and Iqbal, 1980). In any case, as would be expected of its effect on calmodulin, TFP blocks axoplasmic transport in the desheathed nerve preparation as shown in Fig. 12.11.

The effective concentration required to block was relatively high for the desheathed nerve preparation, suggesting a low membrane permeability of the agent.

Trifluoperizine is a member of the chlorpromazine class of agents and chlorpromazine had been reported to also block axoplasmic transport (Edström, Hansson, and Norström, 1973a,b), an action which at the time could not be accounted for. These agents are likely to have the same action as TFP and act by blocking calmodulin (cf. Kanje *et al.*, 1981).

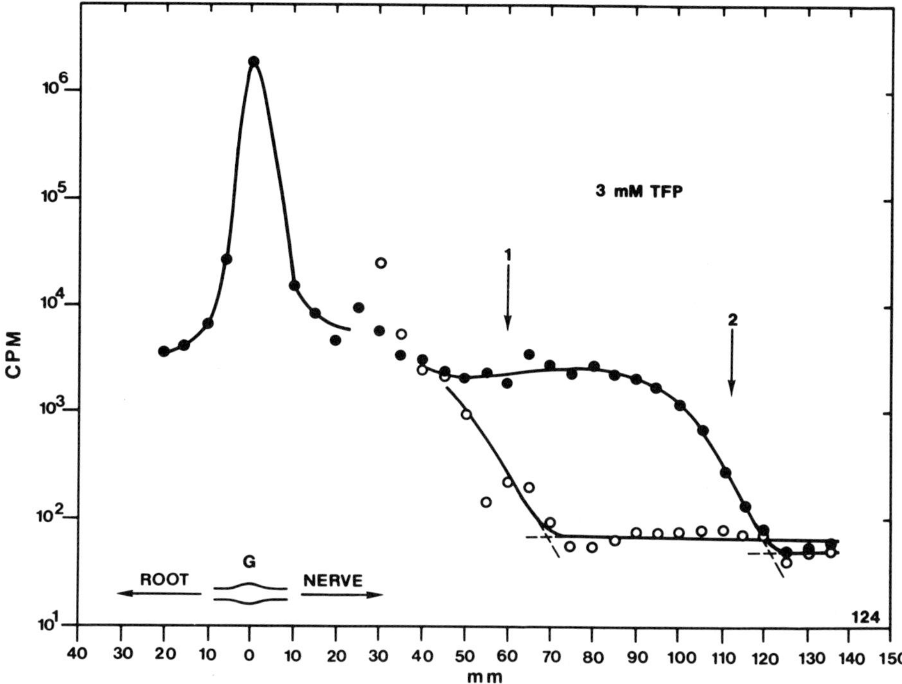

Fig. 12.11 The effect of trifluoperizine (TFP) to block transport. The nerve in which labeled proteins have been transported *in vivo* is removed from the animal and the desheathed nerve preparation is placed in a medium containing 3 mM of TFP. The desheathed branch (○) shows a block of transport (arrow 1) with little effect seen in the sheathed branch (●) (arrow 2). From Ochs and Iqbal (unpublished experiments).

2. Transport Block by High Intra-Axonal Ca^{2+}

As has been noted in Chapter 8, exposure of desheathed nerves to a high level of Ca^{2+} in the incubation medium causes a block of transport. With high levels of free Ca^{2+} in the axon a block of transport occurs and if sufficiently high and present for long enough times, gives rise to neuropathological changes as a result of proteolysis. Such changes can come about not only as a result of exposure to high external concentrations of Ca^{2+}, but also as a result of the release of Ca^{2+} from storage sites within the axon under certain conditions.

When the level of free Ca^{2+} in the axon rises too high, the neuropathological changes seen include a disruption of mitochondria, ER, and the disassembly of microtubules and neurofilaments. Such morphological effects were seen by Schlaepfer (1971) in the fibers of rat nerves cut transversely at small lengths of 0.3 to 0.5 mm to allow a ready access of Ca^{2+} to their interiors. Those fibers, when exposed to an incubation medium containing as little as 1 mM Ca^{2+}, showed those morphological changes. In the desheathed nerve preparation the intact axolemma allows very much less Ca^{2+} to gain entry to the axons and much higher levels of Ca^{2+} need to be present in the medium, concentrations of 60 mM, present for over 4–6 hr before similar axonal changes are seen (Ochs *et al.*, 1978; Chan *et al.*, 1980a; Ochs and Jersild, unpublished observations). When the Ca^{2+}-regulatory mechanisms are overcome, proteolytic enzymes are activated (cf. Schlaepfer, 1971), or calmodulin-activated enzymes brought to threshold for the disruption of axonal organelles.

Transport is blocked in nerves exposed to high levels of Ca^{2+} well before that point is reached, as shown by the reversal of the block when the nerves are placed back in a lower Ca^{2+} medium (Chapter 8). This finding indicates that a high level of free Ca^{2+} may produce a block of transport by some other means before the irreversible disruption of structural organelles takes place. The Ca–Mg ATPase associated with tubulin shows a maximal activity in a Ca^{2+} range of 10^{-6}–10^{-7} M (Ochs and Iqbal, 1980) with a lower activity at 10^{-3} M Ca^{2+}. Possibly at the higher levels of Ca^{2+} transport may be blocked by a depression of Ca–Mg ATPase activity.

Brady *et al.* (1980) claimed that transport can continue in a high Ca^{2+} medium after a disruption of microtubules is produced. Those experiments were done in nerves which remained sheathed when exposed to high Ca^{2+}. As described in Chapter 8, the sheath is an effective barrier to Ca^{2+}. Kanje, Edström and Ekström (1982) were also unable to confirm the claim of Brady *et al.* (1980) that high concentrations of Ca^{2+} destroys microtubules without impairing axoplasmic transport in studies using sheathed frog nerves. These investigators pointed out that the permeability of the perineurium to Ca^{2+} in this species while low, is greater than that of the rat or cat. In any case, in the frog nerve they found transport to be blocked only when most of the microtubules in the fibers were destroyed. R. S. Smith studying rat sciatic nerve found the fibers not to be affected by an incubation of the sheathed nerves in 75 mM Ca^{2+} for 6 hr (quoted by Ellisman, 1981). In this species, as in the cat, the perineurial permeability to Ca^{2+} is relatively low. Further, in the desheathed preparation, even after exposure to 60 mM Ca^{2+} for 4 hr, a time beyond that effective in blocking transport (Chan *et al.*, 1980), a good proportion, perhaps 70–80% of the fibers, were seen to contain microtubules in near-normal densities (Ochs and Jersild, unpublished observations). These nerves can recover when replaced in a low Ca^{2+} medium after 3 hr of exposure to such high levels of Ca^{2+} (Fig. 8.10). After a longer-lasting exposure of desheathed nerves to 60 mM Ca^{2+} for 6 hr, however, few fibers retaining microtubules are found. The need for microtubules in nerve fibers for transport to continue has been supported by these and other observations (cf. Brimijoin, 1982b).

Increased levels of free Ca^{2+} can appear in the axoplasm of axons secondarily to its release from the mitochondria and/or the ER, organelles in which relatively high levels of Ca^{2+} are sequestered as part of their Ca^{2+}-regulatory function (Chapters 8 and 10). A prolonged interference with oxidative phosphorylation such as, for example, occurs during a prolonged period of anoxia, could cause a release of Ca^{2+} from the mitochondria. This could explain the "dissociation" in the recovery of axoplasmic transport and excitability following the longer periods of anoxia (Chapter 7). Friede and Ho (1977) found that a prolonged exposure of nerve pieces *in vitro* to cyanide (CN) caused not only the expected block of transport as measured by accumulation of mitochondria at the cut ends, but a marked decrease in microtubular density as well. In those studies the long-maintained anoxia or some non-specific action of CN might cause the release of excess Ca^{2+} from the mitochondria in amounts sufficient to activate mechanisms leading to a disassembly of microtubules. Also to be considered is the increased pCO_2 during prolonged anoxia and pH changes which might also affect Ca^{2+} regulation (Baker and Honerjager 1978). Possibly some of the neurotoxic agents noted above, and the disease entities to be discussed in the following section, might involve such Ca^{2+} releases from sequestration sites in neurons.

I. NEUROPATHOLOGICAL STATES AND ANIMAL MODELS WITH TRANSPORT DEFECTS

Dyck (1975) grouped neuropathological entities as (a) those giving rise to Wallerian degeneration, (b) primary segmental demyelination, and (c) axonal atrophy. Another classification of neuropathological conditions of peripheral nerve might be (a) those in which synthesis in the cell bodies is altered in various ways with changes related to one or other neural function, including an indirect effect on Schwann cells, (b) those in which the transport mechanism *per se* is affected with a profound disruption of form and function, i.e., Wallerian degeneration, and (c) primary Schwann cell defects. At present, studies of transport in relation to neurological defects have been few and essentially exploratory in nature. Two patients with the Charcot–Marie–Tooth type of distal peripheral neuropathy and one case of hypertrophic neuropathy of the Dejerine–Sottas type were shown to have a decreased accumulation of DBH in nerves taken for biopsy (Brimijoin *et al.*, 1973). It is not known if the decrease in accumulation seen represented a defect of the transport mechanism or an altered production of DBH.

In the class of degenerative motor neuron diseases characterized as the "dying-back" type (cf. Section E above), the spasticity of varying severity, different ages of onset, and the different muscles afflicted, tends to defy a valid and completely agreed upon classification. Spencer and Schaumberg (1976) consider three possible etiological mechanisms: (a) a general metabolic defect of the perikaryon leading to a deficiency in the amount or an alteration in types of materials transported, (b) a defect of the transport mechanism itself, and (c) a direct toxic action on the distal part of the axon. Spencer *et al.* (1979) and Sabri, Moore, and Spencer (1979) proposed as a common feature in the dying-back pattern produced by a number of agents an interference with the supply of glycolytic enzymes. The resulting failure of an adequate supply of ATP to the transport system would impair transport (Chapter 7). This would lead to a degeneration when needed components are no longer available to maintain the nerve, first more distally in the nerve, and then at successively more proximal levels. Another possible explanation for dying-back noted above with respect to the action of BPAU is that a failure of turn-around might lead to a "plugging" of the distal part of the axon and a loss of supply of needed components (Brimijoin, 1981).

Diabetes is a metabolic disease with neuropathy as one of its sequelae (Gregersen, 1967). Schmidt *et al.* (1975) reported a decreased accumulation of AChE and choline acetylase in the nerves of rats made diabetic by injection of streptozotocin, an indication of a defect of transport. In studies where ^{3}H-leucine or ^{14}C-glucosamine was injected as a precursor to assess axoplasmic transport in streptozotocin-diabetic rats, Sidenius and Jakobsen (1979) found no apparent change in the rate of anterograde fast transport. Controls and the streptozotocin-diabetic rats alike showed rates of transport which were calculated to be about 380 mm/day, a figure not significantly different from the 410 mm/day rate usually found in mammalian nerves (Chapter 2). There was, however, a small increased lag time seen between the injection of the precursor ^{3}H-leucine and the appearance of fast-transported radioactivity in the diabetic animals, as well as a 40% reduction in the amount transported. Glycoproteins labeled with ^{14}C-glucosamine on the other hand were not much affected. The authors consider these changes in the amount transported to be due to a reduction in axon caliber, a change associated with the reduction in the conduction velocity of the peripheral nerves of the streptozotocin-diabetic rats. However, axonal changes are likely to be secondary to metabolic shifts in the nerve cell bodies, in turn affecting the amount of transported components. The decreased axonal diameter in itself would not likely affect the rate of fast axoplasmic transport (Chapter 2).

A somewhat different result was found by Bisby (1980a). Two to five weeks after injecting streptozotocin, rats showed no difference in their fast rate of transport, either *in vitro* or *in vivo* as measured by the wave fronts of ^{3}H-protein. Slow transport *in vivo* or the regeneration rate of the fibers was also not changed in the streptozotocin-diabetic animals. However, a longer delay in the onset of nerve regeneration was seen after a crush (3.05 ± 0.5 vs 1.71 ± 0.24 days), was seen. This could be due to an effect on the process of turn-around of materials at the distal ligation for retrograde transport (Chapter 6), or a diminished response of the cell to the ascending trigger causing regenerative changes to begin. Sidenius and Jakobsen (1980a) have reported a reduction in the amount of materials accumulated after a turn-around and rectrograde transport in streptozotocin-diabetic rats using the technique described by Bisby. An effect on slow transport was also found by Jakobsen and Sidenius (1980).

The metabolic disturbance produced by excess galactose results in a neurotoxic effect in chicks, a lag in outflow from the nerve cell bodies, and a reduced amount of material fast-transported (Knull, Lobert, and Wells, 1974). A defect in retrograde transport was observed by Sidenius and Jakobsen (1980b) in galactose-fed rats and the authors consider that this agent produces a picture similar to that of diabetes.

A disease which could be due to a failure of proper synthesis in the cell bodies of components needed by their axons is amyotrophic lateral sclerosis (ALS). This disease, established by Gowers in the last quarter of the 19th century, is seen as a progressive muscular atrophy, though spasticity may be the most visible sign appearing if the lateral columns are affected predominantly. The recognition of the disease as a primary abnormality of motoneurons at various levels of the neuraxis has led to a search for an animal model of the disease (Andrews and Andrews, 1976). The murine paralytic disease seen in an unbred strain of mice known as "wobbler" was investigated for a possible defect of transport, insofar as the affected mice show spontaneous lower motor degeneration (Andrews et al., 1974). The results so far are conflicting. Bird et al. (1971) have reported some defect in slow transport while no changes at all were seen in wobbler nerves by Bradley and Jaros (1973). Hopefully, other animal models may be developed (Andrews, Johnson, and Brazier, 1976) which will allow a further examination of this disease from the point of view of a defect in transport.

The clinical picture of muscular dystrophy could be brought about by the lack of some specific component carried down the nerve into the muscle (McComas et al. 1977), possibly a trophic substance controlling the synthesis of some component(s) by the muscle (Chapter 14), or an involvement of the transport mechanism itself. The latter possibility was suggested by reports of a decreased rate of transport in dystrophic mice (Komiya and Austin, 1974; Tang et al., 1974; Yu et al., 1974). However, the decrease found was of a superfast rate of transport which may not represent a true transport (Chapter 2). Bradley and Jaros (1973) reported no difference in the rate of transport in the nerves of dystrophic animals as compared to controls. Though a twofold increased rate of transport was claimed in an ischemic model of transport (Wood and Boegman, 1975) and a still greater increase in the dystrophic-like condition produced by pargyline (Boegman, Wood, and Pinaud, 1975), the rate in the nerves of pargyline-treated animals was determined to be the same as that in normal animals by Ranish and Dettbarn (1977). Fast axoplasmic transport in the dystrophic chickens investigated by injecting ^{3}H-leucine or ^{3}H-fucose into their spinal cords, revealed no difference in the rate or amount of labeled materials transported in their radial nerves as compared to normal animals (Stromska and Ochs, 1977; Stromska, Ochs, and Muller, 1981). No difference in the rate was found in the dystrophic hamster compared to normals (Boegman and Wood, 1981) nor in the retrograde transport of HRP in the nerves of dystrophic animals as compared to control animals (DeSantis et al., 1977). Thus, it appears from these studies that the mechanism of axoplasmic transport itself is unaffected in muscular dystrophy.

This leaves as a remaining possibility a defect in the synthesis of some needed component transported in the nerve. The reduction in the transport of choline acetyltransferase in the dystrophic mice reported by Jablecki and Brimijoin (1974), points in that direction. A defect in the supply of a trophic substance required for normal muscle function (Chapter 14) could account for the diseased state. While experiments involving limb-bud transplantation (Linkhart *et al.*, 1975) and parabiosis did not support this hypothesis (Douglas, 1975; Law *et al.*, 1976), Hironaka and Miyata (1975) gave evidence for a trophic component in dystrophy. The subject of neurotrophic substances in the nerve affecting muscle is taken up in Chapter 14.

Experimental allergic neuritis (EAN) is considered an animal model for Guillain-Barré Syndrome. While demyelination is the most prominent defect, some axonal involvement also occurs. Rabbits injected with peripheral nerve homogenate to induce EAN showed an effect on ACHe transport studied with a double-ligation technique (Oikarinen, Molnar, and Riekkinen, 1980). The anterograde transport rate of 413 mm/day and its mobile portion (6%) found in controls (cf. Section A, Chapter 4), were not changed over a period of 8–14 days after administration of EAN. The amount of retrograde transport which was normally 4.7% was decreased as a result of EAN to 3.2%, a highly significant change. The neuronal change in EAN and presumably Guillain-Barré Syndrome appears to entail some change in protein synthesis in the cell body and apparently in the processes involved in the turn-around of materials in the nerve fibers (cf. Section B4, Chapter 6).

J. TRANSPORT IN OLD ANIMALS

Neurons as a cell type are unique in that after early development no further cell divisions occur in the mammal. The neurons, therefore, must serve for the life of the organism. It has been speculated that changes in neurons during senescence are the cause of the regressive changes seen in the body tissues, for example, the decline in neuromuscular transmission and muscle function (Gutmann and Hanzlíková, 1972). Some of the circumstantial evidence for this view is that the atrophy of muscles (Yiengst, Barrows, and Shock, 1959), and the decrease of muscle strength (Norris and Shock, 1960), are not completely reversed by exercise (Gutmann and Hanzlíková, 1972). A possible change in the rate of fast axoplasmic transport in the nerves of a group of aged cats and dogs was looked for, but no statistically significant changes in comparison to young adults were found (Ochs, 1973). Two Dalmatian dogs in that group had a genetic defect analogous to premature aging, with changes mimicking the age-related ceriod-lipofuscinosis group of diseases (Koppang, 1966). The corrected rates of axoplasmic transport in the sciatic nerves of those animals were, however, similar to those of normal young adult animals.

Obtaining an adequate population of cats and dogs of advanced age to carry out such studies is difficult at best. Rats have a life span of about 3 years with a maximum of about 3.5 years (Gutmann and Hanzlíková, 1972) and are ideal for this purpose. In a recent study of axoplasmic transport in rats up to 3.5 years old, no statistically significant difference in the rate of fast transport was found compared to young adult rats except in their last half year when a slight decrease in rate was found (Stromska and Ochs, 1982). Concomitant diseases in these aged animals need to be ruled out before concluding that there is an altered transport mechanism at that age. The reported decreases in the rate of slow transport as a function of age (Komiya, 1980) could, on the basis of the unitary hypothesis (Chapter 11), be interpreted as a decrease in the amount of materials transported in the older animal, rather than a change in the transport mechanism itself. A study of possible changes in the composition of neuronal materials transported in the older animals may lead to a better understanding of the aging process.

13

Nerve Terminal Processes In Peripheral And Hypothalamo-Neurohypophysial Systems

The transport of specific neurotransmitters and transmitter-related enzymes in adrenergic and cholinergic nerves has been discussed in Chapter 4. In this chapter we will relate axoplasmic transport to the mechanism whereby vesicles containing neurotransmitter come to release their contents from the nerve terminals. The uptake of neurotransmitter or neurotransmitter precursors into the terminals, with the synthesis of neurotransmitter and its repackaging into vesicles for subsequent re-release from the nerve terminals, is a related topic.

Comparable to the release of neurotransmitters is the release of neurohormones into the circulation to act at distant cellular receptor sites. Transport in the axons of the hypothalamo-neurohypophysial system (HNS), and the release of its neurohormones from its terminals, will be described in the last part of this chapter. As will be shown, transport in the HNS neurons shares the same properties found in other neurons.

A. NERVE TERMINALS

1. Morphology of the Nerve Terminal in Relation to Transport

In the most widely held theory of neurotransmission, that of quantal release (Katz, 1962, and cf. Eccles, 1964, 1977), vesicles containing neurotransmitter release their contents from the nerve terminals by exocytosis. When the nerve is depolarized, the vesicles move toward and attach to the inner surface of the nerve terminal membrane, the two membranes merge, and the open vesicles release their contents into the synaptic cleft. Essentially the same mechanism is found in peripheral neurons and in the neurons of the HNS system, as shown in Fig. 13.1.

After exocytosis, the vesicles are reformed. This may occur by various

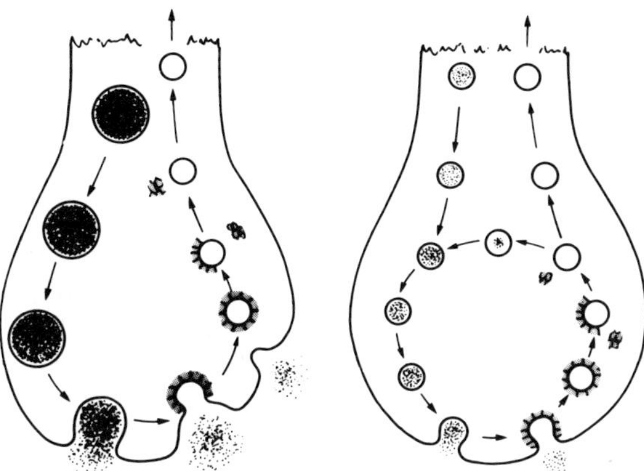

Fig. 13.1 Exocytosis and membrane recapture mechanisms. Comparison of (a) a neurosecretory nerve terminal and (b) an adrenergic or cholinergic neuron terminal is shown with respect to exocytosis and membrane recapture. The possibility that the smaller vesicles are initially derived from larger vesicles transported down from the perikaryon is indicated. From Douglas (1973).

routes: (a) the partially opened vesicle may reclose and move to the interior of the nerve terminal, (b) a new vesicle may reform from the original vesicle membrane materials, or (c) membrane materials of the synaptic terminal may be used to reform new vesicles. In this reformation of vesicles, or endocytosis, the special protein clathrin appears to play a key role (Schook *et al.*, 1979; Goldstein *et al.*, 1979). The clathrin protein network layers onto the inner surface of the membrane and then by its basket-like folding results in the formation of vesicles coated by the clathrin (Fig. 13.1). The clathrin-coated vesicles can interact with F-actin or α-actin (Schook *et al.*, 1979) and in the presence of Ca^{2+} with calmodulin (Linden *et al.*, 1981). No doubt these interactions are important in the mechanical events required in the folding of clathrin in the formation of the vesicles in the course of the endocytotic process occurring in the terminals, a concept depicted in the model of Kanaseki and Kadota (1969). Using centrifuged contents of the synaptosomes, they clearly demonstrated that a basket-like structure covers the vesicles obtained from the terminals and inferred that the folding of the protein basket material on the membrane forms the coated vesicle. Clathrin has been reported to be slow-transported in SCb in nerve (Garner and Lasek, 1981), and an interaction with the cytoskeleton was proposed. In any case, we would expect that clathrin would be carried into the nerve terminals where it is required for the endocytotic process. A portion of the newly reformed vesicles which are interiorized in the terminal are restocked with neurotransmitter. The source of that neurotransmitter may be a synthesis from precursors taken up into the terminal, or transmitter which has been previously released. The restocked vesicles can undergo a further release of neurotransmitter with several cycles of release, repackaging, and release occurring until, with the irreversible loss of some components from the vesicles, the restocking of vesicles with neurotransmitter fails. The unuseable vesicles or parts of vesicles are carried back to the cell body by retrograde transport, presumably for catabolism and the reuse of their constituent components.

The process of exocytosis is not unique to the release of neurotransmitters from nerve terminals. Exocytosis not associated with neurotransmitter release takes place in both the pre-and post-synaptic neuronal membranes of the toad spinal ganglion cell (Rosenblueth and Wissig, 1964). It is also seen in the terminals of sensory nerve fibers and in non-neural cells.

Considering the transport of vesicles by the transport filaments (Chapter 10) down into the nerve terminals, the question arises as to how the vesicles are then carried from the microtubules to the release sites at the terminal membrane surface. The relative closeness of vesicle transport in the axon to the release site was indicated by the EM pictures of Smith *et al.* (1970) where vesicles are pictured associated with the microtubules in the nerve fibers of the lamprey larvae near a synapse (Fig. 13.2).

The nearness of the vesicles a short distance from a synaptic site on the axolemma suggests a relationship to neurotransmission (Smith, 1971; Smith *et al.*, 1970). A connecting or guiding structure, however, is not apparent in those pictures, nor are they usually seen within the presynaptic nerve endings in EM sections taken from brain and spinal cord. By the addition of albumin to fixing solutions in the course of preparing synaptosomes for EM observation, Gray (1976, 1978) was able to better preserve microtubules within synaptosomal nerve endings. With this method microtubules were seen to curve down to, and in some cases insert into, dense areas of the inner side of the presynaptic membrane to which synaptic vesicles are apposed, and which appear to represent release sites (Fig. 13.3).

The inner surface of the presynaptic nerve membrane may also show ridges which appear to channel the vesicles to specific release sites on the inside of the presynaptic membrane (Akert *et al.*, 1972). In other cases the microtubules were seen to sweep around the whole of the inside of the synaptosome with synaptic vesicles present all along their course (Bird, 1976).

An association of vesicles with microtubules ending in proximity to the presynaptic membrane was found in the nerve terminal of the neuromuscular junction (Fig. 13.4).

The sites at which the microtubules terminate on the inner face of the presynaptic membrane have a definite relationship to the post-junctional folds of the muscle membrane seen just under the nerve terminal. At the active sites, the neurotransmitter receptors are most dense where there is a functional linkage of these elements in the nerve and muscle. This morphological feature could subserve the exchange of materials between the corresponding nerve and muscle elements where such an apposition exists.

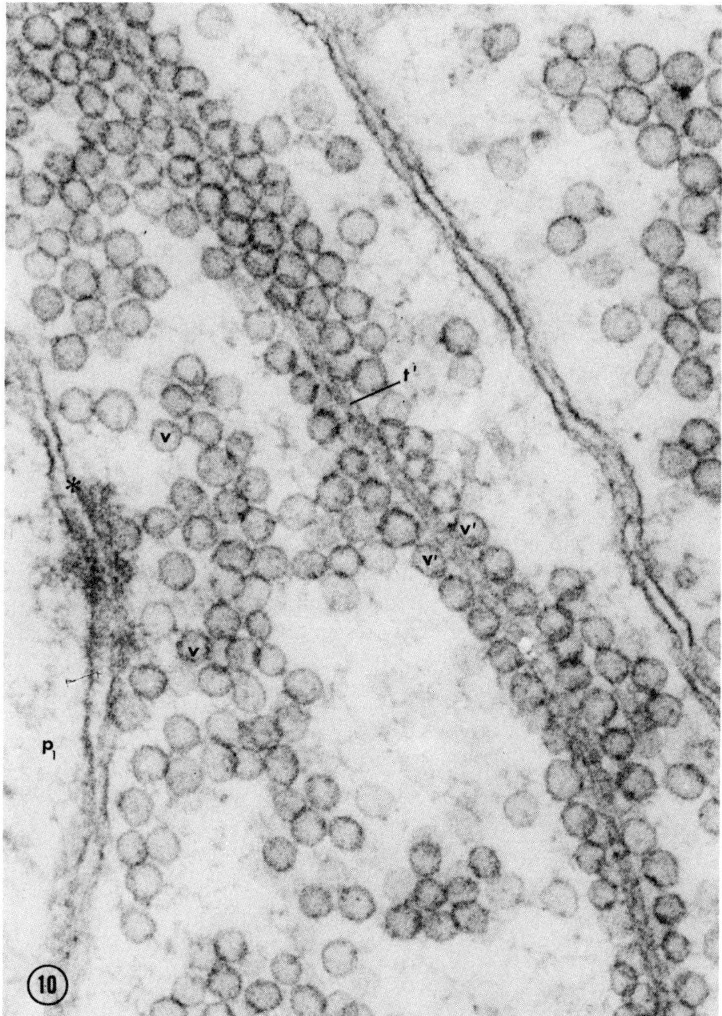

Fig. 13.2 Association of vesicles with microtubules. A section from the spinal cord of the lamprey larva shows an axon with synaptic vesicles (V′) lined up along the microtubules (t′). Vesicles (V) are also seen a short distance from microtubules near the axolemma at a synaptic site (P₁). From D. S. Smith *et al.* (1970).

Microtubules serve to bring vesicles down to their release site and for the retrograde transport of various components (Chapter 3) on the basis of the transport filament model (Chapter 10). It is also possible that they insert into the membrane for direct transport to the site of exocytosis. Alternatively, some other guiding element acts to bring the vesicles and other components from the transport system to the membrane (Figs. 13.5A and 13.5B). Thus, transport filaments carrying vesicles (and trophic components) could, after

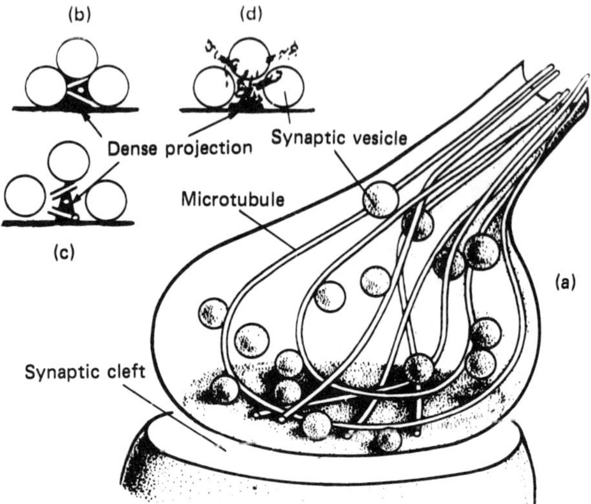

Fig. 13.3 Microtubules in nerve terminals and associated vesicles. (a) The schematization derived from EM sections shows the association of synaptic vesicles to microtubules and to the inner face of the membrane. Loops and a termination of microtubules in the membrane near dense projections at the inner face are also shown with vesicles (b, c, d). From Gray (1976).

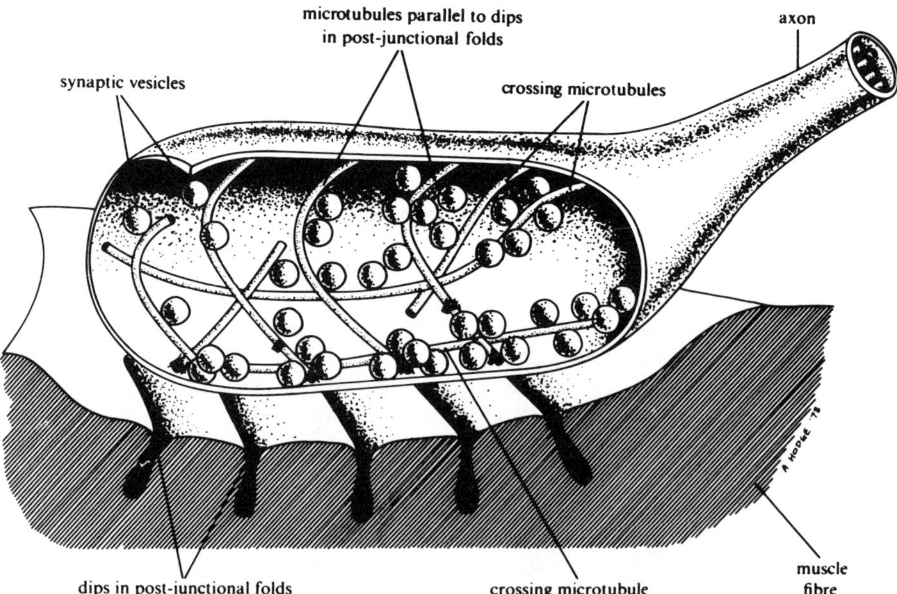

Fig. 13.4 Microtubules in the nerve terminal of the neuromuscular junction. The microtubules are seen to sweep around the inside of the motoneuron terminal to end at sites in the membrane at a position over the folds of the post-junctional membrane of the muscle. Synaptic vesicles are seen along the microtubules and in clusters over the region of postsynaptic folds. From Gray (1978).

leaving the anterogradely directed set of microtubules, divest themselves of vesicles which then find their way to the membrane or onto the ascending microtubules for retrograde transport. Microfilaments may act as an intervening guiding element, one of the possibilities indicated schematically in Figs. 13.5A and 13.5B.

It is interesting with regard to these possibilities that a curving of the

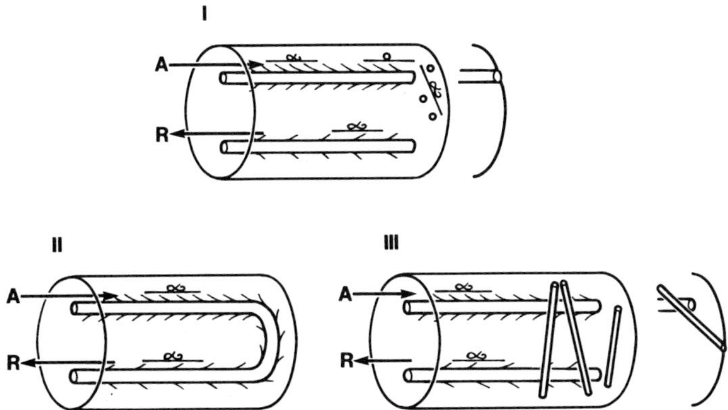

Fig. 13.5A Microtubules and hypothetical transport at the terminals. (I) Movement of transport filaments along one set of microtubules subserving anterograde (A), and another set retrograde (R), transport are shown with transport filaments leaving one set to find their way onto the other. (II) A looping with an ascent of filament on microtubules is indicated. (III) Bridging elements which could be microfilaments are shown acting as guides between the two sets of microtubules or the membrane. From Gray (1976).

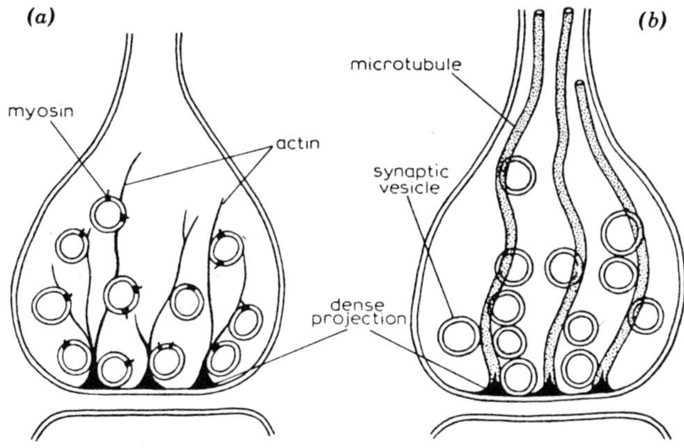

Fig. 13.5B Hypothesis of transport elements to the terminal membrane. In the schematization shown at the left, (a) actin is guiding myosin-coated vesicles to the terminal membrane. In (b) microtubules are shown acting as the guides for the synaptic vesicles to the membrane. From Gray (1976).

microtubules within nerve terminals has been found in a substantial number of cases by Chan and Bunt (1978). The close relation between the mitochondria and microtubules noted in Chapter 10, was also seen in the synaptosomes where the mitochondria appear to be bent into a horseshoe shape as they followed the contour of the looped microtubules (Fig. 13.6).

Vinblastine induces crystalloid precipitates in isolated synaptosomes (Kadota, Kadota, and Gray, 1976). This observation is in accord with other direct morphological evidence that microtubules are present, but much more needs to be known of the properties of microtubules in the nerve terminals, particularly their lability in the process of fixation. Neurofilaments may also be seen in the synaptic terminals of cold-adapted goldfish (Roots and Bondar, 1977). In the Rhesus monkey, the enucleation of the eye is followed by swollen boutons in the lateral geniculate containing a high density of neurofilaments (Ghetti, Haroupian, and Wiśniewski, 1975). The optic nerves survive for a somewhat longer time, up to 10 days following severance from their somas, compared to peripheral neurons (Chapter 6). This would allow the neurofilament subunits to be carried to the nerve terminals to there assemble and give rise to the increase in neurofilaments seen.

The lability of microtubules and neurofilaments appears to apply as well to other organelles in the nerve terminals. Gray (1976) had at one point considered the possibility that the vesicles often seen associated with microtubules might represent a breakdown of smooth ER in the course of tissue preparation for EM. However, this cannot explain the similar close association of DCVs to the microtubules in adrenergic fibers (Chapter 4) and the DCVs in the hypothalamo-neurohypophysial axons and terminals (see Part B below). We may consider as a possibility that the smooth ER structures seen in the terminals under certain circumstances might represent a merging of vesicles, a phenomenon augmented by some of the methods used for tissue preparation.

2. Ca^{2+} and the Mechanism of Release

Calcium is a necessary element in the release of neurotransmitter from the nerve terminals as formalized in the "calcium hypothesis" (Katz and Miledi,

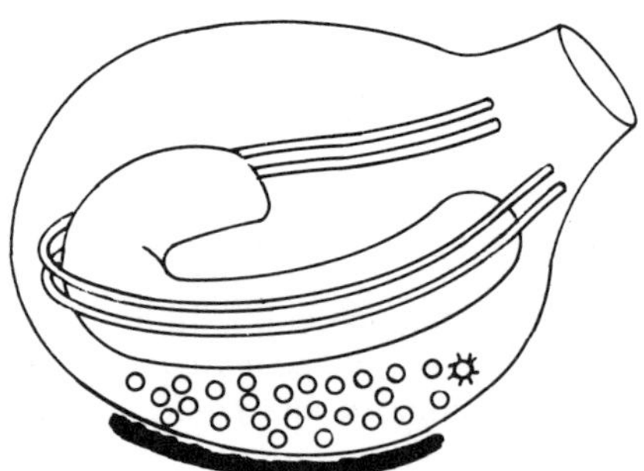

Fig. 13.6 Relation of mitochondria to microtubules in a nerve terminal. The mitochondrion is seen to be curved into a horseshoe-shaped structure as it apparently conforms to the course of microtubules in the terminal. Vesicles at synaptic sites in the membrane are also shown. From Chan and Bunt (1978).

(1967; 1969; Hubbard, 1970). When the action potential sweeps over the nerve terminal to depolarize it, there is an associated increase in the Ca^{2+} permeability of the membrane. The entry of Ca^{2+} into the terminal somehow causes the vesicles to move toward and fuse with the presynaptic membrane, this followed by exocytosis and the release of vesicular contents into the synaptic cleft. The mediation of Ca^{2+} is essential, but how it brings about the merging of vesicles with the membrane and exocytosis is at present unknown (cf. Llinas and Heuser, 1977; Blaustein, 1979).

Possibly the entry of Ca^{2+} activates calmodulin in the nerve terminals and the activator in turn a protein kinase or other enzyme to bring about the exocytosis. A fairly large amount of calcium-binding protein (CaBP) has been found present in the nerve endings (Iqbal and Ochs, 1978), the CaBP most likely including the regulatory protein calmodulin (Iqbal and Ochs, 1979). Baker, Knight, and Whitaker (1980), using a system where the membrane has been made leaky by exposure to brief high-voltage discharges, have shown that exocytosis has a specific requirement for Mg-ATP and is activated by micromolar quantities of Ca^{2+}. Most intriguing is Baker's observation that the exocytosis is blocked by TFP, as would be expected of an intermediation of calmodulin (cf. Chapter 8).

Alternatively, most of the CaBP in the nerve terminals may be part of the mechanism of Ca^{2+} regulation acting to keep the level of Ca^{2+} low in the face of the Ca^{2+} entering the terminal as a result of activity. If Ca^{2+} were not well regulated, the concentration of free Ca^{2+} would rise to the point where release of neurotransmitter could no longer be triggered. The electron-dense particles found present in the nerve terminals with EM and Ca^{2+} precipitation methods (Blaustein, Ratzlaff, and Kendrick, 1978a; McGraw, Somlyo, and Blaustein, 1980), most likely represent sites of Ca^{2+} binding to CaBP, these similar to the oxalate or pyroantimonate electron-dense particles in the axon proper (Chapter 8). Mechanisms by which the level of Ca^{2+} within the nerve terminals are regulated are likely to be similar to those present in axons (Chapter 8), as schematized in Fig. 8.11.

A model advanced to explain the mechanism of exocytosis invokes an actomyosin-like process (Berl, Puszkin, and Nicklas, 1973). When an adequate level of Ca^{2+} is present, neurin (actin) present on the vesicle membrane and stenin (myosin) on the inner surface of the nerve terminal, meet at the inside of the terminal. The actomyosin so formed can then utilize the ~P of ATP to open up the vesicular membrane and allow the release of its content of neurotransmitter (Fig. 13.7).

Further information has served to amplify this concept. Puszkin and Schook (1979) found α-actinin, a protein able to bind to F-actin and stimulate actomyosin Mg-ATPase, to be associated with the vesicles. Thus, the microfilaments in the terminals might be involved in the movement of the vesicles to the inner surface of the synaptosomal membrane by its interaction with α-actinin. This most likely represents the clathrin contribution to the coated vesicles described above (and cf. Kadota, Kadota, and Grey, 1976) in relation to endocytosis.

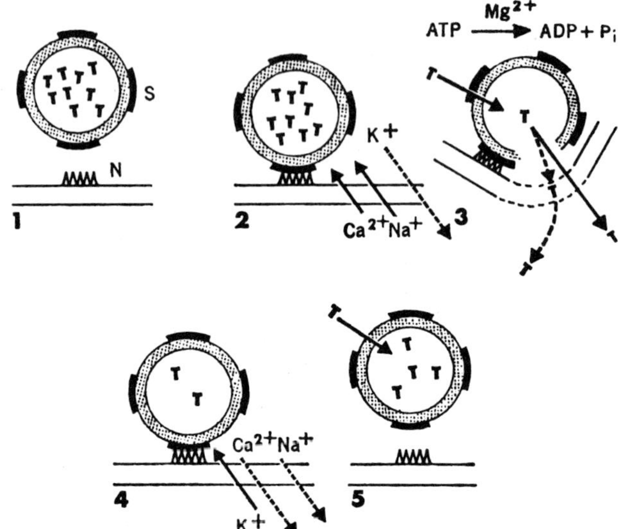

Fig. 13.7 A model of actomyosin-controlled exocytosis. A model for the release of transmitter vesicles is schematized. Neurin (N) (myosin) is shown associated with the inner face of the synaptic membrane and stenin (S) (actin) with the vesicle membranes. Stage 1, the synaptic vesicle is distant from the presynaptic membrane. Stage 2, the synaptic vesicle is in intimate contact with the presynaptic membrane as a result of the influx of Ca^{2+} following activation of the membrane. Stage 3, the Ca^{2+} triggers an interaction between the neurin and stenin with, in turn, a conformational change in the synaptic membrane and an opening of the membranes allowing transmitter (T) to be released into the synaptic cleft. Stage 4, the action is terminated by efflux of Ca^{2+} from the terminal. Stage 5, the reformed vesicle separates from the membrane. From Berl, Puszkin, and Nicklas (1973).

In the model of vesicular release proposed by Rasmussen (1970), when Ca^{2+} is present in adequate amount, cyclic AMP activates a microtubule "contractile" system which guides the vesicles down to the synaptic membrane. Just how the cyclic nucleotides operate with respect to movement, either alone or in relation to calmodulin (Cheung, 1980), remains unknown. It is parenthetically of interest in this regard, that in a number of exocrine cells release is blocked by colchicine or vinca alkaloids (Dustin, 1978).

3. Cholinergic Nerve Terminals

The neurotransmitter ACh is not only carried down within the cholinergic axon by axoplasmic transport (Chapter 4), but it is also locally resynthesized in the nerve terminals. When released in the process of neurotransmission in sympathetic ganglia or at the neuromuscular junction, ACh is hydrolyzed by the AChE present in the synaptic cleft and the choline moiety taken back up into the nerve terminals (Collier and MacIntosh, 1969) where it becomes acetylated. A calcium-dependent transfer of mitochondrial acetyl-coenzyme A supplies the acetate for the formation of vesicle Ach (Benjamin and Quastel, 1981). The newly synthesized ACh repackaged into vesicles is preferentially released on a subsequent activation of the nerve terminals. This was shown by first allowing ^{3}H-labeled choline to be taken up by the terminals and synthesized into ACh. Then, on subsequent activation of the nerves, a relatively high level of the labeled ACh was seen to be released (Birks and MacIntosh, 1961) as expected if the newly synthesized transmitter is preferentially released on a subsequent excitation (Collier, 1969). Several such cycles of re-uptake of choline with its resynthesis into ACh and release from the vesicles are possible before the vesicles can no longer be utilized for such repackaging and release, presumably after the loss of some irreplacable component from the vesicles (cf. Section A4). How much of the vesicular ACh present in the nerve terminals is contributed by anterograde axoplasmic transport and how much by a local resynthesis of ACh in the terminals is unknown.

The dependence of neurotransmission on the continued supply by axoplasmic transport of some component necessary for the release of vesicular contents was shown by Miledi and Slater (1970). Miniature end plate potential (MEPP) activity was recorded from muscle fibers in the end-plate region of rat diaphragm at various times after cutting its phrenic nerve either far from the muscle in the neck or at a point close to the muscle. After a latency of some 12 hr, a decrease in MEPP activity was seen, the decrease occurring earlier in those cases where the nerve had been cut closer to the muscle (Fig. 13.8).

The difference in the times at which the MEPPs fell to 50% of their control level after transecting phrenic nerves close to and far from the muscles and

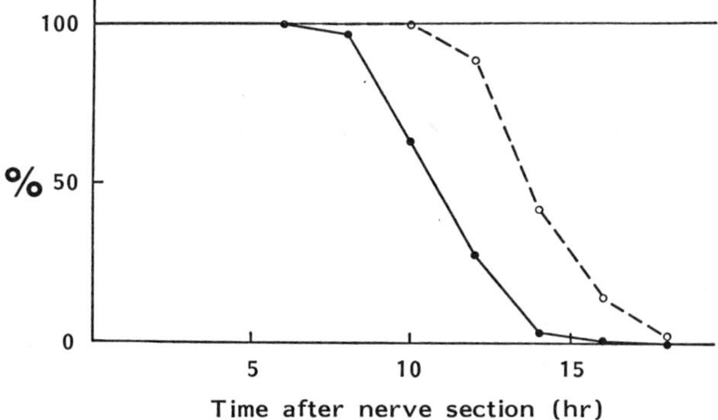

Fig. 13.8 Decrease in MEPP activity following denervation. The time course of failure of MEPP discharge seen after cutting the phrenic nerve supply at the indicated distances close to (●) and far from the muscle (○) of the diaphragm is shown. From Miledi and Slater (1970).

the distance between the cuts was used to derive an estimated rate of transport of 360 mm/day, one reasonably close to the rate of 410 mm/day found for fast transport. The decline in the MEPP discharge and resultant failure of neurotransmission was not due to the Wallerian degeneration of the nerve fibers. That process begins later, starting in the thinnest nerve fibers some 22 hr after nerve transection (cf. Fig. 6.4). The finer fibers appear to have less of a stock of some needed component, this accounting for their failing sooner than the thicker fibers (Chapter 6). The nerve fibers are very thin near their terminals and they could well be most vulnerable on this ground with the loss of neurotransmission appearing as its first sign of deficiency.

Cholinergic nerve fibers also contain the enzyme AChE, with a portion of it carried down within the nerve fibers at the usual fast rate of axoplasmic transport (Chapter 4). Possibly the AChE in the motor nerve fibers may serve to regulate its level of ACh, especially so within the nerve terminal (Birks and McIntosh, 1961; Potter, 1970). Another possibility is that AChE is secreted from the motor nerve terminal into the synaptic cleft (Ochs, 1974b) where it can serve its functional role in hydrolyzing ACh to terminate its synaptic action.

Acetylcholinesterase had been assumed to be integrally present on the post-synaptic membrane in the synaptic region until Hall and Kelley (1971) showed that a mild protease treatment could release AChE without damaging the membrane (cf. Skau and Brimijoin, 1978). EM studies using stains for AChE showed it to be associated with the basal lamina in the cleft within the neuromuscular junction (McMahan, Sanes, and Marshall, 1978). The various AChE species first identified in nerve and muscle by Hall (1973) are of special interest in this regard. Using sucrose gradients, velocity sedimentation of muscle extracts showed three peaks of AChE present, those of 4S, 10S, and 16S (cf. also Fernandez, Duell, and Festoff, 1979; Brimijoin, 1979). The location of these AChE species in the muscle showed a virtual absence of 16S AChE outside of the end-plate regions as compared to the two other species. In denervated muscle, AChE decreased to a greater extent, to 21%, in the end-plate region as compared to the end-plate free region where it fell to 59%. Most of the AChE activity remaining was of the 4S species, the next largest species remaining the 10S, and the 16S species showed the least retention. Brimijoin (1979), using the stop-flow technique (Chapter 4), found a greater percentage of 16S AChE accumulating in the nerve at the cold block, suggesting that this is the major species transported, though the other AChE forms were seen to accumulate as well to some extent. Other evidence for the transport of the 16S species had been given by DiGiamberardino and Couraud (1978). The level of 16S AChE in the muscle supplied by the obturator nerve was seen to fall faster when the obturator nerve was cut close to the muscle than when the nerve was cut further away, the fall taking place over a 4–5 day period (Fernandez *et al.*, 1979). In a later study, the 16S species of AChE was found to be confined to the motor fibers where it was estimated to represent 3–4% of the total

AChE present by Fernandez *et al.* (1980). From its rate of increase at nerve ligations, the authors concluded that this AChE species is transported at a fast rate. The rate of transport of the 20S species (one likely to be similar to the 16S species) was in fact found to be 411 mmm/day when using a double-ligation technique (Couraud and DiGiamberardino, 1980). The information available is thus in accord with a fast transport of the 16S species of AChE and its likely translocation into the synaptic cleft. A remaining unresolved conflict with respect to this concept is that the AChE found released from the nerve terminals by Skau and Brimijoin (1978) on stimulation was of the 10S species.

4. Adrenergic Nerve Terminals

a. Release mechanisms. The pattern of accumulation of noradrenaline (NA) in nerve segments proximal to ligations and stop-flow studies have shown NA to be carried down within the adrenergic nerve fibers by fast axoplasmic transport (Chapter 4). In the nerve terminals NA is released from DCVs by exocytosis in the course of neurotransmission (Smith and Winkler, 1972; Paton, 1979). Unlike ACh, the synaptic action of NA is terminated by its diffusion from the receptor, the degradation by catechol-o-methyltransferase and monamine oxidase, and by its reuptake into the nerve terminals (Iversen, 1967). The NA which reenters the terminals is repackaged into the synaptic vesicles and this NA is then preferentially released from the nerve terminals on a subsequent excitation of the nerve. As in the case of cholinergic neurons, the proportion of NA transported down the axons with respect to that taken back into the nerve terminal and reutilized in neurotransmission remains unknown (Haggendal, 1974).

The adrenergic neuron has a large number of terminal beads at which the neurotransmitter is released (Fig. 13.9).

Two populations of DCVs are seen present in the adrenergic neuron. A larger, 800–1000 Å vesicle is generally found in cell bodies, in the fibers and some in the terminal beads, along with smaller vesicles 400–500 Å in diameter, which are characteristically present in this site (Fig. 13.10).

The larger DCVs are found to accumulate above nerve ligations, evidence pointing to them as the NA-containing organelle transported down the fibers (Kapeller and Mayor, 1967; Geffen and Ostberg, 1969). These may be converted into the smaller DCVs of the terminals after reuptake of NA (Kopin and Silberstein, 1972) as indicated in Fig. 13.10. As can be noted in this figure, in addition to the reuptake of NA by the terminal, there is an extensive synthesis of NA from tyrosine. The pool of catecholamines in the terminals is therefore complex in its origin; the amount carried down by axoplasmic transport, the NA taken back up into the terminals after exocytosis, and the transmitter synthesized in the terminals (cf. Nagatsu, 1973).

The large DCVs are shown guided to the membrane of the bead along microtubules in this schematization. After exocytosis a portion of the NA

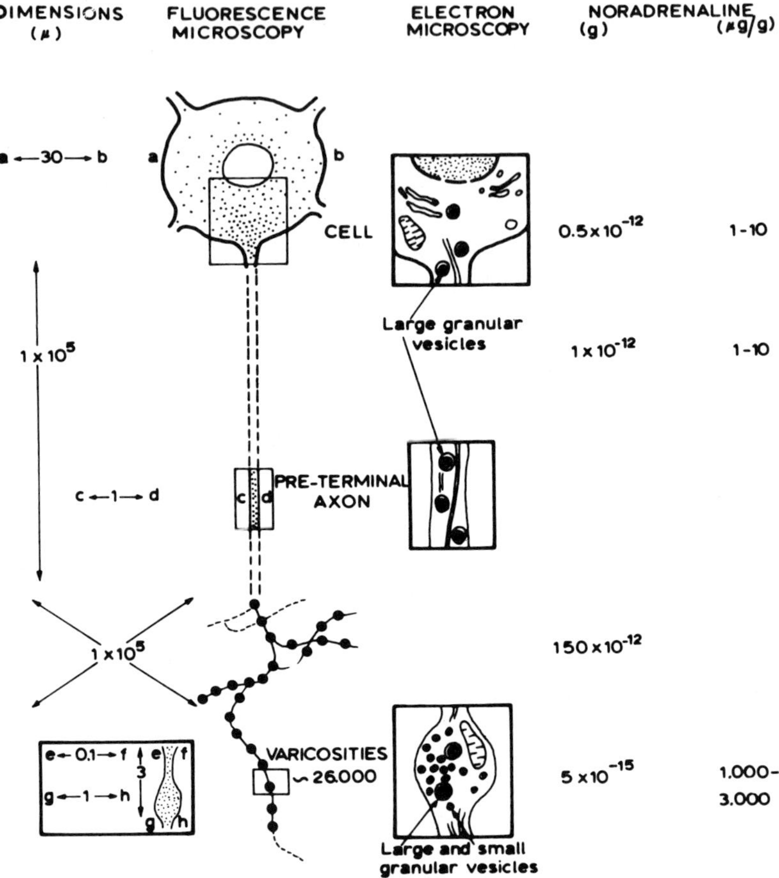

Fig. 13.9 Representation of vesicle transport in an adrenergic neuron. Granular vesicles (dense core vesicles, DCVs) are shown in cell body, axon, and terminal varicosities. Their noradrenaline (NA) contents are indicated at the right in total grams (g) and concentrations (μg per gram weight). Note the multiple-beaded synaptic endings along the terminals. In the drawings representing electron microscopy sections, large and small granular vesicles appear in the beaded endings and only large granules appear in the cell and axon. From Livett (1973).

released binds to the effector cell, part diffuses away from the synaptic region, and some of it taken back into the nerve terminal to form the smaller NA-containing vesicle. Additionally, as shown in Fig. 13.10, NA may be locally synthesized in the terminals from tyrosine supplied by the circulation. The tryosine is converted to NA by the enzyme path described in Chapter 4 (cf. Fig. 4.7). This process may also occur all along the fiber in the DCVs as they are being transported, with the NA formed in the DCVs constantly being released from the vesicles to be catabolized by the monoamine oxidase (MAO) of the mitochondria (cf. Fig. 4.20). The turnover of NA within the

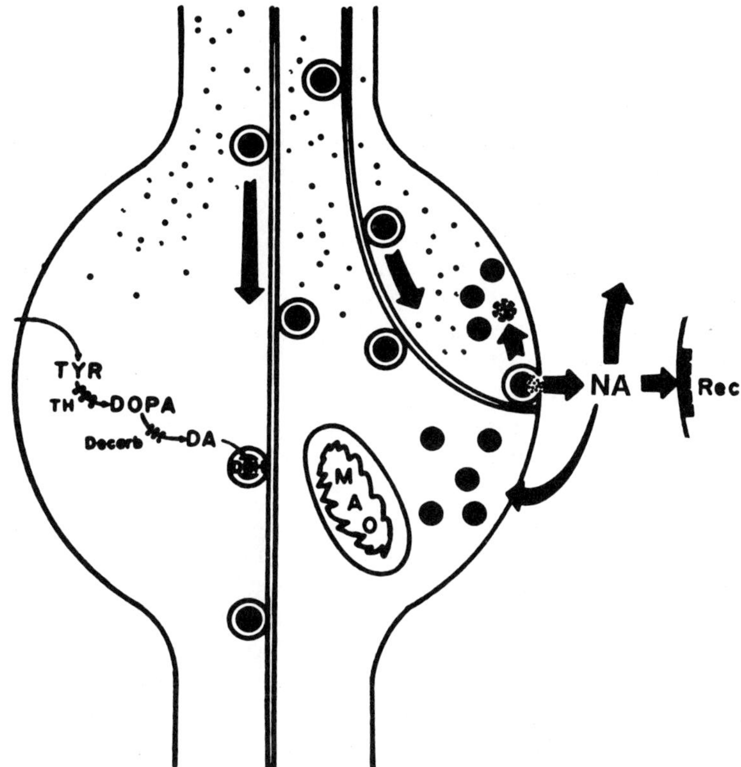

Fig. 13.10 Schematic representation of a varicosity in adrenergic nerve terminals. The dense core vesicles (DCVs) are shown in the course of their transport along microtubules, ending in the varicosity at the membrane where exocytosis (arrow) of noradrenaline (NA) is shown. The arrow indicates its binding to the post-synaptic receptor (Rec) site, another arrow indicating diffusion and reuptake into the terminal for repackaging and reuse. At the left, the synthesis of dopamine from tyrosine (TYR → DOPA → DA) (cf. Fig. 4.7) and the packaging of DA into the DBH-containing vesicle (cf. Fig. 4.20) is indicated. From Kopin and Silberstein (1972).

DCVs as they are transported down the nerve fibers assumes a greater relative importance in the terminal varicosities where a much higher concentration of DCVs is present than in the axon. As noted in Chapter 4 (Section A2), dopamine (DA) is also fast-transported in adrenergic nerve fibers in the course of the steady state turnover of amines in the DCVs.

In addition to NA, DA, and DBH, the protein chromogranin A contained within the DCVs appears to play a role in the storage of the amines. Also present in the DCV is a Mg-ATPase, the enzyme possibly participating in the restocking of NA into the vesicles. A small amount of DBH and chromogranin A was found to be released along with NA from the nerve terminals on excitation (Geffen *et al.*, 1969; Smith *et al.*, 1970; and de Potter *et al.*,

1969). The loss of these substances along with the neurotransmitter is in accord with an exocytosis of NA through the nerve membrane, the relatively large size of the DBH and chromogranin molecules militating against their simple diffusion through the membrane.

Haggendal (1974) discussed some of the possible ways in which NA is released from the vesicles in the process of exocytosis, whether all of it as expected of the quantal theory, or by a partial release of vesicular contents, and considers a partial release of NA from the DCVs to be best in accord with the known findings. A determination of the exact proportion of NA supplied by axoplasmic transport as compared to its local resynthesis in the nerve terminals should help establish whether a partial or a complete exocytosis of all the contents of the vesicle occurs. After several cycles of excytosis and repackaging of NA into the vesicles, most of the DBH and chromogranin is lost with part of the DBH remaining in the spent vesicule carried back to the cell body by retrograde transport (Chapter 4).

b. Uptake of noradrenaline and transmitter pools in the terminals. The amount of ^{3}H-NA taken up by adrenergically innervated tissues is directly related to the amount of adrenergic nerve terminals present (Iversen, 1967). Uptake does not occur if the adrenergic nerve has been cut and enough time allowed for Wallerian degeneration to occur (Hertting *et al.*, 1961). After an uptake of ^{3}H-NA by adrenergically innervated tissue, autoradiography and histochemical fluorescence shows the labeled NA to be located within the nerve fiber terminals (Wolfe *et al.*, 1962). The DCVs containing ^{3}H-NA can be separated and isolated as a granular fraction after tissue homogenization and differential centrifugation (Potter and Axelrod, 1963). The neurotransmitter NA appears to be present in two pools within the terminals, a smaller pool from which an immediate release of NA occurs, and a larger, more stable pool, which acts as a storage depot. The behavior of the two pools was shown using ^{3}H-NA. After ^{3}H-labeled NA is taken up by the terminals, tyramine caused an exocytosis of the vesicles with a release of ^{3}H-NA which depended on when the ^{3}H-NA had been taken up by the nerve terminals beforehand. If the tyramine had been given soon after an uptake of ^{3}H-NA, most of the ^{3}H-NA was released. If tyramine release was initiated 24 hr later, most of the ^{3}H-NA appears to have entered the stable pool with little of it released (Iversen and Whitby, 1963; Potter *et al.*, 1962). Similarly, nerve stimulation causes the release of much of the ^{3}H-NA soon after it was taken up (Kopin *et al.*, 1968) and less released when the nerves were stimulated at later times. Thus, there is a shift with time as the NA taken up into the nerve terminals is first present in a readily releasable ''labile'' pool and then later in a more ''stable'' or ''storage'' pool not as accessible to release (Fillenz, 1977).

Reserpine can deplete NA from both the labile and storage pools. This is followed by a slow recovery of the level of NA in the nerve terminals over a period of several days. The recovery process most likely results from the resupply by the cell bodies of new vesicles containing NA which are carried down to the terminals by axoplasmic transport (Haggendal and Dahlström, 1971). A somal origin of the new supply of NA after a reserpine depletion was indicated by the sequential return of NA toward normal levels; first in the cell body after some 12–14 hr, then more distally in the nearby portion of the axon at 15 hr, in a still more distal portion of the axon after 12-18 hr, and finally in the distal nerve terminals after 24-36 hr.

5. Nerve Terminal Uptake of Exogenous Materials

In addition to neurotransmitter and neurotransmitter precursors, other substances may be taken up by the nerve terminals. One such is nerve growth factor (NGF), of particular importance in the maturation of adrenergic and sensory neurons (Chapters 3 and 14). Additionally, a number of foreign substances may be taken up by the nerve terminals. One of these exogenous materials is horseradish peroxidase (HRP), a protein widely used to trace nerve fiber paths following its uptake and retrograde transport (Chapter 3). Tetanus toxin, cholera toxin, lectins, and certain antibodies by their uptake characteristics reveal the presence of several different uptake mechanisms in the terminals. A relatively high concentration of HRP in the vicinity of the terminals is required for its uptake (Heuser and Reese, 1973; Litchy, 1973; Teichberg *et al.*, 1975), and the uptake is most avid when the nerve is activated. This suggests that the mechanism of HRP uptake is related to exocytosis. After the vesicles have released their content of neurotransmitter, HRP and other substances of this class can enter the open vesicles before they close off. The reformed vesicles interiorized into the nerve terminal are then found to have HRP present in them (Ceccarelli *et al.*, 1973).

A second class of substances does not depend on neuronal activity and exocytosis for its uptake by the nerve terminals (Stöckel *et al.*, 1977). In this latter group are found NGF (Hendry *et al.*, 1974), tetanus toxin (Erdmann *et al.*, 1975; Price *et al.*, 1975; Stöckel *et al.*, 1975b), cholera toxin, wheat germ agglutin (WGA) (Stöckel *et al.*, 1977) and antibody to DBH (Fillenz *et al.*, 1976). These substances appear to have a high affinity for some specific glycolipid and/or glycoprotein binding site in the nerve terminal membrane from which, after binding, they are taken up into the nerve terminal (Dumas *et al.*, 1979).

The uptake and retrograde transport of a member of this second class of substances, tetanus toxin, was shown by coating colloidal gold particles with the toxin (Schwab and Thoenen, 1978). When injected into the anterior chamber of the eye, the coated particles taken up by the adrenergic nerve terminals innervating the iris are carried by retrograde transport into the cell bodies in the superior cervical ganglion where their presence was shown by the electron-dense gold particles in EM sections.

The avidity with which various agents are taken up from the terminals is determined by comparing the amounts deposited in the cell bodies at a fixed time of 14 hr after their injection into the eye, at the time when accumulation is determined to be maximal (Stöckel *et al.*, 1975a), as shown in Fig. 3.13. The uptake characteristics of different materials appears to be due in part to the nature of binding to receptor sites on the nerve terminal membrane. Some substances such as wheat germ agglutinin (WGA) bind with a low affinity but with a high capacity, while others bind with a high affinity but with a low capacity. There appear to be several binding sites with more than one substance sharing a common binding site, as indicated by the competition in the uptake and transport of WGA and NGF. Charge is also important, as indicated by the uptake of cationic ferritin (Olsson and Kristensson, 1981).

The investigation of binding sites for their uptake capabilities is of additional interest in that by coupling substances to agents capable of being taken up into the terminals, those substances may be thus carried back into the cell bodies to there exert specific actions. This technique has great potential both for theoretical and for practical ends as a method of modifying the synthesis and functional capabilities of neurons.

6.   Trans-Synaptic Transport

a.   Neurochemical junctions.   Labeled proteins and polypeptides transported into the nerve terminals appear for the most part to remain there for a considerable period of time, as judged by autoradiography. Some portion, however, can leave the nerve terminals and may enter the post-synaptic cell. This was first indicated by Korr *et al.* (1967), who injected the nucleus of the glossopharyngeal nerve in the dog with ^{32}P or with ^{14}C-labeled amino acids. After allowing 5–9 days for slow transport to carry labeled components to the tongue muscle, it was removed, sectioned, and prepared for autoradiography. Grains of radioactivity appeared to be located in the muscle fibers, with other grains apparently located in the motor nerve fibers between the muscle fibers. The better resolution possible with EM autoradiography should help decide the precise location and show an intramuscular locus. In a further study, neuronal proteins were reported to reach the tongue muscle in four waves (Korr and Appeltauer, 1974), the waves differing in their protein composition (Appeltauer and Korr, 1975). This suggested a transfer of components to the muscles at different times, although such waves could also reflect differences in the turnover of labeled components in the fibers and their redistribution as accounted for by the unitary hypothesis (Chapter 11).

Further evidence for a trans-synaptic transfer was advanced by Miani (1971). After arranging for the uptake by cell bodies in the salivary nucleus of ^{3}H-lysine and ^{3}H-leucine, the ganglion cells in the otic ganglion on which the fibers of the salivary nucleus cells synapse were subsequently found to be labeled. Transfer of labeled materials from the nerve fiber terminals of the salivary neuron to the otic ganglion neuron was not direct, the spread from the pre-synaptic terminals occurring first to the nearby satellite cell and from the latter to the postsynaptic neuron. The nucleus of the post-synaptic neuron seemed to be the target for at least some of the transferred radioactive materials found localized in that site in autoradiographs.

Morphological signs of trans-synaptic changes in the visual system (Ghetti, Horoupian, and Wiśniewski, 1975) indicative of trans-synaptic spread were paralleled by evidence of a trans-synaptic passage of labeled materials by Grafstein (1971). Mouse eyes were injected with ^{3}H-proline and incorporated radioactivity was subsequently found in the striate cortex after its trans-synaptic passage through the lateral geniculate. The geniculate cell bodies received labeled components passed to them from the optic nerve fibers which synapse on them, the labeled components carried in the fibers of the geniculate neurons to their specific sites of termination in the striate cortex. The protein, the TCA insoluble fraction, extracted from the striate cortex was labeled along with a smaller amount of TCA-soluble activity (some part of which could be ^{3}H-leucine). In those experiments a leakage of ^{3}H-leucine into the circulation gave rise to a non-specific labeling of all cortical regions (Chapter 3). The striate cortex, however, contained more radioactivity than the other cortical regions and the differential labeling of the striate cortex with respect to other cortical areas increased with time as expected of a trans-synaptic passage.

Subsequent studies further showed the reality of such trans-synaptic transfer (Casagrande and Harting, 1975; Dräger, 1974; Grafstein and Laureno, 1973; Specht and Grafstein, 1973). The question remains as to the specificity of the transfer. Grafstein has pointed out that the phenomenon might represent a nonspecific transfer of labeled materials from the nerve terminals in that radioactive labeling was seen to spread well beyond the visual cortex area, even into the underlying hippocampus. Such spreads could represent a diffusion of degradation products from the nerve terminals, with some of those materials released taken up and resynthesized by the nearby cellular elements. Thus, an unknown portion of the components transferred trans-synaptically may be non-specific in this sense with another portion specifically transferred.

In any case, the phenomenon has been found useful in showing the organization of afferents in the striate cortex subserving ocular dominance (Wiesel, Hubel, and Lam, 1974; Hubel, Wiesel, and LeVay, 1975). On injecting one eye of the macaque monkey with ^{3}H-fucose and ^{3}H-proline, the incorporated radioactively labeled proteins carried by the optic nerve to the lateral geniculate were, after trans-synaptic transfer, subsequently found in the geniculate fibers ending in layer IV-C of visual cortical area 17. Autoradiographs of the visual cortex revealed alternating stripes of labeling corresponding to the manner in which the lateral geniculate fiber terminations from each eye ending as adjoining columnar bands throughout the visual area, the "ocular dominance bands." These show a fairly restricted localization. The technique of using axoplasmic transport and autoradiography to display the ocular dominance bands has been useful in studying the effects of monocular deprivation which acts to alter the band patterns (Shatz and Stryker, 1978). Changes in their pattern which occur in early development were also studied by this technique (LeVay, Stryker, and Shatz, 1978).

A trans-synaptic transfer of labeled nucleotides in neurons of the visual system which appears to form part of a feedback system of control was reported by Schubert and Kreutzberg (1974). Following application of ^{3}H-uridine or ^{3}H-adenosine to localized areas of the striate region of the rabbit cortex, the transport of the nucleotides in axons terminating on the lateral geniculate cells was studied by autoradiography. When ^{3}H-uridine was used as the precursor, grains of radioactivity were found concentrated in the nucleus of the geniculate cells, while with ^{3}H-adenosine, labeling was seen in both the nucleus and cytoplasm of the cells. The location of these nucleotides following their post-synaptic transfer suggests their possible function as regulatory or trophic substances (Chapter 14).

Trans-synaptic transfer of retrogradely transported material has also been reported. Following an uptake of labeled tetanus toxin by motor nerve fibers, the toxin was seen to have moved up into the motoneuron cells by retrograde transport to cross their membrane, and become localized in the presynaptic terminals synapsing on the motoneuron cell bodies (Schwab and Thoenen, 1976). The labeled material appears in the pre-synaptic nerve terminals just at the time when classical signs of tetanus develop along with electrophysiological evidence of an impaired release of the inhibitory neurotransmitter (Brooks *et al.*, 1957; Curtis and de Groat, 1968; Curtis *et al.*, 1973; Osborne and Bradford, 1973). The correlation between the time of the appearance of the labeled toxin in the terminals of inhibitory interneurons with signs of tetanus is in accord with the concept that a block by the toxin of the release of inhibitory neurotransmitter from the terminals on the motoneurons is the basis for the tetanus (Eccles, 1964).

Another interesting example of retrograde trans-synaptic transport was given by Schwab and Thoenen (1977). Following the injection of ^{131}I-NGF into the anterior chamber of the rat eyes, grains of radioactivity were seen in the cell bodies of the adrenergic neurons innervating the iris, these neurons amounting to 15–25% of the total population present in the superior cervical ganglion. In contrast to this localization, when ^{125}I-tetanus toxin was injected into the eye, the cholinergic pre-synaptic terminals ending on the cell bodies were labeled. The different localizations of these two molecular species thus indicates a specificity in retrograde trans-synaptic transfer.

b. Electrical junctions. At junctions where electrical transmission occurs, different requirements for the transport of components exist. At the gap junction (Revel and Karnovsky, 1967; Revel *et al.*, 1967), a fusion of pre- and post-synaptic membrane occurs with a narrow space of only 20 Å between them. Open channels are present between the two apposed membranes which allows the passage of relatively large molecular species across the junction without leakage out into the gap (Payton *et al.*, 1969). The underlying mechanism of transport of materials in and through the channels (Bennett, 1977) has yet to be determined.

B. HYPOTHALAMO-NEUROHYPOPHYSIAL SYSTEM (HNS)

Historically, the study of the transport of hormones in the nerve fibers of the hypothalamic-neurophypophysial system (HNS) had been carried out with little reference made to studies in the peripheral nervous system and vice-versa (Ochs, 1977b). Only recently has it been generally recognized that the two systems share the same transport properties. Scharrer and Scharrer (1940) noted the marked similarity in the staining appearance of the large cell bodies in the hypothalamus to those of secretory cells, and proposed that the hypothalamus cells, even though they are neurons, also secrete hormones. These large cells are located in the supraoptic nucleus (SON) and paraventricular nuclei (PVN) of the hypothalmus, their axons passing via the pituitary stalk into the posterior pituitary gland. Using Bargmann's chrome-alum-hematoxylin stain, Hild (1951) showed an accumulation of stained material in the fibers above transections made in the stalk with its depletion in the fibers and in the pars nervosa (posterior lobe) below the transections (Fig. 13.11).

Those findings (and cf. Hild, 1954) were consistent with the concept that the stained neurosecretory materials are transported from the cells out into the HNS axons to the posterior lobe of the pituitary where they are stored for later release into the blood stream. The term "neurosecretion" was coined to refer to this process (Bargmann and Scharrer, 1951). One of the hormones isolated from the posterior lobe, vasopressin, is an antidiuretic hormone which also acts to increase blood pressure. The other hormone present is oxytocin, the milk-releasing factor which also controls uterine contractions in parturition.

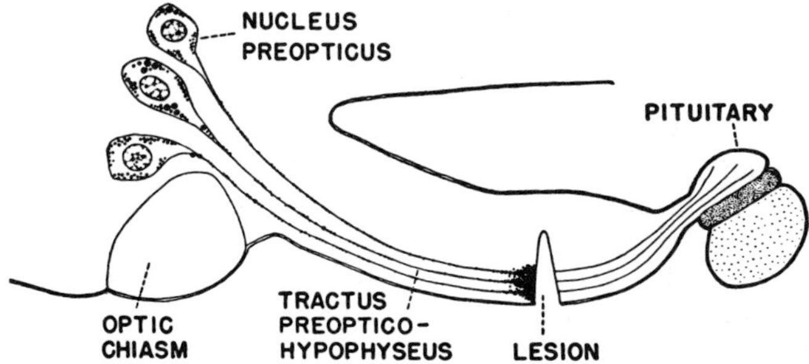

Fig. 13.11 Transport in the hypothalamo-neurohypophyseal system (HNS). After cutting the tract, Hild (1951) found an accumulation of chromatic-staining neurosecretory material above the interruption of the fibers, indicating transport of hormones in the fibers. The work of Hild is schematized in this figure of Bargmann and Scharrer (1951).

1. Vesicular Changes During Transport

The posterior pituitary hormones are present in the electron-dense granules within vesicles 1100–1900 Å in diameter, the DCVs shown in Fig. 13.12.

The granules within the DCVs exhibit different sizes when observed in EMs taken from the cell body, the fiber tract, and the nerve endings of the posterior lobe (Zambrano and DeRobertis, 1966), the overall size of the

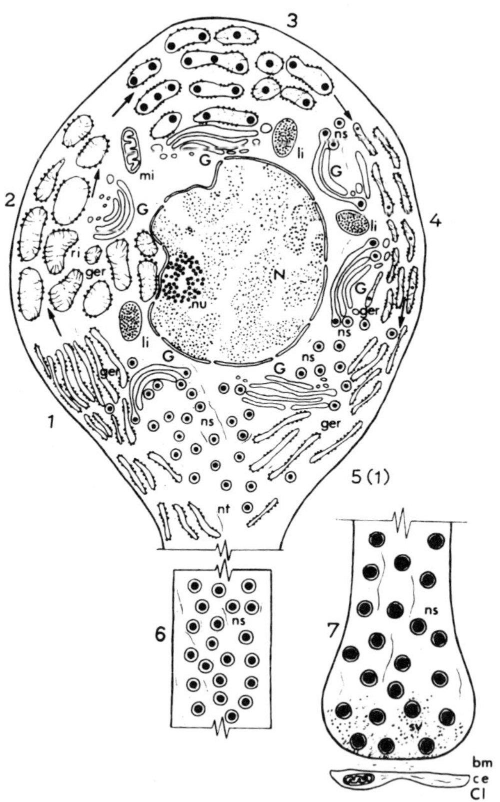

Fig. 13.12 Secretory granules and their transport in HNS neurons. The schematization shows different stages of secretory granules in the supraoptic neuron of the HNS. Stage 1 is a storage or resting phase in a type 1 neuron. Stage 2 shows an enhanced protein synthesis (at the ribosome level) with intracisternal accumulation of molecular material. Stage 3 corresponds to a type II neuron in which intracisternal vesicles are produced at a lower rate of synthesis. Stage 4 is an intermediary stage in which there is both intracisternal material and neurosecretion at the Golgi region (Golgi stage). Stage 5 is similar to stage 1 with storage of neurosecretion elements. Stage 6 represents a section of an axon which indicates an increased size of the granules. In stage 7, the granules reach their maximum size at the nerve ending proper. From Zambrano and De Robertis (1966).

vesicle itself remaining unchanged. One possibility earlier advanced for the variation in the size of the granules was that granular material continues to be synthesized in the vesicles as they move down within the axons. Another explanation was that the granular contents of the vesicles undergo an internal transformation during transport with, as a result, a change in size (Sachs, 1969). The granules contain proteins to which the hormones are attached, a change in the properties and binding of the hormones to the proteins accounting for the increase in granule size in the course of transport. These granular proteins, called neurophysins, have a MW close to 10,000 daltons and they exist in two closely related forms. One, neurophysin A, binds to vasopressin, the other, neurophysin B, to oxytocin. The neurophysins in combination with their hormones are referred to as pro-vasopressin and pro-oxytocin. These, in the course of their transport down the axons, split into their respective hormones and neurophysins (Fig. 13.13).

Changes in the physical properties of the neurophysin in the course of their separation from the hormone give rise to changes in the size of the granule which vary depending on the procedure used for the histological preparation of the tissue (Cannata and Morris, 1973). Recently, vasopressin was shown to be carried down as part of a 20,000 dalton precursor, a pro-pressophysin with a glycoprotein bound to the pressophysin to form the precursor (Brownstein, Russell and Gainer, 1980). Separation of the vasopressin from its binding to the glycoprotein, and perhaps other alterations of the prohormone,

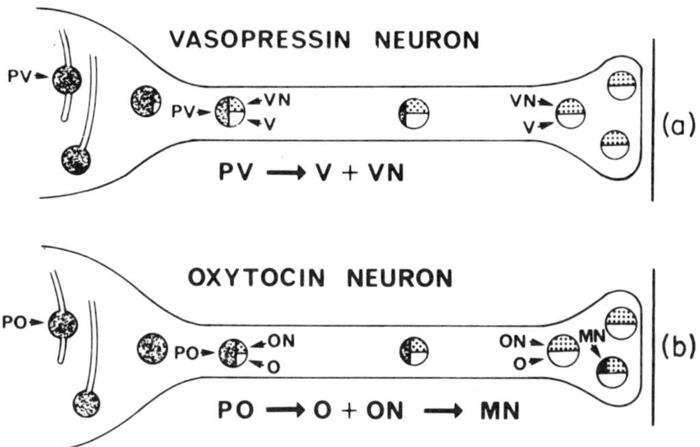

Fig. 13.13 Transformation of secretory granules in the course of transport. Vasopressin (a) and oxytocin (b) neurons are shown with maturation of their neurosecretory-containing granules. PV = pro-vasopressin, V = vasopressin, VN = vasopressin-neurophysin, PO = pro-oxytocin, O = oxytocin, ON = oxytocin-neurophysin, and MN = minor neurophysin. The proportion of bound and unbound hormone in the granules changes in the course of their transport down the axons. From Cross *et al.* (1975).

may account for the change in the size of the vesicle granule in the course of its transport in the axon.

2. The Rate of Transport of Neurophysins and Hormones in HNS Neurons

Sloper *et al.* (1960) used ^{35}S-cysteine and ^{35}S-methionine as hormone precursors to study axoplasmic transport in the HNS. After subarachnoid injection, an early uptake of these precursors into the cells of the SON was followed some 9½ hr later by increased radioactivity appearing in the posterior lobe, this indicating a slow transport of the hormone in the fibers at an apparent rate of several mm/day. Similarly, Flament and Durand (1961) injected ^{35}S-cysteine and ^{35}S-DL methionine into the posterior cisternal region of rats and found the SON and PVN cells to be labeled within a half hour, radioactivity appearing in the posterior lobe 10 hr later. Dehydrated animals showed a greater uptake of labeled activity into the SON cell bodies followed by a greater downflow of labeled activity into the neurohypophysis, a result consistent with the greater production of vasopressin known to occur in the dehydrated animal.

More recent studies revealed a much faster transport rate of the hormones in the fibers of the HNS (Alvarez-Buylla *et al.*, 1973). Norström (1975) injected ^{35}S-cysteine into the SON region of the rat hypothalmus which had been bled to stimulate a greater output of vasopressin, and found labeled activity appearing in the posterior lobe after a delay of only 2 hr. Most of that lag time is involved in the uptake of the labeled precursor, its synthesis, and possibly as well a delay in the movement of the labeled pro-vasopressin from the cell body into the axons. Taking all those factors into account, a transport rate of 190 mm/day in HNS axons was estimated for the labeled hormone. Pickering *et al.* (1975) also found an early accumulation of labeled neurophysin appearing in the neural lobe of the rat following cell body incorporation of the precursor with a similar fast transport rate.

Fast transport of vasopressin and oxytocin was also shown using the precursor ^{3}H-tyrosine injected into the cisterna magna of rats (Jones and Pickering, 1970, 1972; Pickering and Jones, 1971) with the same rate as that of labeled neurophysins (Gainer *et al.*, 1977). This is to be expected in that the neurophysin proteins and their associated hormones are bound together in the pro-hormonal form within the dense granules of the vesicles. Thus, in spite of some difficulties of estimating their exact rates of transport, the evidence for the fast axoplasmic transport of the hormones and their proteins at a rate close to 190 mm/day is reasonably good. Brownstein *et al.*, (1980) found the rate to be > 200 mm/day.

After injecting ^{3}H-choline intracisternally, the vesicular membrane fraction isolated from the posterior lobe, as indicated by isolation of labeled phosphatidyl choline, was also found to be rapidly transported (Swann and Pickering, 1976). Phosphatidyl choline constitutes a major component of the membranes (cf. Chapter 5). Similarly, glycoproteins, another membrane component, were found to be fast-transported in HNS axons when using the precursor ^{3}H-fucose (Norström, 1975). At least a part of the glycoproteins transported are those bound to the pressophysin to form the pro-pressophysin (Brownstein, *et al.*, 1980).

3. Microtubules and Transport in HNS

Flament-Durand and Dustin (1972) showed that colchicine did not prevent the uptake and incorporation of ^{35}S-cysteine by the PVN neurons while it blocked the axoplasmic transport of the hormone, an action similar to that seen in peripheral nerve (Chapter 12). Vincristine also blocked axoplasmic transport in the HNS (Flament-Durand *et al.*, 1975). Microtubular density in the fibers did not decrease until high concentrations of vinca alkaloid were used (Chapter 12). The local anesthetic tetracaine and chlorpromazine also blocked rapid transport of neurophysin in the HNS without causing any apparent change in the microtubule density of the fibers (Edström *et al.*, 1973b, and cf. Chapter 12).

One interesting feature found in HNS neurons which has not been observed in peripheral nerve is that different functional states of the neurons may alter the number and diameter of the microtubules in the HNS axons. Grainger and Sloper (1974) reported that the HNS fibers normally average 23 microtubules per fiber with the microtubules having a mean diameter of 331 Å. In saline-treated rats, an average of 12.5 microtubules per fiber was found with their mean diameter measuring 372 Å. Rats with congenital diabetes insipidus were found to average 23 microtubules per fiber with their microtubules having a mean diameter of 485 Å. This phenomenon suggests a labile control over microtubule assembly (cf. Section A5, Chapter 9).

4. Terminal Release Mechanisms in HNS Terminals

Fawcett, Powell, and Sachs (1968) found that both neurophysins and hormones are released from the HNS terminals, evidence they took for exocytosis from the vesicles insofar as the neurophysin molecule is too large to diffuse through the terminal membrane. The reasoning is the same as that given for the case of neurotransmitter release from adrenergic nerve terminals (Section A4 above). As in peripheral nerve, exocytosis also occurs in the HNS nerve endings of the posterior lobe when an action potential invades and depolarizes the terminals and when Ca^{2+} is present in the external medium (Douglas and Poisner, 1964a,b). The Ca^{2+} enters the terminal to cause, in some as yet unknown manner, the vesicles to merge with the membrane of the nerve terminal and to release their contents by exocytosis (Douglas, Nagasawa, and Schulz, 1971). The smaller electron-lucid vesicles seen in the terminals are considered to represent an endocytosis from the membrane, a process necessary to prevent the surface area of the membrane in the terminals from expanding inordinately as vesicle membrane is contin-

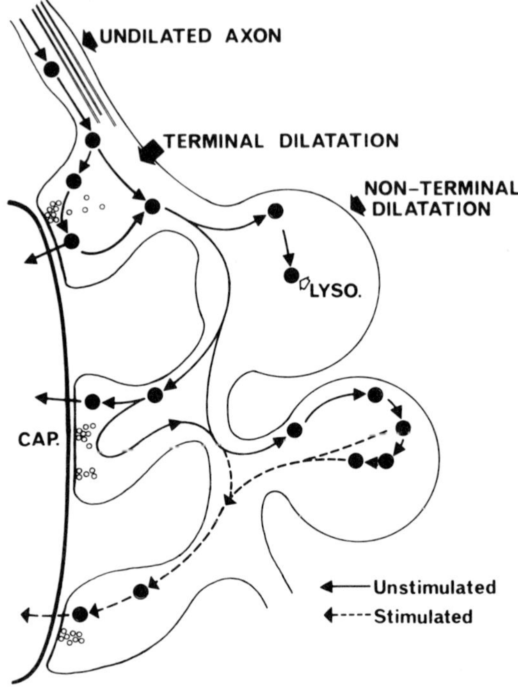

Fig. 13.14 Storage and release sites for neurosecretory granules in HNS terminals. The postulated movements of the secretory granules between readily releasable sites and storage pools is shown diagrammatically. The hormone granules are stored in the nonterminal dilatations and may later be destroyed by lysosomes. Under conditions of stimulation the stored granules may be conveyed to the terminal dilatations where the hormone is discharged into the circulation. From Cross *et al.* (1975).

ually being added to the the nerve terminal membranes in the process of exocytosis. As evidence for such endocytosis, HRP infused into the extracellular space was subsequently found present within the small vesicles.

The vesicles which do not release their hormonal content immediately on their arrival at the terminals appear to be stored in dilated regions of the nerve terminals of the posterior lobe, presumably for a later release of the hormone. This storage concept was developed on the basis that the true nerve ending contains both small electron-lucid vesicles and DCVs while the dilated region contains only granulated vesicles (Pickering *et al.*, 1975). The presumed course of movement of vesicles containing neurosecretory granules into and out of the two regions of the nerve terminals is indicated in Fig. 13.14.

We would expect that the spent vesicles or vesicle membranes are returned to the cell body for reuse by retrograde transport. As yet this retrograde transport has not been directly demonstrated in the HNS. That a retrograde transport of some materials exists may be inferred from the chromatolysis of the cell bodies seen following section of the pituitary stalk. As discussed in Chapter 6, chromatolysis may be initiated by some "signal" having a control over the level of protein synthesis in the perikaryon, a signal carried to the cells by retrograde transport.

14

Neurotrophic Regulation

The observation that the cutting or crushing of nerve leads to muscle wasting, has long been connected with the idea that the nerve has some kind of a trophic (nutritive) influence on muscle. The possibility that there are special trophic nerves (Samuel, 1860; Wyburn-Mason, 1950) has in recent years been set aside, and the term today is defined as "the regulation of muscle and other tissues by nerve not mediated by activity" (Gutmann, 1962, 1976; Miledi, 1963b; Guth, 1968; Harris, 1974). Such regulation is suggested by the early appearance of depolarization, fibrillation, increased presence of neurotransmitter receptors, evidence of altered sodium channels in the muscle membrane, and ionic and metabolic changes, with finally atrophy of the muscle fibers following nerve transection. Much effort in recent years has been expended to determine if some or all of these alterations may be explained by the decrease in muscular activity resulting from denervation rather than by the loss of some trophic substance supplied by the nerve to the muscle. As will be described, some of the changes seen in muscle are in fact due to its diminished activity, others to the loss of a neurotrophic factor (Guth and Albuquerque, 1978). We will first outline in general terms the muscle changes seen following denervation, and then more closely assess the relative contributions of activity and trophic factors. Some of the changes in sensory receptors seen after nerve transection which clearly indicate a loss of a neurotrophic factor will also be described. In the last section some studies aimed at identifying the biochemical nature of the trophic factors will be presented.

A. MUSCLE

1. Effects of Denervation

a. Depolarization and MEPPs. The earliest change recorded in muscle fibers after sectioning their motor nerve supply is a depolarization (Albuquerque *et al*, 1971), one which has been appears within a few hours first in the endplate region (Fig. 14.1).

The depolarization then spreads out from the end-plate region into the rest

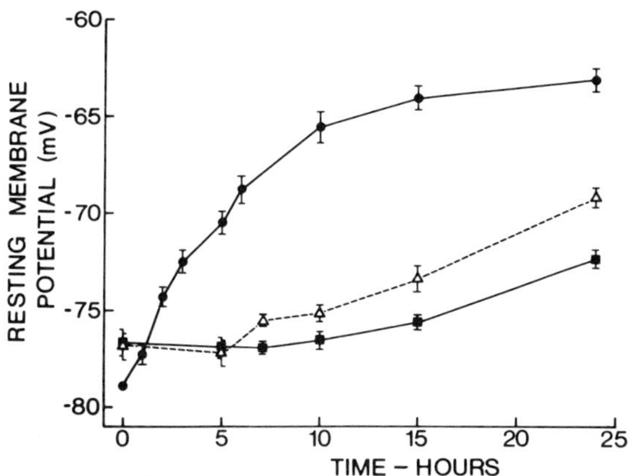

Fig. 14.1 Effect of nerve transection on resting muscle membrane potentials. Mean resting membrane potentials of surface fibers of the extensor muscles of the rat are shown at different times after crushing the motor nerves 0.4–1.0 mm (●) from the nerve's entrance into the muscle, or sectioning the motor nerve 2 cm (△) or 4 cm (■) from the muscle. The values found after crushing the motor nerve were obtained by averaging the membrane potential regardless of the distance of the fiber from the site of nerve crush. From Albuquerque *et al.* (1974).

of the muscle fiber with time until eventually the resting membrane potential (RMP) level falls to approximately 20 mV over its whole length. When the nerves were cut close to the muscle, the decrease in RMP had an earlier onset than when nerves were cut further away, a finding in line with the loss of supply of a needed neurotrophic substance from the nerve rather than a change in muscle activity. Similar observations were made by Deshpande *et al.* (1976) and Bray *et al.* (1976). Possibly the decrease of the RMP on denervation may come about through an effect on the sodium pump. Locke and Solomon (1967) found the level of depolarization produced by ouabain to be similar to that following denervation. Treating denervated muscle with ouabain produced no further fall in RMP.

A later onset of depolarization in muscles after the sectioning their motor nerves was seen by Redfern and Thesleff (1971) and by Card (1977). In rat extensor digitorum longus muscles denervated as close as 1 mm from their synaptic terminations, the RMP began to fall only after a lag time of 10–15 hr. The miniature end plate potentials (MEPPs) increased and then disappeared 12–13 hr after denervations were made, followed several hours later by a fall of the RMP (Fig. 14.2).

When nerves were cut 7 mm from the muscle end-plates, the fall of MEPP activity occurred at a still later time, a finding consistent with the participation of a neurotrophic factor. Card (1977) considers that the failure of MEPP activity following nerve transection signals the loss of a neurotrophic factor which is released from the MEPPs along with ACh, this leading to the fall in RMP. Stimulating the distal nerve stump for an hour at 10 pps caused a faster

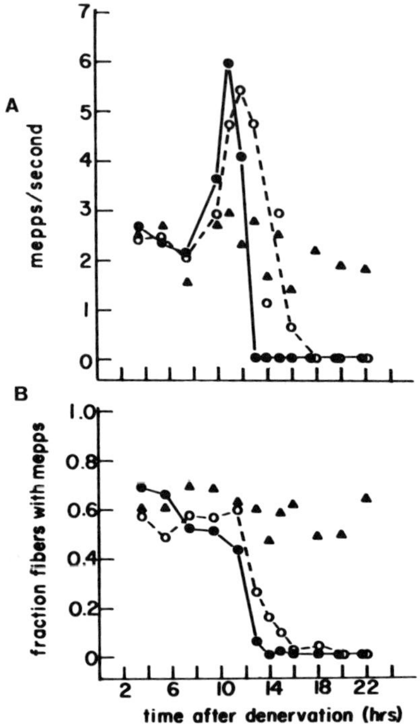

Fig. 14.2 Effect of denervation on resting muscle potentials. (A) Mean frequency of MEPPs recorded less than 1 mm from the site of its denervation (●) and at 6–7 mm (○) from the nerve transection at the various times afterward. The muscles were denervated by crushing or by transection. Discharges from control muscles on the contralateral side are shown (▲). Each point represents 40 muscle fibers. Standard deviations are ± 1 to 1.5 mV. (B) MEPPs recorded divided by the total number of fibers measured at times after denervation, i.e., the fraction of synapses with MEPPs remaining. Symbols given are similar to those in part A. From Card (1977).

depletion of MEPPs and hastened the fall of RMP in accord with the hypothesis. In the experiments of Albuquerque *et al.* (1974) the fall of RMP occurred well before the failure of MEPP discharge, indicating that the trophic factor is not associated with vesicular discharge. The neurotrophic factor on their view could diffuse from the nerve terminals or perhaps be released by vesicles other than those containing ACh. Further studies in other preparations on the sequence in which the MEPP failure and the fall in RMP occurs after denervation may help clarify the relation of the neurotrophic factor released from the nerve terminals to affect muscle RMP.

b. Acetylcholine receptors. Denervated muscles become abnormally sensitive to ACh. This was shown by the injection of a solution of ACh into blood vessels close to the muscle they supply. Normally this has little effect, but it causes large contractions in denervated muscles (Brown, 1937). The increased excitability appearing some time after denervation was referred to as "denervation supersensitivity" by Cannon and Rosenblueth (1949). The mature muscle fiber is normally highly sensitive to ACh only in the immediate vicinity of the end-plate (Miledi, 1960a; Kuffler and Yoshikami, 1975). Denervation supersensitivity is brought about by an increase in the area of the muscle membrane responding to ACh, the area of increased sensitivity spreading slowly from the end-plate region to eventually occupy the whole length of the muscle fiber (Ginetzinski and Shamarma, 1942; Axelsson and

Thesleff, 1959; Miledi, 1960b). This was shown by the changes in depolarization produced by localized brief iontophoretic pulse applications via micropipettes of small amounts of ACh along the length of the muscle fibers (Fig. 14.3).

In the normal muscle fiber shown in the lower row, only the end-plate region responds to the iontophoretically released pulse of ACh with an "ACh depolarization," one having a time course similar to that of the end-plate potential. As can be seen in the upper row, ACh depolarizations were elicited at all points tested along the length of the denervated muscle fiber.

The denervated muscle appears to revert to its immature state at birth where the muscle fibers are seen to be receptive to ACh over their whole length (Diamond and Miledi, 1962). In the course of maturation taking place over a period of a few weeks to a month after birth, the area of the muscle receptive to ACh shrinks to become restricted to the end-plate region, the pattern characteristic of the adult muscle. The mechanisms controlling the restriction of the ACh receptor to the end-plate region in the adult muscle fiber are not yet understood (Jansen, van Essen, and Brown, 1976).

The protein nature of the receptor was determined by the use of ^{125}I-abeled α-bungarotoxin, a toxin which binds irreversibly to the receptor and thus allows the subsequent isolation of the protein by gel chromatography (Miledi, Molinoff, and Potter, 1971; Potter and Molinoff, 1972). When adult muscles

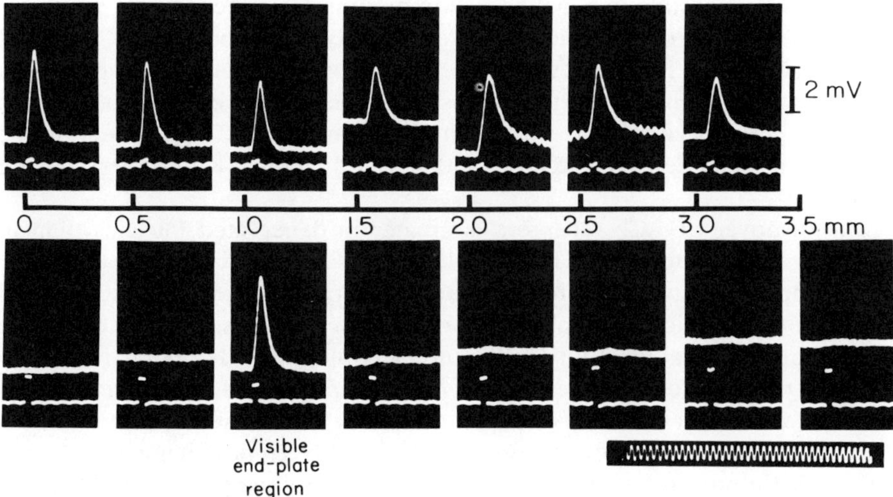

Fig. 14.3 Spread of ACh responses in muscle after denervation. The response of the muscle fiber of cat tenuissimus muscle to micro-iontophoretically applied ACh is recorded in the upper traces and the current is passed through the pipette shown in the lower traces of each record. In an innervated fiber (lower set of records), only the visible end-plate region responded with an ACh potential. In a 14-day denervated fiber (upper set of records) ACh potentials were obtained all along the membrane. From Thesleff (1960a).

are cultured *in vitro*, the loss of nerve supply results in the gradual spread of ACh receptor sensitivity over the length of the fibers. The addition of the protein synthesis blocking agent actinomycin D to the *in vitro* medium prevents the synthesis of the extra-junctional ACh receptors (Grampp, Harris, and Thesleff, 1972). This suggests that the neurotrophic factor normally supplied by the nerve acts to supress the synthesis of ACh receptors by the muscle. The control over the synthesis of ACh receptors most likely is exerted at the DNA–RNA level, as indicated by the action of actinomycin D to block protein synthesis at this point.

The turnover of ACh receptors in the membrane was shown using ^{125}I-α-bungarotoxin. By taking portions of the muscle for counting of ^{125}I activity after uptake of the labeled bungarotoxin, an index of the number of ACh receptors present in the membrane could be determined (Fambrough, 1974). The accumulation of newly formed ACh receptors was calculated as the difference between the receptors incorporated in the membrane and those "hidden" within the fiber, those newly synthesized or undergoing degradation (Devreotes and Fambrough, 1975). The newly synthesized ACh receptor protein is incorporated into membrane-bound vesicles which move toward the membrane to fuse with it and insert its complement of new ACh receptors. Later, the membranes containing receptors which have ended their life cycle are taken back by endocytosis into the cell for degradation. The "hidden" receptors are those in the process of moving to and from the membrane are thus not accessible to α-bungarotoxin binding from the external medium. The hidden receptors, however, represent only a minor part of the total population of ACh receptors so that the uptake of ^{131}I-labeled α-bungarotoxin can be used as a reasonably good measure of the number of receptors. These receptors are not diffusely spread in the membrane. They appear to be clustered as high density patches 1–30 μm in diameter (Ko, Anderson, and Cohen, 1977).

The turnover of ACh receptors is greater in denervated muscles than in normal muscles (Berg and Hall, 1974). The ACh receptors in the end-plate region appear to have a relatively greater stability compared to those inserted into the extrajunctional membrane of the denervated muscle fiber. Using the disappearance of radioactivity following labeling with ^{131}I-α-bungarotoxin to measure their turnover, the end-plate ACh receptors were computed to have a half-life of about 7.5, days while the half-life of the ACh receptors in the extra-junctional membrane was estimated to be only 19 hr (Chang and Huang, 1975). Actinomysin D was seen not only to depress the generation of receptors, but interestingly, to decrease the degradation of receptors as well.

c. Acetylcholinesterase activity.   In the adult muscle, junctional AChE was found to fall to approximately 50% of control levels within several days after denervation, and then to remain at this reduced level for some time (Guth, Albers, and Brown, 1964; Guth, Brown, and Watson, 1967). The nerve thus appears to regulate the synthesis of some part of the AChE associated with the muscle. However, it is important to differentiate between a trophic control over AChE synthesis in the muscle and the possibility that some species of AChE transported within the motor nerve fibers are released from the nerve to become localized within the end-plate region (Section A3, Chapter 13). Ranish and Dettbarn (1978) found a fall in nerve AChE to precede that in the muscle by 24 hr. However, when nerves were sectioned close to and far from the muscle, little evidence for a length-related decrease of AChE was seen as would be expected of its transport in the fibers. A similar finding was reported by Guth *et al.*, (1967). However, those studies did not differentiate between AChE and butyrylcholinesterase activity or between regions of the muscle more and less densely innervated. When those factors were taken into account, it was shown that there was a length-dependent loss of AChE activity in rat diaphragm muscle (Ranish, Dettbarn, and Wecker, 1980). The apparent length-dependent rate of loss was 300 mm/day, a rate fitting well with the usual fast transport rate of 410 mm/day. Davey and Younkin (1978) had also found a length-related decrease of AChE. The fall in the level of the enzyme began 15–48 hr after denervation and the level was significantly greater when a length of nerve was left connected to the muscle. A differentiation was also made between activity-independent end-plate AChE seen when a length of the nerve was left remaining from background AChE which did not change in relation to the length of the nerve remaining. Further evidence relating to those species of AChE transported has been discussed in Section A3, Chapter 13.

d. Tetrodotoxin (TTX)-insensitive action potentials. Denervation causes a change in the properties of the sodium channels of the muscle membrane. Whereas TTX normally will block the Na^+ channels, and thus by its presence prevent an excitation of the action potential, the Na^+ channels of denervated muscles are insensitive to the action of TTX. The onset of TTX-insensitivity begins approximately 2 days after denervation, appearing first in the end-plate region and then spreading out into the remainder of the muscle fiber over a period of a week or so (Redfern and Thesleff, 1971). Both "fast" and "slow" muscles show the presence of TTX-insensitive action potentials after denervation. The time at which TTX-insensitivity develops depends on the length of the transected nerve left connected to the muscle, the effect appearing sooner when nerves are cut closer to the muscle (Fig. 14.4).

This finding of a length-related change in the appearance of TTX-insensitivity is support for the concept that a neurotrophic factor is responsible for this change in the muscle following transection of its nerve supply.

It is not known if the phenomenon is due to a primary change in the Na^+ receptor protein itself, or if some more general membrane change induced by denervation brings about an alteration of the properties of the Na^+ receptor protein inserted into the changed membrane (Edwards, 1979).

e. Fibrillation. Muscle fibrillation, which characteristically occurs following denervation, is seen as a spontaneous irregular twitching or quivering of the muscle associated with irregular large-amplitude potentials recorded

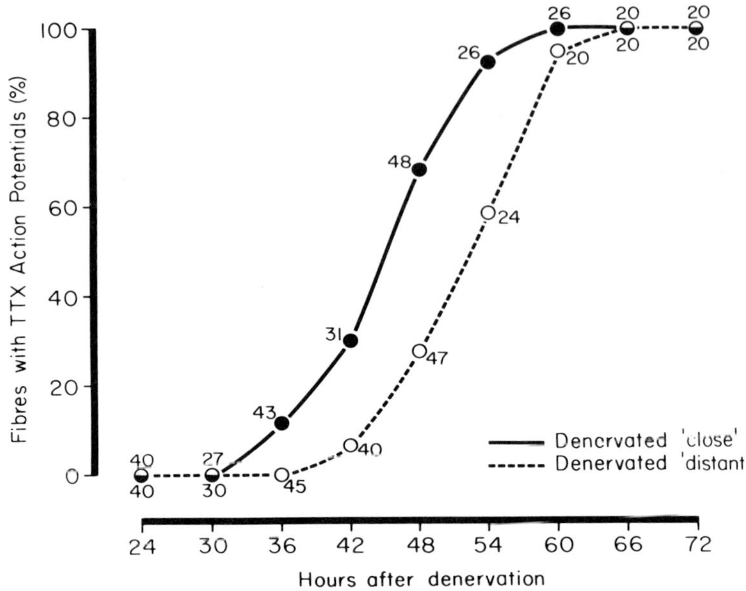

Fig. 14.4 Development of TTX-resistant action potentials in muscle fibers after denervation. Nerves innervating the EDL muscles in the rat were cut close (●) and far (○) from the muscles. The percentage of muscle fibers with TTX-resistant action potentials have an earlier appearance when the nerve stump length was cut approximately 3 cm closer to the muscle. Results from 48 muscles were pooled; the figures indicate the number of fibers examined. From Thesleff (1974).

electromyographically. Fibrillation may begin as early as two days after denervation, the time depending on the length of the nerve left connected with the muscle (Luco and Eyzaguirre, 1955). The onset of fibrillation recorded in the tenuissimus muscle spreads out in the muscle as a function of the time after transection, and the level at which the nerve is cut, as was first shown in the classical studies of Luco and Eyzaguirre. Irregular fibrillation potentials appear first close to the region of muscle entry and then later over the whole length of the muscle, the onset delayed when a longer length of the transected nerve was left attached to the muscle. A similar relation was seen by Salafsky, Bell, and Prewitt, (1968). Landmesser (1971) found that a functionally different nerve such as the sympathetic can form synapses on denervated frog skeletal muscle. When innervation was established the fibrillation of the denervated muscle stopped, indicating that a neurotrophic substance needed to prevent fibrillation was being provided by the foreign nerve fibers innervating the muscle. A neurotrophic influence does not require that a functional synaptic union to be made. This was shown by the reversal of signs of denervation in the muscles even before the restoration of synaptic function (Miledi, 1960b, 1963a,b; Bennett, Pettigrew, and Taylor, 1973; Dennis and Miledi, 1974a,b; Gutmann and Young, 1944). A further discussion of changes appearing on reinnervation will be taken up in Chapter 15.

f. General metabolic changes. Changes in carbohydrate (Wagner and Max, 1979) and protein metabolism are found in the denervated muscle (cf. Gutmann, 1962), these preceded by changes in nucleic acid metabolism (Dresden, 1969). An early increase in nucleic acid metabolism (Politoff and Blitz, 1978), including that of RNA polymerase isolated from nuclei of denervated muscle (Held, 1978), and changes in muscle mRNA following denervation (Metafora *et al.*, 1980), indicates an effect at the level of genetic transcription. The control of metabolism by a neurotrophic factor was indicated by the earlier occurrence of metabolic changes found in muscles when nerve transections were made closer to than farther from the muscles (Hajek *et al.*, 1964). Both the synthesis and the degradation of proteins are depressed in the denervated muscle. Over a long period of time there is a relatively greater decrease in synthesis as compared to degradation, accounting for the atrophy usually seen to follow denervation. Evidence for an increased degradation was an augmentation of autolysis in muscle seen to peak 10–15 days after block of transport by colchicine (Ramírez, 1980). Within a month after its denervation the tibial muscle of the rabbit may be reduced in mass by approximately one-third, with a more gradual further decrease over the next eight months (Gutmann and Hník, 1962). On the other hand, some muscles may show hypertrophy rather than atrophy following denervation, for example, chicken muscle. The denervated diaphragm muscle of the rat shows an early stage of hypertrophy followed later by atrophy (Stewart and Martin, 1956).

g. Cyclic nucleotides. Cyclic nucleotides have been implicated as second messengers for a number of metabolic control mechanisms in various cells (Robison *et al.*, 1971). An increased cyclic AMP has been seen by Carlsen (1975) in rat gastrocnemius muscle appearing 1 to 7 days after nerve transection. On the other hand, little change in the level of cyclic AMP was seen following denervation of frog sartorius muscles (Allen *et al.*, 1978). Instead, a marked rise in cyclic GMP was observed over a 1–5 week period with the increase in cyclic GMP located mainly in the end-plate region. The longer time-course in the development of denervation changes in the amphibian muscle as compared to the mammal is in part due to the lower body temperatures maintained in the frog, in part the lower rate of metabolism of the tissue. The significance of changes in cyclic nucleotides following denervation is difficult to assess until more is known regarding their role in the muscles. Recent studies indicate an interrelation between cyclic nucleotides and calmodulin (Cheung, 1980; Means and Dedman, 1980), and further advances along those lines should lead to a better understanding of the role of cyclic nucleotides in relation to a trophic influence on muscle.

2. Nerve Activity and Denervation Effects

The question as to whether a reduction of muscle activity or the loss of some neurotrophic factor accounts for the denervation changes seen in muscle following nerve transection was a question first experimentally addressed by Denny-Brown and Brenner (1944). They used a clip to produce a partial compression of the sciatic nerves of cats, to the degree where the transmission of nerve impulses through the compressed region was blocked without interrupting the continuity of the nerve fibers. This condition, referred to clinically as a neurapraxia, is to be differentiated from neurotmesis, the complete severance of the fiber (cf. Sunderland, 1978). The muscles innervated by the partially compressed nerves did not show the usual onset of fibrillations typical of a denervated muscle. The authors concluded that the normal state of the muscle depends on the continual supply of some neurotrophic factor supplied by nerve to the muscles, rather than on nerve impulses and muscle activity. In a refinement of that technique, Robert and Oester (1970) applied silastic cuffs in which a local anesthetic had been incorporated, with a hole the size of the sciatic nerves in it so that when placed around the nerves, the anesthetic diffusing from the cuff blocked the transmission of nerve impulses. Such cuffs kept on nerves for several days did not produce the usual signs of denervation. The authors therefore arrived at a similar conclusion, namely, that a neurotrophic supply was essential for normal muscle function.

However, the studies of Lømo and Rosenthal (1972) using silastic cuffs impregnated with the local anesthetic marcaine led them to an opposite conclusion. Those cuffs, when placed around rat sciatic nerves for 5 days, caused signs of denervation to appear in the soleus and EDL muscles. Muscle fibrillation was seen but neurotransmission was preserved and MEPP activity was found remaining in the muscles innervated by those nerves. The MEPPs were actually increased in amplitude, most likely a result of the higher input impedance which is a characteristic of denervated muscle fibers (Hubbard, 1963). A spread of ACh receptors also appeared over the length of the muscle fibers.

The results obtained by Lømo and Rosenthal might be explained if the marcaine, in addition to blocking nerve membrane excitability, is able to enter the nerve fibers and interfere to some degree with axoplasmic transport. To study this possibility, a number of local anesthetics were incorporated into silicone cuffs applied chronically to rat sciatic nerves by Bisby (1975) and an effect on axoplasmic transport was looked for. Transport was assessed either by a block of accumulation of catecholamines or AChE at ligations below the cuffed regions (Chapter 4), or a block of the transport of labeled proteins through the cuffed nerve regions following injections of the L5 motoneuron region with ^{3}H-leucine. After two days of nerve cuffing, axoplasmic transport was seen to be blocked by marcaine (buvicaine), lidocaine, mepivacaine, and procaine but not by benzocaine (cf. Chapter 12 Section F2).

A complete block of transport, however, cannot be the explanation of Lomo and Rosenthal's results because that would result in Wallerian degeneration and it would not account for the MEPP activity and neurotransmission found remaining in their studies. We might suppose that the local anesthetics in their studies had a weaker effect on axoplasmic transport, one allowing materials to still be transported able to maintain the viability of the nerve and neurotransmission, while neurotrophic factors supplying the muscle are blocked. Variability in the degree of block is known to occur when using the silastic cuff technique, such differences as have been reported from different laboratories, most likely due to the differing amounts of local anesthetic diffusing into the nerve trunk, differences which depend on the size of the nerve and the permeability of the perineurial sheaths of the particular nerve investigated (Chapter 8 and cf. Section A3 below).

To obviate an internal action of local anesthetics on the transport mechanism, TTX was employed to chronically block the conduction of nerve impulses. Tetrodotoxin offers a particular advantage for this study. It is very effective in blocking nerve action potentials, by entering and obstructing the Na^+ channels from the outside, and this appears to be its sole action. Tetrodotoxin was shown not to affect axoplasmic transport in *in vitro* studies (Ochs and Hollingsworth, 1971; Chapter 12). With TTX incorporated into silastic cuffs placed around rat sciatic nerves and left *in situ* for up to 7 days, the muscles innervated by the TTX-cuffed nerves showed some of the signs associated with denervation, namely, an increase in ACh receptors as measured by a greater degree of ^{125}I-α-bungarotoxin binding (LaVoie, Collier, and Tenenhouse, 1976). However, the increased amount of ^{125}I-α-bungarotoxin binding to the muscles of TTX-blocked nerves was less than that seen in muscles which had been denervated by transecting their nerve supply. The conclusion drawn from those results was that the inactivity of the muscle incident to the block of nerve activity by TTX can give rise to some of the signs of denervation, but that in addition there is a supply of some neurotrophic substance by the nerve which is interrupted by nerve transection. Essentially the same conclusion was reached by Pestronk, Drachman, and Griffin (1976) using repeated injections of TTX subperineurially into sciatic nerves so as to maintain a chronic block of nerve conduction over a period of 7 days. An increase in the amount of ACh receptors was shown by the increased amount of ^{125}I-α-bungarotoxin binding found in those muscles, with an additional increase of binding seen in muscles which had their nerve supply cut. Pestronk *et al.* (1976) also assessed fast axoplasmic transport in the TTX-blocked nerves by injecting either ^{3}H-leucine or ^{3}H-fucose into the ventral horn of the L5 lumbar cord and found labeled proteins and glycoproteins transported through the TTX-treated region of nerve with the same fast rate of 385 mm/day and with the usual form seen in control nerves. Thus, there was little overall change in transport in nerves chronically treated with TTX, confirming the earlier findings obtained in nerves *in vitro*.

Studies made using TTX were further extended by Bray, Hubbard, and Mills (1979) using a capillary implant technique which made it possible to maintain a TTX block for as long as 21 days. Over this longer period of time, the fall in resting membrane potential was again found to be lower in denervated soleus muscles than in muscles with their nerve activity blocked by TTX (Fig. 14.5).

Additionally, the ACh receptor increase measured by ^{125}I-α-bungarotoxin

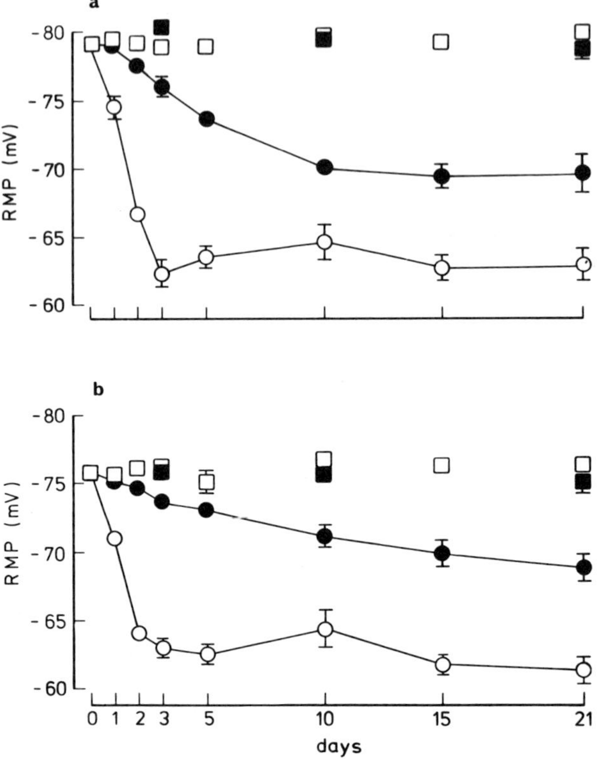

Fig. 14.5 Comparison of TTX block of nerve conduction and denervation on resting membrane potentials. The EDL (a) and soleus (b) muscles are shown. The unfilled squares indicate potentials from control muscles, the filled squares indicate muscles in which a control capillary was implanted in the sciatic nerve. After implanting a TTX-containing capillary, a decrease in muscle membrane potentials was seen (●) with a greater fall after nerve transection (○). Each symbol indicates the mean resting membrane potential of 6 to 11 muscles; a minimum of 15 implacements were made in each muscle. Bars indicates the S.E. of the mean. The difference in potentials seen after TTX block of nerve and denervation was significant at all times represented. From Bray *et al.* (1979).

binding was greater following denervation than after the inactivity produced by a TTX block of the nerves innervating the muscle (Fig. 14.6).

Additional support for a neurotrophic influence came from studies of nerve compression made in the lower limb of baboons using narrow cuff compressions, to the degree adequate to cause a persistent block of activity without causing Wallerian degeneration (Gilliatt *et al.*, 1978). After 28 days of such compression, an increase in the spread of extra-junctional ACh-sensitivity was found in the muscles innervated by those nerves, but with no sign of muscle fibrillation. The authors concluded that a neurotrophic substance supplied by nerve acts to suppress fibrillation, while the loss of muscle activity is responsible for the increased spread of ACh receptors in the muscle fibers.

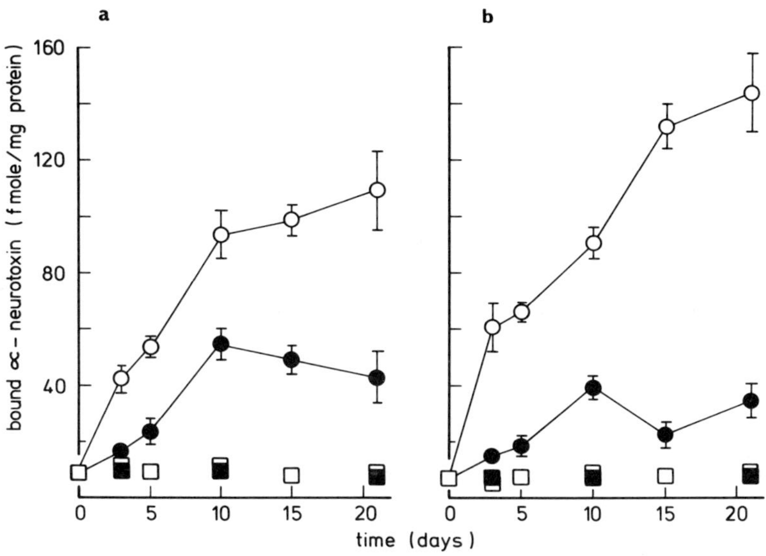

Fig. 14.6 Comparison of TTX nerve block and denervation on ACh receptor density. The EDL (a) and soleus (b) muscles are shown with the same symbols as in Fig. 14.5. The mean binding of ^{125}I-labeled α-neurotoxin (f-mole/mg protein) of 6 to 11 muscles is shown with the bars indicating the S.E. of the means. The difference between TTX-nerve block and denervated muscles was significant at the $P < 0.001$ level in all cases, except for the 10-day e.d.l. values where $P < 0.02$. From Bray *et al.* (1979).

3.   Block of Axoplasmic Transport and Denervation Effects

The inverse of the above experiments, namely, a block of axoplasmic transport with a retention of nerve conduction, was studied by Albuquerque *et al.* (1974). Silastic cuffs containing either vinblastine or colchicine were applied to the sciatic nerves of rats to block axoplasmic transport without affecting the conduction of nerve impulses (Chapter 9). Although some changes in the form of the myelin of the fibers were seen after 10 days, the conduction of impulses through the cuffed regions remained present in those nerves. The block of axoplasmic transport caused denervation-like changes to appear in the muscles innervated by those nerves, a decrease of RMP and the appearance of TTX-insensitive action potentials. Along with those signs of denervation, MEPP activity remained present, the potentials somewhat increased in rate and amplitude. Those changes most likely are due to the increased muscle membrane impedance seen to follow denervation (Hubbard, 1963). Albuquerque and his colleagues concluded that the denervation effects in the muscles were produced by an interference in the supply of a neurotrophic substance to the muscle. Changes in the activity of the muscles would not be causal insofar as the conduction of nerve impulses and activation of muscles remained present.

The inference drawn in the previous section with respect to local anesthetics pertains to the studies of Albuquerque; namely, that colchicine or vinblastine did not effect a complete block of transport. The absence of Wallerian degeneration and the maintained viability of the axons and MEPP discharges indicates that the tubulin-binding agents produced a partial effect on transport, one on the downflow of neurotrophic substances. A more profound effect on transport was seen in other experiments where colchicine was directly applied to the nerve (Perisic and Cuenod, 1972). In those studies a complete block of neurotransmission was produced, one similar to nerve transection. A similar correspondence between colchicine application and the effect of transection was also demonstrated by Pilar and Landmesser (1972) in ciliary nerves. The difference in the results obtained in the latter two experiments compared to those of Albuquerque and his colleagues may be explained by differences in the permeability of the different nerves exposed to the agents used to block axoplasmic transport. The nerves studied by Persic and Cuenod and by Pilar and Landmesser were finer and perhaps also lacked a perineurial sheath, this allowing a greater uptake of colchicine, and as a result a complete block of transport.

As was pointed out above, differences in the exact dimensional relation of the silastic cuff to the nerve may give rise to variations in the uptake of the agent by the nerves or a mechanical effect might be produced which in itself could lead to signs of denervation. Cangiano *et al.*, (1977) used cuffs with somewhat smaller holes and found this to cause a compression of the nerves with resulting signs of denervation; namely, a decrease in RMP, fibrillation potentials, and TTX-resistance action potentials, and considered this to be the main causal factor. Warnick *et al.* (1977), however, pointed out that in their own cuffing experiments with tubulin-binding agents, such compressions were not likely to have occurred because they did not observe muscle fibrillation. To obviate the possibility of a mechanical effect, cotton wetted with colchicine solution was applied to sciatic nerves by Tiedt *et al.*, (1977). The block of axoplasmic transport they found was generally similar to that obtained with silastic cuffs containing tubulin-binding agents, observations taken to support the conclusion of Warnick *et al.* (1977) that mechanical compression in itself cannot explain the results found using silastic cuffs containing tubulin-binding agents.

Yet another methodological objection has been raised to the use of colchicine or vinca alkaloids applied either via cuffs or directly to nerves; namely, that these agents gain entry to the circulation and, carried to the muscle, can act directly on it (Cangiano and Fried, 1977). As evidence for this possibility, Cangiano and Fried injected small amounts of colchicine subperineurally into the nerve on one side of animals and found denervation-like changes appearing 4–5 days later in the contralateral muscle, as well as in the muscle innervated by that nerve. This could only have come about if the agent had gained access to the contralateral muscle through the circulation. However, Kaufmann *et al.* (1974) did not find the same degree of leakage into the circulation when using ^{3}H-colchicine impregnated silastic cuffs; its effects were found limited to the muscle innervated by the cuffed nerve. Additionally, when DeCoursey *et al.* (1978) applied cotton soaked with colchicine to nerve trunks, they found a decreased Cl⁻ conduction, a change correlated with the increased membrane impedance indicative of a denervation (Camerino and Bryant, 1976; DeCoursey, Younkin, and Bryant, 1978) only in the muscle innervated by that nerve.

4. Muscle Activity Counteracting the Effects of Denervation

The possibility that muscle inactivity in itself can cause some of the effects ascribed to denervation, was supported by the finding that a direct stimulation of muscles after nerve transection is able to prevent or reverse the denervation effects (Lømo and Westgaard, 1975). The spread of ACh receptors into the extra-junctional region, typically seen in denervated muscles, was blocked or reversed in muscles directly stimulated with a pattern of electrical pulses close to the rates of discharge normally occurring in the muscle. The block by stimulation of receptor spread was seen even when the program of direct stimulation was begun 5 days after denervation. Jones and Vrbová (1974) also found that direct stimulation of denervated muscles could prevent the spread of ACh receptors into extra-junctional regions, but only when such stimulation was initiated within the first 2–3 days following nerve transection. When stimulation was begun after that time it was only able to reduce the spread of sensitivity to ACh to a slight degree. The explanation for the differences seen by the two groups with respect to the time at which the onset of stimulation was effective in reversing the spread of ACh sensitivity remains unresolved.

Direct stimulation appears able to reverse some other muscle changes which appear after denervation; the ability to accept foreign innervation (Jansen *et al.*, 1973), fibrillation (Lømo and Rosenthal, 1972; Purves and Sakmann, 1974), the decrease of RMP and other changes in the electrical properties of the membrane (Lømo and Westgaard, 1975). However, some denervation effects cannot be explained by a diminished muscle activity. When the doubly-innervated frog sartorius muscle is deprived of one of its nerve supplies, only the denervated portion of the muscle fiber becomes hypersensitive to ACh, in spite of the fact that the whole length of the muscle is still activated through the remaining nerve supply (Katz and Miledi, 1964a).

A further complexity to the problem is that some factor arising from denervated nerve fibers has been proposed to cause denervation effects in normal muscle fibers. Cangiano and Lutzemberger (1977) reported that when rat extensor digitorum longus (EDL) muscle is partially denervated by cutting the L5 root, TTX-resistant action potentials are seen in nearly all the fibers of the muscle 84 hr later, i.e., those which remain innervated by fibers from the other roots. This does not explain the earlier experiments of Cangiano *et al.* (1977) using narrow bore silastic nerve cuffs where, as a result of partial nerve damage, TTX-resistant action potentials were found in muscle fibers lying adjacent to those with normal responses to TTX. Products of nerve degeneration were also considered by Lømo and Westgaard (1975) to cause a transient ACh hypersensitivity in soleus muscles. These may well be non-specific effects in that similar denervation-like changes were produced in muscles by simply placing threads on them, the resulting transient hypersensitivity lasting several days (Jones and Vrbová 1974).

Neurotrophic substances carried from the innervated muscles back to the neurons was suggested by a series of studies by Kuno and his colleagues. Czéh *et al.* (1978) transected thoracic spinal cord to reduce the activation of the lower motoneurons and in turn the muscles. Within 8 days, the soleus motoneuron cells showed a reduction in the duration of their after-hyperpolarization following an action potential, a change indicative of a partial denervation (Huizar *et al.*, 1977). Additionally, a reduction in muscle weight occurred. Sensory nerve deprivation by cutting the dorsal roots had no such effect (Kuno *et al.*, 1974). While block of the soleus nerve with TTX did not produce such an effect in the muscle, electrical stimulation of the nerve below the TTX block prevented the effects of cord transection from appearing, and stimulation above the TTX block did not. The results taken together indicate that the normal activity of the innervated muscle causes some neurotrophic factor to be carried by retrograde axoplasmic transport from the muscles to their motoneuron cell bodies to maintain their usual properties. Watson (1970) used the rat hypoglossal nerve transplanted into the ipsilateral sternomastoid muscle as an experimental model. When the spinal accessory nerve innervating the muscle was cut, metabolic changes were found in the hypoglossal nerve cells. This was prevented by injecting botulinum toxin locally into the muscle at the same time. Those results suggested that the metabolic changes in the cells followed upon some factor from the denervated muscles reaching the nerve terminals via collateral nerve sprouts, these carried by retrograde transport to the hypoglossal cell bodies.

All these results taken together suggest that muscle fibers can exhibit denervation-like phenomena in response to several factors: diminished activity, specific and non-specific external influences, as well as neurotrophic factors. We can no longer view all the changes seen in muscle after partial or complete block of transport as due wholly to a loss of a neurotrophic factor or to the lack of activity. The nerve and the muscles they innervate appear to constitute an interrelated system in which neurotrophic factors act as communicating signals in both directions, both to the muscles and from the muscles to the nerve cell bodies. Activity is a factor in such interactions.

5. Cross-Reinnervation

Fast and slow skeletal muscles in the mammal are differentiated by their mechanical properties: the speed of muscle contraction or development of tension, rate of relaxation following a single contraction, and the fusion of tetanic contractions or tension at a sufficiently high rate of repetitive stimulation (Buller and Lewis, 1965a,b; Close, 1967). In the newborn kitten all muscles are slow, with those destined to become fast muscles taking on their adult characteristics in the immediate post-natal period 3 to 4 weeks after birth (Denny-Brown, 1929).

Isometric contraction durations are measured by the time from the beginning to the peak of muscle tension increase. Relaxation times (half-relaxation time) are measured from the peak of tension to the time where it falls to half its peak amplitude. In the slow muscles of the kitten examined 1 day after birth, the contraction time was typically 80 msec and for the half-relaxation time 89 msec (Buller, Eccles, and Eccles, 1960a,b). By day 29, fast muscles, for example the flexor hallucis longus (FHL), have developed their adult pattern of shorter isometric contraction times, of about 25 msec, with a half-relaxation time of about 16 msec. The slow muscles remain slow-contracting.

To demonstrate the influence of nerve on the rate of contraction of muscle, Buller *et al.*, (1960a,b) used the technique of cross-reinnervation. In this procedure the nerves of the fast flexor (FHL) muscle, and the slow muscle (SOL) are cut, crossed and re-sutured so that the fast nerve was able to regenerate down within the distal nerve stump to reinnervate the slow muscle and similarly the slow nerve to reinnervate the fast muscle (cf. Chapter 15). Sufficient time was allowed for the nerves to reinnervate their "foreign" muscles before testing for any resulting changes in the contraction properties of those muscles (Fig. 14.7).

The cross-reinnervated muscles are seen to take on the contraction properties characteristics of their new nerve supply. The slow soleus muscles when cross-reinnervated by fast nerves now gives rise to isometric muscle contractions typical of fast muscles and the fast muscles cross-reinnervated by slow nerves those of the slow muscle. These changes are not simply due to the effect of the nerve transection and the regeneration of nerves into the muscles to different rates of repetitive stimulation showed a correspondence in the fusion of their contractions. Fusion occurs at the lower stimulation rates of about 15–20 pps in slow muscles, while only an incomplete fusion muscles to different rates of repetitive stimulation showed a correspondence to the fusion of their contractions. Fusion occurs at the lower stimulation rates of about 15–20 pps in slow muscles, while only an incomplete fusion is seen in fast muscles at this rate of stimulation. When cross-reinnervated by a slow nerve, fast muscle became changed in its tetanic response and showed a considerable amount of fusion at the low frequency of stimulation as expected of a slow muscle. The cross-reinnervated slow muscle became shifted in its response to that typically expected of a fast muscle with less fusion seen at the low frequency of stimulation.

Biochemical processes fundamental to the contraction process also change as a result of cross-reinnervation. Mommaerts (1969) has shown that the level of actomyosin ATPase activity (or myosin ATPase which parallels it) of muscle can be correlated with its speed of its contraction (Bárány, 1967). This follows from the control exerted by the ATPase over the rate of utiliza-

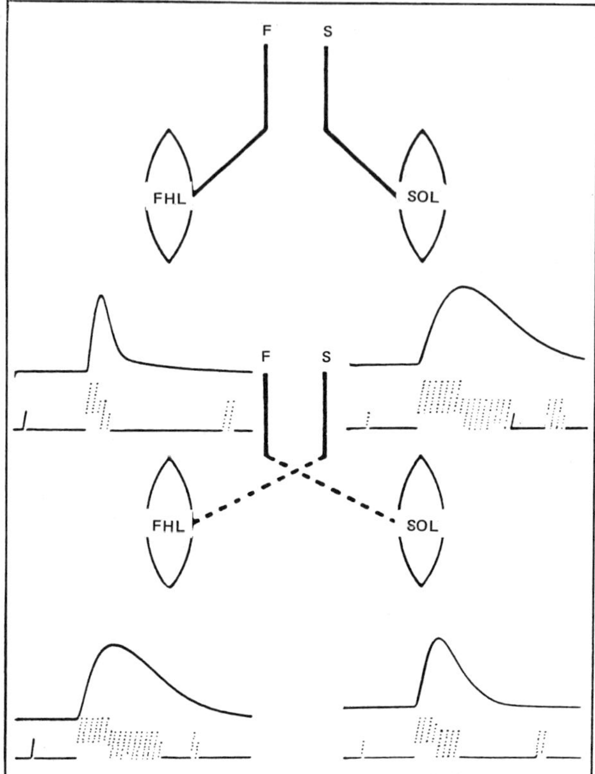

Fig. 14.7 Effect of cross-reinnervation on muscle contractions. The changes in the duration of isometric muscle contraction are shown before (upper curves) and following (lower curves) cross-reinnervation. The normal duration of the muscle responses of flexor hallucis longus (FHL) is short and that of soleus (SOL) muscles is long. The responses of the cross-reinnervated FHL and soleus muscles are changed, with the flexor muscle now giving a longer, and the soleus muscle a short-lasting, contraction. From Buller (1970).

tion of ATP in contraction as the filaments slide along one another (Huxley, 1969). The high level of activity of myosin ATPase characteristic of fast muscles was seen to fall in muscles after cross-reinnervation, to the lower level typical of slow muscles, and conversely the low level of myosin ATPase characteristic of slow muscles to rise after cross-reinnervation to approach, though not reach, the high levels usually found in fast muscles (Buller *et al.*, 1969). Cross-reinnervation also alters the ability of isolated microtubular vesicles to take up Ca^{2+} (Mommaerts *et al.*, 1969), another process related to the speed of muscle contraction, with a change conforming to the change imposed on the muscle by its new innervation.

The earlier results of Buller, Eccles, and Eccles (1960a,b) indicated some asymmetry in the effect of cross-reinnervation. The slow cat soleus muscle when cross-reinnervated by the fast flexor digitorium longus (FDL) nerve showed a shortening of its twitch duration by only 10–15% rather than the expected complete shift in its response to the duration characteristic of a fast muscle. On the other hand, a complete transformation was seen in the rat when those same muscles were taken for cross-reinnervation (Luff, 1975). The relatively small change found in the cat as compared to the rat was traced to a difference in the FDL muscles of these two species. A relatively large number of slow muscle fibers were found present in the fast FDL muscle of the cat, accounting for the small change when it was cross-reinnervated by the slow soleus muscle. When the nerve of the fast extensor digitorum longus (EDL) muscle of the cat, which contains only a few slow motor units, was used to cross-reinnervate with the nerve of the slow soleus muscle, a complete shift in the response of the soleus muscle toward that typical of a fast muscle was seen (Luff, 1975).

The cross-reinnervation experiments can be explained two ways. Specific neurotrophic materials transported in the fast and slow nerve fibers could confer their special character on the muscles they innervate, normally or after cross-reinnervation. Alternatively, the changes seen in the muscles following cross-reinnervation could result from the different pattern and rate of nerve impulses imposed on the muscle by its new motoneuron supply. In support of the latter hypothesis, a period of direct stimulation of fast muscles at a slow pulse rate converted them to slow muscles, as shown by the appearance of the longer-duration contractions typical of slow muscles (Salmons and Vrbová, 1969). The effect of different patterns of stimulation of muscles with fast and slow patterns of stimulation was further studied by Lømo *et al.* (1974). Brief trains of 100 pps were used to induce fast, and 10 pps slow muscle changes. The denervated soleus muscles of the rat were shifted in their responses to correspond to the rate of stimulation used, i.e., they become fast muscles with the fast rate of stimulation and remained slow muscles with a slow rate of stimulation. Other properties associated with slow and fast muscles were similarly shifted by the stimulation pattern (Lømo *et al.*, 1980). A higher level of alkaline-resistant actomyosin ATPase typical of fast muscles appears in slow soleus muscles following their cross-reinnervation with a fast nerve, and this was also seen to be the case when the soleus muscle was directly activated by a fast pattern of nerve stimulation (Guth *et al.*, 1970).

Shifts in muscle properties brought about by an imposed pattern of muscle stimulation appear to have a relation to changes in human muscle seen following a given regime of exercise. Brief bursts of activity favor the development of fast muscle properties, a performance required for a fast sprint, while long-maintained activity, such as occurs during long-distance running, the development of slow muscle responses.

6. Neurotrophic Influence on Regeneration of Minced Muscle

Studitsky (1964) discovered that minced muscle can regenerate into more or less normal-appearing functional muscles (Gutmann, 1976; Carlson, 1973).

This was shown by removing the gastrocnemius, plantaris, and soleus muscles from the back of the hindlimb of adult rats, mincing the muscles and placing the mince back into the hollowed-out place in the leg. Those fragments were found to become transformed, first into myoblasts, then myotubes, and eventually to reform muscles. The nerve must be present in the regenerating field for the minced graft to progress beyond the first stages of development (Zhenevskaya, 1960, 1961). Nerve fibers grow into the graft over a 10–14-day period, during which time the regenerating muscle fibers begin to develop cross-striations and then later primitive motor end-plates, the latter typically after 30 days. If the limb is denervated, muscle regeneration ceases with only narrow bands of connective tissue seen remaining (Hsu, 1971). A few myotubes and fibers with cross-striations were found, indicating that the failure of regeneration occurs just after the myotube stage. Hsu also showed that motor fibers are required for the muscle to develop, sensory or autonomic fibers being ineffective in this regard. Furthermore, fast nerves organized the minced muscle preparation into fast muscles and slow nerves into slow muscles.

7. Influence of Nerve on Limb Regeneration in Lower Vertebrates

If the leg of a newt or salamander is amputated, a whole new limb can regenerate as long as the nerve is present. If the nerve has been removed, the limb does not regenerate, a phenomenon observed in the frog, newt, and salamander (Rose, 1948). Both sensory and motor fibers are effective in inducing limb regeneration, the degree of regeneration depending on the total amount of nerve tissue present rather than any specific type of nerve (Singer, 1952). Of the possible explanations for the phenomenon, a neurotrophic factor supplied by the nerve fibers best fits with the evidence (Singer, 1959). Studies bearing on the identification of those neurotrophic factors will be described below in Section C.

B. NEUROTROPHIC CONTROL OF SENSORY ORGANS

The organization of non-neural cells around the nerve terminals to form sensory organs such as the taste buds, lateral line organs, and muscle stretch receptors, indicates the action of a neurotrophic substance (Zelená, 1964; Guth, 1969), one unrelated to mechanical activity—as appears to be the case, at least in part, in muscle.

1. Sensory Organ Denervation

a. Tastebuds.   Early evidence showing a neurotrophic effect on taste buds was their degeneration upon cutting the glossopharyngeal nerve supply to the tongue muscle of dogs (von Vintschgau and Hönigschmied, 1877). This phenomenon has since been verified in a number of species (Olmsted, 1920). The cells of the taste bud, a flask-shaped structure with its epithelial cells organized around the nerve terminal, undergoes a constant turnover with a complete renewal of cells taking place over a period of several days (Beidler, 1963). The sensory organ requires its nerve supply to maintain that turnover. A neurotrophic influence was indicated by the finding that cutting of the nerve to the taste buds of the barbils in the catfish close to the sensory organ caused it to degenerate sooner than when the nerve was cut some distance from it (Torrey, 1934). Taste buds reappear when the sensory nerve fibers regenerate to reinnervate the cells in that region. The neurotrophic substance appears to require a specific target cell. Sensory nerve fibers will not cause taste buds to appear in nongustatory cells, such as for example, epithelial cells taken from some other site (Zalewski, 1972). This was shown by transplanting sensory ganglia and various epithelial cells to the anterior chamber of the rat eye where the specificity of the neurotrophic effect could be assessed. Either the peripheral or central process of the sensory neuron was able to induce a regeneration of the sensory organ, indicating some generality in the neurotrophic substance supplied by those neurons in their daughter branches (Zalewski, 1969).

b. Lateral-line organs.   The lateral-line organs found distributed anteroposteriorially in placodes along the body surface of fish act as receptors for electrical potentials (Bennett, 1971). The lateral-line organ develops by the budding and the migration of cells to the surface of their sensory placodes, the sensory nerve fibers terminating at the base of the hair cells in the fully formed organ. The neurotrophic influence of the nerve on the lateral-line organ was shown by the failure of the organ to develop when its nerve fiber supply had been cut (DeVillers, 1948). A further indication of a neurotrophic factor was that the time at which the lateral-line organ degenerated after denervation depended on the length of the transected nerve stump left remaining in connection with the organ (Parker, 1932).

In some species the lateral-line organ appears not to depend on neural influences (Stone, 1937). However, as Zelená (1964) has shown, the presence of the nerve is necessary at an early stage where it acts as an inducer for the later development of the organ. After the induction period, the further development of the organ is independent of its nerve supply.

c. Muscle Receptors.   The studies of Tower (1932) and Zelená (1957) focused attention on the influence of nerve on the development of the sensory receptors of muscle. The muscle spindle receptor is a composite organ. A large sensory nerve fiber ending constitutes the annulospiral part of the receptor in the center with its muscular portions present at either end. In the course of its histogenesis, the spindle proceeds from the myoblastic stage to the myotubular stage with its chain of nuclei lined up in the longitudinal axis. In the fully differentiated spindle the nuclei are accumulated in the center where it receives the sensory fiber. If the developing spindle organ of the rat is deprived of its sensory nerve at the myotube stage, the spindle does not develop beyond the myotube stage and it degenerates.

Small (gamma) motor fibers innervate the polar (muscular) regions of the spindle organ. On cutting the ventral roots which supplies gamma fibers to the muscular parts of the spindle in the newborn rat, the muscular regions were reduced in size while the central region supplied by the sensory fibers remained relatively unaffected (Tower, 1932; Boyd, 1962a,b). Zelená and Hník (1963) were able to show that when the spindles were subsequently reinnervated, development could begin anew, if this was done within a certain time. The effect was seen in the rat when reinnervation was arranged for in the immediate post-partum stage of life, and was no longer effective in maintaining the receptor 14 days later (Zelená, 1964).

C.   THE NATURE OF THE NEUROTROPHIC FACTOR

1.   Block of Release of Neurotrophic Factor

The toxin of *Clostridium botulinum* acts in very low concentrations to prevent the release of ACh from the motor nerve terminals (Gunderson, 1980). Thesleff (1960b) had earlier demonstrated that botulinum toxin causes a spread of ACh sensitivity in muscle fibers, a sign of denervation. Other denervation-like effects produced by the botulinum toxin were subsequently found, i.e., fibrillation and the formation of new end-plates with an acceptance of foreign innervation in extra-junctional regions (Purves and Lichtman, 1978, 1980). The possibility had been raised by Thesleff that ACh itself could be the neurotrophic substance. Miledi (1960a), however, showed that the increased spread of ACh sensitivity indicative of denervation seen in organ cultures of muscles was not reversed when ACh was added in high concentrations. Also, on reinnervation of denervated muscles, the ACh receptors which had spread outside the end-plate region were seen to shrink back toward the junctional region well before the reappearance of MEPP activity (Miledi, 1960b).

An alternative explanation is that the toxin prevents the release of ACh from the nerve and an accompanying neurotrophic factor as well. Evidence that botulinum toxin may block such a release was the finding that fast-transported proteins become accumulated to a greater extent within the nerve terminals of preparations exposed to the botulinum toxin than in untreated preparations (Bray and Harris, 1975). On the other hand, Miledi and Spitzer (1974) concluded that the toxin may not block the release of all materials from the nerve terminals. Slow muscles normally do not give rise to full action potentials, but they do so when directly stimulated 2 weeks after denervation. The addition of botulinum toxin prevented such elicitations of full action potentials. Miledi and Spitzer inferred from those findings that botulinum toxin fails to block the release of a neurotrophic factor which normally acts to supress the elicitation of full action potentials in the slow muscles.

Beta-bungarotoxin, which prevents the release of ACh from nerve terminals, also produces denervation effects. Thesleff (1974) considers that β-bungarotoxin and botulinum toxin act to block the release of a neurotrophic material from the nerves coupled to ACh release or alternatively, that ACh acts to increase the permeability of the muscle membrane to a neurotrophic factor which is also released from the nerve terminals.

Drachman (1974) concluded that the maintenance of normal muscle function requires the spontaneous release of ACh by MEPP activity, the activity of the muscle and the supply by nerve of some as yet unspecified neurotrophic substance, one which he has referred to as "mysterine." Efforts to reveal the nature of the neurotrophic substances are described in the following section.

2. Characterization of Neurotrophic Factors

A suitable assay system is needed to isolate and characterize neurotrophic factors. Muscles showing characteristic signs of denervation in organ culture have been favored for this purpose. Lentz (1974) used the triceps muscle of the newt (*Triturus vindescens*). In cutting the nerve, the axon terminals are seen to rapidly degenerate while the junctional folds of the end-plate do so more slowly before they too eventually disappear. A gradual fall in the content of AChE also appears in these denervated muscles. When a sensory ganglia explant is added near the muscles in the organ culture, these regressive changes are mostly prevented or reversed. The junctional folds maintain their normal appearance and AChE levels, while they do not revert completely to their control values, are higher than those seen in the cultured denervated muscle without the ganglion present nearby. Newt brain homogenates added to the muscle culture also have a similar neurotrophic effect. Cyclic AMP plus theophylline had a similar influence, as did adenosine. With respect to the latter finding, Kreutzberg and Schubert (1979) pointed out the general significance of the transport and release of nucleotides from nerve terminals, especially adenosine, which acts on the membrane of the post-synaptic cell as a modulator of synaptic activity. McIlwain (1976, 1978) has also, from a theoretical point of view, considered nucleotides as possible modulators of the activity of neurons and perhaps have a neurotrophic influence.

Davey, Younkin, and Younkin (1979) used the denervated extensor digitorum of the rat in which a fall of AChE was measured 5 days later. Such muscles were taken 3 days after denervation and kept *in vitro* as an organ culture for 2 days before assaying them. The influence of trophic substances added to the culture medium was seen by the reversal of the fall of AChE which normally occurs over the period of 5 days. Brain or spinal cord extracts added to the medium reversed some of the decrease of AChE, as did muscle extracts. Of particular importance was the finding that the substances transported and accumulating in nerves just proximal to transections, were also effective. A similar result was seen by Fernandez, Patterson, and Duell (1980), who found some soluble component in a nerve extract (or perhaps the 16S AChE component itself) to augment the level of 16S AChE in cultured denervated rat anterior gracilis muscle. A less well-defined protein released from nerve on stimulation was also found to act as a neurotrophic factor (Musick and Hubbard, 1972; Fernandez and Inestrosa, 1976).

Neurotrophic factors were implicated in the studies of cultured muscles carried out by Peterson and Crain (1970, 1972). Isolated fragments of 18-day fetal rodent skeletal muscle explanted to a culture medium exhibited myotube formation with cross-striations. The cross-striations remained present in some of the fibers after the 2nd week, but they began to atrophy after the 3rd week (Crain and Peterson, 1974). When a spinal cord-ganglion fragment was implanted nearby, the cultured muscles went on to a further stage of maturation which was maintained for months. Furthermore, those muscle fibers could become innervated. Even muscles which had become atrophic took on a normal appearance and became innervated under the influence of a nearby neural implant (Crain, 1976).

Oh, Johnson, and Kim (1972) and Oh (1975) used chick embryonic skeletal muscle cells in culture as an assay and found that extracts of chick brain and spinal cord were able to stimulate morphological development, protein synthesis, and an increased level of AChE activity in the muscle fibers. Soluble extracts of adult chick sciatic nerve were fractionated and a protein of 10,000–50,000 dalton isolated which could stimulate the synthesis of macromolecules, increase the level of AChE activity, and support a long-term maintenance of muscle cells in culture (Oh, 1975). Later studies indicated that the neurotrophic factor is a glycoprotein (Markelonis and Oh, 1978). Using Sephadex G-200 columns, the most active fraction extracted from chicken nerve was found to have a molecular weight of 84,000 (Markelonis and Oh, 1979). Its protein nature was shown by its inactivation with trypsin or protease hydrolysis. In transected adult chicken nerves the trophic material was retained in the distal stumps of transected nerves (Oh *et al.*, 1980), possibly by its sequestration in Schwann cells upon its release from the degenerated axons.

Somewhat similar results indicating a neurotrophic influence were seen using the regenerating limb buds of the newt for assay (Lebowitz and Singer, 1970). When the forelegs are amputated, a new limb becomes regenerated from a bud stage which appears some 10–14 days later, a process requiring the presence of a nerve supply. If the nerve was removed, regeneration was depressed (cf. Section A7 above). Following denervation of the newt limb, an early decline in the incorporation of labeled thymidine into DNA and uridine in RNA was seen. This was followed by a decline of ^{3}H-leucine incorporation into protein 7 hrs after denervation, preceded by a short-lasting period of increased incorporation (Singer and Caston, 1972). Using newts with forelegs amputated and denervated as a test preparation, newt nerves were homogenized and either homogenates or the supernatant taken after centrifugation of the homogenate for 1 hr at 40,000 g were injected into these animals. The effect was to increase protein synthesis in the limb buds, as indicated by an augmentation of ^{3}H-leucine incorporation. Heating the material abolished the neurotrophic activity of these materials, suggesting its protein nature.

When organ-cultured muscle is used to assay a possible neurotrophic substance, an appropriate stage of development of the muscle must be selected. If taken too late, the muscle can develop response requirements other than those present at an earlier stage. Also, undefined culture media introduces considerable variability into the response characteristics of the system. All these factors make it difficult at present to completely predict the action of a given material suspected of being a neurotrophic agent (Peterson and Crain, 1979). However, with adequate precautions, the organ cultures represent a valuable technique for isolation of possible neurotrophic substances.

Davis and Kiernan (1980) have recently reported that denervated extensor digitorum muscles of the adult rat can be used as an assay. The reversal of some of the denervation effects by nerve extracts would indicate a neurotrophic action. In this muscle the cross-sectional area of type IIb fibers decreases to about 40% of control levels following denervation. About half of this decrease was reversed following 7 days of intramuscular injection with a protein isolated from nerve (cf. also Davis and Kiernan, 1981). How proteins with their relatively large molecular weights can enter muscle fibers *in vivo* to restore the deficiencies of denervation remains to be determined. However, these positive results appear to be promising in the search for neurotrophic factors.

Relatively unknown are those neurotrophic substances which might act on the neurons of the CNS. A large literature deals with morphological changes in CNS neurons suggestive of neurotrophic effects, those for example seen in post-synaptic CNS cells following transection of their presynaptic afferent fibers (Canton and Appel, 1974; Cowan, 1970). A number of similar changes have also suggested the operation of a retrogradely transported neurotrophic influence (Ghetti, Horoupian, and Wisniewski, 1975). The trans-synaptic transfer of neurotrophic materials (Chapter 12) may very well be required for the normal development of neurons and the maintenance of their function. The identification of such neurotrophic substances is clearly an important goal in the understanding of neuronal properties underlying behavior.

The CNS is complex and a simpler system which serves as a model for it is the autonomic ganglia. Those cells show a behavior indicating the participation of a trophic factor after axotomy and the reformation of synaptic connections (Purves and Njå, 1978; Purves and Lichtman, 1978). Evidence that nerve growth factor (NGF) is a neurotrophic agent has been obtained in this system. A further discussion of NGF will be taken up in Section A2 of Chapter 15 where its neurotropic action will be described.

15

Regeneration and Reinnervation

In the course of embryonic development some signals are necessary for growing neurites to arrive at their proper target cells: on neurons, muscles, cells of sensory organs, and secretory cells. A similar process may be seen in the mature nerves after interrupting their fibers. From the central part of crushed or transected nerves, fine regenerating nerve fibers grow out to find their way into the distal part of the nerve, with more or less of them able to reinnervate their original target cells. In this chapter we discuss axoplasmic transport in relation to factors affecting the growth of nerve fibers. As will be seen, the evidence indicates that substances arising from the target cells act on the neurites to direct their growth toward them, substances referred to as "neurotropic," the process termed "neurotropism." These require a routing in the neuron so that regeneration can proceed in the appropriate fiber branches, as described in the following section.

A. PERIPHERAL NERVOUS SYSTEM

1. Neurotropism in Relation to Regeneration of Nerve Fibers

At a relatively early stage in the course of normal development, neurites are seen to sprout from the neuron cell bodies with an orientation toward their presumptive target cells, a directivity which is genetically predetermined (Jacobson, 1978). The growing fibers thereafter appear to be guided by signals in their environment, ultimately reaching their target cells. The signals are received by a special structure at the ends of the growing neurite, the growth cone, which has the ability to move and as a result to determine the path taken by the neurite elongating behind it (cf. Section A2 below).

The concept of a neurotropic signal was inferred from the behavior of regenerating peripheral nerve fibers. Within a short time after nerves are interrupted by a cut or a crush, the process of regeneration begins. Sprouting of collaterals may begin from the fibers just above their interrupted ends even as early as 4 hr after crushing a rat sciatic nerve (Zelená, Lubińska, and Gutmann, 1968), but it is more obvious after a day or so. A relatively large number of thin fibers may arise from the nerve fibers, usually from one or

362

several of the nodes above the interruption, as was shown by Perroncito (1905), Marinesco and Minea (1906), and Ramon y Cajal (1928). An example is given in Fig. 15.1.

This figure is to be compared to the detailed picture of regenerating fibers in nerves transected made 4 days beforehand as reconstructed by Friede and Bischausen (1980) from EM serial sections (Fig. 10.11). As can be seen in Fig. 15.1, each sprouting fiber has a growth cone at its end. The cones serve to guide the fine growing fibers toward the distal degenerated part of the nerve (Ramon y Cajal, 1928; Young, 1949b).

Some of the fibers grow between the empty sheaths of the degenerated fibers in the distal stump, others find their way down inside the empty nerve sheaths. To show the course of growth toward the amputated stump, Ramon y Cajal (1928) cut nerves and then stitched the cut nerves in place in various

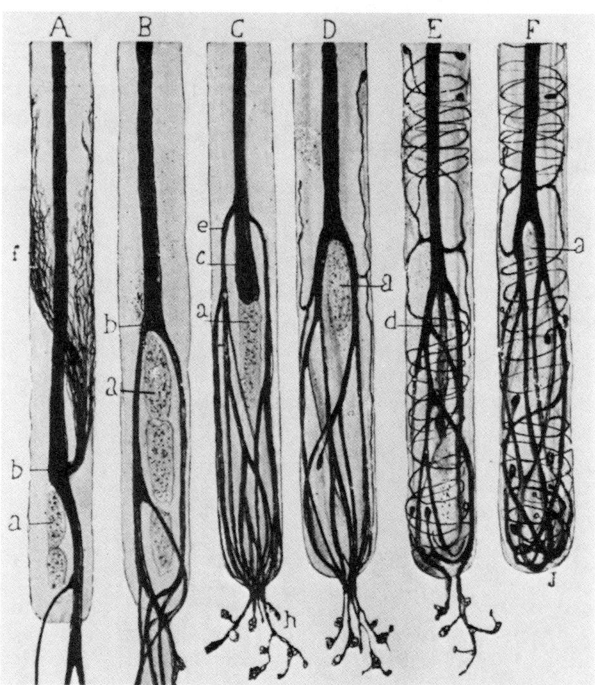

Fig. 15.1 Regenerating fibers arising from the proximal end of cut nerve fibers. Schematic drawings show the varied forms of the regenerating nerve fibers arising from the proximal cut end of fibers three and a half days after section of the sciatic nerve: A, axons with a main ascending and descending branch; B, axon with two main terminal branches; C, axon with collaterals and a sterile tip (c); D, axon similar to B, but with branches giving rise to intratubal recurrent twigs; E, helicoidal apparatus of Perroncito, with emergent branches; F, sterile apparatus of Perroncito, i.e., without emergent branches; a, necrotic portion of the axon; d, sterile segment of the axon in an apparatus of Perroncito. From Ramon y Cajal (1928).

ways, separating the two ends by more or less of a distance. After allowing sufficient time for the regeneration of fibers from the upper portion of the nerve, a neurotropic influence arising from the degenerated fibers in the lower portion of the nerve was inferred by the direction in the growth of regenerating fibers toward it (Fig. 15.2).

In the experiment shown at the left of Fig. 15.2, the upper end of the cut nerve had been turned away from the distal amputated end. The regenerating fibers are seen to have grown down to enter the stump of the degenerated nerve. When the cut ends were more closely approximated, as shown at the right in Fig. 15.2, there was less wandering in the growth of the regenerating fibers suggesting a stronger directive influence on their growth by the closer-placed amputated nerve fibers.

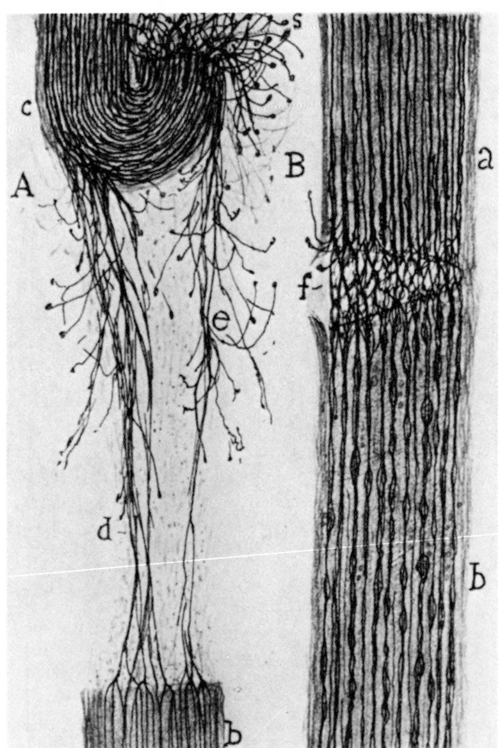

Fig. 15.2 Neurotropic effect on regenerating fibers. (A) A nerve (c) was cut and the upper end turned away from the distal amputated stump and stitched in place. Later, numbers of fine fibers are seen to grow from it down into the distal, degenerated part of the nerve (b). A fiber is seen bifurcating into thick and thin fiber branches (e), the latter being interstitial (d) with clubs and growth into the empty sheath. (B) Only the border of a nerve bundle was cut (f). Upper nerve fibers (a) grow directly down to penetrate the distal segment (b). There is little evidence of lateral outgrowth, presumably because of the closeness of the distal amputated fibers and the neurotropic influence of the distal amputated fibers. From Ramon y Cajal (1928).

The proportion of regenerating fibers directed toward and growing into the amputated distal stump of nerve appears to be greater than can be accounted for on a chance basis. On that basis Ramon y Cajal (1928) considered that a neurotropic substance is released from the degenerated nerves to direct the growth of regenerating fibers. The neurotrophic substance was thought to act on the growth cones, causing them to move toward the degenerated nerve fibers. He visualized the growth cone at the tip of the growing neurite as "a living battering ram, soft and flexible, which advances, pushing aside mechanically the obstacles which it finds in the way, until it reaches the area of its peripheral distribution. This curious terminal club, I christened the growth cone" (Ramon y Cajal, 1937). Further studies using EM have revealed the fine structure of this important entity, as will be described in detail in Section A2 below.

Possibly the neurotropic influence arises from the Schwann cells which increase in number in the denervated nerve (Ramon y Cajal, 1928; Schroder, 1975). The mitosis of Schwann cells in the course of their proliferation is shown by the increased uptake of the DNA precursor ^{3}H-uridine. In the mouse, the increase is seen to peak in the distal amputated stump of nerve 5–10 days after its transection (Asbury, 1967; Bradley and Asbury, 1970). Later, the Schwann cells become progressively encircled by fibroblasts and then by cells of the perineurial type. Then after a much longer time these cells are replaced by connective tissue (Weinberg and Spencer, 1978), their presence acting to hinder reinnervation (Aguayo and Bray, 1980).

It is interesting to note how closely Ramon y Cajal's reconstruction of the growing neurites and their growth cones which he derived from fixed histological preparations corresponds to the behavior of growth cones visualized in the living state. The latter was first observed by Harrison (1910) using organ cultures of the neural tube of frog larvae, of the Rohon–Beard sensory neurons (Fig. 15.3).

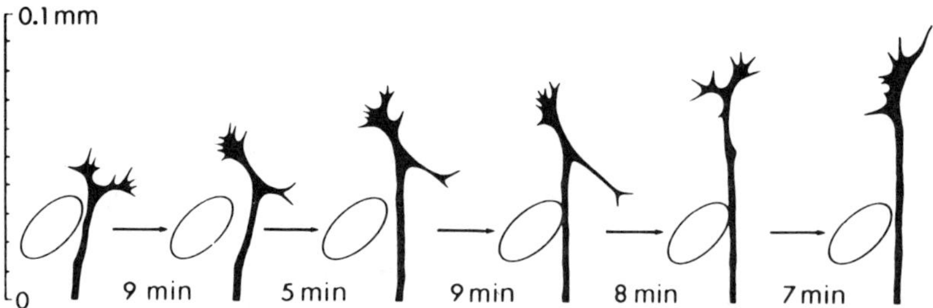

Fig. 15. 3 Successive views of a growing fiber and its growth cone. Six successive views of a growing nerve fiber in tissue culture show changes in shape and rate of growth. Note resorption of side-branch off the main line of growth. The red blood corpuscle, shown in outline, marks a fixed point. The average rate of elongation of the nerve was about 1 mm/min and the total length of the nerve fiber was 800 μm. The observations were made after 4 days of growth on lymph. From Jacobson (1978) as modified from Harrison (1910).

The growing nerve fibers and their growth cones show changes in form as one growth cone and its fiber advances while others retract and undergo involution. The tadpole has a translucent skin allowing visualization of the neurite growth in the tail fin (Speidel, 1941, 1964). Making use of this preparation, Speidel found the growth behavior of the neurites in the living organism to be similar to that observed in tissue culture. The fibers branch and grow in several directions, those branches which advance toward their ultimate destination remain, while aberrant branches involute. An appreciation of the multiplicity of neurite aborizations and their changes in the course of growth may be gained from Speidel's drawings (Fig. 15.4).

Speidel pointed out the salient features of the growing neurons: (a) an overabundance of neurons and of nerve fiber branches early in development,

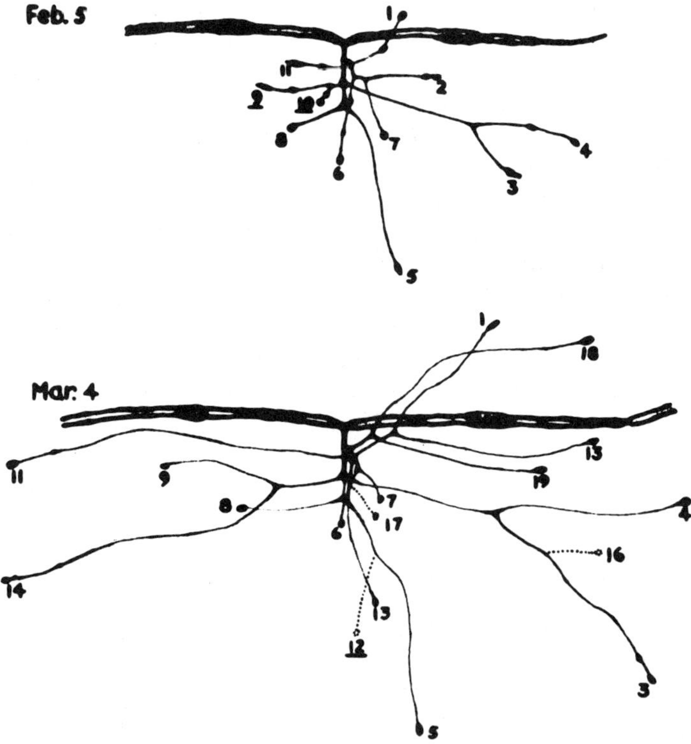

Fig. 15.4 Growth of nerve fibers visualized *in vivo* in a tadpole embryo tail. Multiplicity of the aborizations are seen in the course of a period of 27 days. The sketch on February 5 in the upper panel shows 11 cutaneous endings arising from the side-branch at a node of Ranvier between the two most distal myelin segments of a fiber. Six of the endings were characterized by resting end bulbs, four by retraction clubs, and one by a growth cone. By March 4, the arborization had grown and consisted of 14 endings as shown in the lower panel. Eight new endings had formed and five endings have disappeared in the interim. From Speidel (1964).

(b) an apparent randomness in the growth of most of the neurites with a more direct orientation toward their ultimate target by some, (c) the achievement of the proper target cell by one fiber, (d) a regression and involution of abererrent fiber branches and redundant neurons, and (e) the further growth and thickening of the successful neurite which has innervated their target cells. The latter process has been termed maturation. The innervation of the target cells at first is not precise, the target cells are innervated in a polyneuronal fashion. Then in the course of maturation, the wide-spread innervation becomes progressively reduced as each innervating fiber becomes confined to a smaller number, or even to an individual target cell. This was shown to be the case for skeletal muscle, parasympathetic and sympathetic ganglia and the visual cortex of the CNS (Purves and Lichtman, 1980).

Similar changes appear in the course of regeneration of myelinated nerve fibers in the adult animal. A large number of fine unmyelinated regenerating fibers may be found within the old nerve tubes early in the course of regeneration, these fibers later diminishing in number until only a few fibers remain to become myelinated and thickened (Shawe, 1955). The greatest reduction in the number of proliferated fibers appears to come about when one of the fibers makes contact with a target cell (Young, 1949b). This suggests that a signal substance ascending within the successful fiber by retrograde transport to the cell body acts to augment the further growth of that fiber by increase of transported materials, at the same time bringing about an involution of aberrant fibers. This sequence of events requires a channeling of the transported materials within the nerve branches of the individual neurons, a process preventing substances from spreading to all the many other fiber branches of the regenerating neuron in an indiscriminant fashion. One of the puzzling problems in neurobiology is how such a selective channeling within one or a few of the many branches of a neuron can come about. We can explain this phenomenon on the basis of the routing phenomenon (Chapter 11). Each of the fiber sprouts would have their individual subset of microtubules. Retrograde transport can carry neurotropic materials up one such subset of microtubules with, subsequently, a supply of needed materials carried down by anterograde transport in those same fiber sprouts and not in the others. This process would allow for the growth of one fiber at the expense of the others (Ochs, 1981c).

2. The Growth Cone

Only a few microtubules and rarely neurofilaments are seen at the base of the growth cone in EM (Fig. 15.5).

In the process of nerve fiber elongation, substances are carried by axoplasmic transport along the microtubules of the neurites to the growth cone to supply the materials it requires and as well the growing nerve fiber just behind it. These substances would include, among others, the tubulin subunits of the microtubules, and the triplet protein subunits of neurofilaments

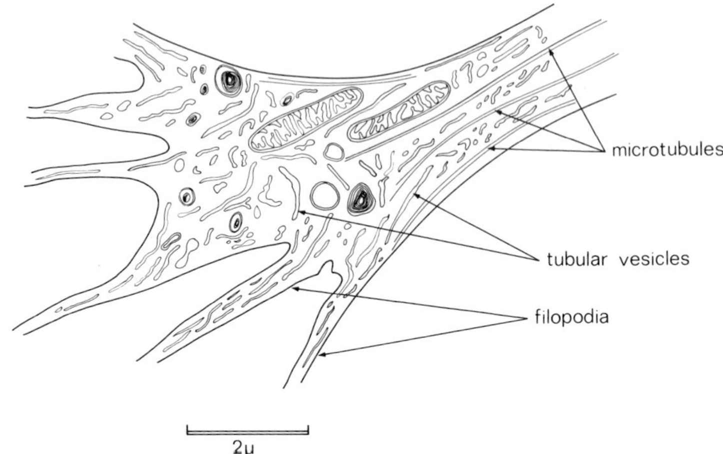

Fig. 15.5 Sketch of growth cones and its filopodia. This sketch represents the organelles seen in EM in the growth cone. Note that the microtubules extend into the base of the growth cone with mitochondria present. Neither of these organelles are seen in the filopodia. From Bray (1973).

(Chapters 9, and 11). From the analysis carried out in Chapter 11, it would appear that these subunits are added to their respective organelles at the base of the growth cones in the course of fiber growth (cf. Bray and Gilbert, 1981). The growth cone not only serves as the assembly site for the neurofilaments and microtubules, but acts as well to direct the growth of the neurite, functions likely to be interrelated.

The static picture given of a growth cone in Fig. 15.5 does not indicate the complicated motion of the filopodia extending over its surface as seen under the microscope in tissue culture. The filopodia extend, wave back, regress. A number of them undergoing this process and the veil-like covering over its surface, the velopedia, gives the appearance of a continual ruffling movement over the surface of the cone.

The filopodial have lengths up to 10–20 μm and diameters of 0.3 μm and internally contain mainly "a finely filamentous matrix" (Tennyson, 1970), the matrix consisting mainly of microfilaments (Yamada *et al.*, 1971). Bray (1973) has emphasized the relation of the ER to the microfilaments within the filopodia. In addition, dense-cored vesicles and cisternae are also seen in the filopodia (Bunge, 1973). Ferritin taken up from the medium is seen in clusters of vesicles in filopodia at their base near the body of the growth cones (Bunge, 1977). These vesicles are present in mound-like protuberances at the base of the filopodia which have been thought to act as an initial packaging depot for the retrograde transport of materials to the perikaryon. The vesicles, however, appear to be produced in the course of preparation of the tissue for EM (Bunge, Johnson, and Argiro, 1982), the actual uptake mechanisms apparently having a different morphology in the living state.

The movement of the filopodia may come about through the action of the actin-containing microfilaments. The agent cytochalasin B disrupts microfilaments, causing the filopodia to "wilt" and shorten so that the growth cone becomes rounded up and collapses in as little as 10 min after exposure to the agent (Yamada *et al.*, 1971). The microtubules and neurofilaments at the base of the growth cone remain unaffected. And, in line with its inefficacy on those organelles, cytochalasin B has little effect on fast axoplasmic transport, as shown by Anderson, Edström, and Mattsson (1972), Crooks and McClure (1972), and Banks, Mayor, and Mraz (1973b). Removing cytochalasin B from the culture medium reverses its effect on the filopodia. Cyclohexaimide does not interfere with the course of recovery from cytochalasin B treament, indicating that protein synthesis is not involved in the recovery process. The microfilaments appear to be part of a special contractile system responsible for the movement of the filopodia and the translocation of materials in them, one subserved by actomyosin. This process is separate from the supply of materials to the growth cones needed for the elongation of neurites subserved by the microtubules, axoplasmic transport.

How the filopodia extend from and regress back into the membrane of the growth cone has been discussed by Bray (1973). Vesicles may move into the filopodia and merge with the membrane so as to increase their length, but other possibilities remain open. In any case, microtubules are not present in the filopodia and their contained microfilaments. Most likely the motility is due to the actomyosin present in or associated with those structures which causes the mechanical response of the growth cone to tension (Bray, 1979).

3.   Nerve growth factor

Nerve growth factor (NGF) is a specific substance in the organism necessary for the maturation of sympathetic and sensory neurons (Levi-Montalcini, 1976; Mobley *et al.*, 1977). We have already referred to NGF with respect to its uptake by the nerve terminals (Chapter 12) and its retrograde transport in fibers (Chapter 3). Here we describe some of its general properties before passing on to a discussion of its neurotropic action.

Nerve growth factor is isolated from snake venom and mouse salivary glands where, for some reason, it is present in large amounts. It is composed of two identical polypeptides of 13,259 daltons (Server and Shooter, 1977; Greene and Shooter, 1980). This dimer (the β NGF subunit) is found associated with two α and two γ subunits. The latter components, though not necessary for the activity of the β subunit, appear to play a role in biosynthesis, storage, and protection of the active NGF moiety (Mobley *et al.*, 1977). When given in excess to newborn rats, NGF causes a marked hypertrophy of the sympathetic ganglia, to as much as 12× their normal size. The ganglion cell bodies are not only enlarged but increased in their number. This occurs because normally there is an oversupply of cells present at birth, a number of them dying in the course of maturation. When NGF is present in excess these cells retain their vitality and growth capabilities.

As part of its growth-stimulating properties, NGF increases the density of microtubules and neurofilaments in the neurons. It also promotes the polymerization of actin and increases myosin ATPase activity (Calissano et al., 1978). In adrenergic neurons, NGF induces as well an increase of the synthetic enzymes TH and DBH (Paravicini et al., 1975). Conversely, if NGF antibodies are injected into newborn rats it will, by combining with NGF, decrease the level of NGF in the organism and cause a marked involution of the sympathetic ganglia and its neurons (Levi-Montalcini and Angeletti, 1968). The adult animal also appears to require NGF for the maintenance of sympathetic neurons in that injection of NGF antibodies in the adult causes some atrophy of the sympathetic neurons with the appearance of neurofibrillary material (Angeletti et al., 1971).

When given in excess to animals at birth, NGF will also cause a similar hypertrophy of neural elements of the peripheral sensory nervous system. The dorsal root ganglia of the chick, specifically the mediodorsal group of cells, increase in size and number in response to NGF given 7 to 10 days post-partum. As in the case of the sympathetic neurons, NGF increases the number of cells in the sensory ganglia by maintaining the viability of cells which normally would die in the course of maturation (Mobley et al., 1977; Harper and Thoenen, 1980).

A marked effect on neurite growth in tissue culture by NGF was found by Roisen and Murphy (1973) using a quantitative assessment of its effect on the number, diameter, length, and degree of arborization of the neurites. Roisen suggested that NGF stimulates membrane-bound adenyl cyclase, the resulting increase in cyclic AMP promoting the assembly of microtubules from an existing pool of tubulin as part of the process of enhanced neurite growth. As evidence for this position, the addition of butyrl cyclic AMP or butyrl cyclic GMP to the medium caused an increased growth of neurites (Roisen, Murphy, and Braden, 1972). Furthermore, the effect of an addition of cytochalasin B or colcemide to cause neurite retraction could be reversed by supplying dibutryl cyclic AMP or NGF to the culture (Roisen and Murphy, 1973).

Evidence that NGF is taken up by the nerve terminals before it exerts its growth promotion was shown by Campenot (1977). A 3-chamber culture system was used to contain the cell bodies of neurons in the central portion, its neurites growing through seals into the two side chambers. By this means NGF could be placed either in the central chamber or into one or other of the side chambers without its leaking to other chambers. The addition of NGF to the central chamber containing the cell bodies did not alter neurite growth. Addition of NGF to either of the side chambers augmented the growth in that chamber, and conversely the removal of NGF from either of the side chambers caused growth on that side to stop while growth in the other side chamber remained unaffected. Thus, the growth of neurites was contingent on the presence of NGF immediately at their terminals where it is taken up and carried to the cell bodies to affect synthetic processes. Such findings can be explained by a routing along one subset of microtubules (see Section A1 above and Chapter 11).

In addition to its growth-promoting function, i.e., its neurotrophic effect, NGF has a neurotropic action. This was suggested by the observation that NGF injected intracerebrally caused neurites of nearby sympathetic neurons to grow aberrantly and in profusion into that region (Levi-Montalcini, 1976). This indicated that the NGF released from target organs taken up by the growth cones directs it and the growth of its neurites toward those target cells as indicated in Fig. 15.6.

The NGF which enters the tips of the growth cone would be carried by retrograde transport to the cell bodies to stimulate it to produce the materials needed for the growth of the fibers, routed back down (Chapter 11) in those specific fiber branches to increase their growth at the expense of the other neurite branches (cf. Section A1 above). Possibly NGF may also act directly within the growth cones to enhance the microtubule and neurofilament assembly needed for neurite growth.

The effect of NGF to direct the growth of neurites was demonstrated in chick dorsal-root explants (Gundersen and Barrett, 1979). The tip of a micropipette containing a very small amount of NGF was placed a short distance from a growth cone at an angle to it and within 9–21 minutes the axon was seen to turn and grow toward the micropipette. Successive repositioning of the pipette away from the growth cone caused similar further growth directed toward the source of NGF with, as a result, a controlled bending of the axon. This experiment supports the concept that NGF has a neurotropic action, one able to modify the direction in which the growth cone moves and its neurite grows.

A glial factor has been found to stimulate growth of neuroblastema cells, (Monard et al., 1973), cells which do not respond to NGF (Monard et al.,

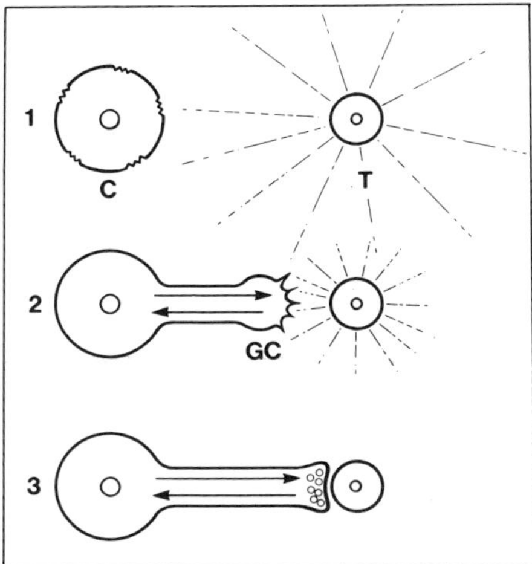

Fig. 15.6 Neurotropic effect proposed for nerve growth factor (NGF). In 1, the target cells (T) are considered to secrete NGF, which, as shown by the radiation, reaches the cell body (C). In 2, a neurite is shown growing toward the target cell along the path of NGF diffusion. The NGF taken up by the growth cone is carried by retrograde transport in the fibers to the cell body, where it augments the directed growth to the target cells. In 3, synaptic contact is made.

1975). It seems most likely that other similar factors will be discovered, those able to stimulate the growth of motoneurons and other types of neurons in the peripheral and central nervous systems not responsive to NGF.

4.   Do Degenerated Nerve Fibers Release a Specific Neurotropic Factor?

The question arises as to whether neurotropic substances are released from degenerating nerve fibers which are capable of attracting a specific type of nerve fiber to it. To test this point, Weiss and Hoag (1946) cut the peroneal and tibial branches of the sciatic nerve of rats and inserted the proximal ends of the cut nerves into a Y-shaped sleeve of blood vessels so as to channel their regenerative growth toward the stem of the Y in which the distal cut end of the tibial nerve was inserted (Fig. 15.7).

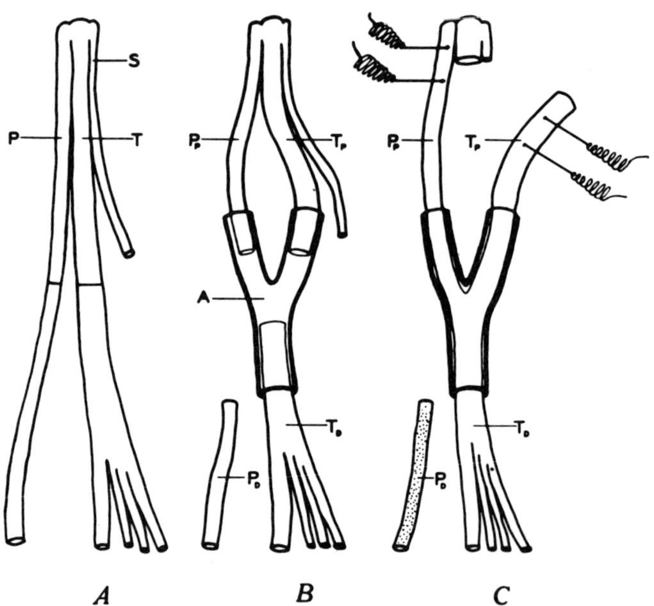

Fig. 15.7 Test for a neurotropic influence from a distal amputated nerve stump. (*A*) The peroneal and tibial branches of rat sciatic nerve were cut and (*B*) inserted into the arms of a Y-shaped blood vessel (A) leading to the distal cut end of the tibial nerve (T_p) and its muscle. In (C), after regeneration of fibers, the proximal ends of the peroneal (P_p) and tibial (T_p) nerves are cut and each is stimulated to determine if there has been a preferential re-innervation of the tibial nerve stump and its muscle by the tibial nerve. From Weiss and Hoag (1946).

If a specific neurotropic factor arises from the distal amputated portion of tibial nerve able to attract the growth of the fibers from the regenerating part of the tibial nerve toward it, we might expect that more fibers of the tibial nerves will have regenerated down into the tibial nerve stump in preference to fibers from the peroneal nerve. This would be shown (cf. Fig. 15.7) if, after allowing sufficient time for regeneration to occur, stimulating the tibial nerve above the site of regeneration gave rise to a greater contractile response of the tibial muscle than stimulating the peroneal nerve. The result of the experiment was that the response of the tibial muscle was the same when stimulating either nerve. The fibers of the peroneal and tibial nerves had thus grown down into the tibial nerve stump on a chance basis rather than preferentially into the degenerated distal portion in response to a neurotropic influence arising from it.

Bernstein and Guth (1961) performed a somewhat similar experiment. In the rat more ventral root fibers of the lumbar 4th spinal cord segment innervate the plantaris muscle than the soleus muscle. The difference in innervation of the two muscles was shown by stimulating the 4th lumbar root and measuring the tensions developed in the two muscles. After crushing the sciatic nerve and allowing regeneration and a subsequent reinnervation of the muscles to take place, the responses of the two muscles to stimulation of the lumbar 4th root were then seen to be similar. This experiment showed that the regenerating nerve fibers had entered the degenerated plantaris and soleus nerves below the crush on a chance basis.

An aspect of this experiment requires some comment. When nerves are crushed, we might expect the nerve tubes to remain, although the integrity of their axons has been lost (cf. Sunderland, 1980). The regenerating fibers could then grow down within the same nerve tubes and thus be guided to their original targets (Young, 1949b). This would result in the same asymmetry of response on stimulating the roots as seen in the normal animal. It would appear that the crushes made in Bernstein and Guth's studies had been sufficiently disruptive as to be the equivalent of a nerve transection, i.e., that the nerve tube continuity was lost, allowing the regenerating fibers to enter the distal degenerated nerve tubes on a chance basis.

Brushart and Mesulam (1980) also demonstrated the randomness of the regeneration of fibers of the rat peroneal and tibial motoneurons following sciatic nerve transection. When the tibial or the peroneal muscle was injected with HRP, the marker substance is carried by retrograde transport into the spinal cord where it was found located in the corresponding peroneal and tibial motoneuron cell bodies. However, after nerve transection followed by regeneration, application of HRP to either one of the muscles was followed by its presence in both tibial and peroneal motoneuron cell body populations.

The apparent lack of a neurotropic influence directing regenerating fibers to their original nerve tubes indicated by the studies described above underlines an important problem in the treatment of nerve lesions in man. Functional recovery after surgical repair in the adult is poor, in large part because of the indiscriminate regeneration of nerve fibers with respect to their old tubes. The chance entry of regenerating motor fibers into the motor nerve tubes of antagonistic muscles or sensory fibers entering motor fiber nerve tubes, and vice versa, would lead to no useful recovery of function (Mark, 1969). If some means could be found to direct the growth of fibers back into their old tubes, and thence to their original target organs, the degree of restitution of function would be markedly enhanced. To this end individual nerve fascicules have been resutured. The procedure is of limited success, however, due to the irregularity of the path taken by the fibers as they pass down within the complex pattern of fascicules present in most peripheral nerves (Sunderland, 1978).

5.   Rate of Growth and Transport in Regenerated Nerve Fibers

The tips of the growing nerve fibers are very sensitive to chemical or mechanical stimulation (Tinel, 1915). The most distal point down the limb at which the animal reflexly responds to a mild nociceptive stimulation of the nerve is found at successively more distal sites as the fibers grow down (Gutmann et al., 1942). The rate of regenerative growth of sensory fibers so determined was 4.5 mm/day, starting several days after nerve interruption. Bisby (1978) found a lag of a little over a day before regenerative nerve growth began with a rate of regenerative growth measured in the rat of close to 3.8 mm/day (Fig. 15.8).

To assess the rate of growth of motor fibers, the sciatic nerve was interrupted at various distances from the leg muscles in a group of animals and the return of motor function determined by the reappearance of reflex toe-spreading (Gutmann et al., 1942). The rate of regeneration found by this means was 3.5–4.5 mm/day.

Another method used to assess the rate of growth of motor fibers is to inject ^{3}H-leucine into the cord and determine the distance to which the front of labeled proteins moves down regenerating fibers to their distal ends (Griffin et al., 1976). The rate of nerve regeneration measured in this way was 3.5 mm/day, a rate similar to that found by other methods (cf. Fig. 15.8).

Considering the marked difference in size between the thin regenerating fibers and the large myelinated fiber from which these fibers arise, it was of interest to determine what differences, if any, may be present in the rate of fast transport in the regenerated nerve fibers as compared to normal nerve fibers. Nerves were crushed and sufficient time allowed for the regeneration of sufficiently long lengths of nerve fibers. Transport in the regenerated fibers was then assessed by injecting ^{3}H-leucine into the spinal cord for motoneuron cell uptake and determining the position of the front of labeled proteins in those fibers (Griffin et al., 1976). The rate of axoplasmic transport found from the advance of the front of radioactively labeled material was seen to be identical in regenerated and normal control nerves (Fig. 15.9).

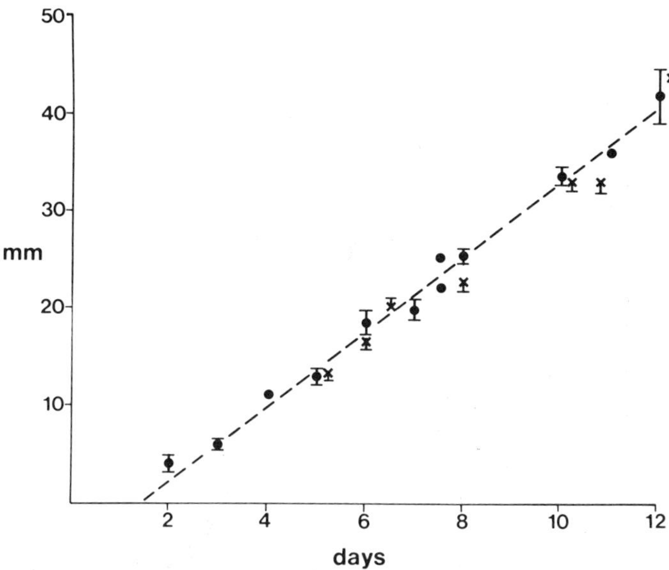

Fig. 15. 8 Rate of regeneration of rat sciatic nerve. The distance from a site of crush lesion to which the regeneration of axons occurs was measured by a pinch-reflex test (●) in 86 animals at different times after crushing. The advance of the front of transported radioactivity in the nerve (×) is also shown for 70 animals. On the abscissa, the days elapsing after making the crush and testing are shown. The points representing the mean ± S.E. are fitted with a regression line calculated by the method of least squares. From Bisby (1978).

The similarity of rates might have been expected, insofar as the rate of axoplasmic transport was found to be the same in unmyelinated fibers as in myelinated fibers of all diameters (Chapter 2). The amplitude of the crest of labeled proteins in the regenerated fibers was larger then that of the control fibers. This is the case because there is a large number of unmyelinated regenerating fibers present early in regeneration. Later on, fewer unmyelinated fibers are left remaining to transport labeled materials (Shawe, 1955).

A similar study of transport in regenerated sensory fibers of the rat made by Bisby (1978) also showed the rate of fast axoplasmic transport to be the same as that of normal fibers. In those studies Bisby observed an early temporary decrease in the amount of labeled activity transported into the regenerating nerves, followed after a few days by a return to control levels. The types of labeled proteins transported into the regenerated nerves also showed little difference from those present in normal nerves, as judged by comparing 23 protein peaks identified by SDS–PAGE. There were only some quantitative differences in several polypeptides (Bisby, 1980c). The small shifts in the peaks found were somewhat different from the minor changes found by Theiler and McClure (1978) in regenerating nerves (Chapter 6). In

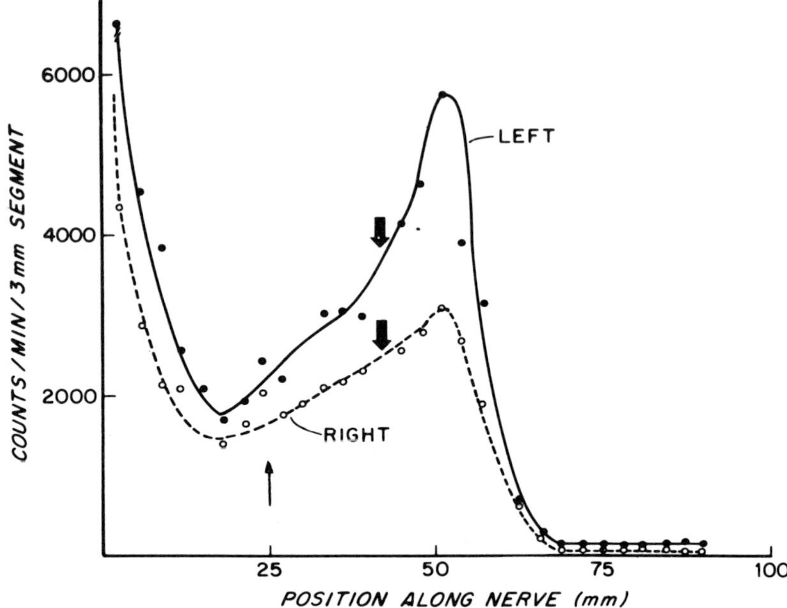

Fig. 15.9 Fast transport in regenerating fibers. Transport was examined 28 days after making a proximal freeze lesion in the nerve on the left. The ventral horn was injected with ³H-leucine and transport was allowed beyond the site at which the lesion was made (thin arrow) down to at least 90 mm distally. Transport in the regenerated nerve fibers (●) had a rate of 402 mm/day with the rate on the normal side (○) 396 mm/day. The ratio of crest sizes on the left (regenerated nerve) to that on the right (control nerve) was 1.86. From Griffin, Drachman, and Price (1976).

the experiments of Theiler and McClure, the nerves were taken above the transections, while Bisby studied components transported within the regenerated fibers themselves. In any case, both studies revealed only minor differences in the proteins of regenerating nerves as compared to normal neurons.

More recently, Willard and his colleagues identified novel proteins present in elevated amounts relative to other proteins during periods of axonal growth (Levine, Skene, and Willard, 1981). They have designated them as growth-associated proteins (GAPs) with molecular weights of 50,000, 43,000, and 23,000–24,000. The 43,000 MW component, GAP 43, appears to be most closely related to axonal growth. It is increased in the rabbit hypoglossal nerve where regeneration occurs and does not increase in the optic fibers which do not regenerate. The GAPs are similar to the proteins showing a minor shift in regenerating rat sciatic nerves (Theiler and McClure, 1978) and in goldfish optic nerve (Benowitz, Shashona, and Yoon, 1981). These findings suggest that the GAPs might play a specific role in axonal growth.

6. Regeneration and the Time of Surgical Repair

There has been some uncertainty in the clinical literature regarding the best time for the repair of transected nerves, whether it should be done soon after a nerve injury, or after a delay of 2 or 3 weeks. Much longer postponement is generally considered to give progressively less satisfactory results, although regeneration is possible even after many months (Sunderland, 1978). The basis for waiting several weeks after nerve interruption is that chromatolysis reaches a peak at about this time, suggesting that the nerve cell bodies might be in a "primed" metabolic state (Ducker *et al.*, 1969). On comparing crest amplitudes of fast transported radioactivity at different times after making cells chromatolytic, little support for an increased amount of transported labeled protein at this time was found (Chapter 5 and Ochs, 1976). However, some change in the properties of the chromatolytic neuron does appear several weeks following a transection. When a second nerve transection was made 2 weeks after a first conditioning crush, the rate of nerve fiber regeneration was found to be increased 27% (McQuarrie and Grafstein, 1973). In those studies the rate of nerve fiber growth was determined morphologically by the use of silver-staining. In a subsequent study (McQuarrie *et al.*, 1977), both the first and second nerve interruptions were made by crushing, and the pinch test was used as a functional means to determine the rate of growth of the sensory axons. The rate of nerve fiber regeneration found after a single crush was 4.3 mm/day, and after a crush following a first conditioning crush made two weeks beforehand, the rate was increased by 23%.

The increase in rate suggests that the preceeding lesion brings about more favorable conditions for the assembly of the fibers, including their microtubules and neurofilaments, at their growing ends in the growth cones (cf. Section A2 above and Chapter 11).

The increase in the rate of regeneration of the fibers would not in itself necessarily give rise to a better clinical outcome. The factors involved in a clinical evaluation of recovery of functon are complex (Sunderland, 1978). As noted above, the fundamental problem in bringing about a satisfactory clinical result is to somehow direct regenerating fibers back into their appropriate nerve tubes in the amputated nerve so that the regenerating fibers can reinnervate their original target cells.

7. Collateral Sprouting Near Target Sites

When a muscle is partially denervated, collaterals sprouting from nearby nerve fibers reinnervate the denervated muscle fibers (Van Harreveld, 1945). This was shown in the rabbit by removing a portion of the nerve supply of the sartorius muscle at some distance above it. By stimulating individually the L4, L5, and L6 roots and measuring the force of muscle contraction, L6 was shown to be the major root segment contributing fibers to the sartorius muscle. The L6 root was then removed by avulsion to prevent its regeneration, and the force of contraction of the sartorius to stimulation of the L5 root determined at intervals over a period of several months. These were seen to increase to normal in a relatively short period of time.

Histological examination showed the muscle fibers to have a normal appearance without the atrophy expected from a partial denervation. Van Harreveld explained the phenomenon by a collateral sprouting of fibers from nearby normal nerves reinnervating the denervated muscle fibers. Such collateral sprouting was shown by Hoffman (1950) and Edds (1953) to arise from the terminal regions of the nearby normal nerve fibers, either from the nodes, or in some cases the internodes. The neurolemmal sheaths of the degenerated axons serve to guide the collateral nerve fibers back to the original end-plates of the denervated muscle fibers (Barker and Ip, 1965). The rapid functional recovery occurs quickly because sprouting begins after only a short delay and the collaterals need grow only a relatively short distance to reinnervate the denervated muscle fibers.

Sprouting was reported to be stimulated by the intramuscular injection of a number of ill-defined substances: ether extracts of egg yolk, muscle, nerve, or brain (Hoffman, 1950), substances to which the term "neurocletins" was given. Similar efforts along this line to find stimulating substances had also been made by Forssmann (1898, 1900) and by Von Muralt (1945), an endeavor analogous to the search for neurotrophic agents (cf. Chapter 14). It would appear that modern biochemical methods would be fruitful. Along that vein Gorio *et al.* (1980) found gangliosides to promote motor nerve sprouting.

Occasionally, collateral sprouts are seen in normal muscles, suggesting that a constant process of renewal of nerve terminals is going on all the time (Barker and Ip, 1965). This may be similar to the constant regenerative renewal which appears to go on in the sensory nerve terminals (Weddel and Zander, 1951). This would further suggest the presence of some regulatory signals between the neuron and the target cells to maintain the pattern of innervation.

Factors influencing the sprouting and regeneration of collateral nerves in relation to their target sites was studied by Aguilar *et al.* (1973) in the salamander (*Amblystoma tigrinum*). In this animal the lower lumbar spinal nerves of segments 15, 16, 17 innervate the peripheral sensory fields of the hind leg. These fields were mapped by lightly touching the skin and recording the resulting action potentials from each of the spinal nerves (*A* of Fig. 15.10) .

In the intact limb nerve 16 was seen to receive its sensory input from a wide region of the limb, with nerves 15 and 17 receiving sensory input from more limited skin regions on either side of the field innervated by nerve 16 (left limb, *A* of Fig. 15.10). When nerve 16 on the right side was cut and the limb mapped 14 days later, the sensory fields of nerves 15 and 17 were

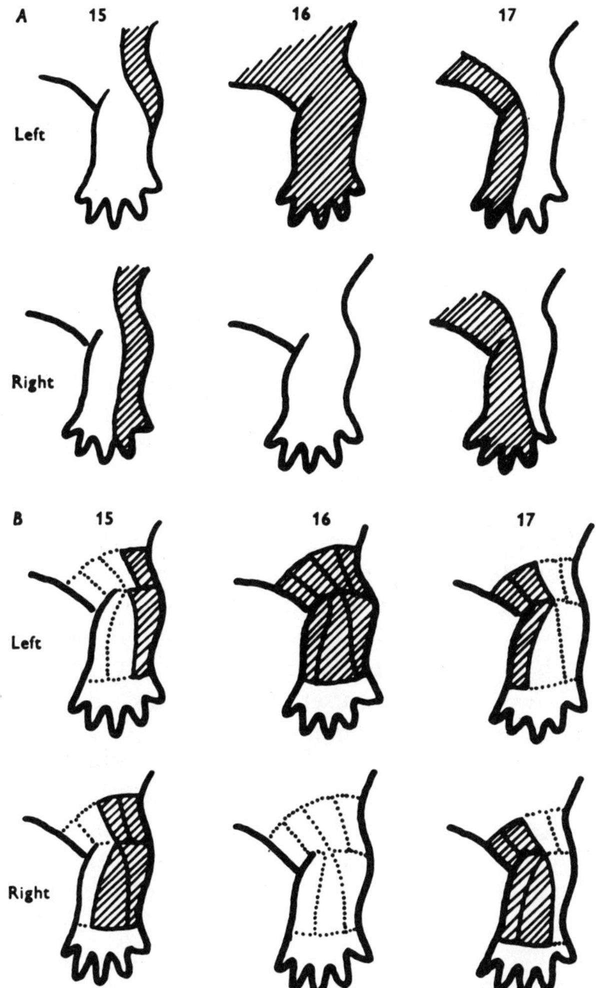

Fig. 15.10 Collateral spread into denervated areas. (A) *Amblystoma* limb sensory nerve territories were mapped using a light touch of the skin while recording the resulting action potentials from each of the spinal nerves. The peripheral fields recorded in nerves 15, 16, and 17 from the left limb are shown on the upper row. Fourteen days after sectioning nerve 16 on the right side, the peripheral fields were mapped. The typical result is a spread of sensory fields 15 and 17 into the areas of skin surface subserved by nerve 16, as shown in the lower row. (B) Motor fields obtained by stimulating nerves and observing or recording from muscles which contracted in response to the stimulation are shown in the upper row taken from the left limb. Fourteen days after section of nerve 16 on the right side, the invasion of its denervated territory by collaterals from nerves 15 and 17 is shown on the bottom row. From Aguilar *et al.* (1973).

seen to have enlarged and to invade the receptive territory originally supplied by nerve 16 (right limb, *A* of Fig. 15.10).

A similar phenomenon was found for the motor fibers. Motor maps were determined in the muscles of the skinned hind limbs by stimulating nerves of segments 15, 16 or 17 and observing muscle contractions or recording EMG responses from the separate muscles of the hind limb (*B* of Fig. 15.10).

In the normal hind limb shown for the left limb, nerve 16 was found to supply most of the limb musculature, and nerves 15 and 17 muscles more limited portions of the musculature on either side of the region supplied by nerve 16. Fourteen days after transecting nerve 16, the motor territories supplied by nerves 15 and 17 were seen to have enlarged to invade the muscle territory normally innervated by nerve 16 (right limb, *B* of Fig. 15.10). If only a partial lesion of nerve 16 had been made, no such collateral spreads from nerves 15 and 17 were seen. The explanation is that the remaining uncut fibers of nerve 16 can more quickly provide collaterals to those portions of the denervated musculature than the more distant nerves 15 and 17.

Exposure of nerve 16 to colchicine for 30 min gave rise to the same result as a complete transection of the nerve, namely, collaterals from nerves 15 and 17 spreading to reinnervate the region normally supplied by nerve 16. The concentration of colchicine used, 0.1 M, was sufficient to block axoplasmic transport completely (Aguilar *et al.*, 1973), as was shown by the failure of accumulation of AChE and catecholamines in the nerves below the site to which the colchicine had been applied (Chapters 4 and 12).

The experiments could be interpreted to mean that normally nerves secrete a substance which inhibits sprouting of nearby fibers. When a nerve is transected the neighboring fibers are released from such inhibition and collaterals sprout to innervate the denervated muscles. Colchicine can block collateral sprouting. This was shown by first cutting nerve 16 and then applying colchicine to nerve 15. This prevents the expected spread of fiber collaterals from nerve 15 to innervate a part of the limb region normally supplied by nerve 16. The block of axoplasmic transport produced by colchicine could prevent the supply of substances needed for the fibers to sprout. Alternatively, the block of axoplasmic transport could interfere with the response of the nerve to a trophic signal arising from the degenerated field which is needed to initiate the supply of materials needed for growth of the collateral fibers.

Collateral sprouting in the plantaris muscle of the rat was also shown by the use of colchicine block (Guth *et al.*, 1980). When the fibers of the L4 segment were transected it was followed by a sprouting of L5 fibers to innervate the denervated muscles. Colchicine placed on the L4 nerves to block transport induced a similar collateral sprouting from L5 nerve fibers. These studies are considered to support the hypothesis (Aguilar *et al*, 1973) that collateral sprouting depends on the transport of some component in nerve fibers.

8. Reinnervation Receptivity of Muscle

Muscle preferentially becomes reinnervated at the site of its original end-plates (Gutmann and Young, 1944; Bennett, McLachlan, and Taylor, 1973a,b; Miledi, 1960a). Other regions of a muscle fiber, however, may become innervated under certain conditions. Frank *et al.* (1975) laid the cut end of the superficial fibular nerve onto the surface of the soleus muscle. Three weeks later, the soleus nerve was crushed and the denervated soleus muscle fiber outside of its original end-plate portion was then seen to become innervated by sprouts from the fibular nerve. Later, collaterals from the fibular nerve grew out to reinnervate the original end-plates and then the fiber branch which had innervated the membrane outside the original end-plate region became resorbed. Thus, while an extrajunctional innervation can occur, this may be only temporary and replaced by a reinnervation of the original synaptic site (cf. Elsberg, 1917).

We can infer from such evidence that the original end-plate region exerts a neurotropic influence directing nerve fibers to it, one which can act over a relatively great distance. The presence of a neurotropic influence arising from the synaptic sites of muscles to act on neurons was further indicated in cultures containing completely dissociated neurons and muscles cells. The neurites grow to form functional synapses on the muscles (Fishbach, 1970) which are seen in EM preparations to have the usual appearance. The possibility that the ACh receptors exert a neurotropic influence (Katz and Miledi, 1964a; Fex *et al.*, 1966) was not supported by observations made in organ culture using pharmacological agents to block the ACh receptors: Curare or α-bungarotoxin were unable to prevent the reinnervation of the synaptic sites (Crain and Peterson, 1974; Van Essen and Jansen, 1974).

In the end-plate region, AChE is bound to the basement membrane within the synaptic cleft (McMahan, Sanes, and Marshall, 1978). When regenerated motor axons reinnervate the end plate region of the denervated skeletal muscle, the fibers are seen to end in contact with the basement membrane. This suggests that AChE or some component associated with it in the basement membrane has a neurotropic influence (Sanes, Marshall, and McMahan, 1978).

Another factor conditioning reinnervation is the activity of the muscle. Direct electrical stimulation of denervated muscle prevents reinnervation (Fex *et al.*, 1966; Jansen *et al.*, 1973). When muscle is reinnervated a further reinnervation, a hyperneurotization of the muscle, is blocked (Aitken, 1950; Fex *et al.*, 1966). This could be the result of the supply of a suppressive factor by the innervating fiber, or perhaps the activity induced in the muscle by its reinnervation. In accord with the latter possibility, botulinum toxin which causes a paralysis of muscles (cf. Section 8), results in a sprouting of nerve fibers to it, sprouting which is prevented by directly stimulating the muscles (Brown, Goodwin, and Ironton, 1977). That sprouting may be induced by a lack of muscle activity was indicated by the use of α-bungarotoxin to bind to ACh-receptors and thus prevent muscle activation via its motor fibers. This procedure led to an increased sprouting of collateral nerve fibers to those muscles (Holland and Brown, 1980). The sprouts making terminations to the inactive muscles have an inceased quantal content and the frequency of discharge is increased as shown by the spontaneous miniature EPPs recorded from those muscles (Snider and Harris, 1979).

Using the superficial fibular nerve as a foreign nerve implanted into the soleus muscle, the reinnervation of the soleus muscle by the fibular nerve after cutting the soleus nerve shows a series of well-defined changes (Lømo, 1980). There is at first, as a result of muscle inactivity, an increase in ACh receptors in the denervated muscle with an induction of nerve sprouts. Spontaneous MEPPs and evoked MEPPs are seen 2.5–3 days later (Lømo and Slater, 1980a). The ACh receptors then accumulate in the muscle membranc to form hot spots, the multiple innervation of those sites later becoming reduced to a single innervation. At 6–7 days after section, AChE activity appears (Lømo and Slater, 1980b). The ectopic junctions formed by the nerve sprouts at the sites of ACh receptors and their mobilization and stabilization into hot spots were not affected by increased muscle activity or by a block of impulse activity in the nerve. These observations point to neurotropic and neurotrophic influences. The ACh sensitivity is influenced by muscle activity and as well as by nerve (Cangiano *et al.*, 1980), the muscle considered to be the source of AChE (Lømo and Slater, 1980b). Its supply by the nerve is not, however, excluded (Lømo, 1980), a point discussed in Chapters 13 and 14.

9.   Reinnervation of Receptor Organs in the Skin

The possibility of a neurotropic influence acting on regenerating sensory nerve fibers was shown for the dome-shaped receptors containing 20 to 50 Merkel cells innervating certain regions of the hairy skin of the cat (Burgess *et al.*, 1974). After crushing the sensory nerve fibers innervating the sensory domes, the fibers regenerate back to form new domes at their original sites with the same pattern of innervation present before denervation. When the nerves were cut rather than crushed, the coincidence of regenerated domes with those of the old sites was not as good, but it was better than would have occurred on a chance basis, this indicating a neurotropic influence.

Sensory fibers end as free nerve endings in the skin, their pattern determining sensation according to Weddel, Palmer, and Pallie (1955) and Sinclair (1967). More recently a reversion to a more deterministic sensory fiber specification for sensation has been supported (Bannister, 1976). It is unknown what directs the fibers to their sensory territories, a distribution which survives the considerable turnover of cellular elements constantly going on.

10.   Axon-Schwann Cell Interaction

Schwann cells engulf a number of unmyelinated fibers and individually cover the larger fibers to myelinate them at specified intervals along their length.

Some kind of signaling is needed to determine how the cells originating from the neural crest come to position themselves along the axons and maintain their function. (Bray, Raminsky and Aguayo, 1981)

The inference has long been drawn that the axon provides some necessary signal to the Schwann cell as shown by the Wallerian degeneration which follows upon axonal interruption (Chapter 6). This was shown also by cutting a predominantly myelinated nerve (sternohyoid) in the rat and cross-anastomosing it to the cut distal stump of a largely unmyelinated (cervical sympathetic) nerve. The increased number of myelinated fibers subsequently found in the distal grafted portion containing the regenerated fibers, is in accord with some signal from those axons initiating their myelination by the Schwann cells in the graft, from Schwann cells which ordinarily progress no further than the single wrapping characteristic of the unmyelinated axons (Weinberg and Spencer, 1975). Similar results suggesting an axon signal for myelination were found by taking the ordinarily myelinated fibers of the phrenic nerve and cross-anastomosing it to the amputated stump of unmyelinated sympathetic nerve trunk (Aguayo *et al.*, 1976). The fibers became myelinated in the graft, reflecting the influence of the axons on the surrounding Schwann cells. Conversely, when unmyelinated fibers were allowed to regenerate into the stumps of myelinated fibers, those fibers remained unmyelinated in the graft. The multipotential capabilities of Schwann cells were assessed in the mutant mice known as "trembler" and "quaking." A lack of myelin is seen in the nerves of the trembler and a less than normal myelination in quaking mice nerves. When pieces of these nerves were grafted to the nerves of normal animals in suitably immune-repressed animals, the fibers regenerated in the grafts lacked myelin or showed little myelin, reflecting the inability of the grafted Schwann cells to respond to the axon signal (Aguayo, Bray, and Perkins, 1979). The Schwann cells of the normal portion of nerve joined to the graft do not invade the graft region nor do the grafted Schwann cells migrate into the nerve territory occupied by normal Schwann cells, suggesting that some signals prevent such migration.

The nature of Schwann cell–neuron interaction was further studied in tissue culture (Bunge, 1975) where it is possible to obtain separate cultures of axons and Schwann cell elements and to recombine them. In such studies it would appear that a mitogenic signal from axons is provided to the Schwann cells to cause them to myelinate the axons. Some signal from the Schwann cells to axons was also suggested in the studies of Aguayo, Bray, and Perkins (1979), who noted that axonal diameters were influenced to some extent by the type of Schwann cells present in a grafted region.

In myelinated nerve fibers the axolemma is differentiated at the nodes where a high density of Na^+ channels is found present (Ritchie and Rogart, 1977). The large number of intramembrane particles seen in the external face of the nodal membrane in freeze-fracture preparations (Rosenbluth, 1976) could represent Na^+ channel proteins. Additionally, there are some special features present as shown by the dense staining of this region with ferric ion-ferrocyanide compared to the internodal axolemma (Waxman and

Quick, 1977). The question arises as to whether the nodal axolemma is differentiated subsequent to the myelination of the fiber or whether some axon membrane differences initiate the nodal changes. The latter was indicated in that intramembrane particles seen using freeze-fracture (Wiley and Ellisman, 1979), and ferric ion-ferrocyanide staining appeared at positions expected of nodes in the immature fiber before its myelination (Waxman and Foster, 1980; Waxman, 1981). The results were formalized as the "axonal demarcation" hypothesis, which holds that some signal from the axon to the Schwann cell cause the latter to position itself on the axon with a subsequent myelination of the internodal position of the fiber, the myelination depending on the multipotential capability of the Schwann cell (cf. Aguayo *et al.*, 1976; Weinberg and Spencer, 1976; Selzer, 1980).

B. REGENERATION IN THE CNS

1. Cell Orientation and the Direction of Neurite Growth

The difference between the genetic (intrinsic) factors which determine the original orientation of neuron cell bodies within the CNS, the direction in which their neurites sprout, and those influences which act on the sprouts to guide their neurite growth towards their target (extrinsic factors) are exemplified by the "improperly" oriented pyramidal cells occasionally seen in the cerebral cortex (Van der Loos 1965, 1976). Typically the pyramidal neuron has its apex and main dendrite oriented toward the surface of the cortex and its axon descending into the underlying white matter. In rare cases of improper orientation, pyramidal cells may be tipped sideways or even inverted. In those disoriented neurons Van der Loos noted that the axons retain their normal downward direction. This indicates that the extrinsic neurotropic factor acts differently on the axons and the dendrites, the latter remaining misdirected.

Unlike the mammal, the power of regeneration of fibers in the CNS is preserved to a large measure in lower forms, e.g., in the urodele amphibia (newts and salamanders) and anuran amphibia (frogs and toads). In these animals a neurotropic influence acting on neurites was shown by implanting extra pieces of neural tissues containing the presumptive Mauthner cells from donor embryos into the gastrula or neurula of host animals, either in their normal (caudo-rostral) orientation, or with a reversed (rostro-caudal) direction (Hibbard, 1965). The pieces were taken at a stage of development when the polarity and the direction of axonal growth of the presumptive cells had already been genetically determined, but when fiber growth had not yet begun. The normal direction of growth taken by the Mauthner cell fiber is to cross the midline and then grow caudally down to the tail region. This can be seen when implants were made in the correct orientation and the fibers took their normal direction of growth (A–D of Fig. 15.11) .

In preparations where the implants were made with a reversed rostro-caudal orientation of their cell bodies, the fibers at first grew rostrally instead of caudally as a result of their intrinsically determined polarity. Then, in response to the extrinsic neurotropic influence acting on them, the neurites turn caudally to grow down to their targets, as shown in *E–H* of Fig. 15.11.

2.   Retino-Tectal Regeneration

The optic fibers in the adult anuran amphibian and the goldfish are able to regenerate and grow back to their usual terminations in the tectum of the brain. Following a transection of the optic tract, the fibers regenerate back to the same cells of the tectum with a resumption of vision (Sperry, 1944).

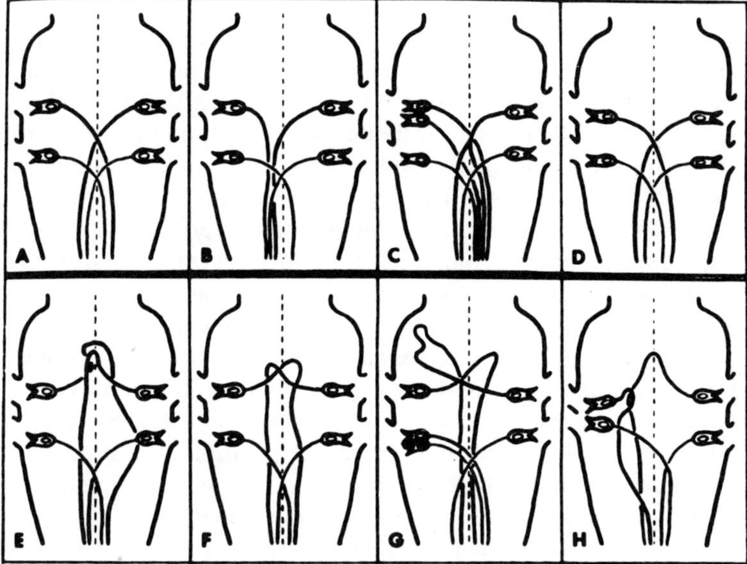

Fig. 15.11 The direction of axonal growth taken by Mauthner cell axons. This figure shows the course of growth taken by neurites of extra Mauthner cells in the medullas after grafting pieces of medulla containing the cells into a host animal. A through D represent four animals viewed from the dorsal side in horizontal sections, with the anterior part at the top and the posterior at the bottom. Mauthner cell grafts were implanted in their correct antero-posterior orientation. The fibers grew out, crossed the midline, and then grew caudally to their targets in normal fashion. E through H represent animals in which the grafts containing extra cells were implanted with a reversed antero-posterior orientation. The initial outgrowth of the axons within the grafted segment was at first toward the original posterior end of the graft, i.e., toward the anterior. After a short period of anterior growth, the neurites reversed their direction of growth, crossed the midline, and grew posteriorly toward their normal targets. The Mauthner cells normally present in the animal below the grafts showed normal downward growth and crossing (except in H). From Hibbard (1965).

Regeneration and reinnervation of the tectum can occur even when the optic nerve had been cut and the eye rotated 180 degrees before being replaced for regeneration. In the case where the eye had been rotated, the regenerated fibers terminate on the same sites on the optic tectum but then give rise to an inverted retino-tectal map of the visual field. Such an inversion of the animal's visual field was shown by its subsequent behavior. In response to a stimulus such as a fly presented in the upper part of the visual field, a stimulus to which the animals normally reflexly react with an upward strike, the visually inverted animals give a reflex strike directed downward and away from the target.

The interplay between the intrinsic genetically determined polarity of the cells of the retina and the extrinsic influences directing the growth and termination of the fibers onto their target cells in the tectum was shown in *Xenopus* by Jacobson (1967, 1978). Before embryonic Stage 28, the eye rudiment, when removed, rotated 180 degrees, and then reimplanted in the animal, gave rise in the adult animal to a normal retino-tectal projection. Thus, at this early stage of growth the polarity of the retinal cells had not yet been genetically determined. After this stage, when the eye rudiment was removed, rotated 180 degrees, and then reimplanted, an inverted retinotectal projection was later found in the adult animal. Various optokinetic tests in addition to a reflex strike toward a target have all shown that the regenerated fibers make their usual retino-tectal connections, even though the path taken by the regenerating fibers from the ganglion cells of the retina had been altered when the optic nerve was cut and the eye rotated.

Attardi and Sperry (1963) gave evidence for the regeneration of fibers back to their original point of termination in the tectum. The optic tract in the goldfish is completely crossed with the fibers from each eye passing to the tectum of the opposite side of the brain where their fibers sweep over the surface of the tectum toward their points of termination. When the dorsal (upper) or ventral (lower) portions of the retina were ablated, the fibers remaining in the tectum were in the lower and upper portions, respectively. Similarly, when the nasal (anterior) or temporal (posterior) halves of the retina were ablated, there was a corresponding loss of the posterior and anterior innervation fields in the tectum, respectively. Sperry (1963) proposed that the termination of the individual retinal fibers at a specific point on the tectum was brought about by the action of two specific neurotropic gradients operating in the tectum, one along the rostro-caudal, the other along the medio-lateral axis. The two gradients together could thus specify a unique point for the termination of each nerve fiber in the surface of the tectum.

The first studies of retino-tectal projections mapped by the use of an electrophysiological recording technique (Jacobson and Gaze, 1964) appeared to confirm the concept of a point-to-point retino-tectal organization (Jacobson and Gaze, 1965). A point source of light presented in the visual field is focused onto a corresponding point of the retina. The light-stimulated retinal receptors excite ganglion cells and impulses are carried in their fibers to the tectum where they discharge a small group of neuronal units. These are recorded with microelectrodes either as single units with an "on", "off", or an "on-off" response, or more often as a grouped discharge. This occurs in a small region of the tectum corresponding to the point-to-point organization indicated by Sperry and his collaborators. Subsequent studies, however, indicated that a greater degree of plasticity in the organization of retinotectal connections is present than had earlier appeared to be the case (Gaze, 1970; Jacobson, 1978). If more time was allowed after making retinal ablations before mapping the tectal responses, the fibers terminating on the tectum were seen to have expanded so as to spread over all of the tectum (Gaze and Sharma, 1970; Yoon, 1971).

Yet another form of plasticity was found when half of the tectum was ablated and a relatively long time was allowed before mapping the responses. The fibers from the retina were then seen to have become compressed into the smaller expanse of tectum left remaining so that all the points of the visual field could be mapped in point-to-point fashion on the smaller portion of tectum.

If the tectum was cut superficially halfway between its caudal and rostral poles so as to interrupt the tectal fibers, and a thin strip of metal or other inert material inserted in the cut in order to prevent the regeneration of fibers into the caudal portion, the fibers remaining in the rostral pole became compressed into the small extent of remaining tectum when a sufficiently long time of regeneration was allowed. Then, later, when the blocking strip was removed and the nerve fibers allowed to regenerate, the fibers were found to regrow so as to terminate in point-to-point fashion over the whole of the expanded tectum thus made available again. A series of such compressions and expansions could be done a number of times, i.e., in "concertina" fashion. The plasticity exhibited indicates that the receptive cells in the tectal field have the power of somehow changing their neurotropic signals and the fibers, the power to change their termination goals so as to pattern themselves in response to such signals.

At present a concept adequate to explain such plastic behavior has yet to be formulated. What seems to be needed is a principle where the operation of the tectal field as a whole somehow determines the pattern of the individual nerve fibers terminating within it. How the cells in the tectum and the neurons in the retina come to have different interactive properties which are at times individualistic and at other times subservient to the organizing influence of the tectal field as a whole is little understood. The reader is referred to the recent studies and discussions of Jacobson (1978), Lund (1978), Edds (1979), and Schmidt (1978) for further insight into this problem.

3. Spinal Cord Regeneration

Unfortunately, the regeneration of axons in the mammalian CNS is feeble compared to that of the lower vertebrates (Ramon y Cajal, 1928; Lee, 1929; Kiernan, 1979). This makes a functional restitution in humans who have suffered spinal transections an as yet unachieved goal. The experimental study of the regeneration of nerve fibers in the spinal cord, and of techniques advanced to augment the power of nerve fiber regeneration has a long history (Mark, 1969; Guth, 1974, 1975; Puchala and Windle, 1977, Windle, 1980). Following an experimental transection of the spinal cord in animals such as the rat, cat, dog, and monkey, nerve fibers in the spinal tracts are seen to sprout, but then after several weeks to stop their regenerative growth. One of the factors believed to prevent growth has been assumed to be the glial formation occurring at the site of transection. Pyromin, a substance which suppresses glial growth, was employed to augment fiber regeneration (Windle *et al.*, 1955), and more recently trypsin and hyaluronidase. Their promotion of regeneration following cord transection of the rat appears equivocal (Kiernan, 1978). Guth *et al.* (1978) recently studied this point using rats with their spinal cord hemisected at C2, just above the level at which descending fibers synapse on the phrenic nerve motoneurons. The diaphragm on that side is paralyzed but respiration adequate to maintain the animal is carried out by the intact phrenic nerve innervating the diaphragmmatic muscle on the other side. The descending fibers transected in the spinal cord need only grow a few mm to make synaptic connections on the phrenic motoneurons, and if they do, the return of function can be readily assessed by the renewed respiratory contractions of the paralyzed diaphragmatic muscle on that side. For 5 to 6 months following hemisection, repeated injections of hyaluronidase and trypsin were made into the cord in the transected region to prevent gliosis. There was no indication of any recovery of respiratory function on the transected side and histological examination of the transected cervical cord showed little evidence of nerve fiber growth across the line of the transection.

Histological study at the sites of spinal cord transection shows cysts or cavities formed near the cut ends of the cord. The cut nerve fibers show terminal clubs without evidence of sprouting, the terminal clubs appearing to enlarge and then become autolyzed with the ruptured site forming cysts or cavitations. Bulbs confined within an enlarged myelin sheath appear at some distance from cavitations (Kao and Chang, 1977; Kao *et al.*, 1977a–c). Kao considered that removing the cavitations and using pieces of nerve to bridge the gap might enhance the regeneration of spinal cord fibers. The use of pieces of peripheral nerve as a bridge was based on the increased number of Schwann cells in degenerated peripheral nerve which appear able to promote the regeneration of peripheral nerve fibers (cf. Section A4 above). Earlier studies using peripheral nerve implants in the CNS seemed to show an enhanced regeneration (Ramon y Cajal, 1928; Sugar and Gerard, 1940) although this was not confirmed by other investigations (Brown and McCough, 1947; Feigen, Geller, and Wolf, 1951). In the studies of Kao and his colleagues, crushes of the cord were made, or the cord cut subpially, and then after a week or two the cavitations, the cysts, and lytic material facing the cut surfaces were removed and pieces of peripheral nerve inserted into the debrided gaps. Using this delayed nerve graft technique, a significant number of axons were seen traversing the nerve graft. Similarly, Richardson, McGuinness, and Aguayo (1980) implanted segments of peripheral nerve into cuts made in rat spinal cords and found regenerating cord fibers to have entered the graft. The fibers bridging the gap were viable, as indicated by injecting HRP either above or below the graft and finding the marker transported through the bridge region. Thus, it appears that the Schwann cells in the nerve graft may promote regeneration, their presence augmenting the normally feeble regenerative powers of central neurons. Other studies have suggested that glia may promote growth under certain circumstances (cf. Liu *et al.*, 1979).

Kiernan (1978) proposed as a hypothesis that the growing tips of the fibers take up some protein which is carried by retrograde transport to the cell body to there stimulate the synthesis of materials required for regeneration. The failure of uptake of such proteins by the fibers in the CNS could explain their feeble power of regeneration as compared to peripheral nerve. Possibly the peripheral nerve grafts may supply an analogous protein.

It is possible for "foreign" fibers to grow into the CNS. In experiments along those lines, the cut ends of the ventral roots of the cat L7 spinal cord segment were grafted to the cut ends of the L7 dorsal roots of the other side, and regenerating ventral roots were found to grow into the spinal cord via the dorsal root (Barnes and Worrall, 1968). The entry of regenerating ventral root fibers into the dorsal root was shown by injecting ^{3}H-leucine into the L7 ventral horn of the cord on the side supplying the ventral roots and tracing an outflow of radioactivity into the dorsal root (Ochs and Barnes, 1969). Barnes and Worrall (1968) reported that such entering fibers could make synaptic connections, as shown by changes in ventral reflexes recorded electrically when stimulating that new input. However, whether the fibers which enter the spinal cord via the dorsal roots do in fact make functional synaptic connections has to be confirmed. This points to another basic problem which must be solved before functional return can occur, namely, that not only must there be a regeneration of fibers, but those fibers must make their proper terminations on the target cells in the CNS, a process requiring neurotropic control, for functional restitution.

Postscript

Over the last several decades the acceptance of axoplasmic transport as a neuronal function has revolutionized what was essentially a static view of the neuron. Sherrington (1906) pictured the flow of action potentials to constitute the integrative action of the nervous system, the changed patterns of electrical activity of the contributing neural elements bringing about the behavior typical of the organism. As can be appreciated from the studies of axoplasmic transport described in this book, the neuron is revealed as a much more dynamic entity. The various materials transported within the axons and dendrites serve not only to maintain the form and function of the neuron itself but materials communicated to other neurons and affector and effector cells can by this means alter their functional capabilities. Conversely, neurons are influenced by signals taken up from muscles and other cells. Such processes could account for the long-lasting neuropathological changes seen following a brief exposure to a chemical or physical agent. Conceivably, transport may participate in the enduring neuronal changes which underlie memory. To understand those processes may now appear to be forbiddingly difficult. However, the history of science has shown that the insurmountable problems of the past have become the commonplace practices of today. No doubt further advances in our understanding of the transport process will help us better understand those basic neuronal functions which are dependent on transport and the altered function characteristic of neuropathological states.

References

Abe, T., Haga, T., and Kurokawa, M. (1973), Rapid transport of phosphatidylcholine occurring simultaneously with protein transport in the frog sciatic nerve, *Biochem. J.* **136:** 731–740.

Abe, T., Haga, T. and Kurokawa, M. (1974), Retrograde axoplasmic transport: its continuation as anterograde transport, *FEBS Lett.* **47:** 272–275.

Abe, T., Haga, T., and Kurokawa, M. (1975), Blockage of axoplasmic transport and depolymerisation of reassembled microtubules by methyl mercury, *Brain Res.* **86:** 504–508.

Abelous, J. E., and Lassalle, H. (1928), Modifications d'excitabilité d'un nerf au cours de la dégénérescence du nerf homologue sectionné. Origine humorale de ces modifications, *Compt. Rend. Soc. Biol.* **98:** 1105–1107.

Adams, W. E. (1942), The blood supply of nerves. II. The effects of exclusion of its regional sources of supply on the sciatic nerve of the rabbit, *J. Anat.* **77:** 243–250.

Aguayo, A. J., and Bray, G. M. (1980), Experimental nerve grafts, Chapter 6, in: *Nerve Repair and Regeneration,* Jewett, D. L., and McCarroll, H. R., Jr. (eds.), C. V. Mosby, St. Louis.

Aguayo, A. J., Bray, G. M., and Perkins, S. C. (1979), Axon–Schwann cell relationships in neuropathies of mutant mice, *Ann. N.Y. Acad. Sci.* **317:** 512–531.

Aguayo, A. J., Epps, J., Charron, L., and Bray, G. M. (1976), Multipotentiality of Schwann cells in cross-anastomosed and grafted myelinated and unmyelinated nerves: quantitative microscopy and radioautography, *Brain Res.* **104:** 1–20.

Aguayo, A., Nair, C. P. V., and Midgley, R. (1971), Experimental progressive compression neuropathy in the rabbit: histologic and electrophysiologic studies, *Arch. Neurol.* **24:** 358–364.

Aguilar, C. E., Bisby, M. A., Cooper, E., and Diamond, J. (1973), Evidence that axoplasmic transport of trophic factors is involved in the regulation of peripheral nerve fields in salamanders, *J. Physiol. (London)* **234:** 449–464.

Aitken, J. T. (1950), Growth of nerve implants in voluntary muscle, *J. Anat.* **84:** 38–49.

Akert, K., Cowan, W. M., Cuénod, M., and Holländer, H. (1975), International symposium on the use of axonal transport for studies of neuronal connectivity, *Brain Res.* **85:** i–v and 201–353.

Akert, K., Pfenninger, K., Sandri, C., and Moor, H. (1972), Freeze etching and cytochemistry of vesicles and membrane complexes in synapses of the central nervous system, pp. 67–86, in: *Structure and Function of Synapses,* Pappas, G. D., and Purpura, D. P. (eds.), Raven Press, New York.

Akert, K., Sandri, C., Weibel, E. R., Peper, K., and Moor, H. (1976), The fine structure of the perineurial endothelium, *Cell Tissue Res.* **165:** 281–295.

Albuquerque, E. X., and Daly, J. W. (1976), Batrachotoxin, a selective probe for channels modulating sodium conductances in electrogenic membranes, pp. 297–338, in: *The Specificity and Action of Animal, Bacterial and Plant Toxins* (Receptors and Recognition, Series B, Volume 1), Cuatrecasas, P. (ed.), Chapman and Hall, London.

Albuquerque, E. X., Schuh, F. T., and Kauffman, F. C. (1971), Early membrane depolarization of the fast mammalian muscle after denervation, *Pflügers Arch.* **328:** 36–50.

Albuquerque, E. X., Warnick, J. E., Sansone, F. M., and Onur, R. (1974), The effects of vinblastine and colchicine on neural regulation of muscle, *Ann. N.Y. Acad. Sci.* **228:** 224–243.

Aldskogius, H., and Arvidsson, J. (1978), Nerve cell degeneration and death in the trigeminal ganglion of the adult rat following peripheral nerve transection, *J. Neurocytol.* **7:** 229–250.

Allen, D. O., Gardner, R. M., Yount, R. A., and Ochs, S. (1978), Increased cyclic GMP in the end-plate region of denervated frog muscle, *J. Neurobiol.* **9:** 445–451.

Allen, R. D., and Kamiya, N. (1964), *Primitive Motile Systems in Cell Biology,* Academic Press, New York.

Alonso, G., Gabrion, J., Travers, E., and Assenmacher, I. (1981), Ultrastructural organization of actin filaments in neurosecretory axons of the rat, *Cell. Tissue Res.* **214:** 323–341.

Alvarez-Buylla, R., Livett, B. G., Uttenthal, L. O., Hope, D. B., and Milton, S. H. (1973), Immunochemical evidence for the transport of neurophysin in the hypothalamo-neurohypophysial system of the dog, *Z. Zellforsch.* **137:** 435–450.

Amos, L. A., Linck, R. W., and Klug, A. (1976), Molecular structure of flagellar microtubules, pp. 847–867, in: *Cell Motility,* Book C, Goldman, R., Pollard, T., and Rosenbaum, J. (eds.), Cold Spring Harbor Laboratory, Cold Spring Harbor, N.Y.

Anderson, K.-E., and Edström, A. (1973), Effects of nerve blocking agents on fast axonal transport of proteins in frog sciatic nerves *in vitro, Brain Res.* **50:** 125–134.

Anderson, K.-E., Edström, A., and Mattsson, H. (1972), Effects of cytochalasin B on uptake of glucosamine, leucine and sulphate into nerve cells; incorporation into glycoproteins and rapid axonal transport, *Brain Res.* **48:** 343–353.

Anderson, L. E., and McClure, W. O. (1973), Differential transport of protein in axons: comparison between the sciatic nerve and dorsal columns of cats, *Proc. Natl. Acad. Sci. USA* **70:** 1521–1525.

Andrews, J. M., and Andrews, R. L. (1976), The comparative neuropathology of motor neuron diseases, pp. 181–216, in: *Amyotrophic Lateral Sclerosis,* Andrews, J. M., Johnson, R. T., and Brazier, M. A. B. (eds.), Academic Press, New York.

Andrews, J. M., Johnson, R. T., and Brazier, M. A. B. (eds.) (1976), *Amyotrophic Lateral Sclerosis,* Academic Press, New York.

Andrews, J. M., Gardner, M. B., Wolfgram, F. J., Ellison, G. W., Porter, D. D., and Brandkamp, W. W. (1974), Studies on a murine form of spontaneous lower motor neuron degeneration— The wobbler *(wr)* mouse, *Am. J. Pathol.* **76:** 63–78.

Angeletti, P. U., Levi-Montalcini, R., and Caramia, F. (1971), Analysis of the effect of the antiserum to the nerve growth factor in adult mice, *Brain Res.* **27:** 343–355.

Appeltauer, G. S. L., and Korr, I. M. (1975), Axonal delivery of soluble, insoluble and electrophoretic fractions of neuronal proteins to muscle, *Exp. Neurol.* **46:** 132–146.

Asbury, A. K. (1967), Schwann cell proliferation in developing mouse sciatic nerve, *J. Cell Biol.* **34:** 735–743.

Attardi, D. G., and Sperry, R. W. (1963), Preferential selection of central pathways by regenerating optic fibers, *Exp. Neurol.* **7:** 46–64.

Austin, L., Bray, J. J., and Young, R. J. (1966), Transport of proteins and ribonucleic acid along nerve axons, *J. Neurochem.* **13:** 1267–1269.

Austin, L., and Morgan, I. G. (1967), Incorporation of ^{14}C-labelled leucine into synaptosomes from rat cerebral cortex *in vitro, J. Neurochem.* **14:** 377–387.

Autilio-Gambetti, L., Gambetti, P., and Shafer, B. (1973), RNA and axonal flow. Biochemical and autoradiographic study in the rabbit optic system, *Brain Res.* **53:** 387–398.

Autilio, L. A., Appel, S. H., Pettis, P., and Gambetti, P.-L. (1968), Biochemical studies of synapses *in vitro,* I. Protein synthesis, *Biochemistry (Washington)* **7:** 2615–2622.

Axelsson, J., and Thesleff, S. (1959), A study of supersensitivity in denervated mammalian skeletal muscle, *J. Physiol. (London)* **147:** 178–193.

Baglivi, G. (1723), *The Practice of Physic Together with New and Curious Dissertation Particularly of the Tarantula and the Nature of Its Poison,* Midwinter *et al.,* London (pp. 372–373, in 1st edition, 1695).

Baker, P. F. (1972), Transport and metabolism of calcium ions in nerve, *Prog. Biophys. Molec. Biol.* **24:** 179–223.

Baker, P. F. (1976), The regulation of intracellular calcium, *Symp. Soc. Exp. Biol.* **30:** 67–88.

Baker, P. F., and Glitsch, H. G. (1973), Does metabolic energy participate directly in the Na^+-dependent extrusion of Ca^{2+} ions from squid giant axons? *J. Physiol. (London)* **233:** 44P–46P.

Baker, P. F., and Honerjager, P. (1978), Influence of carbon dioxide on level of ionized calcium in squid axons, *Nature* **273:** 160–161.

Baker, P. F., Knight, D. E., and Whitaker, M. J. (1980), Calcium and the control of exocytosis, pp. 47–55, in: *Calcium-Binding Proteins: Structure and Function,* Siegel, F. L., Carafoli, E., Kretsinger, R. H., MacLennan, D. H., and Wassermann, R. H. (eds.), Elsevier/North Holland, New York.

Baker, P. F., and McNaughton, P. A. (1976), Kinetics and energetics of calcium efflux from intact squid giant axons, *J. Physiol. (London)* **259:** 103–144.

Balázs, R., and Cremer, J. E. (eds.) (1972), *Metabolic Compartmentation in the Brain,* John Wiley & Sons, New York.

Banks, P., Mangnall, D., and Mayor, D. (1969), The re-distribution of cytochrome oxidase, noradrenaline and adenosine triphosphate in adrenergic nerves constricted at two points, *J. Physiol. (London)* **200:** 745–762.

Banks, P., and Mayor, D. (1972), Intra-axonal transport in noradrenergic neurons in the sympathetic nervous system, *Biochem. Soc. Symp.* **36:** 133–149.

Banks, P., Mayor, D., and Mraz, P. (1973a), Metabolic aspects of the synthesis and intra-axonal transport of noradrenaline storage vesicles, *J. Physiol. (London)* **229:** 383–394.

Banks, P., Mayor, D., and Mraz, P. (1973b), Cytochalasin B and the intra-axonal movement of noradrenaline storage vesicles, *Brain Res.* **49:** 417–421.

Banks, P., and Till, R. (1975), A correlation between the effects of anti-mitotic drugs on microtubule assembly *in vitro* and the inhibition of axonal transport in noradrenergic neurones, *J. Physiol. (London)* **252:** 283–294.

Bannister, L. H. (1976), Sensory terminals of peripherial nerves, Chapter 8, in: *The Peripherial Nerve,* Landon, D. N. (ed.), John Wiley & Sons, New York.

Bárány, M. (1967), ATPase activity of myosin correlated with speed of muscle shortening, *J. gen. Physiol.* **50:** 197–218.

Bargmann, W., and Scharrer, E. (1951), The site of origin of the hormones of the posterior pituitary, *Am. Sci.* **39:** 255–259.

Barker, L. F. (1899), *The Nervous System and Its Constituent Neurons,* Appleton and Co., New York.

Barker, D., and Ip, M. C. (1965), Sprouting and degeneration of mammalian motor axons in normal and de-afferentated skeletal muscle, *Proc. Roy. Soc. B* **163:** 538–554.

Barker, J. L., Neale, J. H., and Bonner, W. M. (1977), Slab gel analysis of rapidly transported proteins in the isolated frog nervous system, *Brain Res.* **124:** 191–196.

Barker, J. L., Neale, J. H., and Gainer, H. (1976), Rapidly transported proteins in sensory, motor and sympathetic nerves of the isolated frog nervous system, *Brain Res.* **105:** 497–515.

Barnes, C. D., and Worrall, N. (1968), Reinnervation of spinal cord by cholinergic neurons, *J. Neurophysiol.* **31:** 689–694.

Barnes, J. F., and Davenport, H. A. (1937), Cells and fibers in spinal nerves. III. Is a 1:1 ratio in the dorsal root the rule? *J. Comp. Neurol.* **66:** 459–469.

Barondes, S. H. (1966), On the site of synthesis of the mitochondrial protein of nerve endings, *J. Neurochem.* **13:** 721–727.

Barondes, S. H. (1968), Further studies of the transport of protein to nerve endings, *J. Neurochem.* **15:** 343–350.

Barondes, S. H. (1973), Neuronal proteins: synthesis, transport and neuronal regulation, pp. 215–226, in: *Proteins of the Nervous System,* Schneider, D. J., Angeletti, R. H., Bradshaw, R. A., Grasso, A., and Moore, B. W. (eds.), Raven Press, New York.

Bastholm, E. (1950), *The History of Muscle Physiology,* Munksgaard, Copenhagen.

Behnke, O. (1975), An outer component of microtubules, *Nature* **257:** 709–710.

Beidler, L. M. (1963), Dynamics of taste cells, pp. 133–148, in: *Olfaction and Taste,* Zotterman, Y. (ed.), *Proceedings of the 1st International Symposium, Vol. 1,* Pergamon Press, Oxford.

Belágyi, J. (1975), Water structure in striated muscle by spin labelling technique, *Acta Biochim. Biophys. Acad. Sci. Hung.* **10:** 63–70.

Benjamin, A. M., and Quastel, J. H. (1981), Acetylcholine synthesis in synaptosomes: mode of transfer of mitochondrial acetyl coenzyme A, *Science* **213:** 1495–1497.

Ben-Jonathan, N., Maxson, R. E., and Ochs, S. (1978), Fast axoplasmic transport of noradrenaline and dopamine in mammalian peripheral nerve, *J. Physiol. (London)* **281:** 315–324.

Ben-Jonathan, N., and Porter, J. C. (1976), A sensitive radioenzymatic assay for dopamine, norepinephrine and epinephrine in plasma and tissue, *Endocrinology* **98:** 1497–1507.

Bennett, M. R., McLachlan, E. M., and Taylor, R. S. (1973a), The formation of synapses in reinnervated mammalian striated muscle, *J. Physiol. (London)* **233:** 481–500.

Bennett, M. R., McLachlan, E. M., and Taylor, R. S. (1973b), The formation of synapses in mammalian striated muscle reinnervated with autonomic preganglionic nerves, *J. Physiol. (London)* **233:** 501–517.

Bennett, M. R., Pettigrew, A. G., and Taylor, R. S. (1973), The formation of synapses in reinnervated and cross-reinnervated adult avian muscle, *J. Physiol. (London)* **230:** 331–357.

Bennett, M. V. L. (1971), Electrolocation in fish, *Ann. N.Y. Acad. Sci.* **188:** 242–269.

Bennett, M. V. L. (1977), Electrical transmission: a functional analysis and comparison to chemical transmission, pp. 357–416, in: *Handbook of Physiology,* Section 1 (Vol. 1, part 1), American Physiological Society, Bethesda, Maryland.

Benowitz, L. I., Shashoua, V. E., and Yoon, M. (1980), Rapidly transported proteins in the regenerating optic nerve of goldfish, *Soc. Neurosci. Abstr.* **6:** 388.

Bentley, F. H., and Schlapp, W. (1943), The effects of pressure on conduction in peripheral nerve, *J. Physiol. (London)* **102:** 72–82.

Berg, D. K., and Hall, Z. W. (1974), Fate of α-bungarotoxin bound to acetylcholine receptors of normal and denervated muscle, *Science* **184:** 473–475.

Bergen, L. G., and Borisy, G. G. (1980), Head-to-tail polymerization of microtubules *in vitro, J. Cell Biol.* **84:** 141–150.

Berl, S., and Puszkin, S. (1970), Mg^{2+}–Ca^{2+}-activated adenosine triphosphatase system isolated from mammalian brain, *Biochemistry (Washington)* **9:** 2058–2067.

Berl, S., Puszkin, S., and Nicklas, W. J. (1973), Actomyosin-like protein in brain, *Science* **179:** 441–446.

Bernstein, J. (1912), *Elektrobiologie,* Vieweg, Braunschweig.

Bernstein, J. J., and Guth, L. (1961), Nonselectivity in establishment of neuromuscular connections following nerve regeneration in the rat, *Exp. Neurol.* **4:** 262–275.

Berthold, C.-H., and Skoglund, S. (1967), Histochemical and ultrastructural demonstration of mitochondria in the paranodal region of developing feline spinal roots and nerves, *Acta Soc. Med.* **72:** 37–70.

Biondi, R. J., Levy, M. J., and Weiss, P. A. (1972), An engineering study of the peristaltic drive of axonal flow, *Proc. Natl. Acad. Sci. USA* **69:** 1732–1736.

Bird, M. M. (1976), Microtubule–synaptic vesicle associations in cultured rat spinal cord neurons, *Cell Tissue Res.* **168:** 101–115.

Bird, M. T., Shuttleworth, E., Jr., Koestner, A., and Reinglass, J. (1971), The wobbler mouse mutant: An animal model of hereditary motor system disease, *Acta Neuropath. (Berlin)* **19:** 39–50.

Birks, R., and MacIntosh, F. C. (1961), Acetylcholine metabolism of a sympathetic ganglion, *Can. J. Biochem. Physiol.* **39:** 787–827.

Bisby, M. A. (1975), Inhibition of axonal transport in nerves chronically treated with local anesthetics, *Exp. Neurol.* **47:** 481–489.

Bisby, M. A. (1978), Fast axonal transport of labelled protein in sensory axons during regeneration, *Exp. Neurol.* **61:** 281–300.

Bisby, M. A. (1980a), Axonal transport of labeled protein and regeneration rate in nerves of streptozocin-diabetic rats, *Exp. Neurol.* **69:** 74–84.

Bisby, M. A. (1980b), Retrograde axonal transport, *Adv. Cell. Neurobiol.* **1:** 69–117.

Bisby, M. A. (1980c), Changes in the composition of labeled protein transported in motor axons during their regeneration, *J. Neurobiol.* **11:** 435–445.

Bisby, M. A., and Bulger, V. T. (1977), Reversal of axonal transport at a nerve crush, *J. Neurochem.* **29:** 313–320.

Bisby, M. A., and Jones, D. L. (1978), Temperature sensitivity of axonal transport in hibernating and nonhibernating rodents, *Exp. Neurol.* **61:** 74–83.

Bischoff, A., and Thomas, P. K. (1975), Microscopic anatomy of myelinated nerve fibers, Chapter 6, in: *Peripheral Neuropathy,* Dyck, P. J., Thomas, P. K., and Lambert, E. H. (eds.), Saunders, Philadelphia.

Black, M. M., and Lasek, R. J. (1978), A difference between the proteins conveyed in the fast component of axonal transport in guinea pig hypoglossal and vagus motor neurons, *J. Neurobiol.* **9:** 433–443.

Black, M. M., and Lasek, R. J. (1979), Axonal transport of actin: Slow component b is the principal source of actin for the axon, *Brain Res.* **171:** 401–413.

Black, M. M., and Lasek, R. J. (1980), Slow components of axonal transport: Two cytoskeletal networks, *J. Cell Biol.* **86:** 616–623.

Blakemore, W. F., and Cavanagh, J. B. (1969), "Neuroaxonal dystrophy" occurring in an experimental "dying back" process in the rat, *Brain* **92:** 789–804.

Blaker, W. D., Toews, A. D., and Morell, P. (1980), Cholesterol is a component of the rapid phase of axonal transport, *J. Neurobiol.* **11:** 243–250.

Blasberg, R., and Lajtha, A. (1966), Heterogeneity of the mediated transport systems of amino acid uptake in brain, *Brain Res.* **1:** 86–104.

Blaustein, M. P. (1974), The interrelationship between sodium and calcium fluxes across cell membranes, *Rev. Physiol. Biochem. Pharmacol.* **70:** 33–82.

Blaustein, M. P. (1979), The role of calcium in catecholamine release from adrenergic nerve terminals, pp. 39–58, in: *The Release of Catecholamines from Adrenergic Neurons,* Paton, D. M. (ed.), Pergamon Press, New York.

Blaustein, M. P., Ratzlaff, R. W., and Kendrick, N. K. (1978a), The regulation of intracellular calcium in presynaptic nerve terminals, *Ann. N.Y. Acad. Sci.* **307:** 195–212.

Blaustein, M. P., Ratzlaff, R. W., Kendrick, N. C., and Schweitzer, E. S. (1978b), Calcium buffering in presynaptic nerve terminals. I. Evidence for involvement of a nonmitochondrial ATP-dependent sequestration mechanism, *J. gen. Physiol.* **72:** 15–41.

Blaustein, M. P., Ratzlaff, R. W., and Schweitzer, E. S. (1978c) Calcium buffering in presynpatic nerve terminals. II. Kinetic properties of the nonmitochondrial Ca sequestration mechanism, *J. gen. Physiol.* **72:** 43–66.

Blobel, G., and Dobberstein, B. (1975), Transfer of proteins across membranes. I. Presence of proteolytically processed and unprocessed nascent immunoglobulin light chains on membrane-bound ribosomes of murine myeloma, *J. Cell Biol.* **67:** 835–851.

Bobillier, P., Petitjean, F., Salvert, D., Ligier, M., and Seguin, S. (1975), Differential projections of the nucleus raphe dorsalis and nucleus raphe centralis as revealed by autoradiography, *Brain Res.* **85:** 205–210.

Bodian, D., and Howe, H. A. (1941), The rate of progression of poliomyelitis virus in nerves, *Bull. Johns Hopkins Hosp.* **69:** 79–85.

Boegman, R. J., and Albuquerque, E. X. (1980), Axonal transport in rats rendered paraplegic following a single subarachnoid injection of either batrachotoxin or 6-aminonicotinamide into the spinal cord, *J. Neurobiol.* **11:** 283–290.

Boegman, R. J., and Riopelle, R. J. (1980), Batrachotoxin blocks slow and retrograde axonal transport *in vivo, Neurosci. Lett.* **18:** 143–147.

Boegman, R. J., and Wood, P. L. (1981), Axonal transport in dystrophic hamsters, *Can. J. Physiol. Pharmacol.* **59:** 202–204.

Boegman, R. J., Wood, P. L., and Pinaud, L. (1975), Increased axoplasmic flow associated with pargyline under conditions which induce a myopathy, *Nature* **253:** 51–52.

Boerhaave, H. (1743), *Academical Lectures on the Theory of Physic.,* Innys, London.

Boesch, J., Marko, P., and Cuénod, M. (1972), Effects of colchicine on axonal transport of proteins in the pigeon visual pathways, *Neurobiology* **2:** 123–132.

Bondy, S. C., and Madsen, C. J. (1971), Development of rapid axonal flow in the chick embryo, *J. Neurobiol.* **2:** 279–286.

Borisy, G. G., Marcum, J. M., Olmsted, J. B., Murphy, D. B., and Johnson, K. A. (1975), Purification of tubulin and associated high molecular weight proteins from porcine brain and characterization of microtubule assembly *in vitro, Ann. N.Y. Acad. Sci.* **253:** 107–132.

Borisy, G. G., Olmstead, J. B., and Klugman, R. A. (1972), *In vitro* aggregation of cytoplasmic microtubule subunits, *Proc. Natl. Acad. Sci. USA* **69:** 2890–2894.

Borle, A. B. (1973), Calcium metabolism at the cellular level, *Fed. Proc.* **32:** 1944–1950.

Bowler, K., and Duncan, C. J. (1966), Actomyosin-like protein from crayfish nerve: A possible molecular explanation of permeability changes during excitation, *Nature* **211:** 642–643.

Boyd, I. A. (1962a), The nuclear-bag fibre and nuclear-chain fibre systems in the muscle spindles of the cat, pp. 185–190, in: *Symposium on Muscle Receptors,* Barker, D. (ed.), Hong Kong University Press, Hong Kong.

Boyd, I. A. (1962b), The structure and innervation of the nuclear-bag muscle fibre system and the nuclear-chain muscle fibre system in mammalian muscle spindles, *Phil. Trans. R. Soc.* **245:** 81–136.

Bradley, W. G., and Asbury, A. K. (1970), Duration of synthesis phase in neurilemma cells in mouse sciatic nerve during degeneration, *Exp. Neurol.* **26:** 275–282.

Bradley, W. G., and Jaros, E. (1973), Axoplasmic flow in axonal neuropathies. II. Axoplasmic flow in mice with motor neuron disease and muscular dystrophy, *Brain* **96:** 247–258.

Bradley, W. G., and Williams, M. H. (1973), Axoplasmic flow in axonal neuropathies. I. Axoplasmic flow in cats with toxic neuropathies, *Brain* **96:** 235–246.

Brady, S. T., Chrothers, S. D., Nosal, C., and McClure, W. O. (1980), Fast axonal transport in the presence of high Ca^{2+}: Evidence that microtubules are not required, *Proc. Natl. Acad. Sci. USA* **77:** 5909–5913.

Brady, S. T., and Lasek, R. J. (1981), Nerve-specific enolase and creatine phosphokinase in axonal transport: Soluble proteins and the axoplasmic matrix, *Cell* **23:** 515–523.

Brattgård, S.-O., Edström, J.-E., and Hydén, H. (1957), The chemical changes in regenerating neurons, *J. Neurochem.* **1:** 316–325.

Brattgård, S.-O., Edström, J.-E., and Hydén, H. (1958), The productive capacity of the neuron in retrograde reaction, *Exp. Cell Res. Suppl.* **5:** 185–200.

Bray, D. (1973), Model for membrane movements in the neural growth cone, *Nature* **244:** 93–95.

Bray, D. (1979), Mechanical tension produced by nerve cells in tissue culture, *J. Cell Sci.* **37:** 391–410.

Bray, D., and Bunge, M. B. (1981), Serial analysis of microtubules in cultured rat sensory axons, *J. Neurocytol.* **10:** 589–605.

Bray, D., and Gilbert, D. (1981), Cytoskeletal elements in neurons, *Ann. Rev. Neurosci.* **4:** 505–523.

Bray, G. M., Rasminsky, M., and Aguayo, A. J. (1981), Interactions between axons and their sheath cells, *Ann. Rev. Neurosci.* **4:** 127–162.

Bray, J. J., and Austin, L. (1968), Flow of protein and ribonucleic acid in peripheral nerve, *J. Neurochem.* **15:** 731–740.

Bray, J. J., and Harris, A. J. (1975), Dissociation between nerve–muscle transmission and nerve trophic effects on rat diaphragm using type-D botulinum toxin, *J. Physiol. (London)* **253:** 53–77.

Bray, J. J., Hawken, M. J., Hubbard, J. I., Pockett, S., and Wilson, L. (1976), The membrane potential of rat diaphragm muscle fibres and the effect of denervation, *J. Physiol. (London)* **255:** 651–667.

Bray, J. J., Hubbard, J. I., and Mills, R. G. (1979), The trophic influence of tetrodotoxin-inactive nerves on normal and reinnervated rat skeletal muscles, *J. Physiol. (London)* **297:** 479–491.

Bray, J. J., Kon, C. M., and Breckenridge, B. McL. (1971), Reversed polarity of rapid axonal transport in chicken motoneurons, *Brain Res.* **33:** 560–564.

Brimijoin, S. (1975), Stop-flow: A new technique for measuring axonal transport, and its application to the transport of dopamine-β-hydroxylase, *J. Neurobiol.* **6:** 379–394.

Brimijoin, S. (1977), A histofluorescence study of events accompanying accumulation and migration of norepinephrine within locally cooled nerves, *J. Neurobiol.* **8:** 251–263.

Brimijoin, S. (1979), Axonal transport and subcellular distribution of molecular forms of acetylcholinesterase in rabbit sciatic nerve, *Mol. Pharmacol.* **15:** 641–648.

Brimijoin, S. (1982a), The role of axonal transport in nerve disease, in: *Peripheral Neuropathy,* Dyck, P. J., Thomas, P. K., and Lambert, E. H. (eds.), 2nd edition, Saunders, Philadelphia. (In press).

Brimijoin, S. (1982b), Microtubules and the ''capacity'' of the system for rapid axonal transport, *Fed. Proc.* **41:** 2312–2316.

Brimijoin, S., Capek, P., and Dyck, P. J. (1973), Axonal transport of dopamine-β-hydroxylase by human sural nerves *in vitro, Science* **180:** 1295–1297.

Brimijoin, S., and Helland, L. (1976), Rapid retrograde transport of dopamine-β-hydroxylase as examined by the stop-flow technique, *Brain Res.* **102:** 217–228.

Brimijoin, S., Lundberg, J. M., Brodin, E. Hökfelt, T., and Nilsson, G. (1980), Axonal transport of substance P in the vagus and sciatic nerves of the guinea pig, *Brain Res.* **191:** 443–457.

Brimijoin, S., Olsen, J., and Rosenson, R. (1979), Comparison of the temperature-dependence of rapid axonal transport and microtubules in nerves of the rabbit and bullfrog, *J. Physiol. (London)* **287:** 303–314.

Brimijoin, S., Skau, K., and Wiermaa, M. J. (1978), On the origin and fate of external acetylcholinesterase in peripheral nerve, *J. Physiol. (London)* **285:** 143–158.

Brimijoin, S., and Wiermaa, M. J. (1977a), Rapid axonal transport of tyrosine hydroxylase in rabbit sciatic nerves, *Brain Res.* **121:** 77–96.

Brimijoin, S., and Wiermaa, M. J. (1978), Rapid orthograde and retrograde axonal transport of acetylcholinesterase as characterized by the stop-flow technique, *J. Physiol. (London)* **285:** 129–142.

Brink, J. J., and Karnovsky, M. L. (1973), Axonal flow of exogenous RNA in the rat optic tectum, *J. Neurochem.* **21:** 1003–1007.

Brinley, F. J., Jr. (1976), Calcium and magnesium metabolism in cephalopod axons, *Fed. Proc.* **35:** 2572–2573.

Brinley, F. J., Jr., and Scarpa, A. (1975), Ionized magnesium concentration in axoplasm of dialyzed squid axons, *FEBS Lett.* **50:** 82–85.

Brooks, V. B., Curtis, D. R., and Eccles, J. C. (1957), The action of tetanus toxin on the inhibition of motoneurones, *J. Physiol. (London)* **135:** 655–672.

Brown, G. L. (1937), The actions of acetylcholine on denervated mammalian and frog's muscle, *J. Physiol. (London)* **89:** 438–461.

Brown, J. O., and McCouch, G. P. (1947), Abortive regeneration in the transected spinal cord, *J. Comp. Neurol.* **87:** 131–137.

Brown, M. C., Goodwin, G. M., and Ironton, R. (1977), Prevention of motor nerve sprouting in botulinum toxin poisoned mouse soleus muscles by direct stimulation of the muscle, *J. Physiol. (London)* **267:** 42P–43P.

Brownstein, M. J., Russell, J. T., and Gainer, H. (1980), Synthesis, transport, and release of posterior pituitary hormones, *Science* **207:** 373–378.

Brunngraber, E. G. (1970), Glycoproteins in neural tissue, Chapter 19, in: *Protein Metabolism of the Nervous System,* Lajtha, A. (ed.), Plenum Press, New York.

Brunngraber, E. G. (1972), Chemistry and metabolism of glycopeptides derived from brain glycoproteins, *Adv. Exp. Med. Biol.* **25:** 17–49.

Brushart, T. M., and Mesulam, M.-M. (1980), Alteration in connections between muscle and anterior horn motoneurons after peripheral nerve repair, *Science* **208:** 603–605.

Bryan, J. (1974), Biochemical properties of microtubules, *Fed. Proc.* **33:** 152–157.

Bryan, J., and Wilson, L. (1971), Are cytoplasmic microtubules heteropolymers? *Proc. Natl. Acad. Sci. USA* **68:** 1762–1766.

Brzin, M., Tennyson, V. M., and Duffy, P. E. (1966), Acetylcholinesterase in frog sympathetic and dorsal root ganglia. A study by electron microscope cytochemistry and microgasometric analysis with the magnetic diver, *J. Cell Biol.* **31:** 215–242.

Buller, A. J. (1970), The neural control of the contractile mechanism in skeletal muscle, *Endeavour* **29:** 107–111.

Buller, A. J., Eccles, J. C., and Eccles, R. M. (1960a), Differentiation of fast and slow muscles in the cat hind limb, *J. Physiol. (London)* **150:** 399–416.

Buller, A. J., Eccles, J. C., and Eccles, R. M. (1960b), Interactions between motoneurones and muscles in respect of the characteristic speeds of their responses, *J. Physiol. (London)* **150:** 417–439.

Buller, A. J., and Lewis, D. M. (1965a), The rate of tension development in isometric tetanic contractions of mammalian fast and slow skeletal muscle, *J. Physiol. (London)* **176:** 337–354.

Buller, A. J., and Lewis, D. M. (1965b), Further observations on mammalian cross-innervated skeletal muscle, *J. Physiol. (London)* **178:** 343–358.

Buller, A. J., Mommaerts, W. F. H. M., and Seraydarian, K. (1969), Enzymic properties of myosin in fast and slow twitch muscles of the cat following cross-innervation, *J. Physiol. (London)* **205:** 581–597.

Bunge, M. B. (1973), Fine structure of nerve fibers and growth cones of isolated sympathetic neurons in culture, *J. Cell. Biol.* **56:** 713–735.

Bunge, M. B. (1977), Initial endocytosis of peroxidase or ferritin by growth cones of cultured nerve cells, *J. Neurocytol.* **6:** 407–439.

Bunge, M. B., Bunge, R. P., and Ris, H. (1961), Ultrastructural study of remyelination in an experimental lesion in adult cat spinal cord, *J. Biophys. Biochem. Cytol.* **10:** 67–94.

Bunge, M. B., Johnson, M. I., and Argiro, V. J. (1982), Certain aspects of growth cone biology and influences of neuronal age on nerve fiber growth in: *Spinal Cord Reconstruction,* Kao, C., Bunge, R. and Reier, P. (eds.) Raven Press, New York.

Bunge, R. P. (1975), Changing uses of nerve tissue culture 1950–1975, pp. 31–42, in: *The Nervous System,* Vol. 1, Brady, R. O. (ed.), Raven Press, New York.

Burdwood, W. O. (1965), Rapid bidirectional particle movement in neurons, *J. Cell Biol.* **27:** 115A.

Burešová, M., and Tuček, S. (1977), Cessation of axonal transport of acetylcholinesterase after administration of cycloheximide, *Brain Res.* **124:** 379–384.

Burgess, P. R., English, K. B., Horch, K. W., and Stensaas, L. J. (1974), Patterning in the regeneration of Type I cutaneous receptors, *J. Physiol. (London)* **236:** 57–82.

Burnstock, G. (1972), Purinergic nerves, *Pharm. Rev.* **24:** 509–581.

Burnstock, G. (1978), Do some sympathetic neurones synthesize and release both noradrenaline and acetylcholine? *Prog. Neurobiol.* **11:** 205–222.

Burnstock, G. (1979), Autonomic innervation and transmission, *Brit. Med. Bull.* **35:** 255–262.

Burton, P. R., and Fernandez, H. L. (1973), Delineation by lanthanum staining of filamentous elements associated with the surfaces of axonal microtubules, *J. Cell Sci.* **12:** 567–583.

Burton, P. R., and Kirkland, W. L. (1972), Actin detected in mouse neuroblastoma cells by binding of heavy meromyosin, *Nature (New Biol.)* **239:** 244–246.

Burton, P. R., and Paige, J. L. (1981), Polarity of axoplasmic microtubules in the olfactory nerve of the frog, *Proc. Natl. Acad. Sci. USA* **78:** 3269–3273.

Byers, M. R. (1974), Structural correlates of rapid axonal transport: Evidence that microtubules may not be directly involved, *Brain Res.* **75:** 97–113.

Byers, M. R., Fink, B. R., Kennedy, R. D., Middaugh, M. E., and Hendrickson, A. E. (1973), Effects of lidocaine on axonal morphology, microtubules, and rapid transport in rabbit vagus nerve *in vitro, J. Neurobiol.* **4:** 125–143.

Calissano, P., Monaco, G., Castellani, L., Mercanti, D., and Levi, A. (1978), Nerve growth factor potentiates actomyosin adenosinetriphosphatase, *Proc. Natl. Acad. Sci. USA* **75:** 2210–2214.

Camerino, D., and Bryant, S. H. (1976), Effects of denervation and colchicine treatment on the chloride conductance of rat skeletal muscle fibers, *J. Neurobiol.* **7:** 221–228.

Campenot, R. B. (1977), Local control of neurite development by nerve growth factor, *Proc. Natl. Acad. Sci. USA* **74:** 4516–4519.

Cancalon, P. (1979a), Influence of temperature on the velocity and on the isotope profile of slowly transported labeled proteins, *J. Neurochem.* **32:** 997–1007.

Cancalon, P. (1979b), Subcellular and polypeptide distributions of slowly transported proteins in the garfish olfactory nerve, *Brain Res.* **161:** 115–130.

Cancalon, P., and Beidler, L. M. (1975), Distribution along the axon and into various subcellular fractions of molecules labeled with (^{3}H) leucine and rapidly transported in the garfish olfactory nerve, *Brain Res.* **89:** 225–244.

Cancalon, P., Elam, J. S., and Beidler, L. M. (1976), SDS gel electrophoresis of rapidly transported proteins in garfish olfactory nerve, *J. Neurochem.* **27:** 687–693.

Cangiano, A., and Fried, J. A. (1977), The production of denervation-like changes in rat muscle by colchicine, without interference with axonal transport or muscle activity, *J. Physiol. (London)* **265:** 63–84.

Cangiano, A., Lømo, T., Lutzemberger, L., and Sveen, O. (1980), Effects of chronic nerve conduction block on formation of neuromuscular junctions and junctional AChE in the rat, *Acta Physiol. Scand.* **109:** 283–296.

Cangiano, A., and Lutzemberger, L. (1977), Partial denervation affects both denervated and innervated fibers in the mammalian skeletal muscle, *Science* **196:** 542–545.

Cangiano, A., Lutzemberger, L., and Nicotra, L. (1977), Non-equivalence of impulse blockade and denervation in the production of membrane changes in rat skeletal muscle, *J. Physiol. (London)* **273:** 691–706.

Cannata, M. A., and Morris, J. F. (1973), Changes in the appearance of hypothalamo-neurohypophysial neurosecretory granules associated with their maturation, *J. Endocrinol.* **57:** 531–538.

Cannon, W. B., and Rosenblueth, A. (1949), *The Supersensitivity of Denervated Structures*, Macmillan, New York.

Carafoli, E., and Crompton, M. (1976), Calcium ions and mitochondria, *Symp. Soc. Exp. Biol.* **30:** 89–115.

Carafoli, E., and Crompton, M. (1978), The regulation of intracellular calcium, pp. 151–216, in: *Current Topics in Membranes and Transport, Membrane Properties: Mechanical Aspects, Receptors, Energetics and Calcium Dependence on Transport,* Vol. 10, Bronner, F., and Kleinzeller, A. (eds.), Academic Press, New York.

Card, D. J. (1977), Denervation: Sequence of neuromuscular degenerative changes in rats and the effect of stimulation, *Exp. Neurol.* **54:** 251–265.

Carlsen, R. C. (1975), The possible role of cyclic AMP in the neurotrophic control of skeletal muscle, *J. Physiol. (London)* **247:** 343–361.

Carlson, B. M. (1973), The regeneration of skeletal muscle—A review, *Am. J. Anat.* **137:** 119–149.

Carlsson, C. A., Bolander, P., and Sjöstrand, J. (1971), Changes in axonal transport during regeneration of feline ventral roots, *J. Neurol. Sci.* **14:** 75–93.

Carton, H. C., and Appel, S. H. (1974), Biochemical studies of transneuronal degeneration: The effects of enucleation on the biochemical maturation of the chick optic tectum, *Brain Res.* **67:** 289–306.

Casagrande, V. A., and Harting, J. K. (1975), Transneuronal transport of tritiated fucose and proline in the visual pathways of tree shrew *Tupaia glis, Brain Res.* **96:** 367–372.

Ceccarelli, B., Hurlbut, W. P., and Mauro, A. (1973), Turnover of transmitter and synaptic vesicles at the frog neuromuscular junction, *J. Cell Biol.* **57:** 499–524.

Chalfie, M., and Thomson, J. N. (1979), Organization of neuronal microtubules in the nematode *Caenorhabditis elegans, J. Cell. Biol.* **82:** 278–289.

Chan, K. Y., and Bunt, A. H. (1978), An association between mitochondria and microtubules in synaptosomes and axon terminals of cerebral cortex, *J. Neurocytol.* **7:** 137–143.

Chan, S. Y., Ochs, S., and Jersild, R., Jr. (1979), Calcium localization in mammalian nerve fibers in relation to its regulation and axoplasmic transport, *Soc. Neurosci. Abstr.* **5:** 59.

Chan, S. Y., Ochs, S., and Worth, R. M. (1980a), The requirement for calcium ions and the effect of other ions on axoplasmic transport in mammalian nerve, *J. Physiol. (London)* **301:** 477–504.

Chan, S. Y., Worth, R., and Ochs, S. (1980b), Block of axoplasmic transport *in vitro* by vinca alkaloids, *J. Neurobiol.* **11:** 251–264.

Chang, C. C., and Huang, M. C. (1975), Turnover of junctional and extrajunctional acetylcholine receptors of the rat diaphragm, *Nature* **253:** 643–644.

Cheung, W. Y. (1980), Calmodulin plays a pivotal role in cellular regulation, *Science* **207:** 19–27.

Chihara, E., and Honda, Y. (1981), Analysis of orthograde fast axonal transport and nonaxonal transport along the optic pathway of albino rabbits during increased and decreased intraocular pressure, *Exp. Eye Res.* **32:** 229–239.

Chou, S.-M., and Hartmann, H. A. (1964), Axonal lesions and waltzing syndrome after IDPN administration in rats, *Acta Neuropath. (Berlin)* **3:** 428–450.

Clarke, E. (1968), The doctrine of the hollow nerve in the seventeenth and eighteenth centuries, pp. 123–141, in: *Medicine, Science and Culture: Historical Essays in Honor of Owsei Temkin,* Stevenson, L. G., and Multhauf, R. P. (eds.), Johns Hopkins Press, Baltimore.

Clarke, E. (1978), The neural circulation. The use of analogy in medicine, *Med. Hist.* **22:** 291–307.

Clarke, E., and Dewhurst, K. (1972), *An Illustrated History of Brain Function,* Sandford Publications, Oxford.

Close, R. (1965), The relation between intrinsic speed of shortening and duration of the active state of muscle, *J. Physiol. (London)* **180:** 542–559.

Close, R. (1967), Properties of motor units in fast and slow skeletal muscles of the rat, *J. Physiol. (London)* **193:** 45–55.

Clouet, D. H., and Waelsch, H. (1961), Amino acid and protein metabolism of the brain. VIII. The recovery of cholinesterase in the nervous system of the frog after inhibition, *J. Neurochem.* **8:** 201–215.

Coggeshall, R. E. (1980), Law of separation of function of the spinal roots, *Physiol. Rev.* **60:** 716–755.

Cohen, S. S. (1971), *Introduction to the Polyamines,* Prentice-Hall, Englewood Cliffs, N.J.

Collier, B. (1969), The preferential release of newly synthesized transmitter by a sympathetic ganglion, *J. Physiol. (London)* **205:** 341–352.

Collier, B., and MacIntosh, F. C. (1969), The source of choline for acetylcholine synthesis in a sympathetic ganglion, *Can. J. Physiol. Pharmacol.* **47:** 127–135.

Cook, D. D., and Gerard, R. W. (1931), The effect of stimulation on the degeneration of a severed peripheral nerve, *Am. J. Physiol.* **97:** 412–425.

Cooper, J. R., Bloom, F. E., and Roth, R. H. (1978), *The Biochemical Basis of Neuropharmacology,* 3rd ed., Oxford University Press, New York.

Cooper, P. D., and Smith, R. S. (1974), The movement of optically detectable organelles in myelinated axons of *Xenopus laevis, J. Physiol. (London)* **242:** 77–97.

Copeland, A. R. (1976a), Axonal transport. I. Diffusion, *Bull. Math. Biol.* **38:** 425–433.

Copeland, A. R. (1976b), Axonal transport. II. Convection, *Bull. Math. Biol.* **38:** 435–444.

Cosens, B., Thacker, D., and Brimijoin, S. (1976), Temperature-dependence of rapid axonal transport in sympathetic nerves of the rabbit, *J. Neurobiol.* **7:** 339–354.

Cotman, C. W., and Taylor, D. A. (1971), Autoradiographic analysis of protein synthesis in synaptosomal fractions, *Brain Res.* **29:** 366–372.

Couraud, J. Y., and Di Giamberardino, L. (1980), Axonal transport of the molecular forms of acetylcholinesterase in chick sciatic nerve, *J. Neurochem.* **35:** 1053–1066.

Cowan, W. M. (1970), Anterograde and retrograde transneuronal degeneration in the central and peripheral nervous system, pp. 217–251, in: *Contemporary Research Methods in Neuroanatomy,* Nauta, W. J. H. and Ebbesson, S. O. E. (eds.), Springer, New York.

Cowan, W. M., and Cuénod, M. (eds.) (1975), *The Use of Axonal Transport for Studies of Neuronal Connectivity,* Elsevier Publishing Co., New York.

Cowan, W. M., Gottlieb, D. I., Hendrickson, A. E., Price, J. L., and Woolsey, T. A. (1972), The autoradiographic demonstration of axonal connections in the central nervous system, *Brain Res.* **37:** 21–51.

Cox, J. A., Malnoë, A., and Stein, E. A. (1981), Regulation of brain cyclic nucleotide phosphodiesterase by calmodulin. A quantitative analysis, *J. Biol Chem.* **256:** 3218–3222.

Coyle, J. T., and Wooten, G. F. (1972), Rapid axonal transport of tyrosine hydroxylase and dopamine-β-hydroxylase, *Brain Res.* **44:** 701–704.

Cragg, B. G. (1970), What is the signal for chromatolysis? *Brain Res.* **23:** 1–21.

Cragg, B. (1979), Overcoming the failure of electronmicroscopy to preserve the brain's extracellular space, *Trends Neurosci.* **2:** 159–161.

Crain, S. M. (1976), *Neurophysiological Studies in Tissue Culture*, Raven Press, New York.

Crain, S., and Peterson, E. R. (1974), Development of neural connections in culture, *Ann. N.Y. Acad. Sci.* **228:** 6–34.

Crescitelli, F. (1951), Nerve sheath as a barrier to the action of certain substances, *Am. J. Physiol.* **166:** 229–240.

Crooks, R. F., and McClure, W. O. (1972), The effect of cytochalasin B on fast axoplasmic transport, *Brain Res.* **45:** 643–646.

Croone, W. (1664), De Ratione Motus in Musculorum, in: Willis, T., *Opera Omnia*, Storti, Coloniae (1694), translated by Pordage, S., London, 1684.

Crosby, E. C., Humphrey, T., and Lauer, E. W. (1962), *Correlative Anatomy of the Nervous System*, Macmillan, New York.

Cross, B. A., Dyball, R. E. J., Dyer, R. G., Jones, C. W., Lincoln, D. W., Morris, J. F., and Pickering, B. T. (1975), Endocrine neurons, *Recent Prog. Hormone Res.* **31:** 243–294.

Csanyi, V., Gervai, J., and Lajtha, A. (1973), Axoplasmic transport of free amino acids, *Brain Res.* **56:** 271–284.

Cuénod, M., and Schonbach, J. (1971), Synaptic proteins and axonal flow in the pigeon visual pathway, *J. Neurochem.* **18:** 809–816.

Curtis, D. R., and De Groat, W. C. (1968), Tetanus toxin and spinal inhibition, *Brain Res.* **10:** 208–212.

Curtis, D. R., Felix, D., Game, C. J. A., and McCulloch, R. M. (1973), Tetanus toxin and the synaptic release of GABA, *Brain Res.* **51:** 358–362.

Czéh, G., Gallego, R., Kudo, N., and Kuno, M. (1978), Evidence for the maintenance of motoneurone properties by muscle activity, *J. Physiol. (London)* **281:** 239–252.

Dahl, D., Bignami, A., Bich, N. T., and Chi, N. H. (1980), Immunohistochemical characterization of neurofibrillary tangles induced by mitotic spindle inhibitors, *Acta Neuropath. (Berlin)* **51:** 165–168.

Dahlström, A. (1968), Effect of colchicine on transport of amine storage granules in sympathetic nerves of rat, *Europ. J. Pharmacol.* **5:** 111–113.

Dahlström, A. (1969), Synthesis, transport, and life-span of amine storage granules in sympathetic adrenergic neurons, *Symp. Int. Soc. Cell Biol.* **8:** 153–174.

Dahlström, A. (1971), Axoplasmic transport (with particular respect to adrenergic neurons), *Phil Trans. R. Soc.* **261:** 325–358.

Dahlström, A. (1973), Aminergic transmission. Introduction and short review, *Brain Res.* **62:** 441–460.

Dahlström, A. B., Evans, C. A. N., Häggendal, C. J., Heiwall, P.-O., and Saunders, N. R. (1974), Rapid transport of acetylcholine in rat sciatic nerve proximal and distal to a lesion, *J. Neural Transmission* **35:** 1–11.

Dahlström, A., and Fuxe, K. (1964a), A method for the demonstration of adrenergic nerve fibres in peripheral nerves, *Z. Zellforsch. Mikrosk. Anat.* **62:** 602–607.

Dahlström, A., and Fuxe, K. (1964b), A method for the demonstration of monoamine-containing fibres in the central nervous system, *Acta Physiol. Scand.* **60:** 293–294.

Dahlström, A., and Fuxe, K. (1964c), Evidence for the existence of monoamine-containing neurons in the central nervous system: 1. Demonstration of monoamines in the cell bodies of brain stem neurons, *Acta Physiol. Scand.* **62:** Suppl. 232, 1–55.

Dahlström, A., and Häggendal, J. (1966), Studies on the transport and life-span of amine storage granules in a peripheral adrenergic neuron system, *Acta Physiol. Scand.* **67:** 278–288.

Dahlström, A., and Häggendal, J. (1967), Studies on the transport and life-span of amine storage granules in the adrenergic neuron system of the rabbit sciatic nerve, *Acta Physiol. Scand.* **69:** 153–157.

Dahlström, A., Häggendal, J., Hielbronn, E., Hiewall, P.-O., and Saunders, N. R. (1974), Proximodistal transport of acetylcholine in peripheral cholinergic neurons, pp. 275–289, in: *Dynamics of Degeneration and Growth in Neurons,* Fuxe, K., Olson, L., and Zotterman, Y. (eds.), Pergamon Press, New York.

Dahlström, A., Häggendal, J., and Larsson, P. A. (1975), On the noradrenaline loading in axonal amine storage granules in rat crushed sciated nerves, *Acta Physiol. Scand.* **94:** 451–458.

Dahlström, A., and Heiwall, P. O. (1975), Intra-axonal transport of transmitters in mammalian neurons, *J. Neural Transmission Suppl.* **XII:** 97–114.

Dahlström, A., and Jonason, J. (1968), Dopa-decarboxylase activity in sciatic nerves of the rat after constriction, *Europ. J. Pharmacol.* **4:** 377–383.

Dale, H. H. (1900), On some numerical comparisons of the centripetal and centrifugal medullated nerve-fibres arising in the spinal ganglia of the mammal, *J. Physiol. (London)* **25:** 196–206.

Davey, B., and Younkin, S. G. (1978), Effect of nerve stump length on cholinesterase in denervated rat diaphragm, *Exp. Neurol.* **59:** 168–175.

Davey, B., Younkin, L. H., and Younkin, S. G. (1979), Neural control of skeletal muscle cholinesterase: A study using organ-cultured rat muscle, *J. Physiol. (London)* **289:** 501–515.

Davies, L. P., Whittaker, V. P., and Zimmermann, N. H. (1977), Axonal transport in the electromotor nerves of torpedo marmorata, *Exp. Brain Res.* **30:** 493–510.

Davis, H. L., and Kiernan, J. A. (1980), Neurotrophic effects of sciatic nerve extract on denervated extensor digitorum longus muscle in the rat, *Exp. Neurol.* **69:** 124–134.

Davis, H. L., and Kiernan, J. A. (1981), Effect of nerve extract on atrophy of denervated or immobilized muscles, *Exp. Neurol.* **72:** 582–591.

DeCoursey, T. E., Younkin, S. G., and Bryant, S. H. (1978), Neural control of chloride conductance in rat extensor digitorum longus muscle, *Exp. Neurol.* **61:** 705–709.

Dennis, M. J., and Miledi, R. (1974a), Electrically induced release of acetylcholine from denervated Schwann cells, *J. Physiol. (London)* **237:** 431–452.

Dennis, M. J., and Miledi, R. (1974b), Characteristics of transmitter release at regenerating frog neuromuscular junctions, *J. Physiol. (London)* **239:** 571–594.

Denny-Brown, D. (1929), The histological features of striped muscle in relation to its functional activity, *Proc. Roy. Soc. B* **104:** 371–411.

Denny-Brown, D., and Brenner, C. (1944), Lesion in peripheral nerve resulting from compression by spring clip, *Arch. Neurol. Psychiat. (Chicago)* **52:** 1–19.

De Potter, W. P., De Schaepdryver, A. F., Moerman, E. J., and Smith, A. D. (1969), Evidence for the release of vesicle-proteins together with noradrenaline upon stimulation of the splenic nerve, *J. Physiol. (London)* **204:** 102P–104P.

DeSantis, M., Hoekman, T., and Limwongse, V. (1977), Retrograde transport of peroxidase in motor neurons innervating slow and fast muscles: Absence of a difference between normal and dystrophic chickens, *Brain Res.* **119:** 454–458.

Descartes, R. (1664), *Treatise of Man,* translated from the French by T. S. Hall, Harvard University Press, Cambridge, 1972.

Deshpande, S. S., Albuquerque, E. X., and Guth, L. (1976), Neurotrophic regulation of prejunctional and postjunctional membrane at the mammalian motor endplate, *Exp. Neurol.* **53:** 151–165.

Devreotes, P. N., and Fambrough, D. M. (1975), Turnover of acetylcholine receptors in skeletal muscle, *Cold Spring Harbor Symp. Quant. Biol.* **40:** 237–251.

Devillers, M. C. (1948), Morphogenèse—La genèse des organes sensoriels latéraux de la Truite, *Compt. Rend. Acad. Sci.* **226:** 354–356.

Diamond, J., and Miledi, R. (1962), A study of foetal and new-born rat muscle fibres, *J. Physiol. (London)* **162:** 393–408.

Dietz, R., (1972), Die assembly-hypothese der chromosomenbewegung und die veränderungen der spindellänge während der anaphase I in spermatocyten von *Pales ferruginea (Tipulidae, Diptera), Chromosoma* **38:** 11–76.

Di Giamberardino, L. (1971), Independence of the rapid axonal transport of protein from the flow of free amino acids, *Acta Neuropath. (Berlin) Suppl.* **V:** 132–135.

Di Giamberardino, L., and Couraud, J. Y. (1978), Rapid accumulation of high molecular weight acetylcholinesterase in transected sciatic nerve, *Nature* **271:** 170–172.

Di Giamberardino, L., Couraud, J. Y., and Barnard, E. A. (1979), Normal axonal transport of acetylcholinesterase forms in peripheral nerves of dystrophic chickens, *Brain Res.* **160:** 196–202.

DiPolo, R. (1978), Ca pump driven by ATP in squid axons, *Nature* **274:** 390–392.

Dismukes, R. K., (and commentators) (1979), New concepts of molecular communication among neurons, *Behav. Brain Sci.* **2:** 409–448.

Donat, J. R., and Wiśniewski, H. M. (1973), The spatio-temporal pattern of Wallerian degeneration in mammalian peripheral nerves, *Brain Res.* **53:** 41–53.

Donoso, J. A., Green, L. S., Heller-Bettinger, I. E., and Samson, F. E. (1977), Action of the *vinca* alkaloids vincristine, vinblastine, and desacetyl vinblastine amide on axonal fibrillar organelles *in vitro, Cancer Res.* **37:** 1401–1407.

Donoso, J. A., Illanes, J.-P., and Samson, F. (1977), Dimethylsulfoxide action on fast axoplasmic transport and ultrastructure of vagal axons, *Brain Res.* **120:** 287–301.

Douglas, W. B. (1975), Sciatic cross-reinnervation of normal and dystrophic muscle in parabiotic mice: Isometric contractile responses of reinnervated tibialis anticus and triceps surae, *Exp. Neurol.* **48:** 647–663.

Douglas, W. W. (1973), How do neurones secrete peptides? Exocytosis and its consequences, including "synaptic vesicle" formation, in the hypothalamo-neurohypophyseal system, *Prog. Brain Res.* **39:** 21–39.

Douglas, W. W., Nagasawa, J., and Schulz, R. (1971), Electron microscopic studies on the mechanism of secretion of posterior pituitary hormones and significance of microvesicles ('synaptic vesicles'): Evidence of secretion by exocytosis and formation of microvesicles as a by-product of this process, *Mem. Soc. Endocrinol.* **19:** 353–383.

Douglas, W. W., and Poisner, A. M. (1964a), Stimulus–secretion coupling in a neurosecretory organ: The role of calcium in the release of vasopressin from the neurohypophysis, *J. Physiol. (London)* **172:** 1–18.

Douglas, W. W., and Poisner, A. M. (1964b), Calcium movement in the neurohypophysis of the rat and its relation to the release of vasopressin, *J. Physiol. (London)* **172:** 19–30.

Drachman, D. B. (1974), The role of acetylcholine as a neurotropic transmitter, *Ann. N.Y. Acad. Sci.* **228:** 160–176.

Dräger, U. C. (1974), Autoradiography of tritiated proline and fucose transported transneuronally from the eye to the visual cortex in pigmented and albino mice, *Brain Res.* **82:** 284–292.

Dresden, M. H. (1969), Denervation effects of newt limb regeneration: DNA, RNA, and protein synthesis, *Develop. Biol.* **19:** 311–320.

Droz, B. (1965a), Fate of newly synthesized proteins in neurons, *Symp. Int. Soc. Cell Biol.* **4:** 159–175.

Droz, B. (1965b), Accumulation de protéines nouvellement synthétisées dans l'appareil de Golgi du neurone; étude radioautoraphique en microscopie electronique, *Compt. Rend. Acad Sci. (Paris)* **260:** 320–322.

Droz, B. (1979), How axonal transport contributes to maintenance of the myelin sheath, *Trends Neurosci.* **2:** 146–148.

Droz, B., and Barondes, S. H. (1969), Nerve endings: Rapid appearance of labeled protein shown by electron microscope radioautography, *Science* **165:** 1131–1133.

Droz, B., and Leblond, C. P. (1962), Migration of proteins along the axons of the sciatic nerve, *Science* **137:** 1047–1048.

Droz, B., and Leblond, C. P. (1963), Axonal migration of proteins in the central nervous system and peripheral nerves as shown by radioautography, *J. Comp. Neurol.* **121:** 325–346.

Droz, B., Rambourg, A., and Koenig, H. L. (1975), The smooth endoplasmic reticulum: Structure and role in the renewal of axonal membrane and synaptic vesicles by fast axonal transport, *Brain Res.* **93:** 1–13.

Duce, I. R., and Keen, P. (1978), Can neuronal smooth endoplasmic reticulum function as a calcium reservoir? *Neuroscience* **3:** 837–848.

Ducker, T. B., Kempe, L. G., and Hayes, G. J. (1969), The metabolic background for peripheral nerve surgery, *J. Neurosurg.* **30:** 270–280.

Dulak, L., and Crist, R. D. (1974), Microtubule properties in dimethylsulfoxide (DMSO), *J. Cell Biol.* **63:** 90A.

Dumas, M., Schwab, M. E., and Thoenen, H. (1979), Retrograde axonal transport of specific macromolecules as a tool for characterizing nerve terminal membranes, *J. Neurobiol.* **10:** 179–197.

Duncan, D. (1948), Alterations in the structure of nerves caused by restricting their growth with ligatures, *J. Neuropath. Exp. Neurol.* **7:** 261–273.

Duncan, D., and Keyser, L. L. (1936), Some determinations of the ratio of nerve fibers to nerve cells in the thoracic dorsal roots and ganglia of the cat, *J. Comp. Neurol.* **64:** 303–311.

Dunlop, D. S., Van Elder, W., and Lajtha, A. (1974), Measurements of rates of protein synthesis in rat brain slices, *J. Neurochem.* **22:** 821–830.

Durham, A. C. H. (1974), A unified theory of the control of actin and myosin in nonmuscle movements, *Cell* **2:** 123–136.

Dustin, P. (1978) *Microtubules,* Springer-Verlag, New York.

Dutton, G. R., and Barondes, S. (1969), Microtubular protein: synthesis and metabolism in developing brain, *Science* **166:** 1637–1638.

Dyck, P. J. (1975), Pathologic alterations of the peripheral nervous system of man, Chapter 15, in: *Peripheral Neuropathy,* Vol. 1, Dyck, P. J., Thomas, P. K., and Lambert, E. H. (eds.), W. B. Saunders, Philadelphia.

Dyck, P. J. and Hopkins, A. P. (1972), Electron microscopic observations on degeneration and regeneration of unmyelinated fibres, *Brain* **95:** 223–234.

Dziegielewska, K. M., Evans, C. A. N., and Saunders, N. R. (1980), Rapid effect of nerve injury upon axonal transport of phospholipids, *J. Physiol. (London)* **304:** 83–98.

Easton, D. M. (1971), Garfish olfactory nerve: Accessible source of numerous long, homogeneous, nonmyelinated axons, *Science* **172:** 952–955.

Eccles, J. C. (1964), *The Physiology of Synapses,* Academic Press, New York.

Eccles, J. C. (1977), *The Understanding of the Brain,* 2nd ed. McGraw-Hill, New York.

Eccles, J. C., and Sherrington, C. S. (1930), Numbers and contraction-values of individual motor-units examined in some muscles of the limbs, *Proc. Roy. Soc. B* **106:** 326–357.

Edds, M. V., Jr. (1953), Collateral nerve regeneration, *Quart. Rev. Biol.* **28:** 260–276.

Edds, M. V., Jr. (1979), Plasticity of retinotectal connections in fishes and amphibians, *Neurosci. Res. Prog. Bull.* **17:** 251–272.

Editorial (1976), Axonal transport and the eye, *Brit. J. Ophthal.* **60:** 547–550.

Edström, A. (1974), Effects of Ca^{2+} and Mg^{2+} on rapid axonal transport of proteins *in vitro* in frog sciatic nerves, *J. Cell. Biol.* **61:** 812–818.

Edström, A., and Hanson, M. (1973a), Retrograde axonal transport of proteins *in vitro* in frog sciatic nerves, *Brain Res.* **61:** 311–320.

Edström, A., and Hanson, M. (1973b), Temperature effects on fast axonal transport of proteins *in vitro* in frog sciatic nerves, *Brain Res.* **58:** 345–354.

Edström, A., Hansson, H.-A., and Norström, A. (1973a), Inhibition of axonal transport *in vitro* in frog sciatic nerves by chlorpromazine and lidocaine, *Z. Zellforsch. Mikrosk. Anat.* **143:** 53–69.

Edström, A., Hansson, H.-A., and Norström, A. (1973b), The effect of chlorpromazine and tetracaine on the rapid axonal transport of neurosecretory material in the hypothalamo-neurohypophysial system of the rat, *Z. Zellforsch. Mikrosk. Anat.* **143:** 71–91.

Edström, A., Hanson, M., Prus, K., and Wallin, M. (1980), Ca^{2+}- or Mg^{2+}-dependent enzymic ATP hydrolysis associated with the microsomal fraction of frog sciatic nerves, *J. Neurochem.* **35:** 297–303.

Edström, A., and Mattsson, H. (1972a), Fast axonal transport *in vitro* in the sciatic system of the frog, *J. Neurochem.* **19:** 205–221.

Edström, A., and Mattsson, H. (1972b), Rapid axonal transport *in vitro* in the sciatic system of the frog of fucose-, glucosamine-, and sulphate-containing material, *J. Neurochem.* **19:** 1717–1729.

Edström, A., and Mattsson, H. (1975), Small amounts of zinc stimulate rapid axonal transport *in vitro*, *Brain Res.* **86:** 162–167.

Edström, A, and Mattsson, H. (1976), Inhibition and stimulation of rapid axonal transport *in vitro* by sulfhydryl blockers, *Brain Res.* **108:** 381–395.

Edwards, C. (1979), The effects of innervation on the properties of acetylcholine receptors in muscle, *Neuroscience* **4:** 565–584.

Ehrenberg, C. G. (1838), Observations on the structure hitherto unknown of the nervous system in man and animals, translated from the German by Craigie, D., in: *Essays on Physiology and Hygiene*, Haswell, Philadelphia.

Eipper, B. A. (1972), Rat brain microtubule protein: purification and determination of covalently bound phosphate and carbohydrate, *Proc. Natl. Acad. Sci. USA* **69:** 2283–2287.

Eisenstadt, M., Goldman, J. E., Kandel, E. R., Koike, H., Koester, J., and Schwartz, J. H. (1973), Intrasomatic injection of radioactive precursors for studying transmitter synthesis in identified neurons of *Aplysia Californica*, *Proc. Natl. Acad. Sci. USA* **70:** 3371–3375.

Elam, J. S. (1979), Axonal transport of complex carbohydrates, Chapter 12, in: *Complex Carbohydrates of Nervous Tissue*, Margolis, R. U., and Margolis, R. K. (eds.), Plenum Press, New York.

Elam, J. S., and Agranoff, B. W. (1971a), Rapid transport of protein in the optic system of the goldfish, *J. Neurochem.* **18:** 375–387.

Elam, J. S., and Agranoff, B. W. (1971b), Transport of proteins and sulfated mucopolysaccharides in the goldfish visual system, *J. Neurobiol.* **2:** 379–390.

Elam, J. S., Goldberg, J. M., Radin, N. S., and Agranoff, B. W. (1970), Rapid axonal transport of sulfated mucopolysaccharide proteins, *Science* **170:** 458–460.

Ellisman, M. H. (1981), Topographic organization of the elements of the "cytoskeleton," *Neurosci. Res. Prog. Bull.* **20:** 79–91.

Ellisman, M. H., and Lindsey, J. D. (1981), The membranous reticulum within myelinated axons is not rapidly transported, *Soc. Neurosci. Abstr.* **7:** 742.

Ellisman, M., and Porter, K. R. (1980), Microtrabecular structure of the axoplasmic matrix: Visualization of cross-linking structures and their distribution, *J. Cell Biol.* **87:** 464–479.

Elsberg, C. A. (1917), Experiments on motor nerve regeneration and the direct neurotization of paralyzed muscle by their own and by foreign nerves, *Science* **45:** 318–320.

Emson, P. C. (1979), Peptides as neurotransmitter candidates in the mammalian CNS, *Prog. Neurobiol.* **13:** 61–116.

Engh, C. A., Schofield, B. H., Doty, S. B., and Robinson, R. A., (1971), Perikaryal synthetic function following reversible and irreversible peripheral axon injuries as shown by radioautography, *J. Comp. Neurol.* **142:** 465–480.

Erdmann, G., Wiegand, H., and Wellhöner, H. H. (1975), Intraaxonal and extraaxonal transport of ^{125}I-tetanus toxin in early local tetanus, *Naunyn-Schmiedeberg's Arch. Pharmacol.* **290:** 357–373.

Erickson, P. F., Seamon, K. B., Moore, B. W., Lasher, R. S., and Minier, L. N. (1980), Axonal transport of the Ca^{2+}-dependent protein modulator of $3':5'$-cyclic-AMP phosphodiesterase in the rabbit visual system, *J. Neurochem.* **35:** 242–248.

Esquerro, E., Garcia, A. G., and Sanchez-Garcia, P. (1980), The effects of the calcium ionophore, A23187, on the axoplasmic transport of dopamine-β-hydroxylase, *Brit. J. Pharmacol.* **70:** 375–381.

Evans, C. A. N., and Saunders, N. R. (1974), An outflow of acetylcholine from normal and regenerating ventral roots of the cat, *J. Physiol. (London)* **240:** 15–32.

Falck, B., Hillarp, N.-Å., Thieme, G., and Torp, A. (1962), Fluorescence of catecholamines and related compounds condensed with formaldehyde, *J. Histochem.* Cytochem. *10:* 348–354.

Fambrough, D. M. (1974), Acetylcholine receptors. Revised estimates of extrajunctional receptor density in denervated rat diaphragm, *J. gen. Physiol.* **64:** 468–472.

Fawcett, C. P., Powell, A. E., and Sachs, H. (1968), Biosynthesis and release of neurophysin, *Endocrinology* **83:** 1299–1310.

Feigin, I., Geller, E. H., and Wolf, A. (1951), Absence of regeneration in the spinal cord of the young rat, *J. Neuropath. Exp. Neurol.* **10:** 420–425.

Feit, H., Slusarek, L., and Shelanski, M. L. (1971), Heterogeneity of tubulin subunits, *Proc. Natl. Acad. Sci. USA* **68:** 2028–2031.

Feldberg, W., and Vogt, M. (1948), Acetylcholine synthesis in different regions of the central nervous system, *J. Physiol. (London)* **107:** 372–381.

Feng, T. P., and Gerard, R. W. (1930), Mechanism of nerve asphyxiation: With a note on the nerve sheath as a diffusion barrier, *Proc. Soc. Exp. Biol. Med.* **27:** 1073–1076.

Fernandez, H. L., Burton, P. R., and Samson, F. E. (1971), Axoplasmic transport in the crayfish nerve cord. The role of fibrillar constituents of neurons, *J. Cell Biol.* **51:** 176–192.

Fernandez, H. L., Duell, M. J., and Festoff, B. W. (1979), Cellular distribution of 16S acetylcholinesterase, *J. Neurochem.* **32:** 581–585.

Fernandez, H. L., Duell, M. J., and Festoff, B. W. (1980), Bidirectional axonal transport of 16S acetylcholinesterase in rat sciatic nerve, *J. Neurobiol.* **11:** 31–39.

Fernandez, H. L., Huneeus, F. C., and Davison, P. F. (1970), Studies on the mechanism of axoplasmic transport in the crayfish cord, *J. Neurobiol.* **1:** 395–409.

Fernandez, H. L., and Inestrosa, N. C. (1976), Role of axoplasmic transport in neurotrophic regulation of muscle end plate acetylcholinesterase, *Nature* **262:** 55–56.

Fernandez, H. L., Patterson, M. R., and Duell, M. J. (1980), Neurotrophic control of 16S acetylcholinesterase from mammalian skeletal muscle in organ culture, *J. Neurobiol.* **11:** 557–570.

Fernandez, H. L., Singer, P. A., and Mehler, S. (1981), Retrograde axonal transport mediates the onset of regenerative changes in the hypoglossal nucleus, *Neurosci. Lett.* **25:** 7–11.

Fex, S., Sonesson, B., Thesleff, S., and Zelená, J. (1966), Nerve implants in botulinum poisoned mammalian muscle, *J. Physiol. (London)* **184:** 872–882.

Fillenz, M. (1977), The factors which provide short-term and long-term control of transmitter release, *Prog. Neurobiol.* **8:** 251–278.

Fillenz, M., Gagnon, C., Stoeckel, K., and Thoenen, H. (1976), Selective uptake and retrograde axonal transport of dopamine-β-hydroxylase antibodies in peripheral adrenergic neurons, *Brain Res.* **114:** 293–303.

Fink, D. J., and Gainer, H. (1980a), Retrograde axonal transport of endogenous proteins in sciatic nerve demonstrated by covalent labeling *in vivo*, *Science* **208**: 303–305.

Fink, D. J., and Gainer, H. (1980b), Axonal transport of proteins. A new view using *in vivo* covalent labeling, *J. Cell Biol.* **85**: 175–186.

Fink, B. R., Kennedy, R. D., Hendrickson, A. E., and Middaugh, M. E. (1972), Lidocaine inhibition of rapid axonal transport, *Anesthesiology* **36**: 422–432.

Fischer, H. A., and Schmatolla, E. (1972), Axonal transport of tritium-labeled putrescine in the embryonic visual system of zebrafish, *Science* **176**: 1327–1329.

Fischbach, G. D. (1970), Synaptic potentials recorded in cell cultures of nerve and muscle, *Science* **169**: 1331–1333.

Flament, J., and Durand, J. (1961), Endocrinologie. Étude des relations hypothalamo-hypophysaires à l'aide de radioisotopes marqués au soufre 35, *Compt. Rend. Acad. Sci. (D) Paris* **252**: 3487–3489.

Flament-Durand, J., and Dustin, P. (1972), Studies on the transport of secretory granules in the magnocellular hypothalamic neurons. I. Action of colchicine on axonal flow and neurotubules in the paraventricular nuclei, *Z. Zellforsch. Mikrosk. Anat.* **130**: 440–454.

Flament-Durand, J., Couck, A.-M., and Dustin, P. (1975), Studies on the transport of secretory granules in the magnocellular hypothalamic neurons of the rat. II. Action of vincristine on axonal flow and neurotubules in the paraventricular and supraoptic nuclei, *Cell Tissue Res.* **164**: 1–9.

Fondy, T. P., and Kaplan, N. O. (1965), Structural and functional properties of the H and M subunits of lactic dehydrogenases, *Ann. N.Y. Acad. Sci.* **119**: 888–904.

Fonnum, F. (1973), Recent developments in biochemical investigations of cholinergic transmission, *Brain Res.* **62**: 497–507.

Fonnum, F., Frizell, M., and Sjöstrand, J. (1973), Transport, turnover and redistribution of choline acetyltransferase and acetylcholinesterase in the vagus and hypoglossal nerves of the rabbit, *J. Neurochem.* **21**: 1109–1120.

Fonnum, F., Frizell, M., and Sjöstrand, J. (1976), Redistribution of choline acetyltransferase and acetylcholinesterase in an isolated nerve segment of the rabbit vagus nerve, *J. Neurochem.* **26**: 427–429.

Fontana, F. (1778), *Treatise on the Venom of the Viper; on American Poisons; and on the Cherry Laurel and Some Other Vegetable Poisons to Which Are Annexed Observations on the Primitive Structure of the Animal Body; Different Experiments on the Reproduction of the Nerves and a Description of the Nerves; and a New Description of a New Canal of the Eye,* two volumes, translated from the French by J. Skinner, Murray, London.

Forel, A. (1887), Einige Hirnanatomische Betrachtungen und Ergebnisse, *Archiv. Psychiat. Nerven.* **18**: 162–198.

Forgue, S. T., and Dahl, J. L. (1978), The turnover rate of tubulin in rat brain, *J. Neurochem.* **31**: 1289–1297.

Forman, D. S., Grafstein, B., and McEwen, B. S. (1972), Rapid axonal transport of (^{3}H) fucosyl glycoproteins in the goldfish optic system, *Brain Res.* **48**: 327–342.

Forman, D. S., McEwen, B. S., and Grafstein, B. (1971), Rapid transport of radioactivity in goldfish optic nerve following injections of labeled glucosamine, *Brain Res.* **28**: 119–130.

Forman, D. S., Padjen, A. L., and Siggins, G. R. (1977a), Axonal transport of organelles visualized by light microscopy: Cinemicrographic and computer analysis, *Brain Res.* **136**: 197–213.

Forman, D. S., Padjen, A. L., and Siggins, G. R. (1977b), Effect of temperature on the rapid retrograde transport of microscopically visible intra-axonal organelles, *Brain Res.* **136**: 215–226.

Forman, D. S., and Shain, W. G., Jr. (1981), Batrachotoxin blocks saltatory organelle movement in electrically excitable neuroblastoma cells, *Brain Res.* **211**: 242–247.

Forssman, J. (1898), Ueber die Ursachen, welche die Wachsthumsrichtung der peripheren Nerven-fasern bie der Regeneration bestimmen, *Beitr. Path. Anat.* **24:** 56–99.

Forssman, J. (1900), Zur Kenntniss des Neurotropismus, *Beitr. Path. Anat.* **27:** 407–430.

Francoeur, J., and Olszewski, J. (1968), Axonal reaction and axoplasmic flow as studied by radioautography, *Neurology* **18:** 178–184.

Frank, E., Jansen, J. K. S., Lømo, T., and Westgaard, R. H. (1975), The interaction between foreign and original motor nerves innervating the soleus muscle of rats, *J. Physiol. (London)* **247:** 725–743.

Frankenhaeuser, B. (1949), Ischaemic paralysis of a uniform nerve, *Acta Physiol. Scand.* **18:** 75–98.

Frankenhaeuser, B. (1957), The effect of calcium on the myelinated nerve fibre, *J. Physiol. (London)* **137:** 245–260.

Friede, [R.] L., and Bischausen, R. (1980), The fine structure of stumps of transected nerve fibers in subserial sections, *J. Neurol. Sci.* **44:** 181–203.

Friede, R. L. (1959), Transport of oxidative enzymes in nerve fibers; a histochemical investigation of the regenerative cycle in neurons, *Exp. Neurol.* **1:** 441–466.

Friede, R. L. (1970), Determination of neurofilament and microtubule density in nerve fibres (what factors control axon calibre?), pp. 209–222, in: *Alzhemer's Disease and Related Conditions,* Wolstenholme, G. E. W., and O'Connor, M. (eds.), J. & A. Churchill, London.

Friede, R. L. (1971), Changes in microtubules and neurofilaments in constricted, hypoplastic nerve fibers, *Acta Neuropath. Suppl. (Berlin)* **V,** 216–225.

Friede, R. L., and Ho, K.-C. (1977), The relation of axonal transport of mitochondria with microtubules and other axoplasmic organelles, *J. Physiol. (London)* **265:** 507–519.

Friede, R. L., and Martinez, A. J. (1970), Analysis of axon–sheath relations during early Wallerian degeneration, *Brain Res.* **19:** 199–212.

Friede, R. L., and Samorajski, T. (1968), Myelin formation in the sciatic nerve of the rat, *J. Neuropath. Exp. Neurol.* **27:** 546–569.

Frizell, M., Hasselgren, P. O., and Sjöstrand, J. (1970), Axoplasmic transport of acetylcholinester-ase and choline acetyltransferase in the vagus and hypoglossal nerve of the rabbit, *Exp. Brain Res.* **10:** 526–531.

Frizell, M., McLean, W. G., and Sjöstrand, J. (1975), Slow axonal transport of proteins; blockade by interruption of contact between cell body and axon, *Brain Res.* **86:** 67–73.

Frizell, M., and Sjöstrand, J. (1974), The axonal transport of slowly migrating (^{3}H) leucine labelled proteins and the regeneration rate in regenerating hypoglossal and vagus nerves of the rabbit, *Brain Res.* **81:** 267–283.

Gainer, H., Barker, J. L., and Wollberg, Z. (1975), Preferential incorporation of extracellular amino acids into neuronal proteins, *J. Neurochem.* **25:** 177–179.

Gainer, H., Sarne, Y., and Brownstein, M. J. (1977), Biosynthesis and axonal transport of rat neurohypophysial proteins and peptides, *J. Cell Biol.* **73:** 366–381.

Gainer, H., Tasaki, I., and Lasek, R. J. (1977), Evidence for the glia–neuron protein transfer hypothesis from intracellular perfusion studies of squid giant axons, *J. Cell Biol.* **74:** 524–530.

Galvani, L. (1791), *Commentary on the Effects of Electricity on Muscular Motion,* translated from the Latin by Margaret G. Foley with notes and a critical introduction by I. Bernard Cohn, Burndy Library, Norwalk, Conn. (1954).

Gambetti, P., Autilio-Gambetti, L. A., Gonatas, N. K., and Shafer, B. (1972), Protein synthesis in synaptosomal fractions, *J. Cell Biol.* **52:** 526–535.

Gambetti, P., Autilio-Gambetti, L., Shafer, B., and Pfaff, L. (1973), Quantitative autoradiographic study of labeled RNA in rabbit optic nerve after intraocular injection of (3H) uridine, *J. Cell Biol.* **59:** 677–684.

Garbarg, M., Barbin, G., Feger, J., and Schwartz, J.-C. (1974), Histaminergic pathway in rat brain evidenced by lesions of the medial forebrain bundle, *Science* **186**: 833–835.

Garner, J. A., and Lasek, R. J. (1981), Clathrin is axonally transported as part of slow component b: The microfilament complex, *J. Cell Biol.* **88**: 172–178.

Gaskin, F., Cantor, C. R., and Shelanski, M. L. (1975), Biochemical studies on the *in vitro* assembly and disassembly of microtubules, *Ann. N.Y. Acad. Sci.* **253**: 133–146.

Gasser, H. S., and Erlanger, J. (1929), The role of fiber size in the establishment of a nerve block by pressure or cocaine, *Am. J. Physiol.* **88**: 581–591.

Gaze, R. M. (1970), *The Formation of Nerve Connections,* Academic Press, New York.

Gaze, R. M. and Sharma, S. C. (1970), Axial differences in the reinnervation of the goldfish optic tectum by regenerating optic nerve fibers, *Exp. Brain Res.* **10**: 171–181.

Geffen, L. (1975), Biochemical and histochemical methods of tracing transmitter-specific neuronal molecules, Chapter 12, in: *The Use of Axonal Transport for Studies of Neuronal Connectivity,* Cowan, W. M., and Cuénod, M. (eds.), Elsevier Publishing Co., New York.

Geffen, L. B., Hunter, C., and Rush, R. A. (1969), Is there bidirectional transport of noradrenaline in sympathetic nerves? *J. Neurochem.* **16**: 469–474.

Geffen, L. B., and Livett, B. G. (1971), Synaptic vesicles in sympathetic neurons, *Physiol. Rev.* **51**: 98–157.

Geffen, L. B., Livett, B. G., and Rush, R. A. (1969), Immunological localization of chromogranins in sheep sympathetic neurones, and their release by nerve impulses, *J. Physiol. (London)* **204**: 58P–59P.

Geffen, L. B., and Ostberg, A. (1969), Distribution of granular vesicles in normal and constricted sympathetic neurones, *J. Physiol. (London)* **204**: 583–592.

Geffen, L. B., and Rush, R. A. (1968), Transport of noradrenaline in sympathetic nerves and the effect of nerve impulses on its contribution to transmitter stores, *J. Neurochem.* **15**: 925–930.

Gerard, R. W. (1932), Nerve metabolism, *Physiol. Rev.* **12**: 469–592.

Geren, B. B. (1954), The formation from the Schwann cell surface of myelin in the peripheral nerves of chick embryos, *Exp. Cell Res.* **7**: 558–562.

Gersch, I., and Bodian, D. (1943), Some chemical mechanisms in chromatolysis, *J. Cell. Comp. Physiol.* **21**: 253–279.

Gerzon, K. (1980), Dimeric catharanthus alkaloids, Chapter 8, in: *Anticancer Agents Based on Natural Product Models,* Academic Press, New York.

Gerzon, K., Ochs, S., and Todd, G. C. (1979), Polarity of vincristine (VCR), vindesine (VDS), and vinblastine (VLB) in relation to neurological effects, *Proc. Am. Assoc. Canc. Res. Abst.* **20**: 46.

Ghetti, B. (1979), Induction of neurofibrillary degeneration following treatment with maytansine *in vivo, Brain Res.* **163**: 9–19.

Ghetti, B. (1980), Experimental studies on neurofibrillary degeneration, pp. 183–198, in: *Aging of the Brain and Dementia,* L. Amaducci, A. N. Davison, and P. Antuono (eds.), Raven Press, New York.

Ghetti, B., Horoupian, D. S., and Wiśniewski, H. M. (1975), Acute and long-term transneuronal response of dendrites of lateral geniculate neurons following transection of the primary visual afferent pathway, *Adv. Neurol.* **12**: 401–424.

Ghetti, B., and Ochs, S. (1978), On the relation between microtubule density and axoplasmic transport in nerves treated with maytansine *in vitro,* pp. 177–186, in: *Peripheral Neuropathies* (Developments in Neurology, Vol. 1), Canal, N., and Pozza, G. (eds.), Elsevier, Amsterdam.

Gilbert, D. S. (1972), Helical structure of *Myxicola* axoplasm, *Nature (New Biol.)* **237**: 195–198.

Gilbert, D. A., Newby, B. J., and Anderton, B. H. (1975), Neurofilament disguise, destruction and discipline, *Nature* **256**: 586–589.

Gilliatt, R. W. (1980), Acute compression block, Chapter 9, in: *The Physiology of Peripheral Nerve Disease,* Sumner, A. J. (ed.), Saunders, Philadelphia.

Gilliatt, R. W., Westgaard, R. H., and Williams, I. R. (1978), Extrajunctional acetylcholine sensitivity of inactive muscle fibres in the baboon during prolonged nerve pressure block, *J. Physiol. (London)* **280:** 499–514.

Ginetzinski, A. G., and Shamarma, N. M. (1942), The tonomotor phenomenon in denervated muscles (in Russian), *Usp. Sovrem. Biol.* **15:** 283–294.

Glisson, F. (1677), *Tractatus de Ventriculo et Intestinis,* Brome, London, Discussion, p. 160, in: Keele (1974).

Goldberg, D. J. (1981), Block of fast axonal transport caused by microinjection of the actin depolymerizer, DNase I, into the axon of an identified neuron, *Soc. Neurosci, Abstr.* **7:** 743.

Goldberg, D. J., Goldman, J. E., and Schwartz, J. H. (1976), Alterations in amount and rates of serotonin transported in an axon of the giant cerebral neurone of *Aplysia californica, J. Physiol. (London)* **259:** 473–490.

Goldberg, D. J., Harris, D. A., Lubit, B. W., and Schwartz., J. H. (1980), Analysis of the mechanism of fast axonal transport by intracellular injection of potentially inhibitory macromolecules: evidence for a possible role of actin filaments, *Proc. Natl. Acad. Sci. USA* **77:** 7448–7452.

Goldberg, D. J., Schwartz, J. H., and Sherbany, A. A. (1978), Kinetic properties of normal and perturbed axonal transport of serotonin in a single identified axon, *J. Physiol. (London)* **281:** 559–579.

Goldberg, S., and Kotani, M. (1967), The projection of optic nerve fibers in the frog *Rana catesbeiana* as studied by radioautography, *Anat. Record.* **158:** 325–332.

Goldman, J. E., Kim, K. S., and Schwartz, J. H. (1976), Axonal transport of (^{3}H) serotonin in an identified neuron of *Aplysia californica, J. Cell Biol.* **70:** 304–318.

Goldman, R., Pollard, T., and Rosenbaum, J. (eds.) (1976), *Cell Motility,* Books A,B,C, Cold Spring Harbor Laboratory, Cold Spring Harbor, N.Y.

Goldstein, J. L., Anderson, R. G. W., and Brown, M. S. (1979), Coated pits, coated vesicles, and receptor-mediated endocytosis, *Nature* **279:** 679–685.

Goodpasture, E. W. (1925), The axis-cylinders of peripheral nerves as portals of entry to the central nervous system for the virus of herpes simplex in experimentally infected rabbits, *Am. J. Path.* **1:** 11–28.

Gorio, A., Carmignoto, G., Facci, L., and Finesso, M. (1980), Motor nerve sprouting induced by ganglioside treatment. Possible implications for gangliosides on neuronal growth, *Brain Res.* **197:** 236–241.

Grafstein, B. (1967), Transport of protein by goldfish optic nerve fibers, *Science* **157:** 196–198.

Grafstein, B. (1971), Transneuronal transfer of radioactivity in the central nervous system, *Science* **172:** 177–179.

Grafstein, B. (1977), Axonal transport: The intracellular traffic of the neuron, pp. 691–717, in: *The Nervous System* Sect. 1, Vol. 1 *Handbook of Physiology.* American Physiological Society, Bethesda, Md.

Grafstein, B., and Forman, D. S. (1980), Intracellular transport in neurons, *Physiol. Rev.* **60:** 1167–1283.

Grafstein, B., Forman, D. S., and McEwen, B. S. (1972), Effects of temperature on axonal transport and turnover of protein in goldfish optic system, *Exp. Neurol.* **34:** 158–170.

Grafstein, B., and Laureno, R. (1973), Transport of radioactivity from eye to visual cortex in the mouse, *Exp. Neurol.* **39:** 44–57.

Grafstein, B., McEwen, B. S., and Shelanski, M. L. (1970), Axonal transport of neurotubule protein, *Nature* **227:** 289–290.

Grafstein, B., and Murray, M. (1969), Transport of protein in goldfish optic nerve during regeneration, *Exp. Neurol.* **25:** 494–508.

Graham, L. T., Jr., Shank, R. P., Werman, R., and Aprison, M. H. (1967), Distribution of some synaptic transmitter suspects in cat spinal cord: Glutamic acid, aspartic acid, γ-aminobutyric acid, glycine and glutamine, *J. Neurochem.* **14:** 465–472.

Grainger, F., and Sloper, J. C. (1974), Correlation between microtubular number and transport activity in hypothalamo-neurohypophyseal secretory neurons, *Cell Tissue Res.* **153:** 101–113.

Grampp, W., Harris, J. B., and Thesleff, S. (1972), Inhibition of denervation changes in skeletal muscle by blockers of protein synthesis, *J. Physiol. (London)* **221:** 743–754.

Gray, E. G. (1976), Problems of understanding the substructure of synapses, *Prog. Brain Res.* **45:** 207–234.

Gray, E. G. (1978), Synaptic vesicles and microtubules in frog nerve endplates, *Proc. Roy. Soc. B* **203:** 219–227.

Gray, E. G., and Guillery, R. W. (1961), The basis for silver staining of synapses of the mammalian spinal cord: A light and electron microscope study, *J. Physiol. (London)* **157:** 581–588.

Graybiel, A. M. (1975), Wallerian degeneration and anterograde tracer methods, Chapter 9, in: *The Use of Axonal Transport For Studies of Neural Connectivity,* Cowan, W. M., and Cuénod, M. (eds.), Elsevier Publishing Co., New York.

Greene, L. A., and Shooter, E. (1980), The nerve growth factor: Biochemistry, synthesis, and the mechanism of action, *Ann. Rev. Neurosci.* **3:** 353–402.

Grégersen, G. (1967), Diabetic neuropathy: Influence of age, sex, metabolic control and duration of diabetes on motor conduction velocity, *Neurology* **17:** 972–980.

Gremo, F., and Marchisio, P. C. (1975), Dynamic properties of axonal transport of proteins and glycoproteins: A study based on the effects of metaphase blocking drugs in the developing optic pathway of chick embryos, *Cell Tissue Res.* **161:** 303–316.

Griffin, J. W., Drachman, D. B., and Price, D. L. (1976), Fast axonal transport in motor nerve regeneration, *J. Neurobiol.* **7:** 355–370.

Griffin, J. W., Hoffman, P. N., Clark, A. W., Carroll., P. T., and Price, D. L. (1978), Slow axonal transport of neurofilament proteins: Impairment of β, β-iminodiproprionitrile administration, *Science* **202:** 633–635.

Griffin, J. W., Price, D. L., and Drachman, D. B. (1976), Impaired regeneration in acrylamide neuropathy: Role of axonal transport, *Neurology* **26:** 350.

Griffin, J. W., Price, D. L., Drachman, D. B., and Morris, J. (1981), Incorporation of axonally transported glycoprotein into axolemma during nerve regeneration, *J. Cell Biol.* **88:** 205–214.

Griffin, J. W., Price, D. L., Hoffman, P. N., and Cork, L. C. (1981), The axonal cytoskeleton: Alterations of organization and axonal transport in models of neurofibrillary pathology, *J. Neuropath. Exp. Neurol.* **40:** 316.

Griffiths, K. F., and McLean, W. G. (1980), A pharmacological comparison of rapid and slow axonal transport in rabbit vagus nerve, *Brit. J. Pharmacol.* **70:** 173P–174P.

Gross, G. W. (1973), The effect of temperature on the rapid axoplasmic transport in C-fibers, *Brain Res.* **56:** 359–363.

Gross, G. W. (1975), The microstream concept of axoplasmic and dendritic transport, *Adv. Neurol.* **12:** 283–296.

Gross, G. W., and Beidler, L. M. (1973), Fast axonal transport in the C-fibers of the garfish olfactory nerve, *J. Neurobiol.* **4:** 413–428.

Gross, G. W., and Beidler, L. M. (1975), A quantitative analysis of isotope concentration profiles and rapid transport velocities in the C-fibers of the garfish olfactory nerve, *J. Neurobiol.* **6:** 213–232.

Gross, G. W., and Kreutzberg, G. W. (1978), Rapid axoplasmic transport in the olfactory nerve of the pike. I. Basic transport parameters for proteins and amino acids, *Brain Res.* **139:** 65–76.

Gundersen, C. B. (1980), The effects of botulinum toxin on the synthesis, storage and release of acetylcholine, *Prog. Neurobiol.* **14:** 99–119.

Gundersen, R. W., and Barrett, J. N. (1979), Neuronal chemotaxis: Chick dorsal-root axons turn toward high concentrations of nerve growth factor, *Science* **206:** 1079–1080.

Gurdon, J. B., and Brown, D. D. (1965), Cytoplasmic regulation of RNA synthesis and nucleolus formation in developing embryos of *Xenopus laevis, J. Mol. Biol.* **12:** 27–35.

Guth, L. (1968), "Trophic" incluences of nerve on muscle, *Physiol. Rev.* **48:** 645–687.

Guth, L. (1969), "Trophic" effects of vertebrate neurons, *Neurosci. Res. Prog. Bull.* **7:** 1–73.

Guth, L. (1974), Axonal regeneration and functional plasticity in the central nervous system, *Exp. Neurol.* **45:** 606–654.

Guth, L. (1975), History of central nervous system regeneration research, *Exp. Neurol.* **48,** Part 2: 3–15.

Guth, L., and Albuquerque, E. X. (1978), The neurotrophic regulation of resting membrane-potential and extrajunctional acetylcholine sensitivity in mammalian skeletal muscle, *Physiologia bohemoslov.* **27:** 401–414.

Guth, L., Albers, R. W., and Brown, W. C. (1964), Quantitative changes in cholinesterase activity of denervated muscle fibers and sole plates, *Exp. Neurol.* **10:** 236–250.

Guth, L., Bright, D., and Donati, E. J. (1978), Functional deficits and anatomical alterations after high cervical spinal hemisection in the rat, *Exp. Neurol.* **58:** 511–520.

Guth, L., Brown, W. C., and Watson, P. K. (1967), Studies on the role of nerve impulses and acetylcholine release in the regulation of the cholinesterase activity of muscle, *Exp. Neurol.* **18:** 443–452.

Guth, L., Samaha, F. J., and Albers, R. W. (1970), The neural regulation of some phenotypic differences between the fiber types of mammalian skeletal muscle, *Exp. Neurol.* **26:** 126–135.

Guth, L., Smith, S., Donati, E. J., and Albuquerque, E. X. (1980), Induction of intramuscular collateral nerve sprouting by neurally applied colchicine, *Exp. Neurol.* **67:** 513–523.

Gutmann, E. (1976), Neurotrophic relations, *Ann. Rev. Physiol.* **38:** 177–216.

Gutmann, E. (ed.) (1962), *The Denervated Muscle,* Publishing House of the Czechoslovak Academy of Sciences, Prague.

Gutmann, E., Guttmann, L., Medawar, P. B., and Young, J. Z. (1942), The rate of regeneration of nerve, *J. Exp. Biol.* **19:** 14–44.

Gutmann, E., and Hanzlíková, V. (1972), *Age Changes in the Neuromuscular System,* Scientechnica, Bristol.

Gutmann, E., and Hník, P. (1962), Denervation studies in research of neurotrophic relationships, Chapter 1, in: *The Denervated Muscle,* Gutmann, E. (ed.), Publishing House of the Czechoslovak Academy of Sciences, Prague.

Gutmann, E., and Hník, P. (eds.) (1963), *The Effect of Use and Disuse of Neuromuscular Functions,* Publishing House of the Czechoslovak Academy of Sciences, Prague.

Gutmann, E., and Holubář, J. (1950), The degeneration of peripheral nerve fibres, *J. Neurol. Neurosurg. Psychiat.* **13:** 89–105.

Gutmann, E., and Young, J. Z. (1944), The re-innervation of muscle after various periods of atrophy, *J. Anat.* **78:** 15–43.

Ha, H. (1970), Axonal bifurcation in the dorsal root ganglion of the cat: A light and electron microscopic study, *J. Comp. Neurol.* **140:** 227–240.

Haak, R. A., Kleinhans, F. W., and Ochs, S. (1976), The viscosity of mammalian nerve axoplasm measured by electron spin resonance, *J. Physiol. (London)* **263:** 115–137.

Haddad, A., Iucif, S., and Cruz, A. R. (1969), Synthesis of RNA in neurons of the hypoglossal nerve nucleus, after section of the axon, in mice, *J. Neurochem.* **16:** 865–868.

Häggendal, J. (1974), Some aspects of the release of the adrenergic transmitter, *J. Neural Transmission Suppl.* **XI:** 135–161.

Häggendal, J., and Dahlström, A. (1971), The recovery of noradrenaline in adrenergic nerve terminals of the rat after reserpine treatment, *J. Pharm. Pharmacol.* **23:** 81–89.

Hahnenberger, R. W. (1978), Effects of pressure on fast axoplasmic flow. An *in vitro* study in the vagus nerve of rabbits, *Acta Physiol. Scand.* **104:** 299–308.

Haimo, L. T., Telzer, B. R., and Rosenbaum, J. L. (1979), Dynein binds to and crossbridges cytoplasmic microtubules, *Proc. Natl. Acad. Sci. USA* **76:** 5759–5763.

Hájek, I., Gutmann, E., and Syrorý, I. (1964), Proteolytic activity in denervated and reinnervated muscle, *Physiologia bohemoslov.* **13:** 32–38.

Hall, T. S. (1972), *Treatise of Man by Rene Descartes,* with translation and commentary, Harvard University Press, Cambridge.

Hall, Z. W. (1973), Multiple forms of acetylcholinesterase and their distribution in endplate and non-endplate regions of rat diaphragm muscle, *J. Neurobiol.* **4:** 343–361.

Hall, Z. W., and Kelly, R. B. (1971), Enzymatic detachment of endplate acetylcholinesterase from muscle, *Nature (New Biol.)* **232:** 62–63.

Hamel, E., del Campo, A. A., Lowe, M. C., and Lin, C. M. (1981), Interactions of taxol, microtubule-associated proteins, and guanine nucleotides in tubulin polymerization, *J. Biol. Chem.* **256:** 11887–11894.

Hammerschlag, R. (1980), The role of calcium in the initiation of fast axonal transport, *Fed. Proc.* **39:** 2809–2814.

Hammerschlag, R., Bakhit, C., Chiu, A. Y., and Dravid, A. R. (1977), Role of calcium in the initiation of fast axonal transport of protein: Effects of divalent cations, *J. Neurobiol.* **8:** 439–451.

Hammerschlag, R., and Bobinski, J. A. (1981), Ca^{2+}- or Mg^{2+}-stimulated ATPase activity in bullfrog spinal nerve: Relation to Ca^{2+} requirements for fast axonal transport, *J. Neurochem.* **36:** 1114–1121.

Hammerschlag, R., Chiu, A. Y., and Dravid, A. R. (1976), Inhibition of fast axonal transport of (^{3}H) protein by cobalt ions, *Brain Res.* **114:** 353–358.

Hammerschlag, R., Dravid, A. R., and Chiu, A. Y. (1975), Mechanism of axonal transport: A proposed role for calcium ions, *Science* **188:** 273–275.

Hammond, G. R., and Smith, R. S. (1977), Inhibition of the rapid movement of optically detectable axonal particles by colchicine and vinblastine, *Brain Res.* **128:** 227–242.

Hanson, M. (1981), Effects of mitosis inhibitors on respiration and fast axonal transport in frog sciatic nerves, *Mol. Pharmacol.* **19:** 291–294.

Hanson, M., and Edström, A. (1977), Fast axonal transport: Effect of antimitotic drugs and inhibitors of energy metabolism on the rate and amount of transported protein in frog sciatic nerves, *J. Neurobiol.* **8:** 97–108.

Hanson, M., and Edström, A. (1978), Mitosis inhibitors and axonal transport, *Int. Rev. Cytol. Suppl.* **7:** 373–402.

Hansson, H.-A. (1973), Uptake and intracellular bidirectional transport of horseradish peroxidase in retinal ganglion cell, *Exp. Eye Res.* **16:** 377–388.

Hare, W. K., and Hinsey, J. C. (1940), Reactions of dorsal root ganglion cells to section of peripheral and central processes, *J. Comp. Neurol.* **73:** 489–502.

Harper, G. P., and Thoenen, H. (1980), Nerve growth factor: Biological significance, measurement, and distribution, *J. Neurochem.* **34:** 5–16.

Harris, A. J. (1974), Inductive functions of the nervous system, *Ann. Rev. Physiol.* **36:** 251–305.

Harris, C. R. S. (1973), *The Heart and the Vascular System in Ancient Greek Medicine from Alcmaeon to Galen,* Clarendon Press, Oxford.

Harris, E. J., and Ochs, S. (1966), Effects of sodium extrusion and local anaesthetics on muscle membrane resistance and potential, *J. Physiol. (London)* **187:** 5–21.

Harrison, R. G. (1910), The outgrowth of the nerve fiber as a mode of protoplasmic movement, *J. Exp. Zool.* **9:** 787–846.

Hartree, W., and Hill, A. V. (1921a), The nature of the isometric twitch, *J. Physiol. (London)* **55:** 389–411.

Hartree, W., and Hill, A. V. (1921b), The regulation of the supply of energy in muscular contraction, *J. Physiol. (London)* **55:** 133–158.

Harvey, W. (1628), *Movement of the Heart and Blood in Animals,* translation from the Latin by K. J. Franklin, Charles C Thomas, Springfield, 1957.

Haschke, R. H., and Fink, B. R. (1975), Lidocaine effects on brain mitochondrial metabolism *in vitro, Anesthesiology* **42:** 737–740.

Haugaard, N. (1968), Cellular mechanisms of oxygen toxicity, *Physiol. Rev.* **48:** 311–373.

Heap, P. F., Jones, C. W., Morris, J. F., and Pickering, B. T. (1975), Movement of neurosecretory product through the anatomical compartments of the neural lobe of the pituitary gland, *Cell Tissue Res.* **156:** 483–497.

Hebb, C. O. (1963), Formation, storage and liberation of acetylcholine, Chapter 3, in: *Handbuch der Experimentellen Pharmakologie,* Vol. 15, Suppl., Eichler, O., and Farah, A. (eds.), Springer-Verlag, Berlin.

Hebb, C. O., and Silver, A. (1961), Gradient of choline acetylase activity, *Nature* **189:** 123–125.

Heidemann, S. R. (1980), Visualization of microtubule polarity, pp. 341–355, in: *Microtubules and Microtubule Inhibitors,* DeBrabander, M., and DeMey, J. (eds.), Elsevier/North-Holland, Amsterdam.

Heidemann, S. R., and McIntosh, J. R. (1980), Visualization of the structural polarity of microtubules, *Nature* **286:** 517–519.

Heilbrunn, L. V. (1952), *An Outline of General Physiology,* 3rd Edition, W. B. Saunders & Co., Philadelphia.

Heinbecker, P., Bishop, G. H., and O'Leary, J. (1932), Nerve degeneration in poliomyelitis. III. Rate of depression and disappearance of components of conducted action potential in severed nerves; correlation with histologic degeneration in groups of fibers responsible for various components, *Arch. Neurol. Psychiat.* **27:** 1421–1435.

Held, I. R. (1978), Stimulation of nuclear RNA synthesis in denervated skeletal muscles, *J. Neurochem.* **30:** 1239–1243.

Held, I., and Young, I. J. (1972), Transport of radioactivity derived from labeled N-acetylglucosamine in mammalian motor axons, *J. Neurobiol.* **3:** 153–161.

Hendrickson, A. E., and Cowan, W. M. (1971), Changes in the rate of axoplasmic transport during postnatal development of the rabbit's optic nerve and tract, *Exp. Neurol.* **30:** 403–422.

Hendry, I. A., Stöckel, K., Thoenen, H., and Iversen, L. L. (1974), The retrograde axonal transport of nerve growth factor, *Brain Res.* **68:** 103–121.

Henkart, M. P., Reese, T. S., and Brinley, F. J., Jr. (1978), Endoplasmic reticulum sequesters calcium in the squid giant axon, *Science* **202:** 1300–1303.

Hertting, G., Axelrod, J., Kopin, I. J., and Whitby, L. G. (1961), Lack of uptake of catecholamines after chronic denervation of sympathetic nerves, *Nature* **189:** 66.

Herzog, W., and Weber, K. (1978), Fractionation of brain microtubule-associated proteins. Isolation of two different proteins which stimulate tubulin polymerization *in vitro, Europ. J. Biochem.* **92:** 1–8.

Heslop, J. P. (1975), Axonal flow and fast transport in nerves, *Adv. Comp. Physiol. Biochem.* **6:** 75–163.

Heslop, J. P., and Howes, E. A. (1972), Temperature and inhibitor effects on fast axonal transport in a molluscan nerve, *J. Neurochem.* **19:** 1709–1716.

Heuser, J. E., and Reese, T. S. (1973), Evidence for recycling of synaptic vesicle membrane during transmitter release at the frog neuromuscular junction, *J. Cell Biol.* **57**: 315–344.

Hibbard, E. (1965), Orientation and directed growth of Mauthner's cell axons from duplicated vestibular nerve roots, *Exp. Neurol.* **13**: 289–301.

Hiebsch, R. R., Hales, D. D., and Murphy, D. B. (1979), Identity and purification of the dynein-like ATPase on cytoplasmic microtubules, *J. Cell Biol.* **83**: 345a.

Hild, W. (1951), Experimentell-morphologische Untersuchungen über das Verhalten der Neurosekretorischen Bahn nach Hypophysenstieldurchtrennungen, Eingriffen in den Wasserhaushalt und Belastung der Osmoregulation, *Virchow's Archiv.* **319**: 526–546.

Hild, W. (1954), Das morphologische, kinetische und endokrinologische Verhalten von hypothalamischem und neurohypophysärem Gewebe *in vitro, Z. Zellforsch.* **40**: 257–312.

Hill, T. J., Field, H. J., and Roome, A. P. C. (1972), Intra-axonal location of herpes simplex virus particles, *J. Gen. Virol.* **15**: 253–255.

Hindelang-Gertner, C., Stoeckel, M.-E., Porte, A., and Stutinsky, F. (1976), Colchicine effects on neurosecretory neurons and other hypothalamic and hypophysial cells, with special reference to changes in the cytoplasmic membranes, *Cell Tissue Res.* **170**: 17–41.

Hinkley, R. E. Jr., and Green, L. S. (1971), Effects of halothane and colchicine on microtubules and electrical activity of rabbit vagus nerves, *J. Neurobiol.* **2**: 97–105.

Hironaka, T., and Miyata, Y. (1975), Transplantation of skeletal muscle in normal and dystrophic mice, *Exp. Neurol.* **47**: 1–15.

His, W. (1887), Zur Geschichte des Menschlichen Rückenmarkes und der Nervenwurzeln, *Abh. Sächs. Ges. akad., Wiss. math-Phys. Klasse,* **13**: 477–514.

His, W. (1889), Die Neuroblasten und deren Entstehung in embryonalen Marke, *Abh. Sächs Ges. akad., Wiss. math-Phys. Klasse,* **15**: 311.

Hodgkin, A. L., and Keynes, R. D. (1955), Active transport of cations in giant axons from *Sepia* and *Loligo, J. Physiol. (London)* **128**: 28–60.

Hoffman, H. (1950), Local re-innervation in partially denervated muscle: A histo-physiological study, *Aust. J. Exp. Biol. Med. Sci.* **28**: 383–397.

Hoffman, P. N., Griffin, J. W., and Price, D. L. (1981), Changes in the axonal transport of the cytoskeleton during development, aging, and regeneration, *Soc. Neurosci. Abstr.* **7**: 743.

Hoffman, P. N., and Lasek, R. J. (1975), The slow component of axonal transport. Identification of major structural polypeptides of the axon and their generality among mammalian neurons, *J. Cell Biol.* **66**: 351–366.

Hoffman, P. N., and Lasek, R. J. (1980), Axonal transport of the cytoskeleton in regenerating motor neurons: Constancy and change, *Brain Res.* **202**: 317–333.

Hökfelt, T. (1969), Distribution of noradrenaline storing particles in peripheral adrenergic neurons as revealed by electron microscopy, *Acta Physiol. Scand.* **76**: 427–440.

Holland, R. L., and Brown, M. C. (1980), Postsynaptic transmission block can cause terminal sprouting of a motor nerve, *Science* **207**: 649–651.

Holmstedt, B. and Toschi, G. (1959), Enzymic properties of cholinesterases in subcellular fractions from rat brain, *Acta Physiol. Scand.* **47**: 280–283.

Horie, H., Takenaka, T., and Inomata, K. (1981), Effects of antimitotic drugs on axoplasmic transport in tissue cultured nerve cells, *Soc. Neurosci. Abstr.* **7**: 485.

Howe, H. A., and Bodian, D. (1942), *Neural Mechanisms in Poliomyelitis,* Oxford University Press, New York.

Hoy, R. R., Bittner, G. D., and Kennedy, D. (1967), Regeneration in crustacean motoneurons: Evidence for axonal fusion, *Science* **156**: 251–252.

Hsu, L. (1971), The role of nerves in the regeneration of minced muscle in adult anurans, Ph.D. Thesis, University of Michigan.

Hubbard, J. I. (1970), Mechanism of transmitter release, *Prog. Biophys.* **21**: 33–124.

Hubbard, S. J. (1963), The electrical constants and the component conductances of frog skeletal muscle after denervation, *J. Physiol. (London)* **165:** 443–456.

Hubel, D. H., Wiesel, T. N., and LeVay, S. (1975), Functional architecture of area 17 in normal and monocularly deprived macaque monkeys, *Cold Spring Harbor Symp. Quant. Biol.* **40:** 581–589.

Hughes, A. (1953), The growth of the embryonic neurite in tissue culture, *J. Anat.* **87:** 446.

Huizar, P., Kuno, M., Kudo, N., and Miyata, Y. (1977), Reaction of intact spinal motoneurones to partial denervation of the muscle, *J. Physiol. (London)* **265:** 175–191.

Huxley, H.E. (1969), The mechanism of muscle contraction. Recent structural studies suggest a revealing model for cross-bridge action at variable filament spacing, *Science* **164:** 1356–1366.

Huxley, H. E. (1973), Muscular contraction and cell motility, *Nature* **243:** 445–449.

Hydén, H. (1943), Protein metabolism in the nerve cell during growth and function, *Acta Physiol. Scand.* **6** (Suppl. XVII): 5–136.

Hydén, H. (1960), The neuron, Chapter 5, in: *The Cell: Biochemistry, Physiology, Morphology,* Brachet, J., and Mirsky, A. E. (eds.), Vol. 4, Academic Press, New York.

Ingoglia, N. A. (1979), 4S RNA is present in regenerating optic axons of goldfish, *Science* **206:** 73–75.

Ingoglia, N. A., Grafstein, B., McEwen, B. S., and McQuarrie, I. G. (1973), Axonal transport of radioactivity in the goldfish optic system following intraocular injection of labelled RNA precursors, *J. Neurochem.* **20:** 1605–1615.

Ingoglia, N. A., and Sturman, J. A. (1978), Axonal transport of putrescine, spermidine, and spermine in the goldfish visual system: Speculation on the association of axonally transported spermidine and tRNA in regenerating optic nerves, *Adv. Polyamine Res.* **2:** 169–182.

Ingoglia, N. A., Sturman, J. A., and Eisner, R. A. (1977), Axonal transport of putrescine, spermidine and spermine in normal and regenerating goldfish optic nerves, *Brain Res.* **130:** 433–445.

Ingoglia, N. A., Weis, P., and Mycek, J. (1975), Axonal transport of RNA during regeneration of the optic nerves of goldfish, *J. Neurobiol.* **6:** 549–563.

Inoué, S., and Sato, H. (1967), Cell mobility by labile association of molecules. The nature of mitotic spindle fibers and their role in chromosome movement, *J. gen. Physiol.* **50:** 259–292.

Iqbal, Z. (1979), Calmodulin: Is it involved in axoplasmic transport? *Trends Neurosci.* **2:** 311–312.

Iqbal, Z., and Ochs, S. (1975), Fast axoplasmic transport of calcium binding components in mammalian nerve, *Soc. Neurosci. Abstr.* **1:** 802.

Iqbal, Z., and Ochs, S. (1978), Fast axoplasmic transport of a calcium-binding protein in mammalian nerve, *J. Neurochem.* **31:** 409–418.

Iqbal, Z., and Ochs, S. (1979), Characterization of a calcium regulator protein in mammalian nerve, *Abstracts of the 7th Meeting of the International Society for Neurochemistry,* p. 391.

Iqbal, Z., and Ochs, S. (1980a), Calmodulin in mammalian nerve, *J. Neurobiol.* **11:** 311–318.

Iqbal, Z., and Ochs, S. (1980b), Fast axoplasmic transport of calmodulin in mammalian nerve: Possible involvement in axoplasmic transport, *Ann. N.Y. Acad. Sci.* **356:** 389–390.

Iqbal, Z., and Ochs, S. (1980c), Uptake of *vinca* alkaloids into mammalian nerve and its subcellular components, *J. Neurochem.* **34:** 59–68.

Isenberg, G., Schubert, P., and Kreutzberg, G. W. (1980), Experimental approach to test the role of actin in axonal transport, *Brain Res.* **194:** 588–593.

Ishikawa, H., Bischoff, R., and Holtzer, H. (1969), Formation of arrowhead complexes with heavy meromyosin in a variety of cell types, *J. Cell Biol.* **43:** 312–328.

Iversen, L. L. (1967), *The Uptake and Storage of Noradrenaline in Sympathetic Nerves,* University Press, Cambridge.

Iversen, L. L., and Whitby, L. G. (1963), The subcellular distribution of catecholamines in normal and tryamine-depleted mouse hearts, *Biochem. Pharmacol.* **12:** 582–584.

Jablecki, C., and Brimijoin, S. (1974), Reduced axoplasmic transport of choline acetyltransferase activity in dystrophic mice, *Nature* **250:** 151–154.

Jacobson, M. (1967), Retinal ganglion cells: Specification of central connections in larval *Xenopus laevis, Science* **155:** 1106–1108.

Jacobson, M. (1978), *Developmental Neurobiology,* 1st Edition, Chapman and Hall, United Kingdom.

Jacobson, M., and Gaze, R. M. (1964), Types of visual response from single units in the optic tectum and optic nerve of the goldfish, *Quart. J. Exp. Physiol.* **49:** 199–209.

Jacobson, M., and Gaze, R. M. (1965), Selection of appropriate tectal connections by regenerating optic nerve fibers in adult goldfish, *Exp. Neurol.* **13:** 418–430.

Jahn, T. L., and Bovee, E. C. (1969), Protoplasmic movement within cells, *Physiol. Rev.* **49:** 793–862.

Jakobsen, J., and Brimijoin, S. (1981), Axonal transport of enzymes and labeled proteins in experimental neuropathy induced by *p*-bromophenylacetylurea, *Brain Res.* **229:** 103–122.

Jakobsen, J., and Sidenius, P. (1980), Decreased axonal transport of structural proteins in streptozotocin diabetic rats, *J. Clin. Invest.* **66:** 292–297.

James, K. A. C., and Austin, L. (1969), The binding *in vitro* of colchicine to axoplasmic proteins from chicken sciatic nerve, *Biochem. J.* **117:** 773–777.

James, K. A. C., and Austin, L. (1970), The effect of DFP on axonal transport of protein in chicken sciatic nerve, *Brain Res.* **18:** 192–194.

James, K. A. C., Bray, J. J., Morgan, I. G., and Austin, L. (1970), The effect of colchicine on the transport of axonal protein in the chicken, *Biochem. J.* **117:** 767–771.

Jankowska, E., Lubińska, L., and Niemierko, S. (1969), Translocation of AÇhE-containing particles in the axoplasm during nerve activity, *Comp. Biochem. Physiol.* **28:** 907–913.

Jansen, J. K. S., Lømo, T., Nicolaysen, K., and Westgaard, R. H. (1973), Hyperinnervation of skeletal muscle fibers: Dependence on muscle activity, *Science* **181:** 559–561.

Jansen, J. K. S., Van Essen, D. C., and Brown, M. C. (1976), Formation and elimination of synapses in skeletal muscles of rat, *Cold Spring Harbor Symp. Quant. Biol.* **40:** 425–434.

Jarrott, B., and Geffen, L. B. (1972), Rapid axoplasmic transport of tryosine hydoxylase in relation to other cytoplasmic constituents, *Proc. Natl. Acad. Sci. USA* **69:** 3440–3442.

Jeffrey, P. L., and Austin, L. (1975), Axoplasmic transport, *Prog. Neurobiol.* **2:** 207–255.

Jeffrey, P. L., James, K. A. C., Kidman, A. D., Richards, A. M., and Austin, L. (1972), The flow of mitochondria in chicken sciatic nerve, *J. Neurobiol.* **3:** 199–208.

Jersild, R. A., Jr., and Ochs, S. (1982), Variations in cation content of pyroantimonate precipitates within individual subcellular compartments determined by X-ray microanalysis, *Anat. Rec.* (in press).

Johnson, J. L. (1970), Changes in acetylcholinesterase, acid phosphatase and beta glucuronidase proximal to a nerve crush, *Brain Res.* **18:** 427–440.

Johnson, J. L. (1974a), An analysis of the compartmentation and proximo-distal convection of the glutamate–glutamine system in the dorsal sensory neuron: Comparison with the motoneuron and cerebral cortex, *Brain Res.* **67:** 489–502.

Johnson, J. L. (1974b), Glutamine in the dorsal sensory neuron, *Brain Res.* **69:** 366–369.

Johnson, J. L. (1977), Glutamic acid as a synaptic transmitter candidate in the dorsal sensory neuron: Reconsiderations, *Life Sci.* **20:** 1637–1644.

Johnson, R. T. (1974), Pathophysiology and epidemiology of acute viral infections of the nervous system, *Adv. Neurol.* **6:** 27–40.

Jones, C. W., and Pickering, B. T. (1970), Rapid transport of neurohypophysial hormones in the hypothalamo-neurohypophysial tract, *J. Physiol. (London)* **208:** 73P–74P.

Jones, C. W., and Pickering, B. T. (1972), Intra-axonal transport and turnover of neurohypophysial hormones in the rat, *J. Physiol. (London)* **227:** 553–564.

Jones, R., and Vrbová, G. (1974), Two factors responsible for the development of denervation hypersensitivity, *J. Physiol. (London)* **236:** 517–538.

Joseph, B. S. (1973), Somatofugal events in Wallerian degeneration: A conceptual overview, *Brain Res.* **59:** 1–18.

Kadota, T., Kadota, K., and Gray, E. G. (1976), Coated-vesicle shells, particle/chain material, and tubulin in brain synaptosomes, *J. Cell Biol.* **69:** 608–621.

Kalix, P. (1971), Uptake and release of calcium in rabbit vagus nerve, *Pflügers Arch.* **326:** 1–14.

Kanaseki, T., and Kadota, K. (1969), The "vesicle in a basket." A morphological study of the coated vesicle isolated from the nerve endings of the guinea pig brain, with special reference to the mechanism of membrane movements, *J. Cell Biol.* **42:** 202–220.

Kanje, M., and Edström, A. (1981), Effects of Ca^{2+} on axonal transport in the frog sciatic nerve, *Abstr. Internatl. Soc. Neurochem.* **8:** 315.

Kanje, M., Edström, A., and Ekström, P. (1982), The role of Ca^{2+} in rapid axonal transport In: *Axoplasmic Transport* Weiss, D. (ed.) Springer-Verlag, Heidelberg. (in press).

Kanje, M., Edström, A., and Hanson, M. (1981), Inhibition of rapid axonal transport *in vitro* by the ionophores X-537 A and A 23187, *Brain Res.* **204:** 43–50.

Kao, C. C., and Chang, L. W. (1977), The mechanism of spinal cord cavitation following spinal cord transection. Part 1. A correlated histochemical study, *J. Neurosurg.* **46:** 197–209.

Kao, C. C., Chang, L. W., and Bloodworth, J. M. B., Jr. (1977a), The mechanism of spinal cord cavitation following spinal cord transection. Part 2: Electron microscopic observations, *J. Neurosurg.* **46:** 745–756.

Kao, C. C., Chang, L. W., and Bloodworth, J. M. B., Jr. (1977b), The mechanism of spinal cord cavitation following spinal cord transection. Part 3: Delayed grafting with and without spinal cord retransection, *J. Neurosurg.* **46:** 757–766.

Kao, C. C., Chang, L. W., and Bloodworth, J. M. B., Jr. (1977c), Axonal regeneration across transected mammalian spinal cords: An electron microscopic study of delayed microsurgical nerve grafting, *Exp. Neurol.* **54:** 591–615.

Kapeller, K., and Mayor, D. (1967), The accumulation of noradrenaline in constricted sympathetic nerves as studied by fluorescence and electron microscopy, *Proc. Roy. Soc. B* **167:** 282–292.

Karlsson, J.-O. (1977), Is there an axonal transport of amino acids? *J. Neurochem.* **29:** 615–617.

Karlsson, J.-O. (1979), Proteins of axonal transport: interaction of rapidly transported proteins with lectins, *J. Neurochem.* **32:** 491–494.

Karlsson, J.-O., Hansson, H. A., and Sjöstrand, J. (1971), Effect of colchicine on axonal transport and morphology of retinal ganglion cells, *Z. Zellforsch. Mikrosk. Anat.* **115:** 265–283.

Karlsson, J.-O., and Sjöstrand, J. (1969), The effect of colchicine on the axonal transport of protein in the optic nerve and tract of rabbit, *Brain Res.* **13:** 617–619.

Karlsson, J.-O., and Sjöstrand, J. (1971a), Electrophoretic characterization of rapidly transported proteins in axons of retinal ganglion cells, *FEBS Lett.* **16:** 329–332.

Karlsson, J.-O., and Sjöstrand, J. (1971b), Rapid intracellular transport of fucose-containing glycoproteins in retinal ganglion cells, *J. Neurochem.* **18:** 2209–2216.

Karlsson, J.-O., and Sjöstrand, J. (1971c), Synthesis, migration and turnover of protein in retinal ganglion cells, *J. Neurochem.* **18:** 749–767.

Kasa, P. (1968), Acetylcholinesterase transport in the central and peripheral nervous tissue: The role of tubules in the enzyme transport, *Nature* **218:** 1265–1267.

Kataoka, K., Nakamura, Y., and Hassler, R. (1973), Habenulo-interpeduncular tract: A possible cholinergic neuron in rat brain, *Brain Res.* **62:** 264–267.

Kater, S. B., and Nicholson, C. (eds.) (1973), *Intracellular Staining in Neurobiology,* Springer-Verlag, New York.

Katz, B. (1962), The transmission of impulses from nerve to muscle, and the subcellular unit of synaptic action, *Proc. Roy. Soc. B* **155:** 455–477.

420   References

Katz, B. (1966), *Nerve, Muscle and Synapse,* McGraw-Hill, New York.

Katz, B., and Miledi, R. (1964a), The development of acetylcholine sensitivity in nerve-free segments of skeletal muscle, *J. Physiol. (London)* **170:** 389–396.

Katz, B., and Miledi, R. (1964b), The measurement of synaptic delay, and the time course of acetylcholines release at the neuromuscular junction, *Proc. Roy. Soc. B* **161:** 483–495.

Katz, B., and Miledi, R. (1967), The timing of calcium action during neuromuscular transmission, *J. Physiol. (London)* **189:** 535–544.

Katz, B., and Miledi, R. (1969), Spontaneous and evoked activity of motor nerve endings in calcium Ringer, *J. Physiol. (London)* **203:** 689–706.

Kauffman, F. C., Warnick, J. E., and Albuquerque, E. X. (1974), Uptake of (^{3}H) colchicine from silastic implants by mammalian nerves and muscles, *Exp. Neurol.* **44:** 404–416.

Keele, K. D. (1974), Physiology. Chapter 7, in: *Medicine in Seventeenth Century England,* Debus, A. G. (ed.), University of California Press, Berkley.

Kennedy, R. D., Fink, B. R., and Byers, M. R. (1972), The effect of halothane on rapid axonal transport in the rabbit vagus, *Anesthesiology* **36:** 433–443.

Khan, M. A., and Ochs, S. (1974), Magnesium or calcium activated ATPase in mammalian nerve, *Brain Res.* **81:** 413–426.

Khan, M. A., and Ochs, S. (1975), Slow axoplasmic transport of mitochondria (MAO) and lactic dehydrogenase in mammalian nerve fibers, *Brain Res.* **96:** 267–277.

Kidwai, A. M., and Ochs, S. (1967), Separation of labeled protein and peptide components in motor axons following intracord injection of ^{3}H-leucine and axoplasmic flow, *Physiologist* **10:** 220.

Kidwai, A. M., and Ochs, S. (1969), Components of fast and slow phases of axoplasmic flow, *J. Neurochem.* **16:** 1105–1112.

Kiernan, J. A. (1978), An explanation of axonal regeneration in peripheral nerves and its failure in the central nervous system, *Med. Hypoth.* **4:** 15–26.

Kiernan, J. A. (1979), Hypotheses concerned with axonal regeneration in the mammalian nervous system, *Biol. Rev.* **54:** 155–197.

Kim, H., Binder, L. I., and Rosenbaum, J. L. (1979), The periodic association of MAP_2 with brain microtubules *in vitro, J. Cell Biol.* **80:** 266–276.

Kirkpatrick, J. B., Bray, J. J., and Palmer, S. M. (1972), Visualization of axoplasmic flow *in vitro* by Nomarski microscopy. Comparison to rapid flow of radioactive proteins, *Brain Res.* **43:** 1–10.

Kirkpatrick, J. B., Hyams, L., Thomas, V. L., and Howley, P. M. (1970), Purification of intact microtubules from brain, *J. Cell Biol.* **47:** 384–394.

Kirkpatrick, J. B., and Palmer, S. M. (1972), Ionic requirements for rapid particulate axoplasmic flow, *Am. J. Path.* **66:** 4A.

Kirkpatrick, J. B., Stern, L. Z., and Palmer, S. M. (1973), Axoplasmic flow in human peripheral nerves, *J. Neuropath. Exp. Neurol.* **32:** 169.

Kirschner, M. W., and Williams, R. C. (1974), The mechanism of microtubule assembly *in vitro, J. Supramol. Struct.* **2:** 412–428.

Kisch, B. (1954), Forgotten leaders of medicine: Valentine, Gruby, Remak, Auerbach, *Trans. Amer. Phil. Soc. (N.S.)* **44:** 139–192, 227–296.

Klatzo, I., Wiśniewski, H., and Streicher, E. (1965), Experimental production of neurofibrillary degeneration. I. Light microscopic observations, *J. Neuropath. Exp. Neurol.* **24:** 187–199.

Knull, H. R., Lobert, P. F., and Wells, W. W. (1974), Galactose neurotoxicity in chicks: Effects on fast axoplasmic transport, *Brain Res.* **79:** 524–527.

Knull, H. R., and Wells, W. W. (1975), Axonal transport of cations in the chick optic system, *Brain Res.* **100:** 121–124.

Ko, P. K., Anderson, M. J., and Cohen, M. W. (1977), Denervated skeletal muscle fibers develop discrete patches of high acetylcholine receptor density, *Science* **196**: 540–542.

Koda, L. Y., and Partlow, L. M. (1976), Membrane marker movement on sympathetic axons in tissue culture, *J. Neurobiol.* **7**: 157–172.

Koenig, E. (1978), Molecular biology of the Mauthner axon, pp. 167–182, in: *Neurobiology of the Mauthner Cell,* Faber, D. S. and Korn, H. (eds.), Raven Press, New York.

Koenig, E. (1979), Ribosomal RNA in Mauthner axon: Implications for a protein synthesizing machinery in the myelinated axon, *Brain Res.* **174**: 95–107.

Koenig, E., and Koelle, G. B. (1961), Mode of regeneration of acetylcholinesterase in cholinergic neurons following irreversible inactivation, *J. Neurochem.* **8**: 169–188.

Koike, H., Eisenstadt, M., and Schwartz, J. H. (1972), Axonal transport of newly synthesized acetylcholine in an identified neuron of *Aplysia, Brain Res.* **37**: 152–159.

Koike, H., and Nagata, Y. (1979), Intra-axonal diffusion of (^{3}H) acetylcholine and (^{3}H) γ-amino-butyric acid in a neurone of *Aplysia, J. Physiol. (London)* **295**: 397–417.

Komiya, Y. (1980), Slowing with age of the rate of slow axonal flow in bifurcating axons of rat dorsal root ganglion cells, *Brain Res.* **183**: 477–480.

Komiya, Y., and Austin, L. (1974), Axoplasmic flow of protein in the sciatic nerve of normal and dystrophic mice, *Exp. Neurol.* **43**: 1–12.

Komiya, Y., and Kurokawa, M. (1978), Asymmetry of protein transport in two branches of bifurcating axons, *Brain Res.* **139**: 354–358.

Kopin, I. J., Breese, G. R., Krauss, K. R., and Weise, V. K. (1968), Selective release of newly synthesized norepinephrine from the cat spleen during sympathetic nerve stimulation, *J. Pharmacol.* **161**: 271–278.

Kopin, I. J., and Silberstein, S. D. (1972), Axons of sympathetic neurons: Transport of enzymes *in vivo* and properties of axonal sprouts *in vitro, Pharmacol. Rev.* **24**: 245–254.

Koppang, N. (1966), Familiäre Glykosphingolipoidose des Hundes (Juvenile Amaurotische Idiote), *Ergebn. Allg. Path.* **47**: 1–43.

Korr, I. M., and Appeltauer, G. S. L. (1974), The time-course of axonal transport of neuronal proteins to muscle, *Exp. Neurol.* **43**: 452–463.

Korr, I. M., Wilkinson, P. N., and Chornock, F. W. (1967), Axonal delivery of neuroplasmic components to muscle cells, *Science* **155**: 342–345.

Kosower, E. M., Correa, W., Kinon, B. J., and Kosower, N. S. (1972), Glutathione. VII. Differentiation among substrates by the thiol-oxidizing agent, diamide, *Biochim. Biophys. Acta* **264**: 39–44.

Kosower, E. M., and Kosower, N. S. (1969), Lest I forget thee, glutathione . . . , *Nature* **224**: 117–120.

Kretsinger, R. H. (1980), Structure and evolution of calcium-modulated proteins, *CRC Crit. Rev. Biochem.* **8**: 119–174.

Kreutzberg, G. [W.] (1968), Histochemical demonstration of a colchicine-induced blockage of enzyme transport in axons of peripheral nerves, *Proc. 3rd Int. Cong. Histochem. Cytochem.,* pp. 133–134.

Kreutzberg, G. W. (1969), Neuronal dynamics and axonal flow, IV. Blockage of intra-axonal enzyme transport by colchicine, *Proc. Natl. Acad. Sci. USA* **62**: 722–728.

Kreutzberg, G. W., and Emmert, H. (1980), Glucose utilization of motor nuclei during regeneration: A [^{14}Cl-2-deoxyglucose study, *Exp. Neurol.* **70**: 712–716.

Kreutzberg, G. W., and Gross, G. W. (1977), General morphology and axonal ultrastructure of the olfactory nerve of the pike, *Esox lucius, Cell Tissue Res.* **181**: 443–457.

Kreutzberg, G. W., and Schubert, P. (1971), Changes in axonal flow during regeneration of mammalian motor nerves, *Acta Neuropath. (Berlin) Suppl.* **V**: 70–75.

Kreutzberg, G. W., and Schubert, P. (1979), Adenosine transport, release and possible actions, Chapter 60, in: *The Neurosciences: Fourth Study Program*, Schmitt, F. O., and Worden, F. G. (eds.) The MIT Press, Cambridge, Mass.

Kristensson, K. (1970), Transport of fluorescent protein tracer in peripheral nerves, *Acta Neuropath. (Berlin)* **16:** 293–300.

Kristensson, K., Ghetti, B., and Wiśniewski, H. M. (1974), Study on the propagation of *Herpes Simplex* virus (type 2) into the brain after intraocular injection, *Brain Res.* **69:** 189–201.

Kristensson, K., Lycke, E., and Sjöstrand, J. (1971), Spread of herpes simplex virus in peripheral nerves, *Acta Neuropath. (Berlin)* **19:** 44–53.

Kristensson, K., and Olsson, Y. (1971), Retrograde axonal transport of protein, *Brain Res.* **29:** 363–365.

Kristensson, K., Olsson, Y., and Sjöstrand, J. (1971), Axonal uptake and retrograde transport of exogenous proteins in the hypoglossal nerve, *Brain Res.* **32:** 399–406.

Krnjević, K. (1955), The distribution of Na and K in cat nerves, *J. Physiol. (London)* **128:** 473–488.

Krogh, A. (1919a), The number and distribution of capillaries in muscles with calculations of the oxygen pressure head necessary for supplying the tissues, *J. Physiol. (London)* **52:** 409–415.

Krogh, A. (1919b), The supply of oxygen to the tissues and the regulation of the capillary circulation, *J. Physiol. (London)* **52:** 457–474.

Kruta, V. (1971a), J. E. Purkyně's contribution to the cell theory, *Clio Medica* **6:** 109–120.

Kruta, V. (1971b), A note on the history of Purkyně cells, pp. 125–136, in: *Jan Evangelist Purkyně 1787–1869*, Kruta, V. (ed.), Universita Jana Evangelisty Purkyně, Brno.

Kuffler, S. W., and Yoshikami, D. (1975), The distribution of acetylcholine sensitivity at the postsynaptic membrane of vertebrate skeletal twitch muscles: Iontophoretic mapping in the micron range, *J. Physiol. (London)* **244:** 703–730.

Kumagai, H., and Nishida, E. (1979), The interactions between calcium-dependent regulator protein of cyclic nucleotide phosphodiesterase and microtubule proteins. II. Association of calcium-dependent regulator protein with tubulin dimer, *J. Biochem.* **85:** 1267–1274.

Kuno, M., Miyata, Y., and Muñoz-Martinez, E. J. (1974), Differential reaction of fast and slow α-motoneurones to axotomy, *J. Physiol. (London)* **240:** 725–739.

Kumara-Siri, M. H. (1979), Batrachotoxin inhibits axonal transport without affecting membrane potential in single neurons of *Aplysia Californica*, *J. Neurobiol.* **10:** 509–512.

Lagercrantz, H. (1976), On the composition and function of the large dense cored vesicles in sympathetic nerves, *Neuroscience* **1:** 81–92.

Lajtha, A. (1964), Protein metabolism of the nervous system, *Int. Rev. Neurobiol.* **6:** 1–98.

Lajtha, A. (1975), Transport and incorporation of amino acids in relation to measurement of axonal flow, Chapter 2, in: *The Use of Axonal Transport for Studies of Neuronal Connectivity*, Cowan, W. M., and Cuénod, M. (eds.), Elsevier Publishing Co, New York.

Landis, E. M., and Pappenheimer, J. R. (1963), Exchange of substances through capillary walls, Chapter 29, in: *Circulation*, Handbook of Physiology, Sect. 2, Vol. 11, American Physiological Society, Washington, D.C.

Landmesser, L. (1971), Contractile and electrical responses of vagus-innervated frog sartorius muscles, *J. Physiol. (London)* **213:** 707–725.

Larsson, L.-I. (1980), Corticotropin and α-melanotropin in brain nerves: Immunocytochemical evidence for axonal transport and processing, *Adv. Biochem. Psychopharm.* **22:** 101–107.

Lasek, R. J. (1970), Axonal transport of proteins in dorsal root ganglion cells of the growing cat: A comparison of growing and mature neurons, *Brain Res.* **20:** 121–126.

Lasek, R. J. (1967), Bidirectional transport of radioactively labelled axoplasmic components, *Nature* **216:** 1212–1214.

Lasek, R. [J.] (1968), Axoplasmic transport in cat dorsal root ganglion cells: As studied with ^{3}H-L-leucine, *Brain Res.* **7**: 360–377.

Lasek, R. J. (1980), Axonal transport: A dynamic view of neuronal structures, *Trends Neurosci.* **3**: 87–95.

Lasek, R. J., Dabrowski, C., and Nordlander, R. (1973), Analysis of axoplasmic RNA from invertebrate giant axons, *Nature (New Biol.)* **244**: 162–165.

Lasek, R. J., Gainer, H., and Barker, J. L. (1977), Cell-to-cell transfer of glial proteins to the squid giant axon, *J. Cell Biol.* **74**: 501–523.

Lasek, R. J., and Hoffman, P. N. (1976), The neuronal cytoskeleton, axonal transport and axonal growth, pp. 1021–1049, in: *Cell Motility,* Book C., Goldman, R., Pollard, T., and Rosenbaum, J. (eds.), Cold Spring Harbor Laboratory, Cold Spring Harbor, N.Y.

Lasek, R. J., Krishnan, N., and Kaiserman-Abramof, I. R. (1979), Identification of the subunit proteins of 10-nm neurofilaments isolated from axoplasm of squid and Myxicola giant axons, *J. Cell Biol.* **82**: 336–346.

La Vail, J. H. (1975), Retrograde cell degeneration and retrograde transport techniques, Chapter 10, in: *The Use of Axonal Transport for Studies on Neuronal Connectivity,* Cowan, W. M., and Cuénod, M. (eds.), Elsevier Publishing Co., New York.

La Vail, J. H., and La Vail, M. M. (1974), The retrograde intraaxonal transport of horseradish peroxidase in the chick visual system: A light and electron microscopic study, *J. Comp. Neurol.* **157**: 303–358.

La Vail, J. H., Rapisardi, S., and Sugino, I. K. (1980), Evidence against the smooth endoplasmic reticulum as a continuous channel for the retrograde axonal transport of horseradish peroxidase, *Brain Res.* **191**: 3–20.

LaVelle, A., and LaVelle, F. W. (1958), The nucleolar apparatus and neuronal reactivity to injury during development, *J. Exp. Zool.* **137**: 285–315.

Lavoie, P.-A., Bolen, F., and Hammerschlag, R. (1979), Divalent cation specificity of the calcium requirement for fast transport of proteins in axons of desheathed nerves, *J. Neurochem.* **32**: 1745–1751.

Lavoie, P.-A., Collier, B., and Tenenhouse, A. (1976), Comparison of α-bungarotoxin binding to skeletal muscles after inactivity or denervation, *Nature* **260**: 349–350.

Law, P. K., Cosmos, E., Butler, J., and McComas, A. J. (1976), The absence of dystrophic characteristics in normal muscles successfully cross-reinnervated by nerves of dystrophic genotype: physiological and cytochemical study of crossed solei of normal and dystrophic parabiotic mice, *Exp. Neurol.* **51**: 1–21.

LeBeux, Y. J., and Willemot, J. (1975a), An ultrastructural study of the microfilaments in rat brain by means of heavy meromyosin labeling. I. The perikaryon, the dendrites and the axon, *Cell Tissue Res.* **160**: 1–36.

LeBeux, Y. J., and Willemot, J. (1975b), An ultrastructure study of the microfilaments in rat brain by means of E-PTA staining and heavy meromyosin labeling. II. The synapses, *Cell Tissue Res.* **160**: 37–68.

Lebowitz, P., and Singer, M. (1970), Neurotrophic control of protein synthesis in the regenerating limb of the newt. *Triturus, Nature* **225**: 824–827.

Lee, F. C. (1929), The regeneration of nervous tissue, *Physiol. Rev.* **9**: 575–623.

Leestma, J. E. (1976), Velocity measurements of particulate neuroplasmic flow in organized mammalian CNS tissue cultures, *J. Neurobiol.* **7**: 173–183.

Lehninger, A. L. (1975), *Biochemistry: The Molecular Bases of All Structure and Function,* 2nd Edition, Worth Publishing Co., New York.

Lehninger, A. L., Carafoli, E., and Rossi, C. S. (1967), Energy-linked ion movements in mitochondrial systems, *Adv. Enzymol.* **29**: 259–320.

Lentz, T. L. (1974), Neurotrophic regulation at the neuromuscular junction, *Ann. N.Y. Acad. Sci.* **228:** 323–337.

Lentz, T. L. (1972), Distribution of leucine-^{3}H during axoplasmic transport within regenerating neurons as determined by electron-microscopic radioautography, *J. Cell Biol.* **52:** 719–732.

Leone, J., and Ochs, S. (1978), Axonic block and recovery of axoplasmic transport and electrical excitability of nerve, *J. Neurobiol.* **9:** 229–245.

LeVay, S., Stryker, M. P., and Shatz, C. J. (1978), Ocular dominance columns and their development in layer IV of the cat's visual cortex: A quantitative study, *J. Comp. Neurol.* **179:** 223–244.

Levi-Montalcini, R. (1976), The nerve growth factor: Its role in growth, differentiation and function of the sympathetic adrenergic neuron, *Prog. Brain Res.* **45:** 235–258.

Levi-Montalcini, R., and Angeletti, P. V. (1968), Nerve growth factor, *Physiol. Rev.* **48:** 534–569.

Levi-Montalcini, R., and Calissano, P. (1979), Nerve growth factors, *Scient. Am.* **240:** 68–77.

Levine, J., Skene, P., and Willard, M. (1981), 'GAPs and fodrin. Novel axonally transported proteins, *Trends Neurosci.* **4:** 273–277.

Levine, J., and Willard, M. (1980), The composition and organization of axonally transported proteins in the retinal ganglion cells of the guinea pig, *Brain Res.* **194:** 137–154.

Lewis, P. R., and Shute, C. C. D. (1967), The cholinergic limbic system: Projections to hippocampal formation, medial cortex, nuclei of the ascending cholinergic reticular system, and the subfornical organ and supra-optic crest, *Brain* **90:** 521–540.

Lewis, T., Pickering, G. W., and Rothschild, P. (1931), Centripetal paralysis arising out of arrested bloodflow to the limb, including notes on a form of tingling, *Heart* **16:** 1–32.

Libet, B. (1948), Adenosinetriphosphatase (ATP-ase) in nerve, *Fed. Proc.* **7:** 72.

Lieberman, A. R. (1976), Sensory ganglia, Chapter 4, in: *The Peripheral Nerve*, Landon, D. N. (ed.), John Wiley & Sons Inc., New York.

Liem, R. K. H. (1982), Simultaneous separations and purification of neurofilament and glial filament proteins from brain, *J. Neurochem.* **38:** 142–150.

Liem, R. K. H., Leterrier, J. S., Keith, C. H., Trenkner, E., and Shelanski, M. (1982), Chemistry and biology of neuronal and glial intermediate filaments, *Cold Spring Harbor Symp. Quant. Biol.* (in press).

Liem, R. K. H., Yen, S.-H., Salomon, G. D., and Shelanski, M. L. (1978), Intermediate filaments in nervous tissues, *J. Cell Biol.* **79:** 637–645.

Lin, C.-T., Dedman, J. R., Brinkley, B. R., and Means, A. R. (1980), Localization of calmodulin in rat cerebellum by immunoelectron microscopy, *J. Cell Biol.* **85:** 473–480.

Linden, C. D., Dedman, J. R., Chafouleas, J. G., Means, A. R., and Roth, T. F. (1981), Interactions of calmodulin with coated vesicles from brain, *Proc. Natl. Acad. Sci. USA* **78:** 308–312.

Lindsey, J. D., and Ellisman, M. H. (1981), Cobalt induced disruption of the golgi apparatus blocks assembly of forming face element, *Soc. Neurosci. Abstr.* **7:** 485.

Linkhart, T. A., Yee, G. W., and Wilson, B. W. (1975), Myogenic defect in acetylcholinesterase regulation in muscular dystrophy of the chicken, *Science* **187:** 549–550.

Litchy, W. J. (1973), Uptake and retrograde transport of horseradish peroxidase in frog sartorius nerve *in vitro*, *Brain Res.* **56:** 377–381.

Litchy, W. J., Brimijoin, S., and Reiter, C. (1981), Velocity and metabolism of axonally transported labeled catecholamines in bullfrog sympathetic nerve, *Soc. Neurosci. Abstr.* **7:** 742.

Liu, H. M., Balkovic, E. S., Sheff, M. F., and Zacks, S. I. (1979), Production *in vitro* of a neurotropic substance from proliferative neurolemma-like cells, *Exp. Neurol.* **64:** 271–283.

Livett, B. G. (1973), Histochemical visualization of peripheral and central adrenergic neurones, *Brit. Med. Bull.* **29:** 93–99.

Livett, B. G. (1976), Axonal transport and neuronal dynamics: Contributions to the study of neuronal connectivity, *Int. Rev. Physiol.* **10:** 37–124.

Liwnicz, B. H., Kristensson, K., Wiśniewski, H. M., Shelanski, M. L., and Terry, R. D. (1974), Observations on axoplasmic transport in rabbits with aluminum-induced neurofibrillary tangles, *Brain Res.* **80:** 413–420.

Llinás, R. R., and Heuser, J. E. (1977), Depolarization-release coupling systems in neurons, *Neurosci. Res. Prog. Bull.* **15:** 556–687.

Locke, S., and Solomon, H. C. (1967), Relation of resting potential of rat gastrocnemius and soleus muscles to innervation, activity and the Na–K pump, *J. Exp. Zool.* **166:** 377–386.

Lockwood, A. H. (1978), Tubulin assembly protein: immunochemical and immunofluorescent studies on its function and distribution in microtubules and cultured cells, *Cell* **13:** 613–627.

Lockwood, A. H. (1979), Molecules in mammalian brain that interact with the colchicine site on tubulin, *Proc. Natl. Acad. Sci USA* **76:** 1184–1188.

Lohmann, K. (1929), Über di pyrophosphatfraktion im Muskel, *Naturwissenschaften* **17:** 624–625.

Lømo, T. (1980), What controls the development of neuromuscular junction? *Trends Neurosci.* **3:** 126–129.

Lømo, T., and Rosenthal, J. (1972), Control of ACh sensitivity by muscle activity in the rat, *J. Physiol. (London)* **221:** 493–513.

Lømo, T., and Slater, C. R. (1980a), Acetylcholine sensitivity of developing ectopic nerve–muscle junctions in adult rat soleus muscles, *J. Physiol. (London)* **303:** 173–189.

Lømo, T., and Slater, C. R. (1980b), Control of junctional acetylcholinesterase by neural and muscular influences in the rat, *J. Physiol. (London)* **303:** 191–202.

Lømo, T., and Westgaard, R. H. (1975), Control of ACh sensitivity in rat muscle fibers, *Cold Spring Harbor Symp. Quant. Biol.* **40:** 263–274.

Lømo, T., Westgaard, R. H., and Dahl, H. A. (1974), Contractile properties of muscle: Control by pattern of muscle activity in the rat, *Proc. Roy. Soc. B* **187:** 99–103.

Lømo, T., Westgaard, R. H., and Engebretsen, L. (1980), Different stimulation patterns affect contractile properties of denervated rat soleus muscles, pp. 297–309, in: *Plasticity of Muscle,* Pette, D. (ed.), Walter de Gryter, New York.

Longo, F. M., and Hammerschlag, R. (1980), Relation of somal lipid synthesis to the fast axonal transport of protein and lipid, *Brain Res.* **193:** 471–485.

Low, F. N. (1976), The perineurium and connective tissue of peripheral nerve, Chapter 3, in: *The Peripheral Nerve,* Landon, D. N. (ed.), John Wiley & Sons, New York.

Lubińska, L. (1956), Outflow from cut ends of nerve fibres, *Exp. Cell Res.* **10:** 40–47.

Lubińska, L. (1964), Axoplasmic streaming in regenerating and in normal nerve fibres, *Prog. Brain Res.* **13:** 1–71.

Lubińska, L. (1977), Early course of Wallerian degeneration in myelinated fibres of the rat phrenic nerve, *Brain Res.* **130:** 47–63.

Lubińska, L., and Niemierko, S. (1971), Velocity and intensity of bidirectional migration of acetylcholinesterase in transected nerves, *Brain Res.* **27:** 329–342.

Luco, J. V., and Eyzaguirre, C. (1955), Fibrillation and hypersensitivity to ACh in denervated muscle: Effect of length of degenerating nerve fibers, *J. Neurophysiol.* **18:** 65–73.

Luduena, R. F., and Woodward, D. O. (1975), α- and β-tubulin: Separation and partial sequence analysis, *Ann. N.Y. Acad. Sci.* **253:** 272–283.

Luff, A. R. (1975), Dynamic properties of fast and slow skeletal muscles in the cat and rat following cross-reinnervation, *J. Physiol. (London)* **248:** 83–96.

Lund, R. D. (1978), *Development and Plasticity of the Brain,* Oxford University Press, New York.

Lundberg, J. M., Hökfelt, T., Änggård, A., Uvnäs-Wallensten, K., Brimijoin, S., Brodin, E., and Fahrenkrug, J. (1980), Peripheral peptide neurons: Distribution, axonal transport, and some aspects on possible function, *Adv. Biochem. Psychopharm.* **22:** 25–36.

Lundborg, G. (1970), Ischemic nerve injury, *Scand. J. Plas. Reconstr. Surg. Suppl.* **6:** 3–113.

Lundborg, G., and Brånemark, P.-I. (1968), Microvascular structure and function of peripheral nerves, *Adv. Microcirc.* **1:** 66–88.

Magid, A., Fischer, H. A., and Schmatolla, E. (1973), Axonal transport—Simple diffusion? *Science* **182:** 180.

Malpighi, M. (1667), An account of some discoveries concerning the brain and the tongue, *Phil. Trans. R. Soc.* **2:** 491.

Marchisio, P. C., Gremo, F., and Sjöstrand, J. (1975), Axonal transport in embryonic neurons. The possibility of a proximo-distal axolemmal transfer of glycoproteins, *Brain Res.* **85:** 281–285.

Marchisio, P. C., and Gremo, F. (1971), The axoplasmic transport of proteins along the optic pathway of newly hatched chicks—The blocking effects of colchicine and vinblastine, *Abstr. Inter. Soc. Neurochem.* **3:** 153.

Marchisio, P. C., and Sjöstrand, J. (1972), Radioautographic evidence for protein transport along the optic pathway of early chick embryos, *J. Neurocytol.* **1:** 101–108.

Marcum, J. M., Dedman, J. R., Brinkley, B. R., and Means, A. R. (1978), Control of microtubule assembly–disassembly by calcium-dependent regulator protein, *Proc. Natl. Acad. Sci. USA* **75:** 3771–3775.

Margolis, R. L., and Wilson, L. (1978), Opposite end assembly and disassembly of microtubules at steady state *in vitro, Cell* **13:** 1–8.

Marinesco, G., and Minea, J. (1906), Précocité des phénomènes de régénérescence des nerfs après leur section, *Compt. Rend. Hebd. Sci. Mem.* **61:** 383–385.

Mark, R. F. (1969), Matching muscles and motoneurones. A review of some experiments on motor nerve regeneration, *Brain Res.* **14:** 245–254.

Markelonis, G. J., and Oh, T. H. (1978), A protein fraction from peripheral nerve having neurotrophic effects on skeletal muscle cells in culture, *Exp. Neurol.* **58:** 285–295.

Markelonis, G., and Oh, T. H. (1979), A sciatic nerve has a trophic effect on development and maintenance of skeletal muscle cells in culture, *Proc. Natl. Acad. Sci. USA* **76:** 2470–2474.

Marsland, D., and Hiramoto, Y. (1966), Cell division: Pressure-induced reversal of the antimeiotic effects of heavy water in the oocytes of the starfish, *Asterias forbesi, J. Cell. Physiol.* **67:** 13–22.

Maupin-Szamier, P., and Pollard, T. D. (1978), Actin filament destruction by osmium tetroxide, *J. Cell Biol.* **77:** 837–852.

McComas, A. J., Brandstater, M. E., Upton, A. R. M., Delbeke, J., DeFaria, C., and Toyonaga, K. (1977), Sick motor neurons and dystrophy: A reappraisal, pp. 180–186, in: *Pathogenesis of Human Muscular Dystrophies,* L. P. Rowland (ed.), Excerpta Medica, Amsterdam.

McEwen, B. S., and Grafstein, B. (1968), Fast and slow components in axonal transport of protein, *J. Cell Biol.* **38:** 494–508.

McEwen, B. S., Forman, D. S., and Grafstein, B. (1971), Components of fast and slow axonal transport in the goldfish optic nerve, *J. Neurobiol.* **2:** 361–377.

McGeer, P. L., Eccles, J. C., and McGeer, E. G. (1978), *Molecular Neurobiology of the Mammalian Brain,* Plenum Press, New York.

McGraw, C. F., Somlyo, A. V., and Blaustein, M. P. (1980), Localization of calcium in presynaptic nerve terminals. An ultrastructural and electron microprobe analysis, *J. Cell Biol.* **85:** 228–241.

MacGregor, W. S. (1967), The chemical and physical properties of DMSO, *Ann. N.Y. Acad. Sci.* **141:** 3–12.

McIlwain, H. (1974), Adenosine in neurohumoral and regulatory roles in the brain, pp. 3–11, in: *Central Nervous System Studies on Metabolic Regulation and Function,* Genazzani, E., and Herken, H. (eds.), Springer, New York.

McIlwain, H. (1976), Translocation of neural modulators. A second category of nerve signal, *Neurochem. Res.* **1:** 351–368.

McIlwain, H. (1978), Synaptic mediators and the structuring of cerebral activity, *Prog. Neurobiol.* **11:** 189–203.

McIntosh, J. R. (1974), Bridges between microtubules, *J. Cell Biol.* **61:** 166–187.

McIntosh, J. R., Hepler, P. K., and Van Wie, D. G. (1969), Model for mitosis, *Nature* **224:** 659–663.

McLane, J. A., and McClure, W. O. (1977), Rapid axoplasmic transport in dystrophic mice, *J. Neurochem.* **29:** 865–872.

McLean, W. G., and Keen, P. (1972), Local synthesis and breakdown of noradrenaline in constricted rat sciatic nerves, *Europ. J. Pharmacol.* **18:** 74–78.

McMahan, U. J., Sanes, J. R., and Marshall, L. M. (1978), Cholinesterase is associated with the basal lamina at the neuromuscular junction, *Nature* **271:** 172–174.

McQuarrie, I. G., Brady, S. T., and Lasek, R. J. (1980), Polypeptide composition and kinetics of SCa and SCb in sciatic nerve motor axons and optic axon of the rat, *Soc. Neurosci. Abstr.* **6:** 501.

McQuarrie, I. G., and Grafstein, B. (1973), Axon outgrowth enhanced by a previous nerve injury, *Arch. Neurol.* **29:** 53–55.

McQuarrie, I. G., Grafstein, B. and Gershon, M.D. (1977), Axonal regeneration in the rat sciatic nerve: Effect of a conditioning lesion and of dbcAMP, *Brain Res.* **132:** 443–453.

McQuarrie, I. G., King, M., and Lasek, R. J. (1981), Transport of cytoskeletal proteins in newly-formed axons of regenerating rat sciatic nerve motoneurons, *Soc. Neurosci. Abstr.* **7:** 38.

Means, A. R., and Dedman, J. R. (1980), Calmodulin—An intracellular calcium receptor, *Nature* **285:** 73–77.

Means, A. R., Tash, J. S., and Chafouleas, J. G. (1982), Physiological implications of the presence, distribution, and regulation of calmodulin in Eukaryotic cells, *Physiol. Rev.* **62:** 1–39.

Mellon, M. and Rhebun, L. (1976), Studies on the accessible sulfhydryls of polymerizable tubulin, pp. 1149–1163, in: *Cell Motility,* Book C, Goldman, R., Pollard, T., and Rosenbaum, J. (eds.), Cold Spring Harbor Laboratory, Cold Spring Harbor, N.Y.

Mendell, J. R., Sahenk, Z., Saida, K., Weiss, H. S., Savage, R., and Couri, D. (1977), Alterations of fast axoplasmic transport in experimental methyl *n*-butyl ketone neuropathy, *Brain Res.* **133:** 107–118.

Mesulam, M.-M. (ed). (1982), *Tracing Neural Connections with Horseradish Peroxidase.* Wiley-Interscience, New York.

Mesulam, M.-M., and Brushart, T. M. (1979), Transganglionic and anterograde transport of horseradish peroxidase across dorsal root ganglia: A tetramethylbenzidine method for tracing central sensory connections of muscles and peripheral nerves, *Neuroscience* **4:** 1107–1117.

Mesulam, M.-M., and Rosene, D. L. (1979), Sensitivity in horseradish peroxidase neurohistochemistry: A comparative and quantitative study of nine methods, *J. Histochem. Cytochem.* **27:** 763–779.

Metafora, S., Felsani, A., Cotrufo, R., Tajana, G. F., Del Rio, A., De Prisco, P. P., Rutigliano, B., and Esposito, V. (1980), Neural control of gene expression in the skeletal muscle fibre: changes in the muscular mRNA population following denervation, *Proc. Roy. Soc. B* **209:** 257–273.

Metuzals, J. (1966), Electron microscopy of neurofilaments, pp. 459–460, in: *Proceedings of the Sixth International Congress for Electron Microscopy, Kyoto,* Vol. 2, Uyeda, R. (ed.), Maruzen Co., Tokyo.

Metuzals, J. (1969), Configuration of a filamentous network in the axoplasm of the squid *(Loligo Pealii L.)* giant nerve fiber, *J. Cell Biol.* **43:** 480–505.

Miani, N. (1963), Analysis of the somato-axonal movement of phospholipids in the vagus and hypoglossal nerves, *J. Neurochem.* **10:** 859–874.

Miani, N. (1971), Transport of S-100 protein in mammalian nerve fibers and transneuronal signals, *Acta Neuropath. (Berlin) Suppl.* **V:** 104–108.

Miani, N., Rizzoli, A., and Bucciante, G. (1961), Metabolic and chemical changes in regenerating neurons. II. *In vitro* rate of incorporation of amino acids into proteins of the nerve cell perikaryon of the C.8 spinal ganglion of rabbit, *J. Neurochem.* **7:** 161–173.

Mildván, A. S. (1970), Metals in enzyme catalsis, Chapter 9, in: *The Enzymes,* Boyer, P. D. (ed.), Vol. II, 3rd Edition, Academic Press, New York.

Miledi, R. (1960a), The acetylcholine sensitivity of frog muscle fibres after complete or partial denervation, *J. Physiol. (London)* **151:** 1–23.

Miledi, R. (1960b), Properties of regenerating neuromuscular synapses in the frog, *J. Physiol. (London)* **154:** 190–205.

Miledi, R. (1963a), Formation of extra nerve–muscle junctions in innervated muscle, *Nature* **199:** 1191–1192.

Miledi, R. (1963b), An influence of nerve not mediated by impulses, pp. 35–40, in: *The Effect of Use and Disuse of Neuromuscular Functions,* Gutmann, E., and Hník, P. (eds.), Publishing House of the Czechoslovak Academy of Sciences, Prague.

Miledi, R., Molinoff, P., and Potter, L. T. (1971), Isolation of the cholinergic receptor protein of *Torpedo* electric tissue, *Nature* **229:** 554–557.

Miledi, R., and Slater, C. R. (1970), On the degeneration of rat neuromuscular junctions after nerve section, *J. Physiol. (London)* **207:** 507–528.

Miledi, R., and Spitzer, N. C. (1974), Absence of action potentials in frog slow muscle fibres paralysed by botulinum toxin, *J. Physiol. (London)* **241:** 183–199.

Mobley, W. C., Server, A. C., Ishii, D. N., Riopelle, R. J., and Shooter, E. M. (1977), Nerve growth factor, *New Engl. J. Med.* **297:** 1096–1104, 1149–1158, 1211–1218.

Mommaerts, W. F. H. M. (1969), Energetics of muscular contraction, *Physiol. Rev.* **49:** 427–508.

Mommaerts, W. F. H. M., Buller, A. J., and Seraydarian, K. (1969), The modification of some biochemical properties of muscle by cross-innervation, *Proc. Natl. Acad. Sci. USA* **64:** 128–133.

Monard, D., Solomon, F., Rentsch, M., and Gysin, R. (1973), Glia-induced morphological differentiation in neuroblastoma cells, *Proc. Natl. Acad. Sci. USA* **70:** 1894–1897.

Monard, D., Stockel, K., Goodman, R., and Thoenen, H. (1975), Distinction between nerve growth factor and glial factor, *Nature* **258:** 444–445.

Morgan, I. G. (1970), Protein synthesis in brain mitochondrial and synaptosomal preparations, *FEBS Lett.* **10:** 273–275.

Mori, H., Komiya, Y., and Kurokawa, M. (1979), Slowly migrating axonal polypeptides. Inequalities in their rate and amount of transport between two branches of bifurcating axons, *J. Cell Biol.* **82:** 174–184.

Morris, C. J. O. R., and Morris, P. (1975), *Separation Methods in Biochemistry,* 2nd Edition, John Wiley & Sons, New York.

Muñoz-Martínez, E. J., and Chavez, B. (1979), Conduction block and function denervation caused by Tullidora *(Karwinskia humboldtiana), Exp. Neurol.* **65:** 255–270.

Muñoz-Martínez, E. J., Núñez, R., and Sanderson, A. (1981), Axonal transport: A quantitative study of retained and transported protein fraction in the cat, *J. Neurobiol.* **12:** 15–26.

Murray, M., and Grafstein, B. (1969), Changes in the morphology and amino acid incorporation of regenerating goldfish optic neurons, *Exp. Neurol.* **23:** 544–560.

Murray, M. R., and Herrmann, A. (1968), Passive movements of Schmidt–Lantermann clefts during continuous observation *in vitro, J. Cell Biol.* **39:** 149a–150a.

Musick, J., and Hubbard, J. I. (1972), Release of protein from mouse motor nerve terminals, *Nature* **237:** 279–281.

Nachmias, V. T., and Huxley, H. E. (1970), Electron microscope observations on actomyosin and actin preparations from *Physarum polycephalum,* and on their interaction with heavy meromyosin subfragment I from muscle myosin, *J. Molec. Biol.* **50:** 83–90.

Nadelhaft, I. (1974), Microtubule densities and total numbers in selected axons of the crayfish abdominal nerve cord, *J. Neurocytol.* **3**: 73–86.

Nagatsu, T. (1973), *Biochemistry of Catecholamines. The Biochemical Method,* University Park Press, Baltimore.

Narahashi, T., Albuquerque, E. X., and Deguchi, T. (1971), Effects of batrachotoxin on membrane potential and conductance of squid giant axons, *J. gen. Physiol.* **58**: 54–70.

Neale, J. H. Forman, D. S., Shibla, D. B., and Shortell, S. A. (1980), Comparative analysis of rapidly transported axonal proteins in sensory neurons of the frog and rat, *J. Neurochem.* **35**: 838–843.

Neuhuber, W., Niederle, B., and Zenker, W. (1977), Somatopetal transport of horseradish peroxidase (HRP) in the peripheral and central branches of dorsal root ganglion cells, *Cell Tissue Res.* **183**: 395–402.

Newton, I., (1730) *Opticks,* 4th Edition, republished, Dover Publications, New York, 1952.

Nicholson, C. (1980), Modulation of extracellular calcium and its functional implications, *Fed. Proc.* **39**: 1519–1523.

Nicholson, C., Ten Bruggencate, G., Stöckle, H., and Steinberg, R. (1978), Calcium and potassium changes in extracellular microenvironment of cat cerebellar cortex, *J. Neurophysiol.* **41**: 1026–1039.

Nittono, K. (1923), On bilateral effects from the unilateral section of branches of the nerves trigeminus in the albino rat, *J. Comp. Neurol.* **35**: 133–161.

Nordlander, R. H., and Singer, M. (1973), Degeneration and regeneration of severed crayfish sensory fibers: An ultrastructural study, *J. Comp. Neurol.* **152**: 175–192.

Norris, A. H., and Shock, N. (1960), *Science and Medicine of Exercise and Sports,* Harper, New York.

Nörstrom, A. (1975), Axonal transport and turnover of neurohypophyseal proteins in the rat, *Ann. N.Y. Acad. Sci.* **248**: 46–63.

Nunn, F., Dixon, K. L., and Lovis, J. D. (1969), The effects of halothane on mitosis, *Anesthesiology* **30**: 348–349.

O'Brien, R. A. D. (1978), Axonal transport of acetylcholine, choline acetyltransferase and cholinesterase in regenerating peripheral nerve, *J. Physiol. (London)* **282**: 91–103.

Ochs, R. L., and Burton, P. R. (1980), Distribution and selective extraction of filamentous components associated with axonal microtubules of crayfish nerve cord, *J. Ultrastructure Res.* **73**: 169–182.

Ochs, S. (1963), Beading phenomena of mammalian myelinated nerve fibers, *Science* **139**: 599–600.

Ochs, S. (1965), Beading of myelinated nerve fibers, *Exp. Neurol.* **12**: 84–95.

Ochs, S. (1966), Axoplasmic flow in neurons, Chapter 3, in: *Macromolecules and Behavior,* Gaito, J. (ed.), Appleton-Century-Crofts, New York.

Ochs, S. (1967), Axoplasmic transport of protein and the beading phenomenon, *Neurosci. Res. Prog. Bull.* **5**: 337–340.

Ochs, S. (1971a), Local supply of energy to the fast axoplasmic transport mechanism, *Proc. Natl. Acad. Sci. USA* **68**: 1279–1282.

Ochs, S. (1971b), Characteristics and a model for fast axoplasmic transport in nerve, *J. Neurobiol.* **2**: 331–345.

Ochs, S. (1971c), The dependence of fast transport in mammalian nerve fibers on metabolism, *Acta Neuropath. (Berlin) Suppl.* **V**: 86–96.

Ochs, S. (1972a), Rate of fast axoplasmic transport in mammalian nerve fibres, *J. Physiol. (London)* **227**: 627–645.

Ochs, S. (1972b), Axoplasmic flow—The fast transport system in mammalian nerve fibers, Chapter 9, in: *Macromolecules and Behavior,* 2nd Edition Gaito, J. (ed.), Appleton-Century-Crofts, New York.

Ochs, S. (1972c), Fast transport of materials in mammalian nerve fibers, *Science* **176:** 252–260.

Ochs, S. (1972d), Block of fast axoplasmic transport by interruption of metabolism with 2-deoxy-glucose, fluoroacetate and azide, *Abstr. Am. Soc. Neurochem.* **3:** 109.

Ochs, S. (1973), Effect of maturation and aging on the rate of fast axoplasmic transport in mammalian nerve, *Prog. Brain Res.* **40:** 349–362.

Ochs, S. (1974a), Energy metabolism and supply of ~P to the fast axoplasmic transport mechanism in nerve, *Fed. Proc.* **33:** 1049–1058.

Ochs, S. (1974b), Systems of material transport in nerve fibers (axoplasmic transport) related to nerve function and trophic control, *Ann. N.Y. Acad. Sci.* **228:** 202–223.

Ochs, S. (1974c), Axoplasmic transport—Energy metabolism and mechanism, Chapter 3, in: *The Vertebrate Peripheral Nervous System,* Hubbard, J. I. (ed.), Plenum Press, New York.

Ochs, S. (1975a), Mechanism of axoplasmic transport and block of transport by pharmacological agents, *Abstracts of the 6th International Congress on Pharmacology (Neurotransmission),* Helsinki, Finland, p. 78; Vol. 2, pp. 161–174.

Ochs, S. (1975b), Waller's concept of the trophic dependence of the nerve fiber on the cell body in the light of early neuron theory, *Clio Medica* **10:** 253–265.

Ochs, S. (1975c), Retention and redistribution of proteins in mammalian nerve fibres by axoplasmic transport, *J. Physiol. (London)* **253:** 459–475.

Ochs, S. (1975d), Axoplasmic transport—A basis for neural pathology, Chapter 12, in: *Peripheral Neuropathy,* Vol. 1., Dyck, P. J., Thomas, P. K., and Lambert, E. H. (eds.), Saunders, Philadelphia.

Ochs, S. (1975e), A unitary concept of axoplasmic transport based on the transport filament hypothesis, pp. 189–194, in: *Third International Congress on Muscle Diseases,* Bradley, W. G., Gardner-Medwin, D., and Walton, J. N. (eds.), Excerpta Medica, Amsterdam.

Ochs, S. (1976), Fast axoplasmic transport in the fibres of chromatolysed neurons, *J. Physiol. (London)* **255:** 249–261.

Ochs, S. (1977a), The early history of nerve regeneration beginning with Cruikshank's observations in 1776, *Med. Hist.* **21:** 261–274.

Ochs, S. (1977b), Axoplasmic transport in peripheral nerve and hypothalamo-neurohypophyseal systems, Chapter 2, in: *Hypothalamic Peptide Hormones and Pituitary Regulation,* Porter, J. C. (ed.), Plenum Press, New York.

Ochs, S. (1979), The early history of material transport in nerve, *Physiologist* **22:** 16–19.

Ochs, S. (1980a), A brief history of nerve repair and regeneration, pp. 1–8, in: *Nerve Repair and Regeneration: Its Clinical and Experimental Basis,* Jewett, D. L., and McCarroll, H. R. (eds.), C. V. Mosby Company, St. Louis.

Ochs, S. (1980b), Calcium requirement for axoplasmic transport and the role of the perineurial sheath, Chapter 7, in: *Nerve Repair and Regeneration: Its Clinical and Experimental Basis,* Jewett, D. L., and McCarroll, H.R. (eds.), C. V. Mosby Company, St. Louis.

Ochs, S. (1981a), Characterization of fast orthograde transport, *Neurosci. Res. Prog. Bull.* **20:** 19–31.

Ochs, S. (1981b), Axoplasmic transport, Chapter 22, in: *Basic Neurochemistry,* 3rd ed., Siegel, G. J., Albers, R. W., Agranoff, B. W., and Katzman, R. (eds.), Little, Brown and Company, Boston.

Ochs, S. (1981c), Regeneration of nerve in relation to axoplasmic transport: An analysis, pp. 115–127, in: *Posttraumatic Peripheral Nerve Regeneration,* Gorio, A., Mingrino, S., and Millesi, H. (eds.), Raven Press, New York.

Ochs, S. (1982) Block of axoplasmic transport by agents interfering with calcium flux: cobalt, nickel, lanthanum, verapamil; and maintenance of transport in calcium-free media by strontium. Soc. Neurosci. Abstr. (8) (In press).

Ochs, S., and Barnes, C. D. (1969), Regeneration of ventral root fibers into dorsal roots shown by axoplasmic flow, *Brain Res.* **15:** 600–603.

Ochs, S., Booker, H., and DeMyer, W. E. (1961), Note on the signal for chromatolysis after nerve interruption, *Exp. Neurol.* **3:** 206–208.

Ochs, S., and Burger, E. (1958), Movement of substance proximo-distally in nerve axons as studied with spinal cord injections of radioactive phosphorus, *Am. J. Physiol.* **194:** 499–506.

Ochs, S., Chan, S.-Y., and Worth, R. (1978), Calcium and the mechanism of axoplasmic transport, pp. 359–367, in: *The Neurobiological Mechanisms in Manipulative Therapy,* Korr, I. M. (ed.), Plenum Press, New York.

Ochs, S., Dalrymple, D., and Richards, G. (1962), Axoplasmic flow in ventral root nerve fibers of the cat, *Exp. Neurol.* **5:** 349–363.

Ochs, S., and Heckaman, P. (1972), Block of nerve excitability and fast axoplasmic transport by increased O_2 and CO_2, *Biophys. Soc. Abstr.* **12:** 192a.

Ochs, S., and Hollingsworth, D. (1971), Dependence of fast axoplasmic transport in nerve on oxidative metabolism, *J. Neurochem.* **18:** 107–114.

Ochs, S., and Iqbal, Z. (1980), Calmodulin and calcium activation of tubulin associated Ca-ATPase, *Soc. Neurosci. Abstr.* **6:** 501.

Ochs, S., and Jersild, R. A., Jr. (1974), Fast axoplasmic transport in nonmyelinated mammalian nerve fibers shown by electron microscopic radioautography, *J. Neurobiol.* **5:** 373–377.

Ochs, S., Erdman, J., Jersild, R. A., Jr., and McAdoo, V. (1978), Routing of transported materials in the dorsal root and nerve fiber branches of the dorsal root ganglion, *J. Neurobiol.* **9:** 465–481.

Ochs, S., and Johnson, J. (1969), Fast and slow phases of axoplasmic flow in ventral root nerve fibres, *J. Neurochem.* **16:** 845–853.

Ochs, S., Johnson, J., and Ng, M.-H. (1967), Protein incorporation and axoplasmic flow in motoneuron fibres following intra-cord injection of labelled leucine, *J. Neurochem.* **14:** 317–331.

Ochs, S., Kachmann, R., and DeMyer, W. E. (1960), Axoplasmic flow rates during nerve regeneration, *Exp. Neurol.* **2:** 627–637.

Ochs, S., and Ranish, N. (1969), Characteristics of the fast transport system in mammalian nerve fibers, *J. Neurobiol.* **1:** 247–261.

Ochs, S., and Ranish, N. (1970), Metabolic dependence of fast axoplasmic transport in nerve, *Science* **167:** 878–879.

Ochs, S., Sabri, M. I., and Ranish, N. (1969), Fast transport system of axoplasmic flow in cat nerve fibers, *Fed. Proc.* **28:** 458.

Ochs, S., Sabri, M. I., and Ranish, N. (1970), Somal site of synthesis of fast transported materials in mammalian nerve fibers, *J. Neurobiol.* **1:** 329–344.

Ochs, S., and Smith, C. (1971a), Effect of temperature and rate of stimulation on fast axoplasmic transport in mammalian nerve fibers, *Fed. Proc.* **30:** 665.

Ochs, S., and Smith, C. B. (1971b), Fast axoplasmic transport in mammalian nerve *in vitro* after block of glycolysis with iodoacetic acid, *J. Neurochem.* **18:** 833–843.

Ochs, S., and Smith, C. (1975), Low temperature slowing and cold-block of fast axoplasmic transport in mammalian nerves *in vitro*, *J. Neurobiol.* **6:** 85–102.

Ochs, S., and Worth, R. (1975), Batrachotoxin block of fast axoplasmic transport in mammalian nerve fibers, *Science* **187:** 1087–1089.

Ochs, S., Worth, R. M., and Chan, S.-Y. (1977), Calcium requirement for axoplasmic transport in mammalian nerve, *Nature* **270:** 748–750.

Oh, T. H. (1975), Neurotrophic effects: Characterization of the nerve extract that stimulates muscle development in culture, *Exp. Neurol.* **46:** 432–438.

Oh, T. H., Johnson, D. D. and Kim, S. U. (1972), Neurotrophic effect on isolated chick embryo muscle in culture, *Science* **178:** 1298–1300.

Oh, T. H., Markelonis, G. J., Reier, P. J., and Zalewski, A. A. (1980), Persistence in degenerating sciatic nerve of substances having a trophic influence upon cultured muscle, *Exp. Neurol.* **67:** 646–654.

Ohnishi, A., and Ikeda, M. (1980), Morphometric evaluation of primary sensory neurons in experimental *p*-bromophenyl-acetylurea intoxication, *Acta Neuropath. (Berlin)* **52:** 111–118.

Oikarinen, R., Molnar, G. K., and Riekkinen, P. J. (1980), Axonal transport of cholinesterase activity in experimental allergic neuritis, *Acta Univ. Oulu A97 Biochem,* **29:** 85–90.

Okada, Y., and McDougal, D. B., Jr. (1971), Physiological and biochemical changes in frog sciatic nerve during anoxia and recovery, *J. Neurochem.* **18:** 2335–2353.

Olmsted, J. M. D. (1920), The nerve as a formative influence in the development of taste-buds, *J. Comp. Neurol.* **31:** 465–468.

Olmsted, J. B., and Borisy, G. G. (1975), Ionic and nucleotide requirements for microtubule polymerization *in vitro, Biochemistry* **14:** 2996–3005.

Olmsted, J. B., Marcum, J. M., Johnson, K. A., Allen, C., and Borisy, G. G. (1974), Microtubule assembly: Some possible regulatory mechanisms, *J. Supramol. Struct.* **2:** 429–450.

Olsson, T., and Kristensson, K. (1981), Neuronal uptake of iron: Somatopetal axonal transport and fate of cationized and native ferritin, and iron-dextran after intramuscular injections, *Neuropath. Appl. Neurobiol.* **7:** 87–95.

Osborne, R. H., and Bradford, H. F. (1973), Tetanus toxin inhibits amino acid release from nerve endings *in vitro, Nature (New Biol.)* **244:** 157–158.

O'Steen, W. K., and Vaughan, G. M. (1968), Radioactivity in the optic pathway and hypothalamus of the rat after intraocular injection of tritiated 5-hydroxytryptophan, *Brain Res.* **8:** 209–212.

Padieu, P., and Mommaerts, W. F. H. M. (1960), Creatine phosphoryl transferase and phosphoglyceraldehyde dehydrogenase in iodoacetate poisoned muscle, *Biochem. Biophys. Acta* **37:** 72–77.

Palade, G. (1975), Intracellular aspects of the process of protein synthesis, *Science* **189:** 347–358.

Pannese, E. (1963), Investigations on the ultrastructural changes of the spinal ganglion neurons in the course of axon regeneration and cell hypertrophy. I. Changes during axon regeneration, *Z. Zellforsch.* **60:** 711–740.

Papasozomenos, S. C., Autilio-Gambetti, L., and Gambetti, P. (1981a), Redistribution of fast axonal transport of proteins in β, β'-iminodipropionitrile (IDPN) intoxication, an ultrastructural autoradiographic study, *Soc. Neurosci. Abstr.* **7:** 486.

Papasozomenos, S. C., Autilio-Gambetti, L., and Gambetti, P. (1981b), Reorganization of axoplasmic organelles following β, β'-iminodipropionitrile administration, *J. Cell Biol.* **91:** 866–871.

Paravicini, U., Stoeckel, K., and Thoenen, H. (1975), Biological importance of retrograde axonal transport of nerve growth factor in adrenergic neurons, *Brain Res.* **84:** 279–291.

Parker, G. H. (1929), The neurofibril hypothesis, *Quart. Rev. Biol.* **4:** 155–178.

Parker, G. H. (1932), On the trophic impulse so-called, its rate and nature, *Am. Naturalist* **66:** 147–158.

Parker, G. H., and Paine, V. L. (1934), Progressive nerve degeneration and its rate in lateral-line nerve of the catfish, *Am. J. Anat.* **54:** 1–25.

Partlow, L. M. Ross, C. D., Motwani, R., and McDougal, D. B., Jr. (1972), Transport of axonal enzymes in surviving segments of frog sciatic nerve, *J. gen. Physiol.* **60:** 388–405.

Paton, D. M. (ed.) (1976), *The Mechanism of Neuronal and Extraneuronal Transport of Catechol-amines,* Raven Press, New York.

Paton, D. M. (ed.) (1979), *The Release of Catecholamines from Adrenergic Neurons,* Pergamon Press, New York.

Paulson, J. C., and McClure, W. O. (1975), Inhibition of axoplasmic transport by colchicine, podophyllotoxin, and vinblastine: An effect on microtubules, *Ann. N.Y. Acad. Sci.* **253:** 517–527.

Payton, B. W., Bennett, M. V. L., and Pappas, G. D. (1969), Permeability and structure of junctional membranes at an electronic synapse, *Science* **166:** 1641–1643.

Penningroth, S. M., Cleveland, D. W., and Kirschner, M. W. (1976), *In vitro* studies of the regulation of microtubule assembly, pp. 1233–1257, in: *Cell Motility,* Book C, Goldman, R., Pollard, T., and Rosenbaum, J. (eds.), Cold Spring Harbor Laboratory, Cold Spring Harbor, N.Y.

Perišić, M., and Cuénod, M. (1972), Synaptic transmission depressed by colchicine blockage of axoplasmic flow, *Science* **175:** 1140–1142.

Perroncito, A. (1905), La rigenerazione delle fibre nervose, *Boll. Soc. Med. Chir. Pavia.* **4:** 434–444.

Perry, G. W., and Wilson, D. L. (1981), Protein synthesis and axonal transport during nerve regeneration, *J. Neurochem.* **37:** 1203–1217.

Pestronk, A., Drachman, D. B., and Griffin, J. W. (1976), Effect of muscle disuse on acetylcholine receptors, *Nature* **260:** 352–353.

Peters, F. E. (1967), *Greek Philosophical Terms. A Historical Lexicon,* New York University Press, New York.

Peters, A., and Vaughn, J. E. (1967), Microtubules and filaments in the axons and astrocytes of early postnatal rat optic nerves, *J. Cell Biol.* **32:** 113–119.

Peterson, E. R., and Crain, S. M. (1970), Innervation in cultures of fetal rodent skeletal muscle by organotypic explants of spinal cord from different animals, *Z. Zellforsch.* **106:** 1–21.

Peterson, E. R., and Crain, S. M. (1972), Regeneration and innervation in cultures of adult mammalian skeletal muscle coupled with fetal rodent spinal cord, *Exp. Neurol.* **36:** 136–159.

Peterson, E. R., and Crain, S. M. (1979), Factors involved in muscle atrophy after surgical denervation of long-term cultures of cord-innervated fibers, pp. 305–309, in: *Muscle Regen-eration,* Mauro, A., *et al.* (eds.), Raven Press, New York.

Phillis, J. W. (1970), *The Pharmacology of Synapses,* Pergamon Press, New York.

Pickering, B. T., and Jones, C. W. (1971), The biosynthesis and intraneuronal transport of neuro-hypophysial hormones: Preliminary studies in the rat, pp. 337–351, in: *Memoirs of the Society for Endocrinology,* No. 19, Heller, H., and Lederis, K. (eds.), Cambridge University Press, London.

Pickering, B. T., Jones, C. W., Buford, G. D., McPherson, M., Swann, R. W., Heap, P. F., and Morris, J. F. (1975), The role of neurophysin proteins: Suggestions from the study of their transport and turnover, *Ann. N.Y. Acad. Sci.* **248:** 15–35.

Pilar, G., and Landmesser, L. (1972), Axotomy mimicked by localized colchicine application, *Science* **177:** 1116–1118.

Plateau, J. [A. F.] (1873), *Statique expérimentale et theoretique des liquides soumis aux seules forces moleculaires,* Gauthier-Villars, Paris.

Pleasure, D. E., Mishler, K. C., and Engel, W. K. (1969), Axonal transport of proteins in experimental neuropathies, *Science* **166:** 524–525.

Pleasure, D. (1980), Axoplasmic transport, pp. 221–237, in: *The Physiology of Peripheral Nerve Disease,* Sumner, A. (ed.), W. B. Saunders Company, Philadelphia.

Politoff, A. L., and Blitz, A. L. (1978), Neurotrophic control of RNA synthesis in amphibian striated muscle, *Brain Res.* **151:** 561–570.

Pollard, T. D. (1976), The role of actin in the temperature-dependent gelatin and contraction of extracts of *Acanthamolba, J. Cell Biol.* **68:** 579–601.

Pollard, T. D., and Weihing, R. R. (1974), Actin and myosin and cell movement, *CRC Crit. Rev. Biochem.* **2:** 1–65.

Pomerat, C. M., Hendelman, W. J., Raiborn, C. W., Jr., and Massey, J. F. (1967), Dynamic activities of nervous tissue *in vitro*, pp. 119–178, in: *The Neuron*, Hydén, H. (ed), Elsevier Publishing Company, Amsterdam.

Porter, K. R. (1966), Cytoplasmic microtubules and their functions, pp. 308–356, in: *Principle of Biomolecular Organization* (CIBA Foundation Symposium), Wolstenholme, G. E. W., and O'Connor, M. (eds.), Little, Brown, and Co., Boston.

Porter, K. R. (1976), Motility in cells, pp. 1–23, in: *Cell Motility*, Goldman, R., Pollard, T., and Rosenbaum, J. (eds.), Cold Spring Harbor Laboratory, Cold Spring Harbor, N.Y.

Porter, K. R., Byers, H. R., and Ellisman, M. H. (1979), The cytoskeleton, pp. 703–722, in: *The Neurosciences: Fourth Study Program*, Schmitt, F. O., and Worden, F. G. (eds.), M.I.T. Press, Cambridge, Mass.

Potter, D. D., Furshpan, E. J., and Landis, S. C. (1981), Multiple-transmitter status and "Dale's Principle," *Neurosci. Newsletter* **12:** 1–9.

Potter, H. D. (1971), The distribution of neurofibrils coextensive with microtubules and neurofilaments in dendrites and axons of the tectum, cerebellum and pallium of the frog, *J. Comp. Neurol.* **143:** 385–410.

Potter, L. T. (1970), Synthesis, storage and release of (^{14}C) acetylcholine in isolated rat diaphragm muscle, *J. Physiol. (London)* **206:** 145–166.

Potter, L. T., and Axelrod, J. (1963), Subcellular localization of catecholamines in tissues of the rat, *J. Pharm. Exp. Therap.* **142:** 291–298.

Potter, L. T., Axelrod, J., and Kopin, I. J. (1962), Differential binding and release of norepinephrine and tachyphylaxis, *Biochem. Pharmacol.* **11:** 254–256.

Potter, L. T., and Molinoff, P. B. (1972), Isolation of cholinergic receptor proteins, in: *Perspectives in Neuropharmacology. A tribute to Julius Axelrod*, Snyder, S. H. (ed.), Oxford University Press, New York.

Price, C. H., McAdoo, D. J., Farr, W., and Okuda, R. (1979), Bidirectional axonal transport of free glycine in identified neurons R3-R14 of *Aplysia, J. Neurobiol.* **10:** 551–571.

Price, D. L., Griffin, J., Young, A., Peck, K., and Stocks, A. (1975), Tetanus toxin: Direct evidence for retrograde intraaxonal transport, *Science* **188:** 945–947.

Price, D. L., and Porter, K. (1972), The response of ventral horn neurons to axonal resection, *J. Cell Biol.* **53:** 24–37.

Puchala, E., and Windle, W. F. (1977), The possibility of structural and functional restitution after spinal cord injury. A review, *Exp. Neurol.* **55:** 1–42.

Purves, D. (1977), The formation and maintenance of synaptic connections, pp. 21–49, in: *Function and Formation of Neural Systems*, Steut, G. S. (ed.), Dahlern Konferenzen, Berlin.

Purves, D., and Lichtman, J. W. (1978), Formation and maintenance of synaptic connections in autonomic ganglia, *Physiol. Rev.* **58:** 821–862.

Purves, D., and Lichtman, J. W. (1980), Elimination of synapses in the developing nervous system, *Science* **210:** 153–157.

Purves, D., and Njå, A. (1978), Trophic maintenance of synaptic connections in autonomic ganglia, pp. 27–47, in: *Neuronal Plasticity*, Cotman, C. W. (ed.), Raven Press, New York,

Purves, D., and Sakmann, B. (1974), The effect of contractile activity on fibrillation and extrajunctional acetylcholine-sensitivity in rat muscle maintained in organ culture, *J. Physiol. (London)* **237:** 157–182.

Puszkin, S., and Schook, W. (1979), The role of cytoskeleton in neuron activity, *Meth. Achiev. Exp. Pathol.* **9:** 87–111.

Quigley, H. A., Guy, J., and Anderson, D. R. (1979), Blockade of rapid axonal transport. Effect of intraocular pressure elevation in primate optic nerve, *Arch. Ophthalmol.* **97:** 525–531.

Rahmann, H. (1971), Different modes of substance flow in the optic tract, *Acta Neuropath. (Berlin) Suppl.* **V:** 162–170.

Raine, C. S., Ghetti, B., and Shelanski, M. L. (1971), On the association between microtubules and mitochondria within axons, *Brain Res.* **34:** 389–393.

Ramírez, B. U. (1980), Neurotrophic regulation of muscle autolytic activity, *Exp. Neurol.* **67:** 257–264.

Rammler, D. H., and Zaffaroni, A. (1967), Biological implications of DMSO based on a review of its chemical properties, *Ann. N.Y. Acad. Sci.* **141:** 13–23.

Ramon y Cajal, S. (1894), La fine structure des centres nerveux, *Proc. Roy. Soc. B* **55:** 444–468.

Ramon y Cajal, S. (1909), *Histologie du système nerveux de l'homme et des vertébrés,* Vol. 1, translated by Azoulay, L.; republished by Consejo Superior de Investigaciones Cientificas, Instituto Ramon y Cajal, Madrid (1952).

Ramon y Cajal, S. (1911), *Histologie du système nerveux de l'homme et des vertébrés,* Vol. 2, translated by Azoulay, L.; republished by Consejo Superior de Investigaciones Cientificas, Instituto Ramon y Cajal, Madrid (1955).

Ramon y Cajal, S. (1928), *Studies on Degeneration and Regeneration of the Nervous System,* translated by May, R. M. (2 Vol), Oxford University Press, Oxford; reprinted by Hafner, New York. 1968.

Ramon y Cajal, S. (1937), *Recollections of My Life,* translated by Craigie, E. H., with Cano, J., American Philosophical Society, Philadelphia.

Ramon y Cajal, S. (1954), *Neuron Theory or Reticular Theory?* Translated by Purkiss, M. U., and Fox, C. A., Consejo Superior de Investigaciones Cientificas, Instituto Ramon y Cajal, Madrid.

Ranish, N., and Dettbarn, W.-D. (1977), The effect of pargyline and serotonin on fast axoplasmic transport in sciatic nerve, *Fed. Proc.* **36:** 560.

Ranish, N. A., and Dettbarn, W.-D. (1978), Nerve stump length and cholinesterase activity in muscle and nerve, *Exp. Neurol.* **58:** 377–386.

Ranish, N. A., Dettbarn, W.-D., Wecker, L. (1980), Nerve stump length-dependent loss of acetylcholinesterase activity in endplate regions of rat diaphragm, *Brain Res.* **191:** 379–386.

Ranish, N., and Ochs, S. (1972), Fast axoplasmic transport of acetylcholinesterase in mammalian nerve fibres, *J. Neurochem.* **19:** 2641–2649.

Ranvier, L. (1875), Des tubes nerveux en T. et de leurs relations avec les cellules ganglionnaires, *Compt. Rend. Acad. Sci.* **81:** 1274–1276.

Rasmussen, H. (1970), Cell communication, calcium ion, and cyclic adenosine monophosphate, *Science* **170:** 404–412.

Rasool, C. G., Schwartz, A. L., Bollinger, J. A., Reichlin, S., and Bradley, W. G. (1981), Immunoreactive somatostatin distribution and axoplasmic transport in rat peripheral nerve, *Endocrinology* **108:** 996–1001.

Rebhun, L. I. (1972), Polarized intracellular particle transport: Saltatory movements and cytoplasmic streaming, *Int. Rev. Cytol.* **32:** 93–137.

Redfern, P., and Thesleff, S. (1971), Action potential generation in denervated rat skeletal muscle. II. The action of tetrodotoxin, *Acta Physiol. Scand.* **82:** 70–78.

Reiner, M. (1949), *Deformation and Flow,* H. K. Lewis & Co., London.

Reis, D. J., Ross, R. A., and Joh, T. H. (1974), Some aspects of the reaction of central and peripheral noradrenergic neurons to injury, pp. 109–125, in: *Dynamics of Degeneration and Growth in Neurons,* Fuxe, K., Olson, L., and Zotterman, Y. (eds.), Pergamon Press, New York.

Reis, D. J., Gilad, G., Pickel, V. M., and Joh, T. H. (1978), Reversible changes in the activities and amounts of tyrosine hydroxylase in dopamine neurons of the substantia nigra in response to axonal injury as studied by immunochemical and immunocytochemical methods, *Brain Res.* **144:** 325–342.

Remak, R. (1838), *Observations anatomicae et microscopicae de systematis nervosi structura,* Dissertation, Reims, Berolini.

Remillard, S., Rebhun, L. I., Howie, G. A., and Kupchan, S. M. (1975), Antimitotic activity of the potent tumor inhibitor maytansine, *Science* **189:** 1002–1005.

Revel, J. P., and Karnovsky, M. J. (1967), Hexagonal array of subunits in intercellular junctions of the mouse heart and liver, *J. Cell Biol.* **33:** C7–C12.

Revel, J. P., Olson, W., and Karnovsky, M. J. (1967), A twenty-angstrom gap junction with a hexagonal array of subunits in smooth muscle, *J. Cell Biol.* **35:** 112A.

Richardson, P. M., McGuinness, U. M., and Aguayo, A. J. (1980), Axons from CNS neurones regenerate into PNS grafts, *Nature* **284:** 264–265.

Ritchie, J. M., and Rogart, R. B. (1977), Density of sodium channels in mammalian myelinated nerve fibers and nature of the axonal membrane under the myelin sheath, *Proc. Natl. Acad. Sci. USA* **74:** 211–215.

Robert, E. D., and Oester, Y. T. (1970), Absence of supersensitivity to acetylcholine in innervated muscle subjected to a prolonged pharmacologic nerve block, *J. Pharmacol. Exp. Therap.* **174:** 133–140.

Roberts, P. J., Keen, P., and Mitchell, J. F. (1973), The distribution and axonal transport of free amino acids and related compounds in the dorsal sensory neuron of the rat, as determined by the dansyl reaction, *J. Neurochem.* **21:** 199–209.

Robison, G. A., Butcher, R. W., and Sutherland, E. W. (1971), *Cyclic AMP,* Academic Press, New York.

Rodríguez-Echandía, E. L., and Piezzi, R. S. (1968), Microtubules in the nerve fibers of the toad *Bufo Arenarum Hensel.* Effect of low temperature on the sciatic nerve, *J. Cell Biol.* **39:** 491–497.

Rodríguez-Echandía, E. L., Zamora, A., and Piezzi, R. S. (1970), Organelle transport in constricted nerve fibers of the toad *Bufo arenarum Hensel, Z. Zellforsch.* **104:** 419–428.

Roisen, F. J., and Murphy, R. A. (1973), Neurite development *in vitro*: II. The role of microfilaments and microtubules in dibutyryl adenosine 3′,5′-cyclic monophosphate and nerve growth factor stimulated maturation, *J. Neurobiol.* **4:** 397–412.

Roisen, F. J., Murphy, R. A., and Braden, W. G. (1972), Neurite development *in vitro*. I. The effects of adenosine 3′5′-cyclic monophosphate (cyclic AMP), *J. Neurobiol.* **3:** 347–368.

Rome-Talbot, D., Andre, D., and Chalazonitis, N. (1978), Hypothermic decrease in microtubule density and birefringence in unmyelinated axons, *J. Neurobiol.* **9:** 247–254.

Roofe, P. G. (1947), Role of the axis cylinder in transport of tetanus toxin, *Science* **105:** 180–181.

Roots, B. I., and Bondar, R. L. (1977), Neurofilamentous changes in goldfish *(Carassius Auratus L.)* brain in relation to environmental temperature, *J. Neuropath. Exp. Neurol.* **36:** 453–464.

Rose, S. M. (1948), The role of nerves in amphibian limb regeneration, *Ann. N. Y. Acad. Sci.* **49:** 818–833.

Rosenblueth, A., and Del Pozo, E. C. (1943), The centrifugal course of Wallerian degeneration, *Am. J. Physiol.* **139:** 247–254.

Rosenbluth, J. (1976), Intramembranous particle distribution at the node of Ranvier and adjacent axolemma in myelinated axons of the frog brain, *J. Neurocytol.* **5:** 731–745.

Rosenbluth, J., and Wissig, S. L. (1964), The distribution of exogenous ferritin in toad spinal ganglia and the mechanism of its uptake by neurons, *J. Cell Biol.* **23:** 307–325.

Rosenbaum, J. L., Binder, L. I., Granett, S., Dentler, W. L., Snell, W., Sloboda, R., and Haimo, L. (1975), Directionality and rate of assembly of chick brain tubulin onto pieces of neurotubules, flagellar axonemes, and basal bodies, *Ann. N.Y. Acad. Sci.* **253:** 147–177.

Rostas, J. A. P., Austin, L., and Jeffrey, P. L. (1979), Selective labelling of two phases of axonal transport of cholesterol in the chick optic system, *J. Neurochem.* **32:** 1461–1466.

Rothschuch, K. (1973), *History of Physiology,* Translated from the German by Risse, G. B., Krieger Publishing Co., Huntington, N.Y.

Rubinson, K. A., and Baker, P. F. (1979), The flow properties of axoplasm in a defined chemical environment: Influence of axions and calcium, *Proc. Roy. Soc. B* **205:** 323–345.

Rydevik, B., McLean, W. G., Sjöstrand, J., and Lundborg, G. (1980), Blockage of axonal transport induced by acute, graded compression of the rabbit vagus nerve, *J. Neurol. Neurosurg. Psych.* **43:** 690–698.

Sabri, M. I., and Ochs, S. (1971), Inhibition of glyceraldehyde-3-phosphate dehydrogenase in mammalian nerve by iodoacetic acid, *J. Neurochem.* **18:** 1509–1514.

Sabri, M. I., and Ochs, S. (1972), Relation of ATP and creatine phosphate to fast axoplasmic transport in mammalian nerve, *J. Neurochem.* **19:** 2821–2828.

Sabri, M. I., and Ochs, S. (1973), Characterization of fast and slow transported proteins in dorsal root and sciatic nerve of cat, *J. Neurobiol.* **4:** 145–165.

Sabri, M. I., Moore, C. L., and Spencer, P. S. (1979), Studies on the biochemical basis of distal axonopathies—I. Inhibition of glycolysis by neurotoxic hexacarbon compounds, *J. Neurochem.* **32:** 683–689.

Sachs, F., and Latorre, R. (1974), Cytoplasmic solvent structure of single barnacle muscle cells studied by electron spin resonance, *Biophys. J.* **14:** 316–326.

Sachs, H. (1969), Neurosecretion, *Adv. Enzymol.* **32:** 327–372.

Sahenk, Z., and Mendell, J. R. (1979), Analysis of fast axoplasmic transport in nerve ligation and adriamycin-induced neuronal perikaryon lesions, *Brain Res.* **171:** 41–53.

Salafsky, B., Bell, J., and Prewitt, M. A. (1968), Development of fibrillation potentials in denervated fast and slow skeletal muscle, *Am. J. Physiol.* **215:** 637–643.

Salmons, S., and Vrbová, G. (1969), The influence of activity on some contractile characteristics of mammalian fast and slow muscles, *J. Physiol. (London)* **201:** 535–549.

Samson, F. E., Jr. (1971), Mechanism of axoplasmic transport, *J. Neurobiol.* **2:** 347–360.

Samson, F. E. (1976), Pharmacology of drugs that affect intracellular movement, *Ann. Rev. Pharmacol. Toxicol.* **16:** 143–159.

Samuel, S. (1860), *Trophischen Nerven,* Wigund, Leipzig.

Samuels, A. J., Boyarsky, L. L., Gerard, R. W., Libet, B., and Brust, M. (1951), Distribution, exchange and migration of phosphate compounds in the nervous system, *Am. J. Physiol.* **164:** 1–15.

Sandberg, A. A., Rosenthal, H., Schneider, S. L., and Slaunwhite, W. R., Jr. (1966), Protein-steroid interactions and their role in the transport and metabolism of steroids, pp. 1–61, in: *Steroid Dynamics,* Pincus, G., Nakao, T. and Tait, J. F. (eds.), Academic Press, New York.

Sanes, J. R., Marshall, L. M., and McMahan, U. J. (1978), Reinnervation of muscle fiber basal lamina after removal of myofibers, *J. Cell Biol.* **78:** 176–198.

Satir, P. (1968), Studies on cilia. III. Further studies on the cilium tip and a "sliding filament" model of ciliary motility, *J. Cell Biol.* **39:** 77–94.

Sawyer, C. H. (1946), Cholinesterase in degenerating and regenerating peripheral nerves, *Am. J. Physiol.* **146:** 246–253.

Schafer, R. (1973), Acetylcholine: Fast axoplasmic transport in insect chemoreceptor fibers, *Science* **180:** 315–317.

Scharrer, E., and Scharrer, B. (1940), Secretory cells within the hypothalamus, *Assoc. Res. Nervous Mental Disease* **20:** 170–194.

Scharrer, E., and Scharrer, B. (1963), *Neuroendocrinology,* Columbia University Press, New York.

Schatzmann, H. J. (1975), Active calcium transport and Ca^{2+}-activated ATPase in human red cells, *Curr. Topics Memb. Transp.* **6:** 125–168.

Schiff, P. B., Fant, J., and Horwitz, S. B. (1979), Promotion of microtubule assembly *in vitro* by taxol, *Nature* **277**: 665–667.

Schiff, P. B., and Horwitz, S. B. (1980), Taxol stabilizes microtubules in mouse fibroblast cells, *Proc. Natl. Acad. Sci. USA* **77**: 1561–1565.

Schlaepfer, W. W. (1971), Experimental alterations of neurofilaments and neurotubules by calcium and other ions, *Exp. Cell Res.* **67**: 73–80.

Schlaepfer, W. W. (1977a), Immunological and ultrastructural studies of neurofilaments isolated from rat peripheral nerve, *J. Cell. Biol.* **74**: 226–240.

Schlaepfer, W. W. (1977b), Structural alterations of peripheral nerve induced by the calcium ionophore A23187, *Brain Res.* **136**: 1–9.

Schlaepfer, W. W., and Freeman, L. A. (1978), Neurofilament proteins of rat peripheral nerve and spinal cord, *J. Cell Biol.* **78**: 653–662.

Schlaepfer, W. W., and Hasler, M. B. (1979), Characterization of the calcium-induced disruption of neurofilaments in rat peripheral nerves, *Brain Res.* **168**: 299–309.

Schmatolla, E., and Fischer, H. A. (1972), Axonal transport in embryonic visual system in zebrafish, *Exp. Brain Res.* **15**: 168–176.

Schmidt, J. T. (1978), Retinal fibers alter tectal positional markers during the expansion of the half retinal projection in goldfish, *J. Comp. Neurol.* **177**: 279–300.

Schmidt, R. E., Matschinsky, F. M., Godfrey, D. A., Williams, A. D., and McDougal, D. B., Jr. (1975), Fast and slow axoplasmic flow in sciatic nerve of diabetic rats, *Diabetes* **24**: 1081–1085.

Schmidt, R. E., Yu, M. J. C., and McDougal, D. B., Jr. (1980), Turnaround of axoplasmic transport of selected particle-specific enzymes at an injury in control and diisopropylphosphorofluoridate-treated rats, *J. Neurochem.* **35**: 641–652.

Schmitt, F. O. (1968), II. The molecular biology of neuronal fibrous proteins, *Neurosci. Res. Prog. Bull.* **6**: 119–144.

Schonbach, J., and Cuénod, M. (1971), Axoplasmic migration of protein. A light microscopic autoradiographic study in the avian retino-tectal pathway, *Exp. Brain Res.* **12**: 275–282.

Schook, W., Puszkin, S., Bloom, W., Ores, C., and Kochwa, S. (1979), Mechanochemical properties of brain clathrin: interactions with actin and α-actinin and polymerization into basketlike structures or filaments, *Proc. Natl. Acad. Sci. USA* **76**: 116–120.

Schroder, J. M. (1975), Degeneration and regeneration of myelinated nerve fibers in experimental neuropathies, Chapter 16, in: *Peripheral Neuropathy,* Dyck, P. J., Thomas, P. K. and Lambert, E. H. (eds.), Saunders and Company, Philadelphia.

Schubert, P., and Kreutzberg, G. W. (1974), Axonal transport of adenosine and uridine derivatives and transfer to postsynaptic neurons, *Brain Res.* **76**: 526–530.

Schubert, P., and Kreutzberg, G. W. (1975), Dendritic and axonal transport of nucleoside derivatives in single motoneurons and release from dendrites, *Brain Res.* **90**: 319–323.

Schubert, P., Kreutzberg, G. W., and Lux, H. D. (1972), Neuroplasmic transport in dendrites: Effect of colchicine on morphology and physiology of motoneurones in the cat, *Brain Res.* **47**: 331–343.

Schubert, P., Lux, H. D., and Kreutzberg, G. W. (1971), Single cell isotope injection technique, a tool for studying axonal and dendritic transport, *Acta Neuropath. (Berlin) Suppl.* **5**: 179–186.

Schultze, M. (1871), The general characters of the structures composing the nervous system, Chapter 3, in: *Human Comparative Histology,* Vol. 1, Stricker, S. (ed.), New Sydenham Society, London.

Schwab, M. E., and Thoenen, H. (1976), Electron microscopic evidence for a transsynaptic migration of tetanus toxin in spinal cord motoneurons: An autoradiographic and morphometric study, *Brain Res.* **105**: 213–227.

Schwab, M., and Thoenen, H. (1977), Selective trans-synaptic migration of tetanus toxin after retrograde axonal transport in peripheral sympathetic nerves: A comparison with nerve growth factor, *Brain Res.* **122:** 459–474.

Schwab, M. E., and Thoenen, H. (1978), Selective binding, uptake and retrograde transport of tetanus toxin by nerve terminals in the rat iris, *J. Cell Biol.* **77:** 1–13.

Schwartz, J. H., Goldman, J. E., Ambron, R. T., and Goldbert, D. J. (1975), Axonal transport of vesicles carrying [^{3}H]serotonin in the metacerebral neuron of *Aplysia californica, Cold Spring Harbor Symp. Quant. Biol.* **40:** 83–92.

Schwarzacher, H. G. (1958), Der Cholinesterasegehalt motorischer Nervenzellen während der axonalen reaktion, *Acta Anat.* **32:** 51–65.

Scott, F. H. (1905), On the metabolism and action of nerve cells, *Brain* **28:** 506–526.

Scott, F. H. (1906), On the relation of nerve cells to fatigue of their nerve fibres, *J. Physiol. (London)* **34:** 145–162.

Selzer, M. E. (1980), Regeneration of peripheral nerve, Chapter 12, in: *The Physiology of Peripheral Nerve Disease,* Sumner, A. (eds.), W. B. Saunders Company, Philadelphia.

Server, A. C., and Shooter, E. M. (1977), Nerve growth factor, *Adv. Prot. Chem.* **31:** 339–409.

Sharp, G. A., Fitzsimons, J. T. R., and Kerkut, G. A. (1978), Microtubular ATPase in *Carcinus* and *Helix* nerve, *Comp. Biochem. Physiol.* **61A:** 297–301.

Shatz, C. J., and Stryker, M. P. (1978), Ocular dominance in layer IV of the cat's visual cortex and the effects of monocular deprivation, *J. Physiol. (London)* **281:** 267–283.

Shawe, G. D. H. (1955), On the number of branches formed by regenerating nerve fibres, *Brit. J. Surg.* **42:** 474–488.

Shecket, G. N., and Lasek, R. J. (1977), A neurofilament associated Mg^{2+}/Ca^{2+} ATPase from squid giant axon, *Trans. Am. Soc. Neurochem.* **8:** 179.

Shelanski, M. L., Gaskin, F., and Cantor, C. R. (1973), Microtubule assembly in the absence of added nucleotides, *Proc. Natl. Acad. Sci. USA* **70:** 765–786.

Shelanski, M. L., Leterrier, J.-F., and Liem, R. K. H. (1981), Evidence for interactions between neurofilaments and microtubules, *Neurosci. Res. Prog. Bull.* **19:** 32–43.

Shelanski, M. L., and Liem, R. K. H. (1979), Neurofilaments, *J. Neurochem.* **33:** 5–13.

Sherline, P., and Schiavone, K. (1977), Immunofluorescence localization of proteins of high molecular weight along intracellular microtubules, *Science* **198:** 1038–1040.

Sherline, P., Schiavone, K., and Brocato, S. (1979), Endogenous inhibitor of colchicine–tubulin binding in rat brain, *Science* **205:** 593–595.

Sherrington, C. S. (1906), *The Integrative Action of the Nervous System,* Yale University Press, New Haven, revised edition, Cambridge University Press, Cambridge, 1947.

Shkolnik, L. J., and Schwartz, J. H. (1980), Genesis and maturation of serotonergic vesicles in identified giant cerebral neuron of *Aplysia, J. Neurophysiol.* **43:** 945–967.

Shute, C. C. D., and Lewis, P. R. (1963), Cholinesterase-containing systems of the brain of the rat, *Nature* **199:** 1160–1164.

Sidenius, P., and Jakobsen, J. (1979), Axonal transport in early experimental diabetes, *Brain Res.* **173:** 315–330.

Sidenius, P., and Jakobsen, J. (1980a), Impaired retrograde axonal transport from a nerve crush in streptozotocin diabetic rats, *Diabetologia* **19:** 222–228

Sidenius, P., and Jakobsen, J. (1980b), Axonal transport in rats after galactose feeding, *Diabetologia* **19:** 229–233.

Siegel, L. G., and McClure, W. O. (1975), Putrescine: Effect on axoplasmic transport, *Brain Res.* **93:** 543–547.

Siegel, R. E. (1968), *Galen's System of Physiology and Medicine,* Karger, New York.

Siegel, R. E. (1973), *Galen on Psychology, Psychopathology, and Function and Diseases of the Nervous System*, Karger, Basel.

Sigerist, H. E. (1948), The story of tarantism, Chapter 4, in: *Music and Medicine*, Schullian, D. M. and Schoen, M. (eds.), Schuman, New York.

Sinclair, D. (1967), *Cutaneous Sensation*, Oxford University Press, London.

Singer, M. (1952), The influence of the nerve in regeneration of the amphibian extremity, *Quart. Rev. Biol.* **27**: 169–183.

Singer, M. (1959), The significance of nerves in regeneration, in: *Regeneration in Vertebrates*, Thornton, C. S. (ed.), University of Chicago Press, Chicago.

Singer, M., and Bryant, S. V. (1969), Movements in the myelin Schwann sheath of the vertebrate axon, *Nature* **221**: 1148–1150.

Singer, M., and Caston, J. D. (1972), Neurotrophic dependence of macromolecular synthesis in the early limb regeneration of the newt, *Triturus*, *J. Embryol. Exp. Morph.* **28**: 1–11.

Singer, M., and Salpeter, M. M. (1966), Transport of tritium-labelled l-histidine through the Schwann and myelin sheaths into the axon of peripheral nerves, *Nature* **210**: 1225–1227.

Singer, P., and Mehler, S. (1980), 2-Deoxy[^{14}C]glucose uptake in the rat hypoglossal nucleus after nerve transection, *Exp. Neurol.* **69**: 617–626.

Singer, S. J., and Nicolson, G. L. (1972), The fluid mosaic model of the structure of cell membranes, *Science* **175**: 720–731.

Sjöstrand, J. (1969), Rapid axoplasmic transport of labelled proteins in the vagus and hypoglossal nerves of the rabbit, *Exp. Brain Res.* **8**: 105–112.

Sjöstrand, J., and Frizell, M. (1975), Retrograde axonal transport of rapidly migrating proteins in peripheral nerves, *Brain Res.* **85**: 325–330.

Sjöstrand, J., Frizell, M., and Hasselgren, P.-O. (1970), Effects of colchicine on axonal transport in peripheral nerves, *J. Neurochem.* **17**: 1563–1570.

Sjöstrand, J., and Karlsson, J.-O. (1969), Axoplasmic transport in the optic nerve and tract of the rabbit: A biochemical and radioautographic study, *J. Neurochem.* **16**: 833–844.

Sjöstrand, J., Rydevik, B., Lundborg, G., and McLean, W. G., (1978), Impairment of intraneural nucrocirculation, blood–nerve barrier and axonal transport in experimental nerve ischemia and compression, pp. 337–355, in: *The Neurobiologic Mechanisms in Manipulative Therapy*, Korr, J. M. (ed.), Plenum Press, New York.

Skau, K. A., and Brimijoin, S. (1978), Release of acetylcholinesterase from rat hemidiaphragm preparations stimulated through the phrenic nerve, *Nature* **275**: 224–226.

Sloboda, R. D., Dentler, W. L., Bloodgood, R. A., Telzer, B. R., Granett, S., and Rosenbaum, J. L. (1976), Microtubule-associated proteins (MAPs) and the assembly of microtubules *in vitro*, pp. 1171–1212, in: *Cell Motility*, Goldman, R., Pollard, T., and Rosenbaum, J. (eds.), Cold Spring Harbor Laboratory, Cold Spring Harbor, N.Y.

Sloper, J. C., Arnott, D. J., and King, B. C. (1960), Sulphur metabolism in the pituitary and hypothalamus of the rat: a study of radioisotope-uptake after the injection of ^{35}S DL-cysteine, methionine, and sodium sulphate, *J. Endocrinol.* **20**: 9–23.

Smith, A. D., DePotter, W. P., Moerman, E. H., and DeSchaepdryver, A. F. (1970), Release of dopamine β-hydroxylase and chromogranin A upon stimulation of splenic nerve, *Tissue and Cell* **2**: 547–568.

Smith, A. D., DePotter, and DeSchaepdryver, A. F. (1969), Subcellular fractionation of bovine splenic nerves, *Arch. Int. Pharm.* **179**: 495–496.

Smith, A. D., and Winkler, H. (1972), Fundamental mechanisms in the release of catecholamines, *Handb. Exp. Pharmacol.* **33**: 538–617.

Smith, D. S. (1971), On the significance of cross-bridges between microtubules and synaptic vesicles, *Phil. Trans. R. Soc.* **261**: 395–405.

Smith, D. S., Järlfors, U., and Beránek, R. (1970), The organization of synaptic axolplasm in the lamprey (*Petromyzon Marinus*) central nervous system, *J. Cell Biol.* **46:** 199–219.

Smith, D. S., Jaïlfors, U., and Cameron, B. F. (1975), Morphological evidence for the participation of microtubules in axonal transport, *Ann. N.Y. Acad. Sci.* **253:** 472–506.

Smith, M. C. (1975), Histological findings after hemicerebellectomy in man: Anterograde, retrograde and transneuronal degeneration, *Brain Res.* **95:** 423–442.

Smith, R. S. (1973), Microtubule and neurofilament densities in amphibian spinal root nerve fibers: Relationship to axoplasmic transport, *Can. J. Physiol. Pharmacol.* **51:** 798–806.

Smith, R. S. (1978), Axonal inclusions in the crab *Hemigrapsus nudus*, *J. Neurocytol.* **7:** 611–621.

Smith, R. S. (1979), Symposium on Physiology of Retrograde Axonal Transport, Forman, D. S. (Chairman), Society for Neuroscience, 8th Annual Meeting, pp. 63–70 (BIS Conference Report #49), Brain Information Service, Los Angeles.

Smith, R. S. (1980), The short term accumulation of axonally transported organelles in the region of localized lesions of single myelinated axons, *J. Neurocytol.* **9:** 39–65.

Smith, R. S., and Koles, Z. J. (1976), Mean velocity of optically detected intraaxonal particles measured by a cross-correlation method, *Can. J. Physiol. Pharmacol.* **54:** 859–869.

Snider, W. D., and Harris, G. L. (1979), A physiological correlate of disuse-induced sprouting at the neuromuscular junction, *Nature* **281:** 69–71.

Snyder, R. E. Reynolds, R. A., Smith, R. S., and Kendal, W. S. (1976), Application of a multiwire proportional chamber to the detection of axoplasmic transport, *Can. J. Physiol. Pharmacol.* **54:** 238–244.

Snyder, S. H., and Childers, S. R. (1979), Opiate receptors and opioid peptides, *Ann. Rev. Neurosci.* **2:** 35–64.

Somjen, G. G. (1979), Extracellular potassium in the mammalian central nervous system, *Ann. Rev. Physiol.* **41:** 159–177.

Souyri, F., Chretien, M., and Droz, B. (1981), "Acrylamide-induced" neuropathy and impairment of axonal transport of proteins. 1. Multifocal retention of fast transported proteins at the periphery of axons as revealed by light microscope radioautography, *Brain Res.* **205:** 1–13.

Špaček, J., and Lieberman, A. R. (1980), Relationships between mitochondrial outer membranes and agranular reticulum in nervous tissue: Ultrastructural observations and a new interpretation, *J. Cell Sci.* **46:** 129–147.

Specht, S., and Grafstein, B. (1973), Accumulation of radioactive protein in mouse cerebral cortex after injection of ^{3}H-fucose into the eye, *Exp. Neurol.* **41:** 705–722.

Speidel, C. C. (1941), Adjustments of nerve endings, *Harvey Lect.* **36:** 126–158.

Speidel, C. C. (1964), *In vivo* studies of myelinated nerve fibers, *Int. Rev. Cytol.* **16:** 173–231.

Spencer, P. S. (1972), Reappraisal of the model for "bulk axoplasmic flow," *Nature (New Biol.)* **240:** 283–285.

Spencer, P. S., and Schaumburg, H. H. (1976), Central-peripheral distal axonopathy—the pathology of dying-back polyneuropathies, *Prog, Neuropathol.* **3:** 253–295.

Spencer, P., and Schaumburg, H. H. (eds.) (1980), *Experimental and Clinical Neurotoxicology,* Williams and Wilkins, Baltimore.

Spencer, P. S., Sabri, M. I., Schaumburg, H. H., and Moore, C. L. (1979), Does a defect of energy metabolism in the nerve fiber underlie axonal degeneration in polyneuropathies? *Ann. Neurol.* **5:** 501–507.

Sperry, R. W. (1944), Optic nerve regeneration with return of vision in anurans, *J. Neurophysiol.* **7:** 57–69.

Sperry, R. W. (1963), Chemoaffinity in the orderly growth of nerve fiber patterns and connections, *Proc. Natl. Acad. Sci. USA* **50:** 703–710.

Springer, A. D., Heacock, A. M., Schmidt, J. T., and Agranoff, B. W. (1977), Bilateral tectal innervation by regenerating optic nerve fibers in goldfish: A radioautographic, electrophysiological and behavioral study, *Brain Res.* **128:** 417–427.

Starkey, R. R., and Brimijoin, S. (1979), Stop-flow analysis of the axonal transport of Dopa Decarboxylase (EC 4.1.1.26) in rabbit sciatic nerves, *J. Neurochem.* **32:** 437–441.

Stebbings, H., and Bennett, C. E., (1975), The sleeve element of microtubules. pp. 35–45, in: *Microtubules and Microtubule Inhibitors,* Borgers, M., and DeBrabander, M. (eds.), North Holland, Amsterdam.

Steno, N. (Stenson) (1664), De musculis et glandis observationum specimen, Discussion, pp. 142–160, in: Bastholm, 1950.

Stewart, D. M., and Martin, A. W. (1956), Hypertrophy of the denervated hemidiaphragm, *Am. J. Physiol.* **186:** 497–500.

Stewart, M. A., Passonneau, J. V., and Lowry, O. H. (1965), Substrate changes in peripheral nerve during ischaemia and Wallerian degeneration, *J. Neurochem.* **12:** 719–727.

Stjärne, L. (1966), Storage particles in noradrenergic tissues. Second International Catecholamine Symposium, *Pharmacol. Rev. (Milan)* **18:** 425–432.

Stoeckel, M. E., Hinerlang-Gertner, C., Dellmann, H.-D., Porte, A., and Stutinsky, F. (1975), Subcellular localization of calcium in the mouse hypophysis. I. Calcium distribution in the adeno- and neurohypophysis under normal conditions, *Cell Tissue Res.* **157:** 307–322.

Stoeckel, K., Schwab, M., and Thoenen, H. (1975a), Specificity of retrograde transport of nerve growth factor (NGF) in sensory neurons: A biochemical and morphological study, *Brain Res.* **89:** 1–14.

Stöckel, K., Schwab, M., and Thoenen, H. (1975b), Comparison between the retrograde axonal transport of nerve growth factor and tetanus toxin in motor, sensory, and adrenergic neurons, *Brain Res.* **99:** 1–16.

Stoeckel, K., Schwab, M., and Thoenen, H. (1977), Role of gangliosides in the uptake and retrograde axonal transport of cholera and tetanus toxin as compared to nerve growth factor and wheat germ agglutinin, *Brain Res.* **132:** 273–285.

Stone, G. C., and Wilson, D. L. (1979), Qualitative analysis of proteins rapidly transported in ventral horn motoneurons and bidirectionally from dorsal root ganglia, *J. Neurobiol.* **10:** 1–12.

Stone, G. C., Wilson, D. L., and Hall, M. E. (1978), Two-dimensional gel electrophoresis of proteins in rapid axoplasmic transport, *Brain Res.* **144:** 287–302.

Stone, L. S. (1937), Further experimental studies of the development of lateral-line sense organs in amphibians observed in living preparations, *J. Comp. Neurol.* **68:** 83–115.

Streit, P., Reubi, J. C., Wolfensberger, M., Henke, H., and Cuénod, M. (1979), Transmitter-specific retrograde tracing of pathways? *Prog. Brain Res.* **51:** 489–496.

Stromska, D. P., Iqbal, Z., and Ochs, S. (1979), Fast axoplasmic transport of tubulin and triad proteins, *Soc. Neurosci. Abstr.* **5:** 63.

Stromska, D., and Ochs, S. (1977), Slow outflow patterns of labeled proteins in cat and rat sensory nerves, *Fed. Proc.* **36:** 485.

Stromska, D. P., and Ochs, S. (1978), Comparison of fast transported protein and glycoprotein downflow patterns in sensory and motor fibers of mammalian sciatic nerve, *Soc. Neurosci. Abstr.* **4:** 37.

Stromska, D. and Ochs, S. (1981), Patterns of slow transport in the sensory nerves, *J. Neurobiol.* **12:** 441–453.

Stromska, D., and Ochs, S. (1982), Axoplasmic transport in aged rats, *Exp. Neurol.* (in press).

Stromska, D., Ochs, S., and Muller, J. (1981), Rate of axoplasmic transport in the dystrophic chicken, *Exp. Neurol.* **74:** 530–547.

Studitsky, A. N. (1964), Free auto- and homografts of muscle tissue in experiments on animals, *Ann. N.Y. Acad. Sci.* **120:** 789–801.

Sugar, O., and Gerard, R. W. (1940), Spinal cord regeneration in the rat, *J. Neurophysiol.* **3:** 1–19.

Summers, K. E., and Gibbons, I. R. (1971), Adenosine triphosphate-induced sliding of tubules in trypsin-treated flagella of sea-urchin sperm, *Proc. Natl. Acad. Sci. USA* **68:** 3092–3096.

Sumner, A., Pleasure, D., and Ciesielka, K. (1976), Slowing of fast axoplasmic transport in acrylamide neuropathy, *J. Neuropath. Exp. Neurol.* **35:** 319.

Sunderland, S. (1978), *Nerves and Nerve Injuries,* 2nd Edition, Williams and Wilkins, Baltimore.

Sunderland, S. (1980), The anatomical basis of nerve repair, Chapter 3, in: *Nerve Repair and Regeneration: Its Clinical and Experimental Basis,* Jewett, D. L., and McCarroll, H. R., Jr. (eds.), C. V. Mosby Co., St. Louis.

Suszkiw, J. B., and Pilar, G. (1976), Selective localization of a high affinity choline uptake system and its role in ACh formation in cholinergic nerve terminals, *J. Neurochem.* **26:** 1133–1138.

Swammerdam, J. (1752), *Bibel der nature,* translated from the Dutch and Latin originals as *The Book of Nature* by Floyd, T., and Boerhaave, H. (eds.), Hill, London (1758).

Swann, R. W., and Pickering, B. T. (1976), Incorporation of radioactive precursors into the membrane and contents of the neurosecretory granules of the rat neurohypophysis as a method of studying their fate, *J. Endocrinol.* **68:** 95–108.

Szentágothai, J. (1963), Discussion IV; question and response, pp. 205–206, in: *The Effect of Use and Disuse on Neuromuscular Functions,* Gutmann, E., and Hník, P. (eds.), Publishing House of the Czechoslovak Academy of Sciences, Prague.

Takenaka, T., Horie, H., and Sugita, T. (1978), New technique for measuring dynamic axonal transport and its application to temperature effects, *J. Neurobiol.* **9:** 317–324.

Takenaka, T., and Ochs, S. (1980), External detection of axoplasmic transport using ^{32}P-ATP as precursor, *J. Neurobiol.* **11:** 571–576.

Takenaka, T., and Inomata, K. (1981), Axoplasmic transport of microtubule-associated proteins in the rat sciatic nerve, *J. Neurobiol.* **12:** 479–486.

Tang, B. Y., Komiya, Y., and Austin, L. (1974), Axoplasmic flow of phospholipids and cholesterol in the sciatic nerve of normal and dystrophic mice, *Exp. Neurol.* **43:** 13–20.

Tani, E., and Ametani, T. (1970), Substructure of microtubules in brain nerve cells as revealed by ruthenium red, *J. Cell Biol.* **46:** 159–165.

Teichberg, S., Holtzman, E., Crain, S. M., and Peterson, E. R. (1975), Circulation and turnover of synaptic vesicle membrane in cultured fetal mammalian spinal cord neurons, *J. Cell Biol.* **67:** 215–230.

Temkin, O. (1973), *Galenism: Rise and Decline of a Medical Philosophy,* Cornell University Press, Ithaca.

Tennyson, V. M. (1970), The fine structure of the axon and growth cone of the dorsal root neuroblast of the rabbit embryo, *J. Cell Biol.* **44:** 62–79.

Terry, R. D., and Wiśniewski, H. (1970), The ultrastructure of the neurofibrillary tangel and the senile plaque, pp. 145–168, in: *Alzheimer's Disease and Related Conditions,* A Ciba Foundation Symposium, Wolstenholme, G. E. W., and O'Conner, M. (eds.), J. & A. Churchill, London.

Theiler, R. F., and McClure, W. O. (1978), Rapid axoplasmic transport of proteins in regenerating sensory nerve fibers, *J. Neurochem.* **31:** 433–447.

Theron, J. J., Meyer, B. J., Boekkooi, S., and Loots, J. M. (1975), Ultrastructural localisation of calcium in peripheral nerves of the rat, *S. African Med. J.* **49:** 1795–1798.

Thesleff, S. (1960a), Effects of the motor innervation of the chemical sensitivity of skeletal muscle, *Physiol. Rev.* **40:** 734–752.

Thesleff, S. (1960b), Supersensitivity of skeletal muscle produced by botulinum toxin, *J. Physiol. (London)* **151:** 598–607.

Thesleff, S. (1974), Physiological effects of denervation of muscle, *Ann. N.Y. Acad. Sci.* **228:** 89–104.

Thoenen, H., Otten, U., and Oesch, F. (1973), Axoplasmic transport of enzymes involved in the synthesis of noradrenaline: Relationship between the rate of transprot and subcellular distribution, *Brain Res.* **62:** 471–475.

Thomas, P. K., and Fullerton, P. M. (1963), Nerve fibre size in the carpal tunnel syndrome, *J. Neurol. Neurosurg. Psychiat.* **26:** 520–527.

Thomas, P. K., King, R. H. M., and Phelps, A. C. (1972), Electron microscope observations on the degeneration of unmyelinated axons following nerve section, *J. Anat.* **113:** 279.

Thompson, D'A. (1942), *On Growth and Form,* Cambridge University Press, Cambridge.

Thureson-Klein, A., Klein, R. L., and Yen, S.-S. (1973), Ultrastructure of highly purified sympathetic nerve vesicles: Correlation between matrix density and norepinephrine content, *J. Ultrastruct. Res.* **43:** 18–35.

Tiedt, T. N., Wisler, P. L., and Younkin, S. G. (1977), Neurotrophic regulation of resting membrane potential and acetylcholine sensitivity in rat extensor digitorum longus muscle, *Exp. Neurol.* **57:** 766–791.

Tieri, O., Polzella, A., Rinaldo, E., and Vecchione, L. (1962), Research on the mechanism of action of iodoacetic action on the retina. I. Trisephosphate-dehydrogenase activity, *Soc. Italian Biol. Sper. Naples Boll.* **38:** 183–185.

Tilney, L. G., and Porter, K. R. (1967), Studies on the microtubules in Heliozoa, II. The effect of low temperature on these structures in the formation and maintenance of the axopodia, *J. Cell Biol.* **34:** 327–343.

Tinel, J. (1915), Le signe du "Fourmillement" dans les lesion des nerfs peripheriques, *Press Med.* **47:** 388–389.

Titeca, J. (1935), Étude des modifications fonctionnelles due nerf au cours de sa dégénérescence Wallérienne, *Arch. Internatl. Physiol.* **41:** 1–56.

Toews, A. D., Goodrum, J. F., and Morell, P. (1979), Axonal transport of phospholipids in rat visual system, *J. Neurochem.* **32:** 1165–1173.

Torrey, T. W. (1934), The relation of taste buds to their nerve fibers, *J. Comp. Neurol.* **59:** 203–220.

Torvik, A., and Heding, A. (1967), Histological studies on the effects of actinomyocin D on retrograde nerve cell reaction in the facial nucleus of mice, *Acta Neuropath. (Berlin)* **9:** 146–157.

Torvik, A., and Skjörten, F. (1971), Electron microscopic observations on nerve cell regeneration and degeneration after axon lesions. I. Changes in the nerve cell cytoplasm, *Acta Neuropath. (Berlin)* **17:** 248–264.

Tower, S. S. (1932), Atrophy and degeneration in the muscle spindle, *Brain* **55:** 77–90.

Tsukita, S., and Ishikawa, H. (1976), Three-dimensional distribution of smooth endoplasmic reticulum in myelinated axons, *J. Electron Microsc.* **25:** 141–149.

Tsukita, S., and Ishikawa, H. (1980), The movement of membranous organelles in axons. Electron microscopic identification of anterogradely and retrogradely transported organelles, *J. Cell Biol.* **84:** 513–530.

Tsukita, S., and Ishikawa, H. (1981), The cytoskeleton in myelinated axons: Serial section study, *Biomed. Res.* **2:** 424–437.

Tuček, S. (1975), Transport of choline acetyltransferase and acetylcholinesterase in the central stump and isolated segments of a peripheral nerve, *Brain Res.* **86:** 259–270.

Tuček, S. (1978), *Acetylcholine Synthesis in Neurons,* John Wiley & Sons, New York.

Tytell, M., Brady, S. T., and Lasek, R. J. (1980), Axonal transport of microtubule-associated proteins known as *Tau* factor, *Soc. Neurosci. Abstr.* **6:** 501.

Van der Loos, H. (1965), The "improperly" oriented pyramidal cell in the cerebral cortex and its possible bearing on problems of neuronal growth and cell orientation, *Bull. Johns Hopkins Hosp.* **117:** 228–250.

Van der Loos, H. (1967), The history of the neuron, Chapter 1, in: *The Neuron*, Hydén, H. (ed.), Elsevier Publishing Company, Amsterdam.

Van der Loos, H. (1976), Neuronal circuitry and its development, *Prog. Brain Res.* **45**: 259–278.

van Essen, D., and Jansen, J. K. S. (1974), Reinnervation of the rat diaphragm during perfusion with α-bungarotoxin *Aact Physiol. Scand.* **91**: 571–573.

van Gehuchten, A. (1903), La degenerescence dite rétrogràde ou degenerescence Wallérienne indirecte, *Le Nevraxe* **5**: 1–108, cited in van Gehuchten, P., The scientific work of Arthur van Gehuchten (privately printed, no date).

van Harreveld, A. (1945), Re-innervation of denervated muscle fibers by adjacent functioning motor units, *Am. J. Physiol.* **144**: 477–493.

van Harreveld, A., and Marmont, G. (1939), The course of recovery of the spinal cord from asphyxia, *J. Neurophysiol.* **2**: 101–111.

van Leeuwenhoek, A. (1717), Letter to Abraham van Bleiswyk (March 2, 1717) printed in "Epistolae physiologicale super compluribus naturae arcanis" (Beman, Delft, 1719), Epistola XXXII, pp. 310–327, and in part translated pp. 32–35, in: *The Human Brain and Spinal Cord*, Clarke, E., and O'Malley, C. D. (eds.), University of California Press, Berkeley, 1968.

Vaughn, J. E., and Peters, A. (1967), Electron microscopy of the early postnatal development of fibrous astrocytes, *Am. J. Anat.* **121**: 131–152.

von Euler, U. S. (1956), *Noradrenaline*, Charles C Thomas, Springfield.

von Euler, U. S., and Hillarp, N.-Å. (1956), Evidence for the presence of noradrenaline in submicroscopic structures of adrenergic axons, *Nature* **177**: 44–45.

von Muralt, A. (1945), *Die Signalubermittlung im Nerven*, Birkhäuser, Basel.

von Vintschgau, M., and Hönigschmied, J. (1877), Nervus Glossopharyngeus und Schmeckbecher, *Arch. Gesam. Physiol.* **14**: 443–448.

Wagner, K. R., and Max, S. R. (1979), Neurotrophic regulation of glucose 6-phosphate dehydrogenase in rat skeletal muscle, *Brain Res.* **170**: 572–576.

Waldeyer, W. (1891), Ueber Einige Neuere Forschungen im Gebiete der Anatomie des Central Nervensystems, *Deutsch. Med. Wochenschr.* **17**: 1213–1218, 1244–1246, 1267–1269, 1287–1289, 1331–1332, 1352–1356.

Waller, A. V. (1852a), *Nouvelle Méthode Anatomique pour l'Investigation du Système Nerveux*, Georgi, Bonn; Waller, A. V. (1852), A new method for the study of the nervous system, *London J. Med.* **43**: 609–625.

Waller, A. V. (1852b), Huitième Mémoire sur le système nerveux, *Compt. Rend. Hebd. Seanc.* **35**: 561–564.

Waller, A. (1861), *The nutrition and reparation of nerves; being the substance of a lecture delivered at the Royal Institution of Great Britain, Friday, May 31, 1861*, Read and Co., London.

Waller, A. D. (1891), *An Introduction to Human Physiology*, Longmans, Green, and Co., New York.

Warnick, J. E., Albuquerque, E. X., and Guth, L. (1977), The demonstration of neurotrophic function by application of colchicine or vinblastine to the peripheral nerve, *Exp. Neurol.* **57**: 622–636.

Wasserman, R. H., Corradino, R. A., Carafoli, E., Kretsinger, R. H., MacLennan, D. H., and Siegel, F. L. (eds.) (1977), *Calcium-Binding Proteins and Calcium Function*, Elsevier/North Holland, New York.

Watson, D. F., Donoso, J. A., Illanes, J. P., and Samson, F. E. (1975), Comparison of transported proteins in the central and peripheral processes of unipolar neurons, *Abstr. Am. Soc. Neurochem.* **6**: 109.

Watson, D. F., Donoso, J. A., O'Neill, R. E., and Samson, F. E. (1980), Axoplasmic transport with velocities induced by pargyline *J. Neurosci. Res.* **5**: 563–578.

Watson, W. E. (1968), Observations on the nucleolar and total cell body nucleic acid of injured nerve cells, *J. Physiol. (London)* **196:** 655–676.

Watson, W. E. (1970), Some metabolic responses of axotomized neurones to contact between their axons and denervated muscle, *J. Physiol. (London)* **210:** 321–343.

Waxman, S. G. (1981), Cytochemical heterogeneity of the axon membrane, *Trends Neurosci.* **4:** 7–9.

Waxman, S. G., and Foster, R. E. (1980), Development of the axon membrane during differentiation of myelinated fibres in spinal nerve roots, *Proc. Roy Soc. B* **209:** 441–446.

Waxman, S. G., and Quick, D. C. (1977), Cytochemical differentiation of the axon membrane in A- and C-fibres, *J. Neurol. Neurosurg. Psychiat.* **40:** 379–385.

Webb, J. L. (1966), *Enzymes and Metabolic Inhibitors,* Academic Press, New York.

Webster, R. E., Henderson, D., Osborn, M., and Weber, K. (1978), Three-dimensional electron microscopical visualization of the cytoskeleton of animal cells: Immunoferritin identification of actin- and tubulin-containing structures, *Proc. Natl. Acad. Sci. USA* **75:** 5511–5515.

Weddell, G., and Zander, E. (1951), The fragility of non-myelinated nerve terminals, *J. Anat.* **85:** 242–250.

Weddell, G., Palmer, E., and Pallie, W. (1955), Nerve endings in mammalian skin, *Biol. Rev.* **30:** 159–195.

Weinberg, H. J., and Spencer, P. S. (1975), Studies on the control of myelinogenesis. I. Myelination of regenerating axons after entry into a foreign unmyelinated nerve, *J. Neurocytol.* **4:** 395–418.

Weinberg, H. J., and Spencer, P. S. (1976), Studies on the control of myelinogenesis. II. Evidence for neuronal regulation of myelin production, *Brain Res.* **113:** 363–378.

Weinberg, H. J., and Spencer, P. S. (1978), The fate of Schwann cells isolated from axonal contact, *J. Neurocytol.* **7:** 555–569.

Weingarten, M. D., Lockwood, A. H., Hwo, S.-Y., and Kirschner, M. W. (1975), A protein factor essential for microtubule assembly, *Proc. Natl. Acad. Sci. USA* **72:** 1858–1862.

Weir, R. L., Glaubiger, G., Chase, T. N. (1978), Inhibition of fast axoplasmic transport by acrylamide, *Envir. Res.* **17:** 251–255.

Weisenberg, R. C. (1972), Microtubule formation *in vitro* in solutions containing low calcium concentrations, *Science* **177:** 1104–1105.

Weisenberg, R. C. (1974), The role of ring aggregates and other structures in the assembly of microtubules, *J. Supramol. Struct.* **2:** 451–465.

Weiss, B., and Levin, R. M. (1978), Mechanism for selectivity inhibiting the activation of cyclic nucleotide phosphodiesterase and adenylate cyclase by antipsychotic drugs, *Adv. Cyclic Nucleotide Res.* Vol. 9, 285–303.

Weiss, D. G., Krygier-Brévart, V., Gross, G. W., and Kreutzberg, G. W. (1978), Rapid axoplasmic transport in the olfactory nerve of the pike: II. Analysis of transported proteins by SDS gel electrophoresis, *Brain Res.* **139:** 77–87.

Weiss, D. G., Schmid, G., and Wagner, L. (1980), Influence of microtubule inhibitors on axoplasmic transport of free amino acids. Implications for the Hypothetical transport mechanism, pp. 31–41, in: *Microtubules and Microtubule Inhibitors,* DeBrabander, M. and DeMey, J. (eds.), Elsevier/North Holland, Amsterdam.

Weiss, H. D., Walker, M. D., and Wiernik, P. H. (1974), Neurotoxicity of commonly used antineoplastic agents, *New Engl. J. Med.* **291:** 127–133.

Weiss, P. A. (1970), Neuronal dynamics and neuroplasmic flow, pp. 840–850, in: *The Neurosciences: Second Study Program,* Schmitt, F. O. (ed.), Rockefeller University Press, New York.

Weiss, P. A. (1972), Neuronal dynamics and axonal flow, V. The semisolid state of the moving axonal column, *Proc. Natl. Acad. Sci. USA.* **69:** 620–623.

Weiss, P. (1961), The concept of perpetual neuronal growth and proximo-distal substance convection, pp. 220–242, in: *Regional Neurochemistry,* Kety, S. S., and Elkes, J. (eds.), Pergamon Press, New York.

Weiss, P. (1969), Neuronal dynamics and neuroplasmic ("axonal") flow, pp. 3–34, in: *Cellular Dynamics of the Neuron,* Barondes, S. (ed.), Academic Press, New York.

Weiss, P., and Davis, H. (1943), Pressure block in nerves provided with arterial sleeves, *J. Neurophysiol.* **6:** 269–286.

Weiss, P., and Hiscoe, H. B. (1948), Experiments on the mechanism of nerve growth, *J. Exp. Zool.* **107:** 315–395.

Weiss, P., and Hoag, A. (1946), Competitive reinnervation of rat muscles by their own and foreign nerves, *J. Neurophysiol.* **9:** 413–418.

Weiss, P., and Holland, Y. (1967), Neuronal dynamics and axonal flow. II. The olfactory nerve as model test object, *Proc. Natl. Acad. Sci. USA* **57:** 258–264.

Weiss, P. A., and Mayr, R. (1971a), Organelles in neuroplasmic ("axonal") flow: Neurofilaments, *Proc. Natl. Acad. Sci. USA.* **68:** 846–850.

Weiss, P., and Mayr, R. (1971b), Neuronal organelles in neuroplasmic "axonal" flow. II. Neurotubules, *Acta Neuropath. (Berlin) Suppl.* **5:** 198–206.

Weiss, P., and Pillai, A. (1965), Convection and fate of mitochondria in nerve fibers: Axonal flow as vehicle, *Proc. Natl. Acad. Sci. USA* **54:** 48–56.

Weller, R. O., Mitchell, J., and Daves, G. D., Jr. (1980), Buckthorn *(Karwinskia Humboldtiana)* toxins, Chapter 23, in: *Experimental and Clinical Neurotoxicity,* Spencer, P. S., and Schaumburg, H. H. (eds.), Williams and Wilkins, Baltimore.

Westgaard, R. H. (1975), Influence of activity on the passive electrical properties of denervated soleus muscle fibres in the rat, *J. Physiol. (London)* **251:** 683–697.

Wettstein, F. O., Noll, H., and Penman, S. (1964), Effect of cycloheximide on ribosomal aggregates engaged in protein synthesis *in vitro, Biochim. Biophys. Acta* **87:** 525–528.

White, H. D., Coughlin, B. A., and Purich, D. L. (1980), Adenosine triphosphatase activity of bovine brain microtubule protein, *J. Biol. Chem.* **255:** 486–491.

Whittaker, V. P., Zimmermann, H., and Dowdall, M. J. (1975), The biochemistry of cholinergic synapses as exemplified by the electric organ of *Torpedo, J. Neural Transmission Suppl.* **12:** 39–60.

Wick, A. N., Drury, D. R., Nakada, H. I., and Wolfe, J. B. (1957), Localization of the primary metabolic block produced by 2-deoxyglucose, *J. Biol. Chem.* **224:** 963–969.

Wiesel, T. N., Hubel, D. H., and Lam, D. M. K. (1974), Autoradiographic demonstration of ocular-dominance columns in the monkey striate cortex by means of transneuronal transport, *Brain Res.* **79:** 273–279.

Wildy, P. (1967), The progression of herpes simplex virus to the central nervous system of the mouse, *J. Hyg. (Cambridge)* **65:** 173–192.

Wiley, C. A., and Ellisman, M. H. (1979), Development of axonal membrane specializations defines nodes of Ranvier and precedes Schwann cell myelin formation, *J. Cell Biol. (Abstr.)* **83:** 83a.

Willard, M., Cowan, W. M., and Vagelos, P. R. (1974), The polypeptide composition of intra-axonally transported proteins: evidence for four transport velocities, *Proc. Natl. Acad. Sci. USA* **71:** 2183–2187.

Willard, M. B., and Hulebak, K. L. (1977), The intra-axonal transport of polypeptide H: Evidence for a fifth (very slow) group of transported proteins in the retinal ganglion cells of the rabbit, *Brain Res.* **136:** 289–306.

Willard, M., and Simon, C. (1981), Antibody decoration of neurofilaments, *J. Cell Biol.* **89:** 198–205.

Willard, M., Wiseman, M., Levine, J., and Skene, P. (1979), Axonal transport of actin in rabbit retinal ganglion cells, *J. Cell Biol.* **81:** 581–591.

Williams, P. L., and Hall, S. M. (1970), *In vivo* observations on mature myelinated nerve fibres of the mouse, *J. Anat.* **107:** 31–38.

Williams, P. L., and Hall, S. M. (1971), Prolonged *in vivo* observations of normal peripheral nerve fibres and their acute reactions to crush and deliberate trauma, *J. Anat.* **108:** 397–408.

Willis, T. (1664), *The Anatomy of the Brain and Nerves,* Feindel, W. (ed.), with a note on Pordage's English translation and a bibliographic survey of *Cerebri Anatome,* 2 Vol., McGill University Press, Montreal (1965).

Wilson, D. L., and Stone, G. C. (1979), Axoplasmic transport of proteins, *Ann. Rev. Biophys. Bioeng.* **8:** 27–45.

Wilson, L. G. (1959), Erasistratus, Galen and the Pneuma, *Bull. Hist. Med.* **33:** 293–314.

Wilson, L., and Meza, I. (1973), The mechanism of action of colchicine. Colchicine binding properties of sea urchin sperm tail outer doublet tubulin, *J. Cell Biol.* **58:** 709–719.

Wilson, L., and Bryan J. (1974), Biochemical and pharmacological properties of microtubules, *Adv. Cell Mol. Biol.* **3:** 21–72.

Wilson, L., Bamburg, J. R., Mizel, S. B., Grisham, L. M., and Creswell, K. M. (1974), Interaction of drugs with microtubule proteins, *Fed. Proc.* **33:** 158–166.

Windle, W. F. (ed.) (1955), *Regeneration in the Central Nervous System,* Charles C Thomas, Springfield.

Windle, W. F. (1980), Neurotropism: Editorial comment, *Exp. Neurol.* **67:** 251–256.

Wiśniewski, H., Karczewski, W., and Wiśniewska, K. (1966), Neurofibrillar degeneration of nerve cells after intracerebral injection of aluminium cream, *Acta Neuropath.* (Berlin) **6:** 211–219.

Wiśniewski, H., and Terry, R. D. (1967), Experimental colchicine encephalopathy. I. Induction of neurofibrillary degeneration, *Lab. Invest.* **17:** 577–587.

Wolfe, D. E., Potter, L. T., Richardson, K. C., and Axelrod, J. (1962), Localizing tritiated norepinephrine in sympathetic axons by electron microscopic autoradiography, *Science* **138:** 440–441.

Wolosewick, J. J., and Porter, K. R. (1976), Stereo high-voltage electron microscopy of whole cells of the human diploid line, WI-38, *Am. J. Anat.* **147:** 303–324.

Wood, P. L., and Boegman, R. J. (1975), Increased axoplasmic flow in experimental ischemic myopathy, *Exp. Neurol.* **48:** 136–141.

Wooten, G. F., and Coyle, J. T. (1973), Axonal transport of catecholamine synthesizing and metabolizing enzymes, *J. Neurochem.* **20:** 1361–1371.

Worth, R. M., and Ochs, S. (1976), The effects of repetitive electrical stimulation on axoplasmic transport, *Soc. Neurosci. Abstr.* **2:** 50.

Wright, G. P. (1955), The neurotoxins of *Clostridium botulinum* and *clostridium tetani*, *Pharmacol. Rev.* **7:** 413–465.

Wuerker, R. B. (1970), Neurofilaments and glial filaments, *Tissue and Cell* **2:** 1–9.

Wuerker, R. B., and Kirkpatrick, J. B. (1972), Neuronal microtubules, neurofilaments and microfilaments, *Int. Rev. Cytol.* **33:** 45–75.

Wyburn-Mason, R. (1950), *Trophic Nerves,* Kimpton, London.

Yamada, K. M., Spooner, B. S., and Wessells, N. K. (1971), Ultrastructure and function of growth cones and axons of cultured nerve cells, *J. Cell Biol.* **49:** 614–635.

Yarmolinsky, M. B., and de la Haba, G. L. (1959), Inhibition by puromycin of amino acid incorporation into protein, *Proc. Natl. Acad. Sci. USA* **45:** 1721–1729.

Yiengst, M. J., Barrows, C. H., Jr., and Shock, N. W. (1959), Age changes in the chemical composition of muscle and liver in the rat, *J. Geront.* **14:** 400–404.

Yoon, M. [G.] (1971), Reorganization of retinotectal projection following surgical operations on the optic tectum in goldfish, *Exp. Neurol.* **33:** 395–411.

Yoon, M. G. (1972), Reversibility of the reorganization of retinotectal projection in goldfish, *Exp. Neurol.* **35:** 565–577.

Young, J. Z. (1936), Structure of nerve fibers in cephalopods and crustacea, *Proc. Roy. Soc. B* **121:** 319–337.

Young, J. Z. (1944a), Contraction, turgor and the cytoskeleton of nerve fibres, *Nature* **153:** 333–335.

Young, J. Z. (1944b), Surface tension and the degeneration of nerve fibres, *Nature* **154:** 521–522.

Young, J. Z. (1945), The history of the shape of a nerve fiber, pp. 41–94, in: *Essays on Growth and Form, Presented to D'Arch Wentworth Thompson,* LeGros, W. E., and Medawar, P. B. (eds.), Clarendon Press, Oxford.

Young, J. Z. (1949a), Narrowing of nerve fibres at the nodes of Ranvier, *J. Anat.* **83:** 55.

Young, J. Z. (1949b), Factors influencing the regeneration of nerves, *Adv. Surg.* **1:** 165–220.

Young, W. S. III, Wamsley, J. K., Zarbin, M. A., and Kuhar, M. J. (1980), Opioid receptors undergo axonal flow, *Science* **210:** 76–78.

Yu, M. K., Wright, T. L., Dettbarn, W.-D., and Olson, W. H. (1974), Pargyline-induced myopathy with histochemical characteristics of Duchenne muscular dystrophy, *Neurology* **24:** 237–244.

Zackroff, R. V., and Goldman, R. D. (1980), *In vitro* reassembly of squid brain intermediate filaments (neurofilaments): Purification by assembly–disassembly, *Science* **208:** 1152–1155.

Zalewski, A. A. (1969), Regeneration of taste buds after reinnervation by peripheral or central sensory fibers of vagal ganglia, *Exp. Neurol.* **25:** 429–437.

Zalewski, A. A. (1972), Regeneration of taste buds after transplantation of tongue and ganglia grafts to the anterior chamber of the eye, *Exp. Neurol.* **35:** 519–528.

Zambrano, D., and De Robertis, E. (1966), The secretory cycle of supraoptic neurons in the rat, *Z. Zellforsch* **73:** 414–431.

Zatz, M., and Barondes, S.H. (1971), Rapid transport of fucosyl glycoproteins to nerve endings in mouse brain, *J. Neurochem.* **18:** 1125–1133.

Zelená, J. (1957), The morphogenetic influence of innervation of the ontogenetic development of muscle-spindles, *J. Embryol. Exp. Morphol.* **5:** 283–292.

Zelená, J. (1964), Development, degeneration, and regeneration of receptor organs, *Prog. Brain Res.* **13:** 175–213.

Zelená, J. (1968), Bidirectional movement of mitochondria along axons of an isolated nerve segment, *Z. Zellforsch. Mikrosk. Anat.* **92:** 186–196.

Zelená, J. (1972), Ribosomes in myelinated axons of dorsal root ganglia, *Z. Zellforsch.* **124:** 217–229.

Zelená, J., and Gutmann, E. (1968), Bidirectional shifting of mitochondria along axons (in Czechoslovakian) *Czech. Fysiol.* **17:** 39–40.

Zelená, J., and Hník, P. (1963), Effect of innervation of the development of muscle receptors, pp. 95–105, in: *The Effect of Use and Disuse on Neuromuscular Functions,* Gutmann, E., and Hník, P. (eds.), Publishing House of the Czechoslovak Academy of Sciences, Prague.

Zelená, J., Lubińska, L., and Gutmann, E. (1968), Accumulation of organelles at the ends of interrupted axons, *Z. Zellforsch. Mikrosk. Anat.* **91:** 200–219.

Zenker, W., and Högl, E. (1976), The prebifurcation section of the axon of the rat spinal ganglion cell, *Cell Tissue Res.* **165:** 345–363.

Zenker, W., and Hohberg, E. (1973), α-Motorishce nervenfaser: axonquerschnittsfläche von stammfaser und endästen, *Z. Anat. Entwick.-Gesch.* **139:** 163–172.

Zenker, W., Mayr, R., and Gruber, H. (1973), Axoplasmic organelles: Quantitative differences between ventral and dorsal root fibres of the rat, *Experientia* **29:** 77–78.

Zenker, W., Mysicka, A., and Neuhuber, W. (1980), Dynamics of the transganglionic movement of horseradish peroxidase in primary sensory neurons, *Cell Tissue Res.* **207:** 479–489.

Zhenevskaya, R. P. (1960), The influence of de-efferentation on the regeneration of skeletal muscle (in Russian), *Arkh. Anat. Gist. Embriol.* **39:** 42–50 (quoted by Carlson, 1973).

Zhenevskaya, R. P. (1961), Restoration of muscle by the method of transplantation of minced muscle tissue under conditions of sensory deprivation (in Russian), *Arkh. Anat. Gistol. Embriol.* **40:** 46–53 (quoted by Carlson, 1973).

Index

Isoelectric point (pI), 113

K^+, *see* Potassium depolarization and transport
α-Ketoglutarate, 158
Krebs cycle, *see* Tricarboxylic acid

1-Lactate reversal of IAA block, 160
Lactic acid dehydrogenase (LDH), 159-161
Lanthanum, 199, 220
Latent period in regeneration, 136
Lateral-line degeneration, 357
Lattice-like structure in axoplasm, *see*
        Microtrabecular network
LDH, *see* Lactic acid dehydrogenase
Lectins, 117, 236
Lidocaine, 299-300, 346
Limb regeneration:
   assay for trophic factors, 360
   nerve requirement for, 356
Lipid phase in electron spin resonance
      measure, 221
Local anesthetics, 299-300, 346
Local entry of ^{3}H-N-SP, 61-62, 264
Local uptake of amino acids, 102-103
Low temperature and rate of transport, 35-38
Lysis of CNS fibers, 388

Magnesium (Mg^{2+}) as substitution for Ca^{2+},
      197-198
Mannose, 116
MAO, *see* Monoamine oxidase
MAPs, *see* Microtubular-associated proteins
Marcaine (buvicaine), 346
Maturation of nerve fibers, 367
Maturation period in regenerating cell bodies,
      137
Maytansine, 284-285, 293
MBK, *see* Methyl-*n*-butyl ketone
Melvonic acid, 119
Membrane:
   active agents, 299-303
   carrier mechanisms, 100
   excitation block and transport, 299
   mobility in, 234-236
   stimulation excess and transport, 303
   turnover, 252
Mepivacaine, 346
MEPPs, *see* Miniature end plate potentials
Mercurobenzine, 161, 293
Meromyosin binding to actin, 219
Mersalyl, 293
Messenger ribonucleic acid (mRNA), 100
Metabolic blocking agents, 152-163
Methyl-mercury (methyl-Hg), 293
Methyl-*n*-butyl ketone (MKB), 294-295

M (muscle) form of lactic acid dehydrogenase,
      160-161
Microfilament bundles, 221
Microfilaments, 200, 218-221, 317, 319
Micro-iontophoresis, 55-56
   dendritic transport, 55-57
   to muscle fibers, 341
Microperistalsis, 245-246
Micro-spectrophotometry of cell body, 136
Microstream hypothesis of transport,
      231-234, 251
Microtrabecular network, 218-221, 239, 277
   critical-point drying, 219
   rigidity of, 223
   transport mechanism, 239
Microtubular-associated proteins (MAPs),
      203-205, 210-215, 228-229, 249
   distribution, 210-215
   turnover in fibers, 211
Microtubules, 200-215, 273, 279, 284-287,
      300, 315, 335, 367
   assembly, 202-211, 300, 368
     directionality of, 206-208
     initiation site, 206
     at interrupted ends, 248
     at low temperatures, 209, 287
     and MAPs, 204-205, 210, 249
     polypeptide inhibiting factors, 205-206
     and 30S substance, 204
     transphosphorylation in, 204
     and tubulin assembly protein (TAP),
       210, 249
     unidirectional, 207
   breaks in, 286-287
   Ca^{2+} effect on, 204
   calmodulin in, 210
   cell division and tubulin-binding agents, 211
   cilia dynein addition to, 213-214
   coating with glycoproteins and mucopoly-
      saccharides, 214-215, 234
   coherent outgrowth, 246-250
   colchicine, 279-280
   in daughter branches, 268
   density, 267-268, 284, 287-288
   dimers, 202-203
   disassembly, 202-205, 208-211, 300
     at low temperatures, 287-291
     at terminals, 246-250
   form at nerve interruptions, 248
   in HNS neurons, 335
   inhibitory substances, 205-206
   polarity, 208
   proportion to neurofilaments, 294
   protofilaments of, 201
   routing, 272-277